PATENT APPLICATION FOR MSAART EVO TECHNOLOGIES THAT SOLVE THE REAL-WORLD PROBLEMS OF CO2, ENERGY, POWER, OXYGEN AND POLLUTION

THE MSAART EVO SYSTEM APPLIED TO A GAS-POWERED ENGINE

DRAFT 518,400 B KMV - PART ONE OF TWENTY

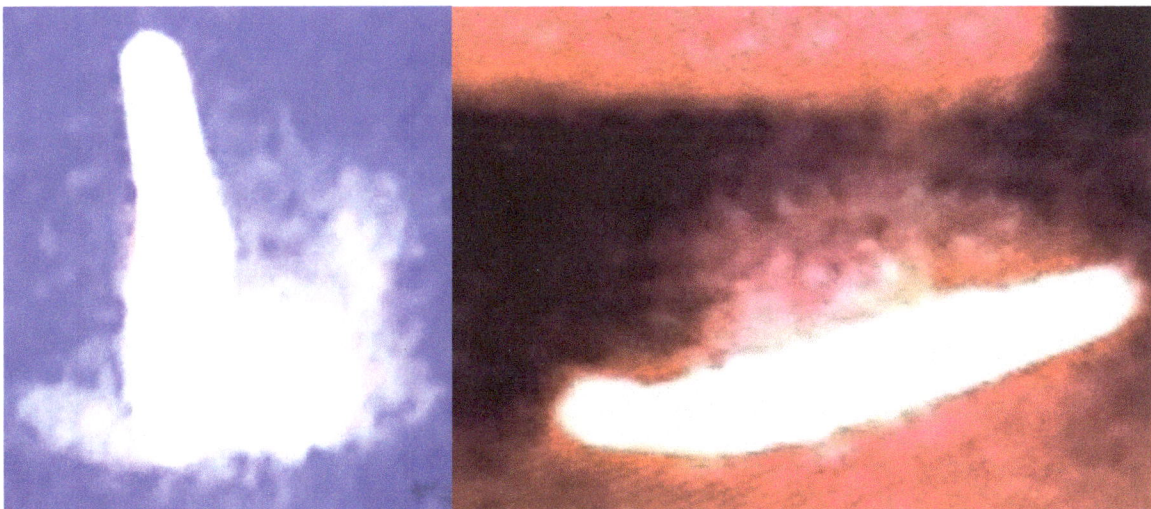

MSAART EVO PHOTOS 1,000 FRAMES PER / SEC WORLDS FIRST DOCUMENTATION OF LIGHT IMPLODING INTO A ZERO POINT

MALCOLM V of SCOTLAND | MALCOLM BENDALL
THURSDAY 22ND SEPTEMBER 2022

APPLICATIONS FOR A MSAART EVO'S FORM AND FUNCTIONS

LIST OF CONTENTS

PART ONE

SUMMARY OF RESULTS

PART TWO

PART THREE

PART FOUR

APPLICATIONS FOR A MSAART EVO'S FORM AND FUNCTIONS

SUMMARY OF RESULTS

APPLICATIONS FOR A MSAART EVO'S FORM AND FUNCTIONS

SUMMARY OF RESULTS

PART NINE

PART TEN

APPLICATIONS FOR A MSAART EVO'S FORM AND FUNCTIONS
SUMMARY OF RESULTS

PART NINE

APPLICATIONS FOR A MSAART EVO'S FORM AND FUNCTIONS

LIST OF DIAGRAMS IN PART ONE – START PAGE 14

APPLICATIONS FOR A MSAART EVO'S FORM AND FUNCTIONS

LIST OF DIAGRAMS IN PART ONE

APPLICATIONS FOR A MSAART EVO'S FORM AND FUNCTIONS

LIST OF DIAGRAMS IN PART ONE

APPLICATIONS FOR A MSAART EVO'S FORM AND FUNCTIONS

LIST OF DIAGRAMS IN PART ONE

END OF THE BEGINNING - PART ONE – PAGE 100

LIST OF DIAGRAMS IN PART TWO -START 101

APPLICATIONS FOR A MSAART EVO'S FORM AND FUNCTIONS

LIST OF DIAGRAMS IN PART TWO

PART TWO - END OF THE BEGINNING – PAGE 140

LIST OF DIAGRAMS IN PART THREE – STARTS 141

APPLICATIONS FOR A MSAART EVO'S FORM AND FUNCTIONS

LIST OF DIAGRAMS IN PART THREE – STARTS 141

APPLICATIONS FOR A MSAART EVO'S FORM AND FUNCTIONS

LIST OF DIAGRAMS IN PART THREE – STARTS 141

END OF THE BEGINNING - PART THREE– PAGE 162

LIST OF DIAGRAMS IN PART FOUR – START 163

APPLICATIONS FOR A MSAART EVO'S FORM AND FUNCTIONS

LIST OF DIAGRAMS IN PART FOUR

APPLICATIONS FOR A MSAART EVO'S FORM AND FUNCTIONS

LIST OF DIAGRAMS IN PART FOUR

APPLICATIONS FOR A MSAART EVO'S FORM AND FUNCTIONS

LIST OF DIAGRAMS IN PART FOUR

APPLICATIONS FOR A MSAART EVO'S FORM AND FUNCTIONS

LIST OF DIAGRAMS IN PART EIGHT – START

APPLICATIONS FOR A MSAART EVO'S FORM AND FUNCTIONS

PART 1

MSAART EVO
TECHNOLOGIES THAT
SOLVE THE REAL-WORLD
PROBLEMS OF CO2,
ENERGY, POWER, OXYGEN,
AND POLLUTION

APPLICATIONS FOR A MSAART EVO'S FORM AND FUNCTIONS

ABSTRACT

THIS PAPER DESCRIBES A METHODOLOGY AND PATHWAY TO CALCULATE AND INDUCE ATOMIC RECONSTRUCTION FROM BOTH ELEMENTAL AND MOLECULAR RESONANCE BY REFERENCING THE MSSART MSAART EVO MODEL OF THE ELEMENTS (PMOE) AND MSAART EVO UNIFICATION MODEL CALCULATOR (PUMC) TABLES TO ENGAGE THE NATURAL PROPERTIES OF THE TOROIDAL MSAART EVO NATURE, GEOMETRY AND FUNCTION INCLUDING AN EXPLANATION OF THE MSAART EVO'S FORM AND STRUCTURE ESPECIALLY AS IT RELATES TO ATOMIC RECONSTRUCTION, COLD FUSION AND LOW ENERGY ATOMIC REACTIONS (LEAR) INDUCED ATOMIC TRANSFORMATIONS.

As all *Matter* is constructed in the mould of *Time* to understand *Matter* one must first understand *Time*. As all calculations of Energy is in terms of work over *Time* if one could reverse or accelerate *Time* one can instead of requiring to do more work may simply reduce the *Time* factor for an increase in Energy. *Time* itself is constructed by equal and opposite mirror numbers. As for every force there is an equal and opposite force, as above so below. *Time's* number is 518,400 which its self is the product of 5 x 1 x 8 x 4 x 4 x 8 x 1 x 5 = 25,600 // 12,800 // 6,400 // 3,200 // 1,600 // 800.

Time's Base number is a product of 1 x 2 x 3 x 4 x 5 x 6 x 8 x 9 x 10 = 518,400. Seven is not in *Time's* Base number sequence as it represents *Direct current (DC)* not *Matter* which is constructed of Frequencies generated by the mould of *Time*. The equal and opposite mirror numbers for *Time's* 518,400 Base Number are as follows.

1 x 2 x 3 x 4 x 5 x 6 x 6 x 5 x 4 x 3 x 2 x 1 = *518,400* and 8 x 9 x 10 x 10 x 9 x 8 = *518,400* both are mirror numbers reflecting the basic symmetry of a Universal truth.

As there are no straight lines in the Universe therefore every line is part of a curve. Every curve is a part of a spiral either imploding clockwise manifesting a Negative charge in its natural state (life force) or exploding anti-clockwise manifesting a Positive Charge (death and destruction force) in its natural state, reflecting *Time's* opposed Numerical Base nature.

This is true as *Time* is the mould within which *Matter* is formed, therefore Elements moulded by *Time* must reflect both equal and opposite numbers, implicit and encoded in their phase change temperatures and equal and opposite positive and Negative charge and clockwise and anti-clockwise spin components.

Therefore because and as a consequence of the above, as the Sun's diameter is *864,000* Miles it must factor out as 8 x 6 x 4 x 4 x 6 x 8 = *36,864*. Whereas Time is the mould in which *Matter* is formed, from 5th dimensional direct current Aether (*5.555*), 6th dimensional Light Energy (6.666) and Time, the 4th dimension (4.444), must logically and absolutely proceed *Matter* being the 3rd dimension (3.333). So as the Sun is the mould in which the pre-curser to *Matter* Protium [Known as Hydrogen (H)] is formed, Protiums Base Mirror Number must again reflect the Sun's and *Time's* Base Numbers. to be subsequently contained within the Sun's self-generated, self-structuring and self-regulated toroidal electromagnetic field, As Protiums Phase Change melting point, where applied external energy intensities reach the critical

ABSTRACT

resonance of Protium its-self causing the implosive atomic focus, is minus - 259.2 Degrees Celsius therefore Protiums Base Mirror Number calculation is as follows.
2 x 5 x 9 x 2 x 2 x 9 x 5 x 2 = *129,600*

129,600 is the Resonant Frequency Energy Unit (RFEU) number representing primal base **Matter** Protium a manifestation of the third dimension. It is therefore a proof that the Suns Diameter *864,000* divided by *129,600* equals 3,333.333. Another proof is that Protium's *129,600* / *36,864* Base Mirror Number = 3.515625 // 7.03125 // 14.0625 // 28.125 // 56.25 // 112.5 // 225 // 450 // 900 // 1,800 // 3,600 // 7,200 // *14,400* // 28,800 // 57,600 // 115,200 // 230,400 // 460,800 // *921,600* // 1,843,200 //3,686,400 [*36,864*] 20 Octaves up.

518,400 Time / *129,600* (RFEU) = 4 as the square represents Matter (AC) and the Circle represents (DC) Aether and *518,400 Time* / *36,864* (Sun's Base Number) = 14.062 (PPOS).

THE MSAART EVO SYSTEM FOR WASTE ENERGY RECOVERY FROM THE OPERATION OF THE INTERNAL COMBUSTION ENGINE ELEMENTS AND MOLECULES INVOLVED WITHIN THE SYSTEM

ABSTRACT – MSAART EVO UNIFICATION MODEL

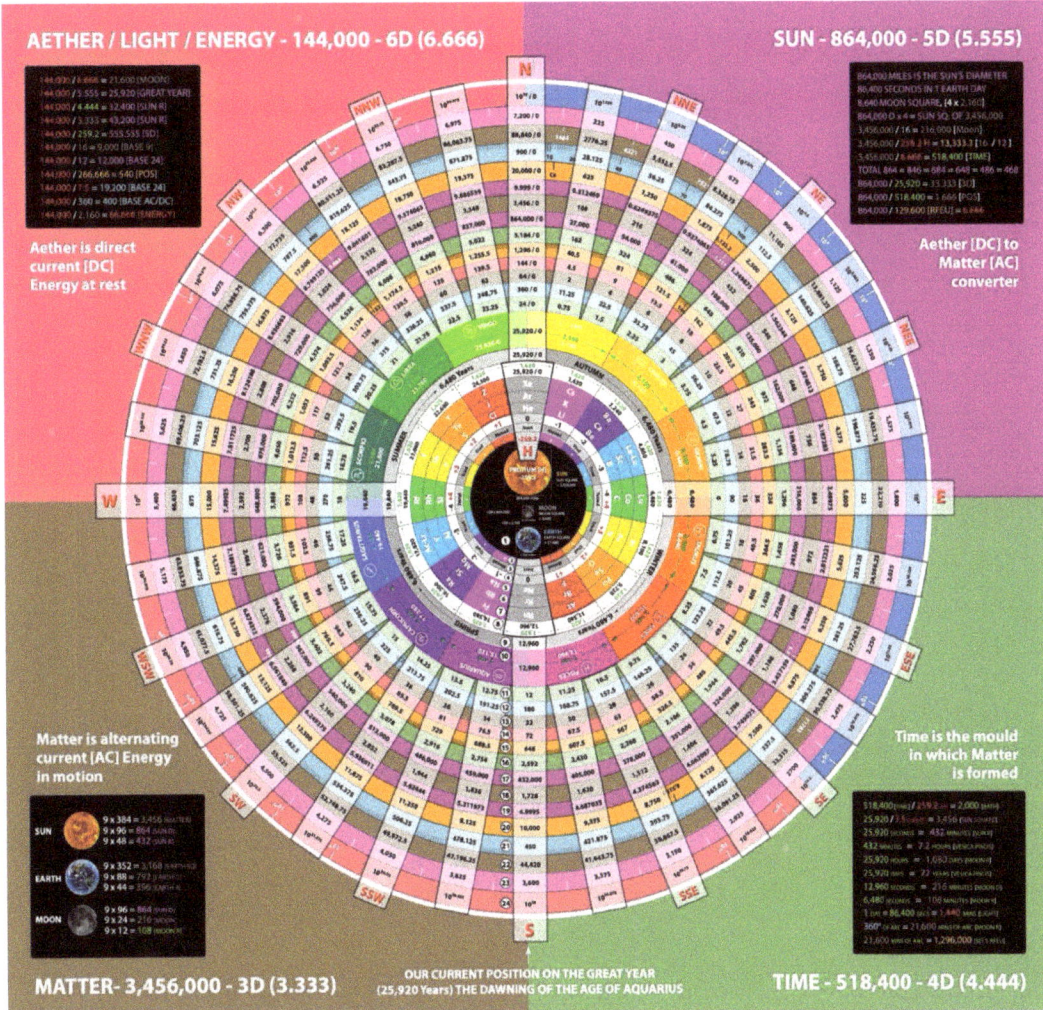

DIAGRAM 1 – MSAART EVO UNIFICATION MODEL.

PLASMOID UNIFICATION MODEL

Plasmoids are doughnut or toroidal shaped clusters of net Protons or net Electrons that once captured and placed into a Toroidal orbit are capable of absorbing, storing and releasing enormous amounts of energy present within their self-generated and structured electro-magnetic containment field. Plasmoids, in effect, function as an atomic battery that can be self-charging due to its ability to convert matter to available clean energy. Plasmoids by their unique geometry cause a consequential electro-magnetic containment field to generate a Zero point naturally and casually, without much effort, have the ability to convert the nuclear Mass of Protium (Atoms) into energy.

The Plasmoid Unification Model (PUM) posits that Plasmoids are epoch-making and that knowledge of them has been hidden in plain sight for centuries. This PUM 'slide rule' reveals the algorithmic relationships between life's elements critical to mankind's existence and development. It starts with Protium [H] which has a melting point of -259.2°C and is the most abundant element in our Solar System. Protium determines the 25,920 Great Year frequency of our Solar System. The resonant frequencies of all other elements can then be calculated when 25,920 years is reduced from years to days, hours and seconds.

The PUM is evidence that the Universe is an intelligent design. That design is in perfect octave harmonic resonance with itself. Therefore, all of creation from Galaxies to Planets to Elements all resonate in unison with a collective chord 'As Above So Below'. This is interconnected with an Energy 'web', the 24 components and laws of which are all based and governed on the same 16 sector Torus Plasmoid precepts shown. The concepts and ruling principles of the PUM can, and have, been applied to make Energy to Matter and Matter to Energy conversions. When applied to the modern hydrocarbon powered internal combustion engine, PUM technology removes exhaust toxic waste products and increases the engine power output by transforming waste energy back into fuel. Plasmoids employed in conjunction with the Plasmoid Toroidal Implosive Turbine provide a new novel Matter to Energy and Energy to Matter propulsion device for water, land, air and space travel.

LEGEND

APPLICATIONS FOR A MSAART EVO'S FORM AND FUNCTIONS

ABSTRACT – AETHER.

The

ABSTRACT – SUN.

The

ABSTRACT – SUN.

The

ABSTRACT – TIME.

The

ABSTRACT – MATTER.

The

ABSTRACT – MSAART EVO SYSTEM.

The CURRIE temperature

ABSTRACT – MSAART EVO SYSTEM – GENERATOR.

The

ABSTRACT – MSAART EVO SYSTEM – CHARGER.

The Currie temperature

APPLICATIONS FOR A MSAART EVO'S FORM AND FUNCTIONS

ABSTRACT – MSAART MSAART EVO IMPLOSIVE TURBINE.

ABSTRACT – MSAART EVO ATOMIC PROTIUM PROTON DRIVE.

ABSTRACT – MSAART EVO PLANETARY POWER PLANT.

ABSTRACT – MSAART EVO ATOMIC OCTAVE RESONANT CAVITY WITH IMPLODED SPHERE SWIRL GUIDE.

APPLICATIONS FOR A MSAART EVO'S FORM AND FUNCTIONS

ABSTRACT

a = 2,160
b = 2,880
c = 3,600
P = 8,640
S = 4,320
K = 3,110,400
ha = 2,880
hb = 2,160
hc = 1,728

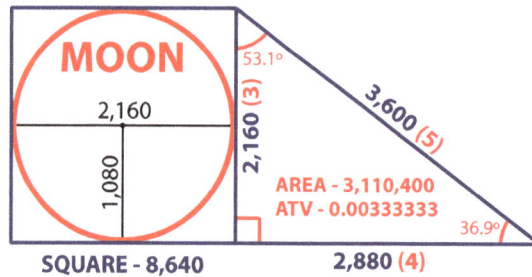

MOON
2,160
1,080
53.1°
2,160 (3)
3,600 (5)
AREA - 3,110,400
ATV - 0.00333333
36.9°
SQUARE - 8,640
2,880 (4)

a = 7,920
b = 10,560
c = 13,200
P = 31,680
S = 15,840
K = 41,817,600
ha = 10,560
hb = 7,920
hc = 6,336

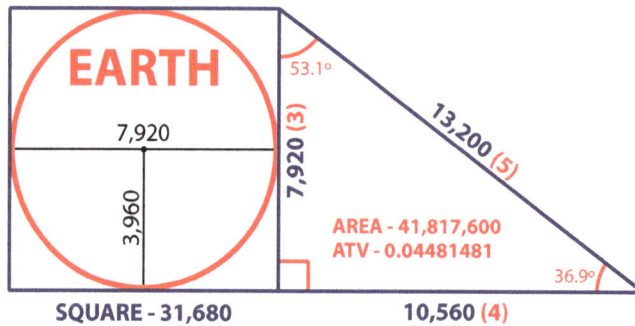

EARTH
7,920
3,960
53.1°
7,920 (3)
13,200 (5)
AREA - 41,817,600
ATV - 0.04481481
36.9°
SQUARE - 31,680
10,560 (4)

a = 864,000
b = 1,152,000
c = 1,440,000
P = 3,456,000
S = 1,728,000
K = 497,664,000,000
ha = 1,152,000
hb = 864,000
hc = 691,200

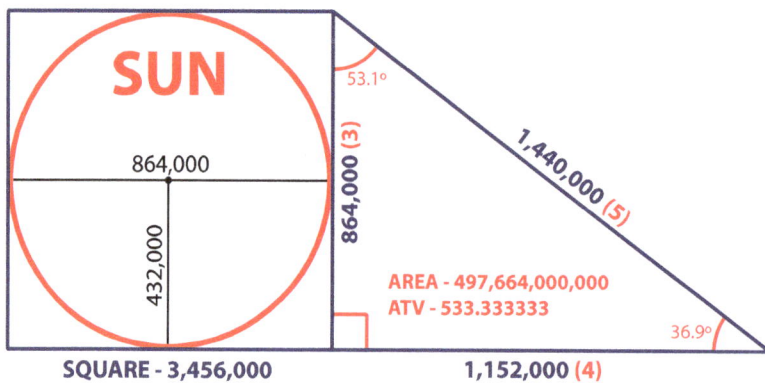

SUN
864,000
432,000
53.1°
864,000 (3)
1,440,000 (5)
AREA - 497,664,000,000
ATV - 533.333333
36.9°
SQUARE - 3,456,000
1,152,000 (4)

DIAGRAM 2 – SUN, MOON, EARTH ATV CALCULATIONS & RIGHT TRIANGLE CALCULATION.

APPLICATIONS FOR A MSAART EVO'S FORM AND FUNCTIONS

ABSTRACT

EXPLANATION OF THE CHANGE IN VARIABLE LIGHT SPEED AND TIME

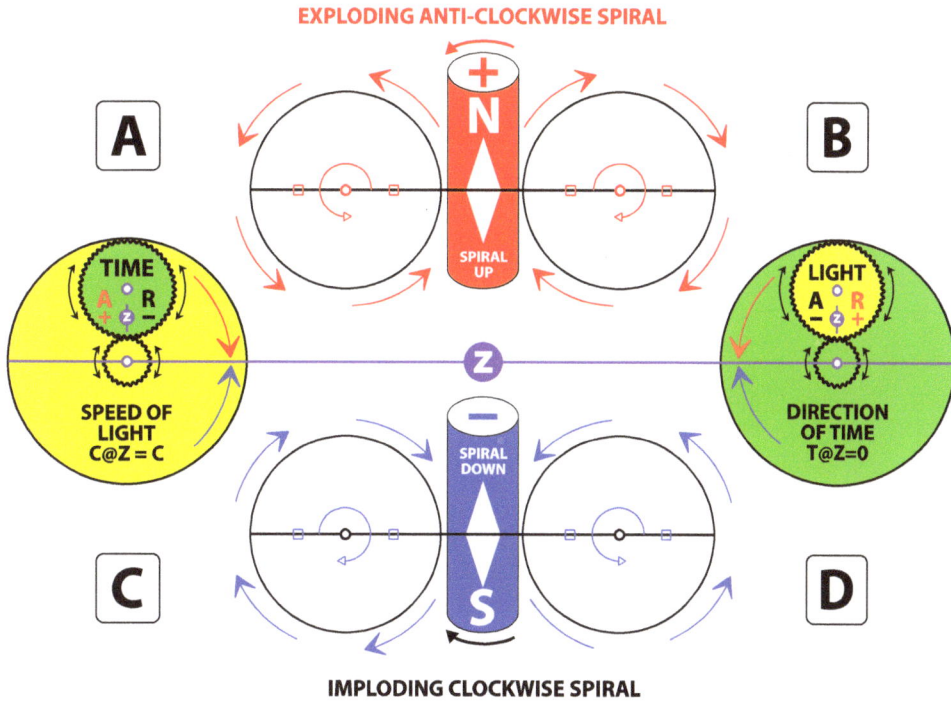

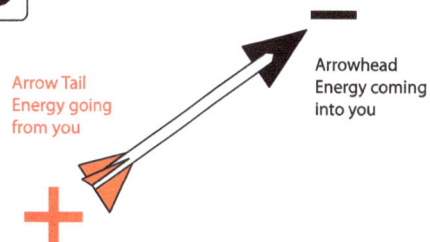

DIAGRAM 3 – VARIABLE LIGHT SPEED AND TIME CALCULATOR.

A

RED A = Time acceleration caused by an anti-clockwise spiral explosive force repelled by the high charge density zero point and attracted to and seeking a lower charge density.

BLUE R = Time reversal caused by a clockwise imploding spiral repelled by a low charge density and attracted to and seeking the high charge density zero point.

B

- As the Negative charge increases, time reverses.
- As the Positive increases, time accelerates
- As the Negative charge increases, the speed of Light increases.
- As the Positive charge increases, the speed of Light decreases.

C

C is only constant at the Zero Point. Where the [AC] left hand clockwise spin [L] is zeroed out by the [AC] right hand anti-clockwise spin [R], thereby creating a zero frequency and consequently a DC point.

Therefore, E only equals M (mass) times the speed of light squared when L/R equals one at zero point.

Therefore $E = MC^2$
$E = L/R (C \times L/R)^2$

D

Arrow Tail Energy going from you

Arrowhead Energy coming into you

APPLICATIONS FOR A MSAART EVO'S FORM AND FUNCTIONS

FORWARD

The results of the treatment of landfill leachate within the Thunderstorm Generator proved that the leachate can be used as a source of Ammonia Hydroxide fuel for the engine as well as a medium by which new Ammonia Hydroxide may be generated in excess of the rate of use by the engine.

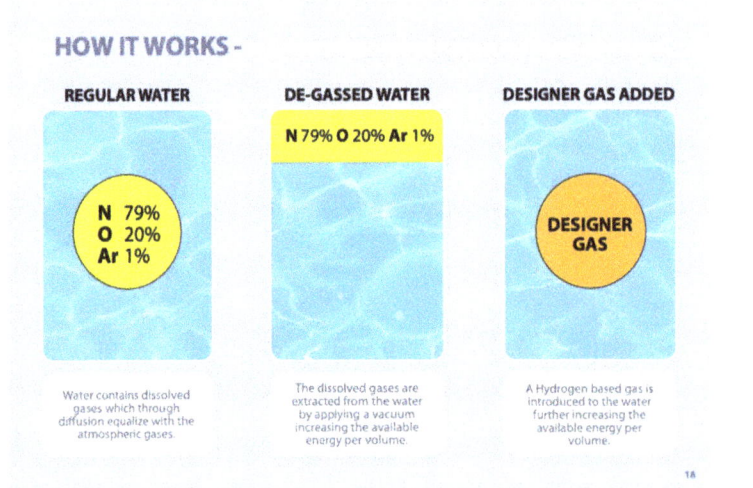

HOW IT WORKS -

REGULAR WATER	DE-GASSED WATER	DESIGNER GAS ADDED
N 79% O 20% Ar 1%	N 79% O 20% Ar 1%	DESIGNER GAS
Water contains dissolved gases which through diffusion equalize with the atmospheric gases.	The dissolved gases are extracted from the water by applying a vacuum increasing the available energy per volume.	A Hydrogen based gas is introduced to the water further increasing the available energy per volume.

DIAGRAM 4 :- PRECONDITIONING WATER FOR THE MSAART EVO GENERATOR.

This has been proven possible through the actions of the technologies applied vacuum, atmospheric air diffused into water (creating symmetrical minute bubbles that are imploded) and the creation, charging and subsequent catastrophic instantaneous discharge of those created MSAART EVO'S in the engine.

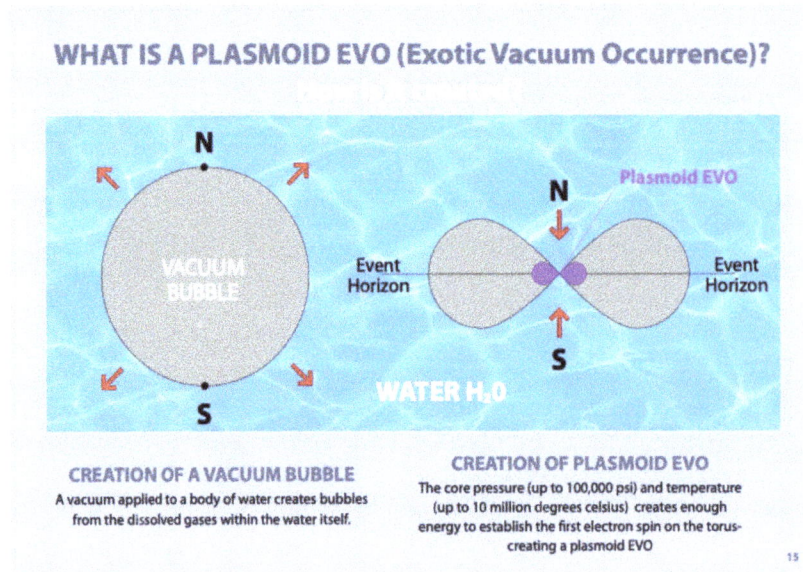

WHAT IS A PLASMOID EVO (Exotic Vacuum Occurrence)?

CREATION OF A VACUUM BUBBLE
A vacuum applied to a body of water creates bubbles from the dissolved gases within the water itself.

CREATION OF PLASMOID EVO
The core pressure (up to 100,000 psi) and temperature (up to 10 million degrees celsius) creates enough energy to establish the first electron spin on the torus-creating a plasmoid EVO

DIAGRAM 5 : - WHAT IS A MSAART EVO AND HOW IS IT CREATED?

APPLICATIONS FOR A MSAART EVO'S FORM AND FUNCTIONS

FORWARD

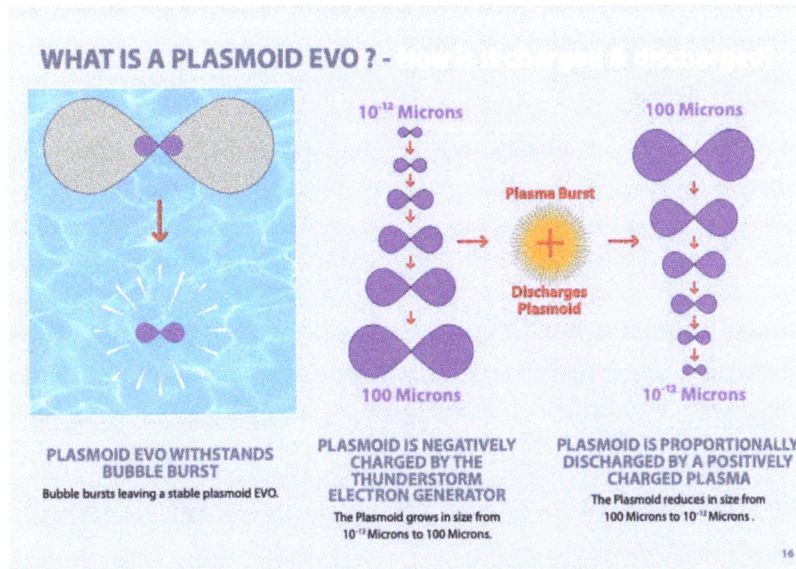

DIAGRAM 6 :- WHAT IS A MSAART EVO – HOW IS IT CHARGED AND DISCHARGED?

Diagrams 1 – 3 above demonstrate the symmetrical imploded bubble MSAART EVO formation, charge and discharge sequences. Diagram 4 below demonstrates the MSAART EVO'S effects on Water, including the separation of Hydrogen and Oxygen, the absorption of free electrons from the disintegrated Hydrogen.

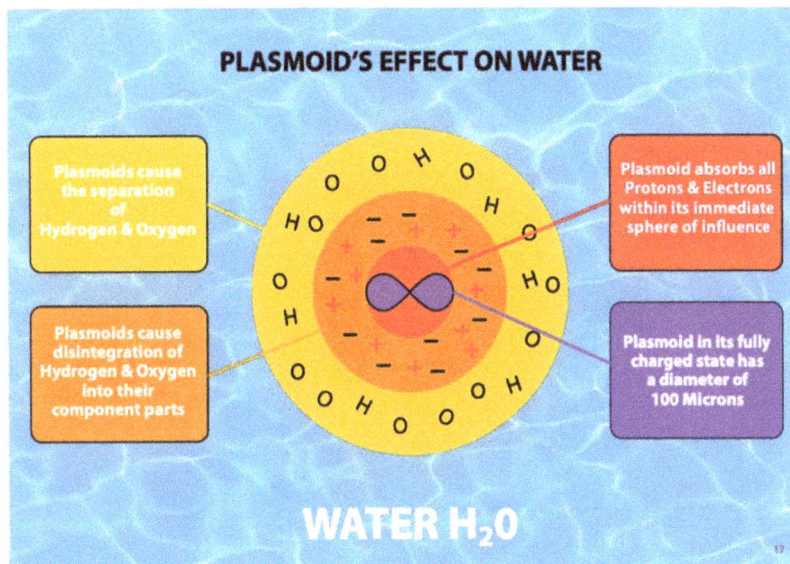

DIAGRAM 7 : - MSAART'S EFFECT ON WATER IS IT SPLITS H2O INTO O AND H2 PULLING APART PROTIUM (H) ATOM.

The MSAART EVO'S induced-flash instantaneous discharge, after its creation and charging within the Catalytic Tornado Resonator (CTR), within the engine's combustion chamber achieves a total clean burn of all fuels resulting in the elimination of all Carbon Monoxide unburnt fuel, and other toxic components.

APPLICATIONS FOR A MSAART EVO'S FORM AND FUNCTIONS

FORWARD

The MSAART EVO swarm displays quantum entanglement effects enabling them to share electrons instantaneously. This means that at the point of creation within the MSAART EVO Creator, the MSAART EVO'S are charging before transferring to the charge area within the Catalytic Tornado Resonator (CRT). After leaving the CTR charge area, MSAART EVO'S can transfer charge back to the MSAART EVO'S within the CTR charge area. A MSAART EVO's ability to store electrons collected within one inch of their container, even though stainless steel. Their capacity to distribute their charge to the swarm has profound implications for the capture use of waste heat. This is not just from the operation of internal combustion engines but any industrial or natural source of heat. The MSAART EVO's ability to controllably discharge its stored charge over time or instantaneously, means it has a wide range of uses as a power or heat source, or as an Atomic Battery. The Thunderstorm Generator Engine's description of normal operational parameters and performance, are augmented by schematics that are documented in Appendix 1. Results are summarised in Diagram 5 below.

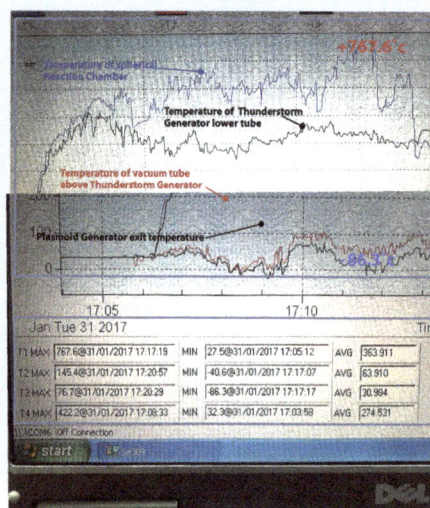

DIAGRAM 8 : - TEST RESULTS – THERMOCOUPLE READINGS.

The Terra Tek analytical results obtained from the treatment of the Leachate over varying time periods, within the Thunderstorm Generator Engine, are documented in Appendix 2 and then are graphed & analysed in Appendix 3. The scientific physics, chemical principles and geometry underpinning the technology are documented in Appendix 7.

APPLICATIONS FOR A MSAART EVO'S FORM AND FUNCTIONS

FORWARD

When one uses alien dimensional and temporal maths, chemistry, physics, music, resonant vibrations and astronomy, you provide the basis to reverse engineer our Sun's function. This can be achieved on a practical, scalable, fractal and safe platform using MSAART EVO'S to create Matter from Energy and Energy from Matter.

A PULSED VACUUM ATMOSPHERIC NITROGEN COMBINED WITH WATER SOURCED FREE PROTIUM TO PRODUCE AMMONIA HYDROXIDE FROM NORLAND'S LANDFILL LEACHATE.

In addition to the actions of the generated MSAART EVO'S on the leachate the applied vacuums influence on PH and therefore the availability of alternate 10,000 times H+ caused by the vacuum pulse and a concurrent reduction of 10,000 (OH-) then Visa -Versa on the pressure pulse provides the conditions precedent to manufacture Ammonia Hydroxide from the Leachate. The paragraphs below describe this process in detail.

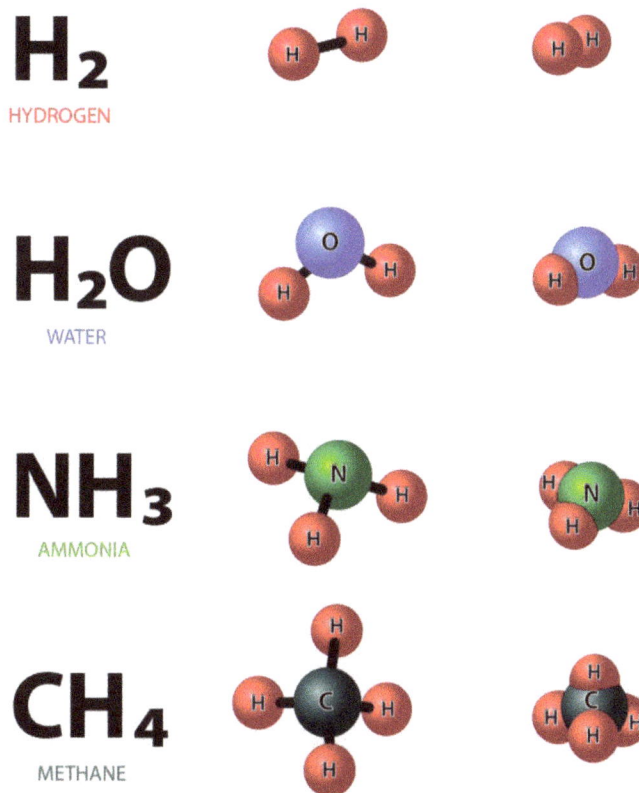

H_2
HYDROGEN

H_2O
WATER

NH_3
AMMONIA

CH_4
METHANE

DIAGRAM 9 – KEY ELEMENTS AND MOLECULES USED BY THE THUNDERSTORM GENERATOR MSAART EVO BASED TECHNOLOGY.

FORWARD

First Nitrogen from the atmosphere N2 is added to three water molecules (3H2O) to produce Ammonia. This is an exothermic reaction meaning that the production of Ammonia fuel in its-self produces available energy from that reaction to enable the production of even more Ammonia Hydroxide.

$$N_{2(g)} + 3H_{2(g)} \rightleftharpoons 2NH_{3(g)} + \Delta H$$

$$\Delta H = -92kJ\ mol^{-1}$$

AMMONIA MOLECULE NH3

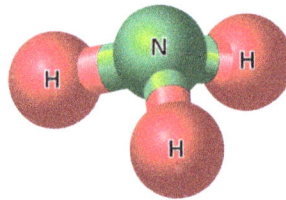

DIAGRAM 10 – AMMONIA MOLECULE NH3.

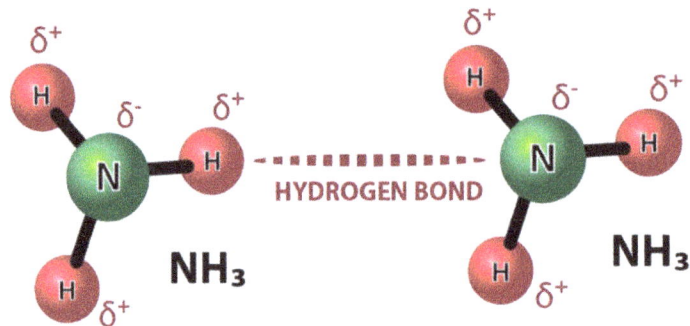

DIAGRAM 11 – AMMONIA'S MECHANISM TO HYDROGEN BOND WITH EACH OTHER.

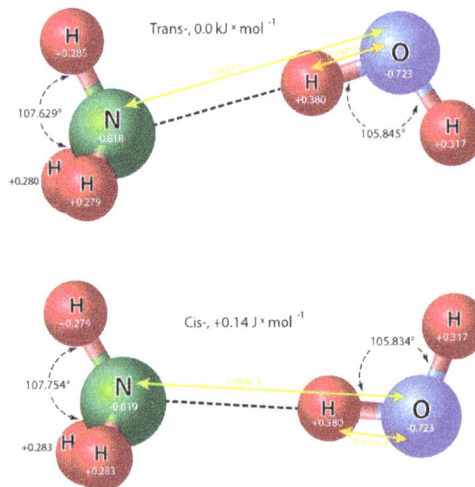

DIAGRAM 12 – AMMONIA'S HYDROGEN BONDING WITH WATER LETTING WATER CONTAIN UP TO 30% NH3.

APPLICATIONS FOR A MSAART EVO'S FORM AND FUNCTIONS

FORWARD

DIAGRAM 10 -

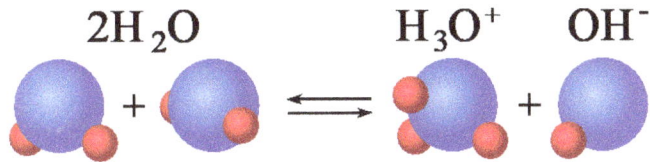

$$2H_2O \qquad H_3O^+ \qquad OH^-$$

DIAGRAM 13 – 2H2O (TWO WATER MOLECULES) TO H30+ + (OH)-.

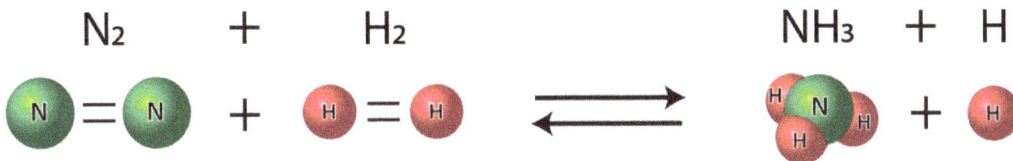

$$N_2 \qquad + \qquad H_2 \qquad\qquad NH_3 \quad + \quad H$$

DIAGRAM 14 – NATURAL N2 AND H2 COMBINED INTO NH3 PLUS A FREE HYDROGEN.

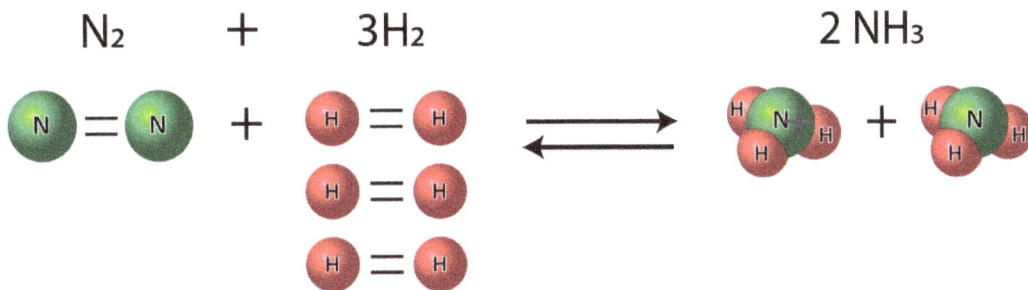

$$N_2 \qquad + \qquad 3H_2 \qquad\qquad 2\,NH_3$$

DIAGRAM 15 -

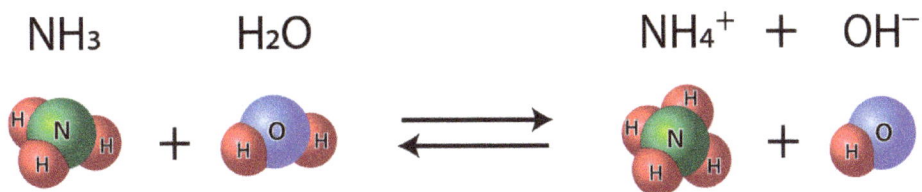

$$NH_3 \qquad H_2O \qquad\qquad NH_4^+ \ + \ OH^-$$

FORWARD

Then Ammonia Hydroxide is formed

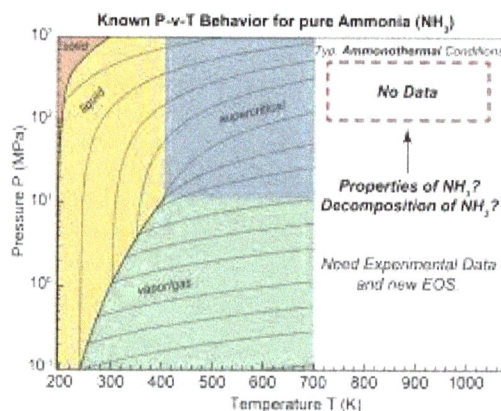

NH3 + H2O = (NH4) NET POSITIVE + (OH-) NET NEGATIVE

DIAGRAM 16-

THE COMBUSTION OF AMMONIA TO NITROGEN AND WATER IS <u>EXOTHERMIC</u>.

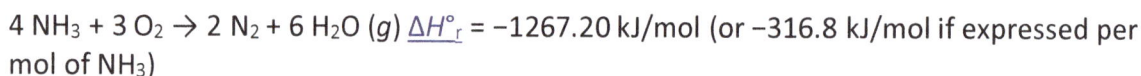

$4\,NH_3 + 3\,O_2 \rightarrow 2\,N_2 + 6\,H_2O$ (*g*) $\underline{\Delta H°_r}$ = −1267.20 kJ/mol (or −316.8 kJ/mol if expressed per mol of NH_3)

The <u>standard enthalpy change of combustion</u>, $\Delta H°_c$, expressed per <u>mole</u> of Ammonia and with condensation of the water formed, is −382.81 kJ/mol. Dinitrogen is the thermodynamic product of combustion: all <u>nitrogen oxides</u> are unstable with respect to N_2 and O_2, which is the principle behind the <u>catalytic converter</u>. Nitrogen oxides can be formed as kinetic products in the presence of appropriate catalysts, a reaction of great industrial importance in the production of <u>nitric acid</u>:

$$4\,NH_3 + 5\,O_2 \rightarrow 4\,NO + 6\,H_2O$$

A subsequent reaction leads to NO_2: $\qquad\qquad 2\,NO + O_2 \rightarrow 2\,NO_2$

The combustion of ammonia in air is very difficult in the absence of a catalyst (such as <u>platinum</u> gauze or warm <u>chromium(III) oxide</u>), due to the relatively low heat of combustion, a lower laminar burning velocity, high auto-ignition temperature, high heat of vaporization, and a narrow flammability range. However, recent studies have shown that efficient and stable combustion of ammonia can be achieved using swirl combustors, thereby rekindling research interest in ammonia as a fuel for thermal power production.[28] The flammable range of ammonia in dry air is 15.15%-27.35% and in 100% relative humidity air is 15.95%-26.55%.[29] For studying the kinetics of ammonia combustion a detailed reliable reaction mechanism is required, however knowledge about ammonia chemical kinetics during combustion process has been challenging.[30]

FORWARD

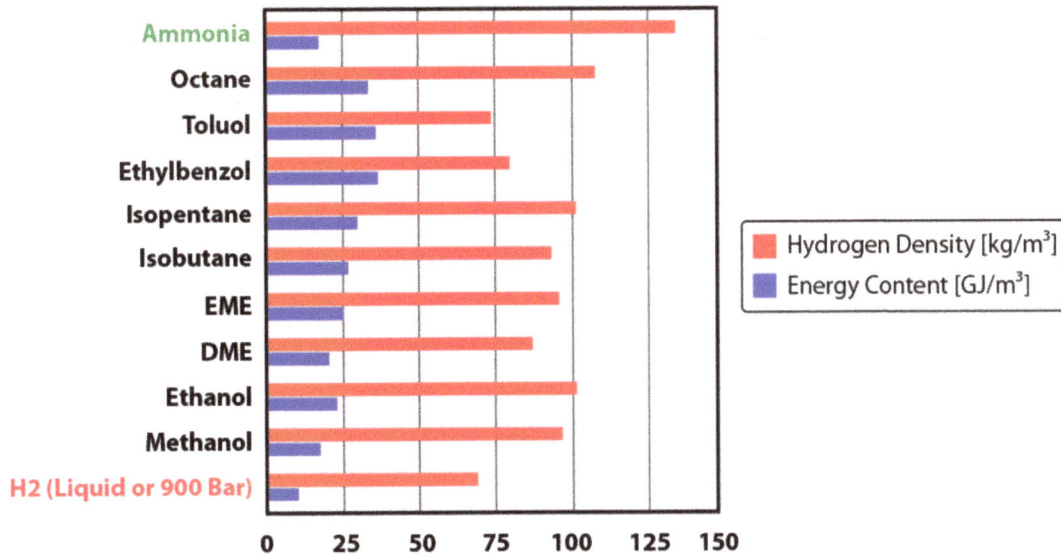

DIAGRAM 17 – AMMONIA COMPARITIVE CHART WITH HYDROCARBON FUELS AND HYDROGEN OF HYDROGEN DENSITY AND ENERGY CONTENT.

The data and operational observations together suggest that equal amounts of fuel used by the engine was equal to that stored within the leachate a conclusion supported by the fact that the amount of Ammonia Hydroxide present at the start of the test was equal to that contained within it at the end of the test. The pulsed intake vacuum and exhaust pressure created, as part the 66% of wasted energy from the normal operation of an internal combustion engine, is critical to enable the low energy creation of Ammonia Hydroxide. The vacuum pulse when applied to the top of the water contained within the MSAART EVO Generator produces a drop in the PH (negative log of free Hydrogen ions) of the water. This drop in PH gives available Hydrogen ions up to 10,000 times more than the Landfill Leachate at normal one Atmosphere of pressure. This means a drop from a recorded 8.8 PH a Strong Alkaline to a PH of 4.8 induced by the application of the vacuum at a negative pressure of minus two atmospheres.

FORWARD

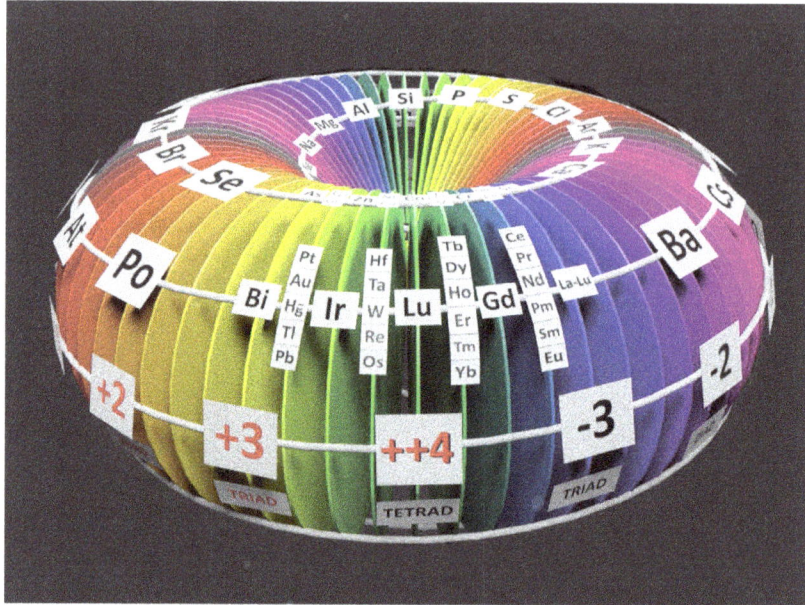

DIAGRAM 18 – MODEL OF THE ELEMENTS.

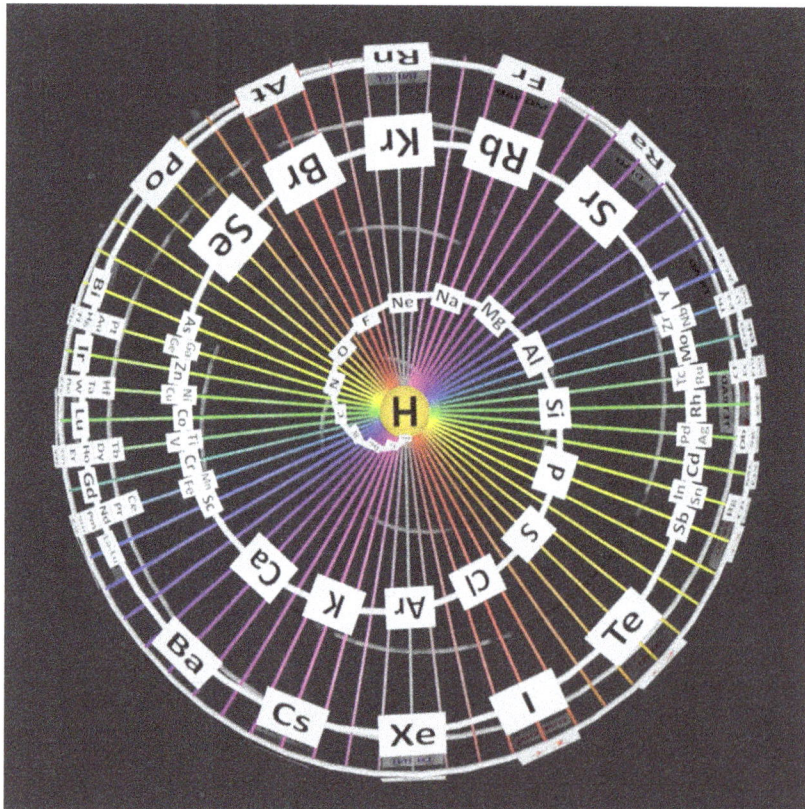

DIAGRAM 19 – ONE SPIRAL OF THE MODEL OF THE ELEMENTS (MOE).

FORWARD

Extraordinarily when the pressure pulse is applied after the vacuum pulse has passed 10,000 times less free Hydrogen ions are available and 10,000 times more Hydroxide (OH-) is available to attach to the two molecules of Ammonia Hydroxide. This production line effect literally "prints" Ammonia Hydroxide at a greater rate than the engine is "spending it".

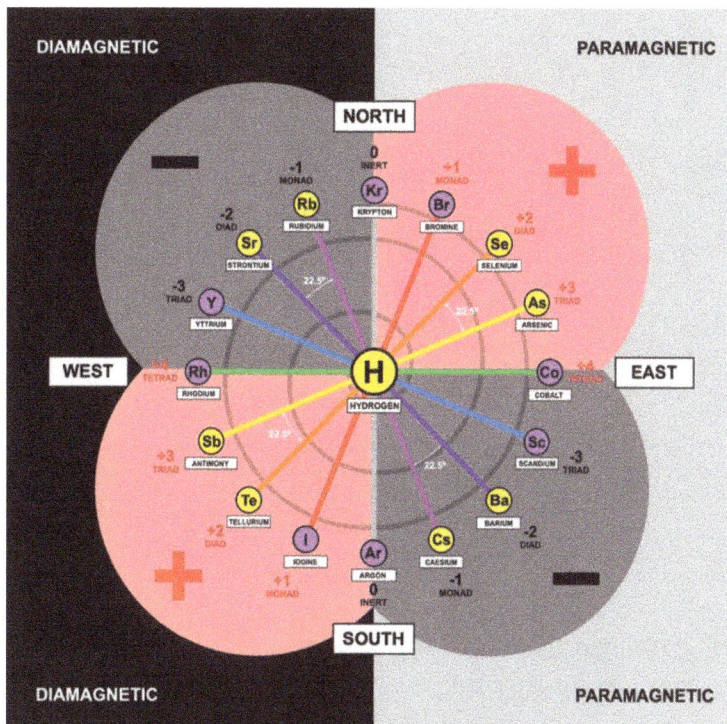

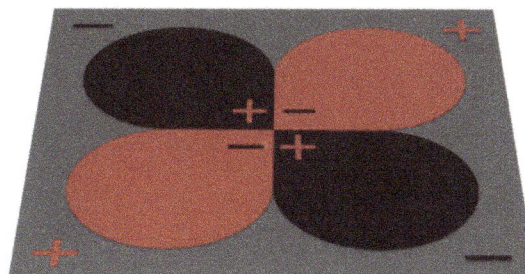

DIAGRAM 20 – ONE SPIRAL OF THE MODEL OF THE ELEMENTS (MOE).

FORWARD

H₂
HYDROGEN

H₂O
WATER

NH₃
AMMONIA

CH₄
METHANE

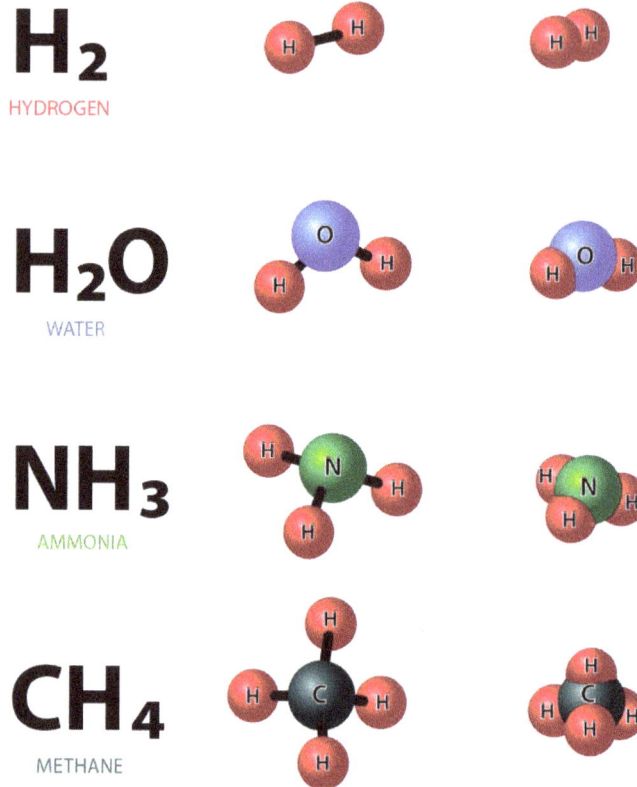

DIAGRAM 21 - KEY ABUNDANT ATMOSPHERIC ELEMENTS AND MOLECULES USED BY THE THUNDERSTORM GENERATOR MSAART EVO BASED TECHNOLOGY.

There is also a complimentary source of energy, in addition to the normal exhaust waste stream, for the generation of Ammonia Hydroxide. That is the burning of Carbon Monoxide , unspent Hydrocarbon fuel and other toxic compounds found within Hydrocarbon fuels. This burning occurs within the Catalytic Tornado Resonator (CTR) component of the Thunderstorm Generator device. The heat generated by this combustion releases a significant amount of free electrons between the 4" outer sphere and the 3" inner sphere. Between the 3" sphere and the 2" sphere a vacuum absorbs a stream of MSAART EVO'S, water in its gas phase, Ammonia and Ammonia Hydroxide. These components all interact with these free protons, electrons and free Hydrogen ions. These interactions induce many changes and transformations within the components present within the engines induced Vacuum's imploding stream.

APPLICATIONS FOR A MSAART EVO'S FORM AND FUNCTIONS

FORWARD

THESE CHANGES HAVE THE EFFECTS LISTED BELOW.

1.) MSAART EVO'S capture and put into orbit the free protons and electrons and thereby grow is size and area of their effect.

2.) The water splits into Hydrogen and oxygen creating fuel.

3.) The proportion of H3O to H2O increases as free Hydrogen ions saturate the highly ionised water molecule. This makes it possible for the fifth state of Flat Water to form on the inside of the 3" sphere due to ionic attraction created by the positive charge being generated by the direction of spin and combusting exhaust gas components creating highly positively charged particles.

4.) Oxygen, Nitrogen, Argon and Carbon Monoxide are all supercharged, ionised and go into the engines combustion chamber fundamentally altered from their precursor normal atmospheric gas low energy states.

5.) Within the combustion chamber the most dramatic difference is from the normal slow burn front of oxidising Hydrocarbon fuels to the instantaneous "flash" burn lightning discharge of the discharging MSAART EVO'S when subjected to the spark plugs positive charge. The MSAART EVO, which in effect is an atomic battery, releases its stored electrons in proportionate response to both the Positive charge pulse of the spark plug but also to the relatively slowly burning hydrocarbons release of positively charged particals.
The end result is that not only do you achieve a perfect complete burn but there is in addition there are no unburnt Hydrocarbons or Carbon Monoxide present within the exhaust stream.

6.)

APPLICATIONS FOR A MSAART EVO'S FORM AND FUNCTIONS

FORWARD

PHASE CHANGE TEMPERATURES FOR ELEMENTS THAT REACH THEIR RESONANT HARMONIC POINT DETERMINED BY THE SUM OF THEIR INTERNAL ANGLES, CHARGE DENSITY, ATV & EXTERNAL PRESSURES.

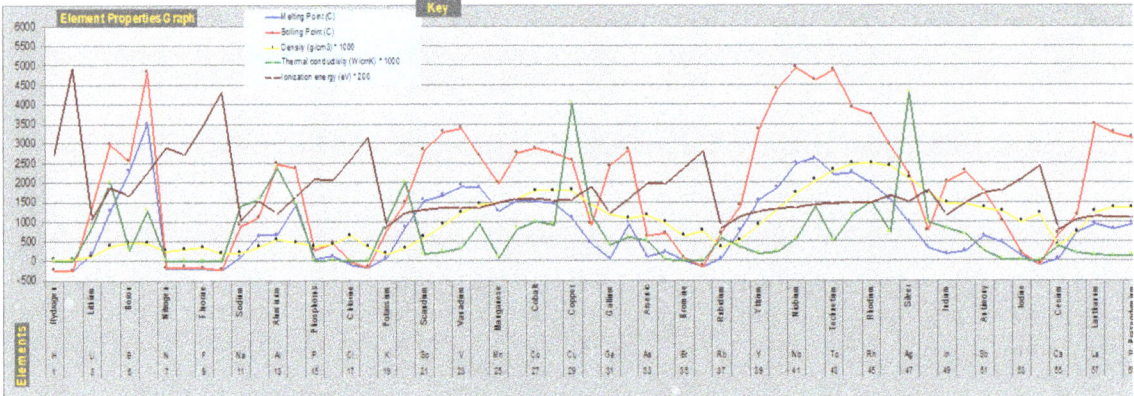

DIAGRAM 22 AND 23 – MODEL OF THE ELEMENTS DATA INCLUDING BOILING POINT, MELTING POINT, DENSITY, IONISATION ENERGY, THERMAL CONDUCTIVITY AND IONISATION ENERGY.

APPLICATIONS FOR A MSAART EVO'S FORM AND FUNCTIONS

FORWARD

THE FIFTH STATE OF WATER - 3D SPHERE
NEGATIVELY CHARGED MSAART EVO NUCLEI CENTRE SPIRAL
WATER MOLECULAR SPHERE
BENDALLS THUNDERSTORM WATER BALLS GEOMETRY.

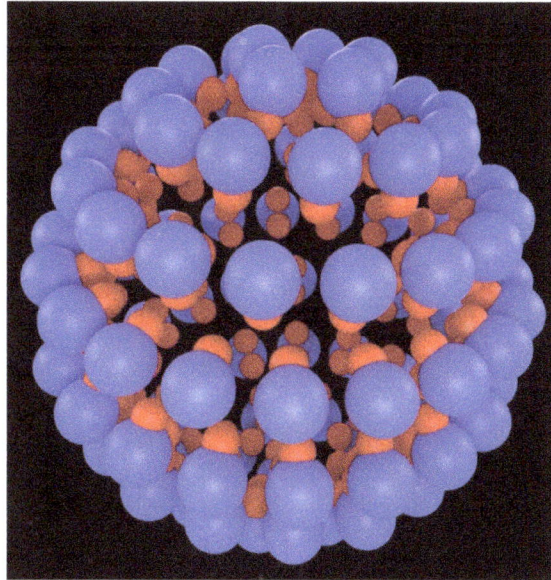

DIAGRAM 24 – A NET NEGATIVELY CHARGED MSAART EVO CENTRED WATER BALL, WITH OXYGEN OUT AND WITH HYDROGEN IN, EFFECTING INCREASED POTENTIAL DIFFERANCE.

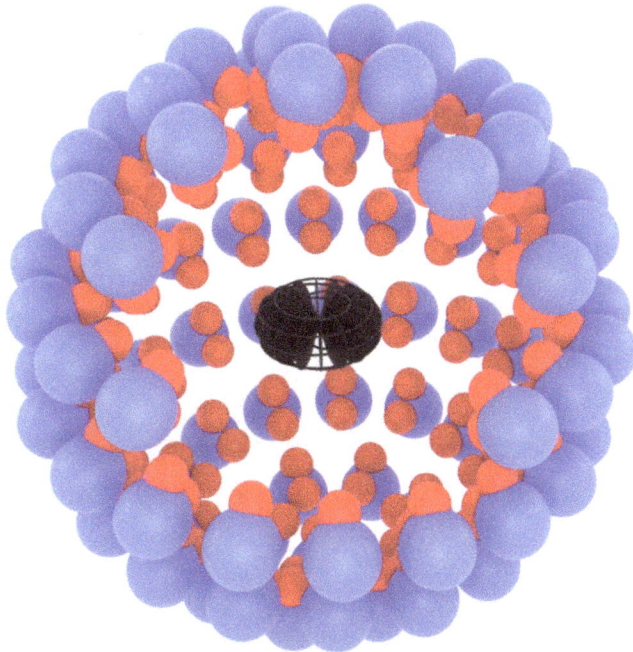

DIAGRAM 25- NET NEGATIVE CHARGE MSAART EVO CENTRED WATER BALL CUT AWAY WITH MSAART EVO.

APPLICATIONS FOR A MSAART EVO'S FORM AND FUNCTIONS

FORWARD

THE FIFTH STATE OF WATER - 3D SPHERE
POSITIVE CHARGE MSAART EVO NUCLEI CENTRES A SPIRAL
WATER MOLECULAR SPHERE
BENDALL'S THUNDERSTORM WATER BALLS GEOMETRY.

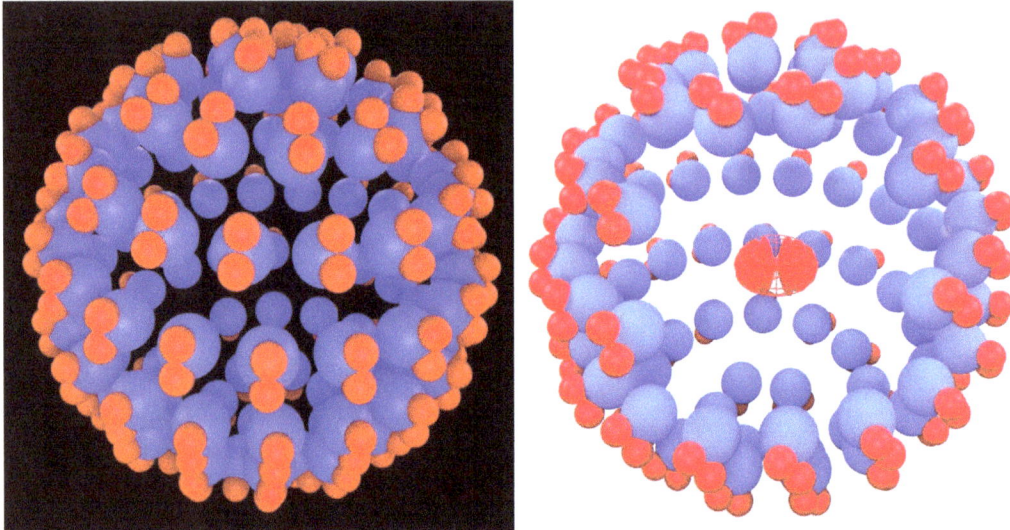

DIAGRAM 26 – NET POSITIVE CHARGE MSAART EVO CENTRED WATER BALL
WITH OXYGEN IN HYDROGEN OUT.

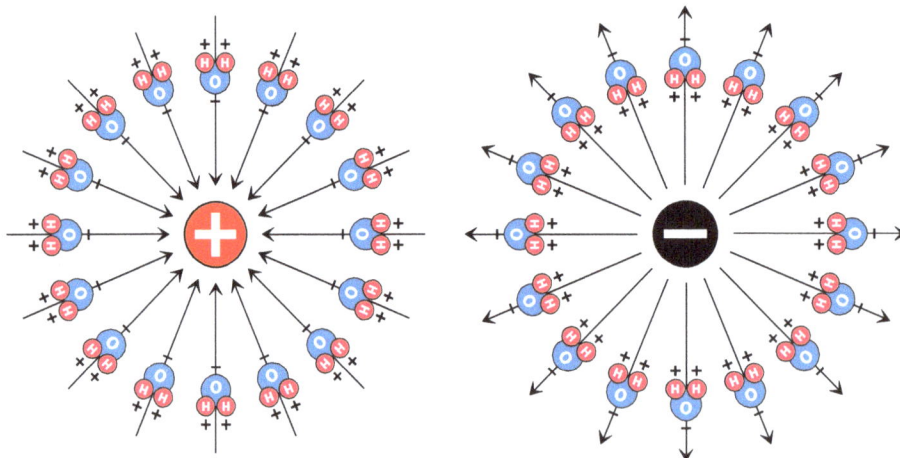

DIAGRAM 27 - NET POSITIVE / NEGATIVE CHARGE MSAART EVO CENTRED WATER BALL.

APPLICATIONS FOR A MSAART EVO'S FORM AND FUNCTIONS

FORWARD

THE FORTH STATE OF WATER – 2D
WOVEN ELECTRON BARRIER SHEETS
CRYSTALINE WATER MOLECULAR SHEETS
KMV'S 2D HEXAELET FLAT SCREEN GEOMETRY PROTON GUN

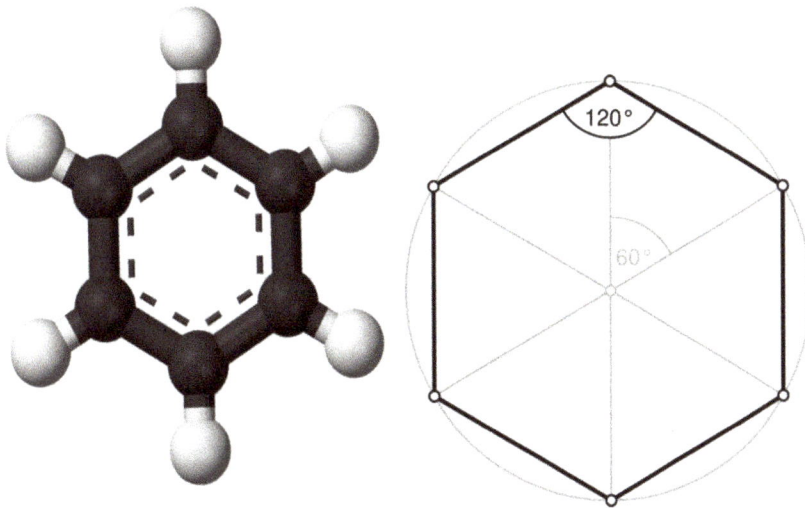

DIAGRAM 28 –

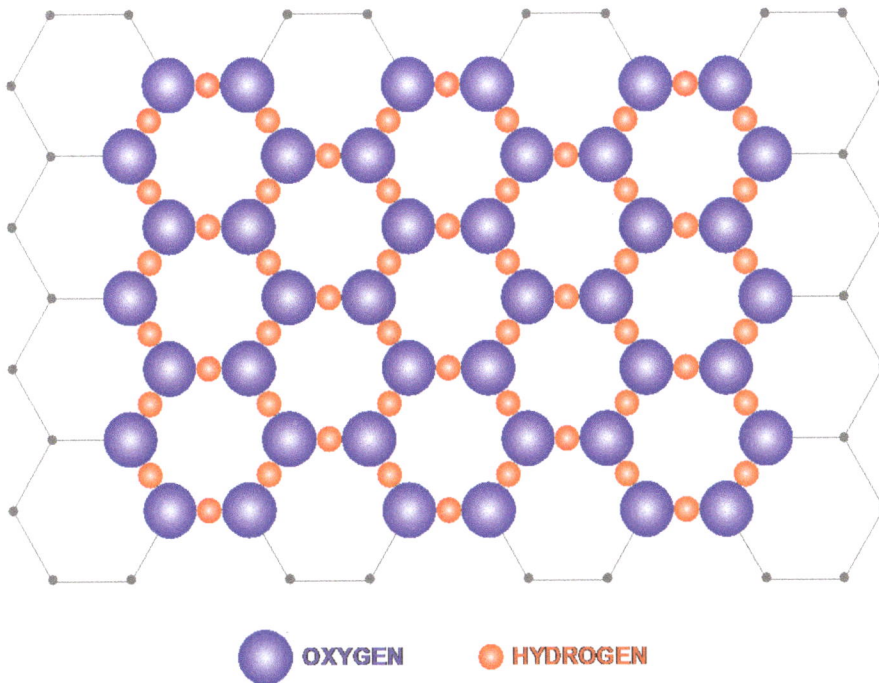

OXYGEN   HYDROGEN

DIAGRAM 29 –

INTRODUCTION

The photo below shows the Norland landfill site that has been sealed and is currently used to extract landfill methane gas for the purpose of generating electricity. As a result of producing the Methane landfill gas from drill holes some leachate fluid, is extracted along with the gas. A sample of this leachate was taken. This sample was treated in the MSAART EVO Creator component of the Thunderstorm Generator. The purpose was to determine the leachates potential use as a source of fuel. And this determines if the Thunderstorm Generators actions could remove harmful toxic components. These include Cyanide from the Leachate. London water currently applies penalties of about 60,000 pounds a year to accept and treat this toxic waste stream.

DIAGRAM 30 – AERIAL PHOTO NORLANDS TIP SITE.

The leachate was known to contain dissolved Methane, Ammonia Hydroxide and other potential fuel components. These may be boiled off by the vacuum generated by the Thunderstorm Generator and used as fuel for the generator its self. Any fuels extracted from the leachate may enhance the power generation capacity of the site. It will also reduce or potentially eliminate any need to pay for the leachate fluid offsite treatment or, at least by removing toxins, reduce the penalties applied by London water.

APPLICATIONS FOR A MSAART EVO'S FORM AND FUNCTIONS

INTRODUCTION

In addition, lower quality gas may benefit from the MSAART EVO's discharge within the engine's combustion chamber. This produces a complete burn of the Methane and also burns other components. This would not otherwise have reached the critical required temperatures for combustion and thereby destroy potentially harmful components. Lower volumes of gas may be required to produce the same amount of electricity. This is due to the influence of the MSAART EVO'S making in effect an additional fuel by treating the leachates of Methane, Ammonia Hydroxide, water disassociation into Hydrogen and Oxygen and the release of energy from the deconstruction of free Hydrogen ions into Protons and Electrons.

This may mean an extended life for the Norlands landfill site as smaller amounts of gas, because of the addition of the waste stream leachate volatiles as an augmenting fuel, may generate the same or increased electricity from the same amount of landfill gas. The Thunderstorm Generator also will remove toxic exhaust components such as Carbon Monoxide, unburnt Hydrocarbons and other unwanted reactive molecules.

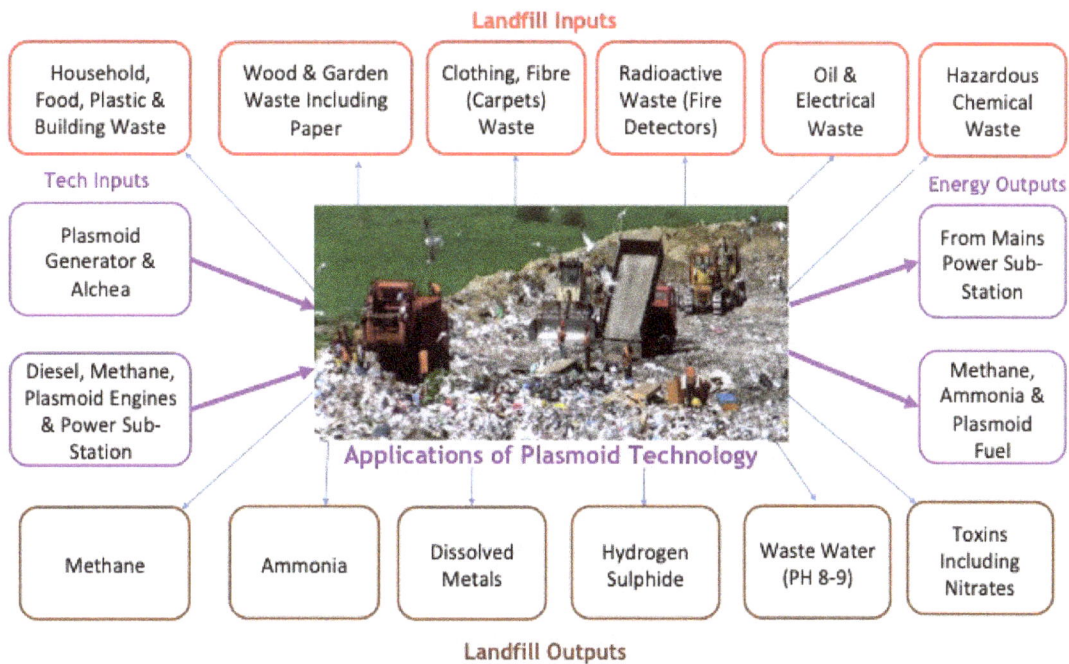

DIAGRAM 31- LANDFILL APPLICATIONS FOR MSAART EVO TECHNOLOGY.

A one litre control Sample was taken of the untreated Norlands landfill leachate fluid (sample A) and of samples treated for 3 minutes (sample C) 5 minutes (sample D) and 8 minutes (sample E) respectively within the Catalytic Tornado Resonator component of the Thunderstorm Generator. The

INTRODUCTION

Thunderstorm Generator apparatus is seen in the schematic of Appendix 1 and the photo below shows the colour changes to the leachate, especially the 3 minute sample, caused by the leachates exposure to a vacuum, catalysts, imploding bubbles and MSAART EVO'S.

DIAGRAM 32 - THE BOTTLES FROM LEFT TO RIGHT ARE THE 5 MINUTE, 3 MINUTE, 1.5 MINUTE AND A ZERO MINUTE CONTROL SAMPLE. THE 3 MINUTE SAMPLE IS LIGHTER THAN THE CONTROL.

These actions are analogous to cavitation bubbles eating away propellers and also to the effects demonstrated by the "The star in the Jar" production of High energy packages including high intensity purple UV light using ultra sonic sound to generate and then collapse bubbles 7,000 times a second.

The samples were then sent to TERRA TEK for analysis. The results of these tests are included below in Appendix 2 and a full analysis of the results are included in Appendix 3. Appendix 4 contains the knowledge of alien dimensional maths, chemistry, physics, numerology, astrology and sacred geometry essential to understand MSAART EVO'S and therefore understand its intrinsic novel Matter to Energy and Energy to Matter capabilities.

APPLICATIONS FOR A MSAART EVO'S FORM AND FUNCTIONS

INTRODUCTION

MSAART EVO'S have a self-generated, self-organising, electro-magnetic containment field and grow in size from its imploded bubble creation size of ten to the minus 12 in size, where it is stable, to in excess of the 100 micron diameter of the human hair, where it is unstable. The MSAART EVO'S imploded sphere geometry creates a direct current (DC) zero point on the toroidal equatorial event horizon plane where positive charge (AC) explodes and negative charge (AC) implodes from that zero point.

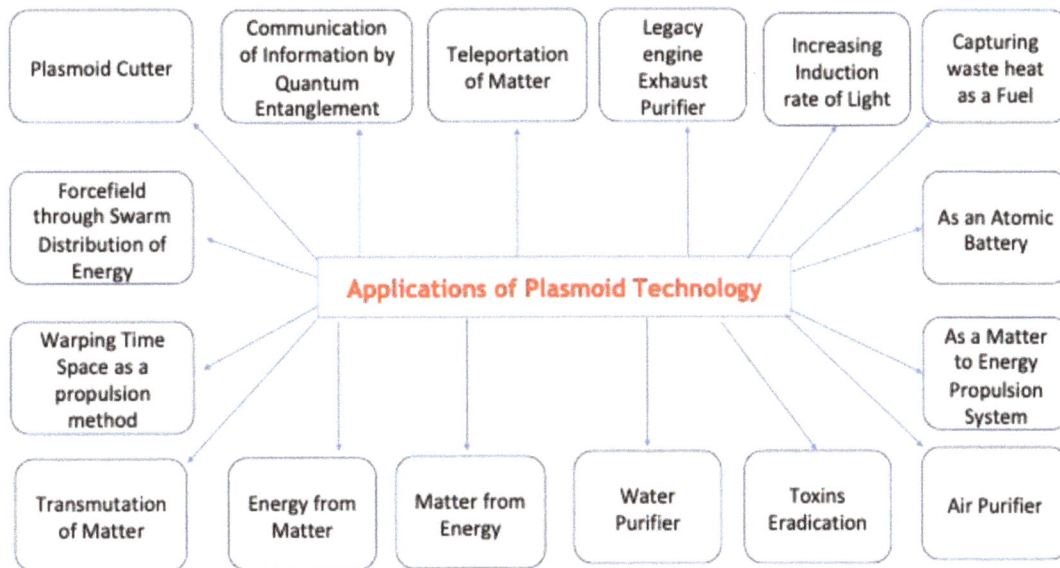

DIAGRAM 33 – SCOPE OF UNIVERSAL APPLICATIONS FOR MSAART EVO TECHNOLOGY.

The Thunderstorm Generator's vacuum and imploding symmetrical bubbles that generate MSAART EVO'S were used to treat landfill liquid leachate, from the London Norland land fill site, within the MSAART EVO Creator bubbler component of the Thunderstorm Generator engine. The leachate was produced as a by-product from the normal operation of Landfill Methane gas extraction by drill holes. Both the TERRA TEK test results and engines improved performance determined that the treated leachate can supply fuel to the engine, from Ammonia Hydroxide and dissolved Methane being boiled off from the leachate into the engines vacuum. The test results indicate that 2/3 of the Ammonia hydroxide present in the leachate was boiled off within the first 3 minutes of operation. However, when the MSAART EVO'S did reach their critical electron mass, after 3 minutes, additional Ammonia Hydroxide fuel was generated

INTRODUCTION

Introducing Plasmoid Technology

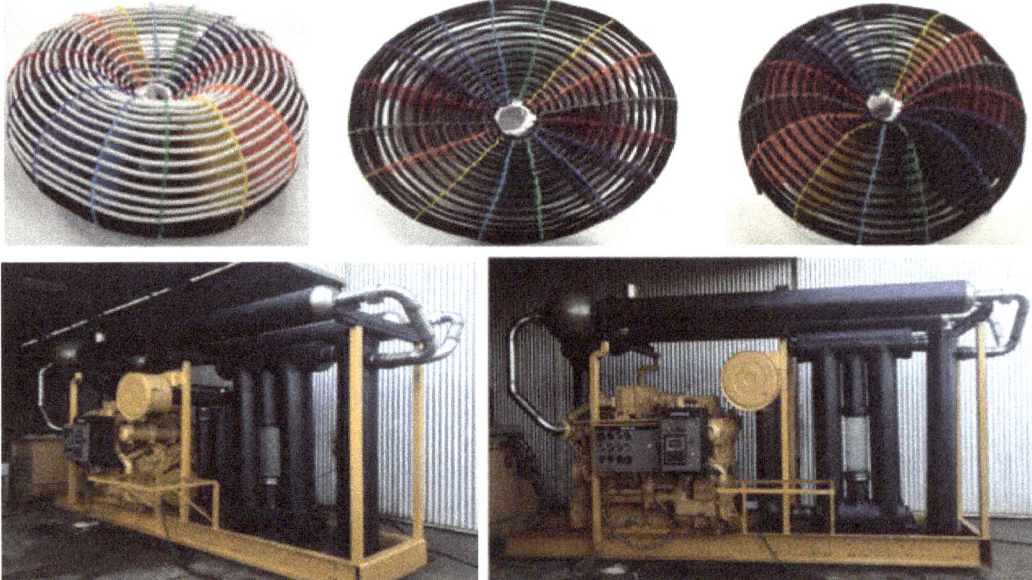

DIAGRAM 34 – ABOVE ARE MSAART EVO MODELS DEMONSTRATING THE FORM THAT ENABLES THEIR FUNCTION. THE SCALING UP TO A CATERPILLAR G3508E PROVED THE MSAART EVO TECHNOLOGY TO BE FRACTAL AND APPLICABLE TO ALL HYDROCARBON FUELS ENHANSING POWER WITH ZERO CARBON MOXIDE OR HYDROCARBON EMISSIONS.

due to the infusion of Nitrogen under 14.7 psi pressure from the atmosphere and the release of free Hydrogen by the action of the UV light, imploding bubbles and MSAART EVO'S. The MSAART EVO'S action upon the water and certain Leachate component parts, including the destroyed Cyanide and other compounds, provide the N, H and O to form Ammonia Hydroxide.

After 8 minutes of treatment the concentration of Ammonia Hydroxide had recovered to levels within the Leachate from the 3 minute sample C level of 576 mg/L to near those at the start of the test, from the zero time control sample A of 1,788 mg/L back to the 8 minute test sample E level of 1,698 mg/L. The engine after the 3 minute period increased in power and needed the idle setting to be wound back to prevent over revving of the motor. Throughout the test the throttle was held hard against the idle adjustment screw, as the Propane fuel from the bottle was only used to start and pre-heat the Catalytic Tornado Resonator component, once the Thunderstorm Generator was engaged the full throttle was moved back to and held in the idle position. These Ammonia and Methane fuels are in addition to the supply of bottled Methane, generated Hydrogen and charged MSAART EVO'S used or produced in the

APPLICATIONS FOR A MSAART EVO'S FORM AND FUNCTIONS

INTRODUCTION

normal operation of the Thunderstorm Generator. The test also determined that the Thunderstorm Generator can remove or reduce both chemical and biological toxins, including Cyanide, from the leachate by the effects of the ionized air, imploding bubbles and MSAART EVO'S on the leachate in the normal operation of the Thunderstorm Generator. The Thunderstorm Generator has also proved capable of removing or reducing the concentration of harmful biological hazards such as mould, fungus, bacteria, viruses and amoeba by the pre-requisite imploded bubbles due to the atomic based Low Energy Nuclear Reaction [LEANR] created by the bubbles. These interactions when reduced to percentage changes of the afore mentioned groups display uniform effects and proportional changes that indicate the validity of the results and prove processes and forces at work which are not classical. The results analysed in Appendices 2 and 3 can only be explained by complex atomic rearrangements of Matter and Energy resulting from forces and transformations not achievable without MSAART EVO influence, reaction and interaction with Matter (mainly Protium) and Energy (mainly High Energy UV light) caused by production of High energy bursts including UV light from the North to South pole implosion of vacuum generated bubbles and MSAART EVO'S. These bubbles also create, at the centre Zero point of the implosion on their equatorial plane event horizon, the energy and geometry for the creation of MSAART EVO'S and the bursts of energy including High energy UV light. The energy created at the bubbles implosive centre is amplified by the prior application of the UV light frequencies by the ioniser component of the Thunderstorm Generator to the air infused by a diffuser prior to the bubbles collapse.

The results contained within Appendices 2 and 3 of this report prove a complex but proportional interaction of the MSAART EVO'S on elements, salts and molecules from the leachate. The orderly time exposure destruction of Cyanide by high energy UV light proves the existence of MSAART EVO'S.

DEFINITION:- WIKIPEDIA MSAART EVO DEFINITION from WINSTON BOSTICK'S 1958 PAPER

"A MSAART EVO IS A COHERENT (TOROIDAL) STRUCTURE OF PLASMA AND MAGNETIC FIELDS. MSAART EVO'S HAVE BEEN PROPOSED TO EXPLAIN NATURAL PHENOMENA SUCH AS BALL LIGHTNING, MAGNETIC BUBBLES IN THE MAGNETOSPHERE AND OBJECTS IN COMET TAILS, IN SOLAR WIND, IN THE SOLAR ATMOSPHERE AND IN THE HELIOSPHERIC SHEET."

APPLICATIONS FOR A MSAART EVO'S FORM AND FUNCTIONS

INTRODUCTION

DIAGRAM 35 - THE THUNDERSTORM GENERATOR TEST SET UP FROM LEFT TO RIGHT IS THE GAS REGULATOR SUPPLYING PROPANE, THEN THE TWO MSAART EVO GENERATORS WITH THE WHITE CAPS AND YELLOW VALVES THEN ABOVE IS THE IONIZER PROVIDING UV TREATED ATMOSPHERIC AIR THEN THE FILLER FUNNEL FOR THE LEACHATE. THE 380CC SINGLE CYLINDER 4 STROKE ENGINE IS IN THE BACKGROUND. PAGE 1 SHOWS THE FRONT VIEW OF THE ENGINE.

DIAGRAM 36 - THE BOSCH EXHAUST GAS ANALYSER MEASURING ZERO EMMISSIONS OF CARBON MONOXIDE AND HYDROCARBONS AFTER TREATMENT OF THE EXHAUST FUMES WITHIN THE CATALYITIC TORNADO RESONATOR (CTR) ASSEMBLY OF AIR PULLED THRU BY THE VACCUM MADE BY THE ENGINE. WHEN THE ENGINE IS JUST RUN NORMALLY ON GAS THE CARBON MONOXIDE LEVELS ARE ABOUT 2% AND THE HYDROCARBON LEVEL IS BETWEEN 1,000 TO 3,000 PPM DEPENDING ON THROTTEL SETTINGS.

APPLICATIONS FOR A MSAART EVO'S FORM AND FUNCTIONS

AIM

The removal of all Carbon Monoxide and Hydrocarbons from the exhaust gas waste stream, as confirmed by our monitoring of the exhaust gas with our Bosche exhaust gas analyser, proved that the Catalytic Tornado Resonator and MSAART EVO generating and charging components are effective in treating the gas phase constituents of water. The aim of this test was to determine among other things whether landfill leachate waste fluid could be used as a fuel for a standard internal combustion engine fitted with the Thunderstorm Generator and still produce zero Carbon Monoxide and Hydrocarbons from the exhaust.

The Thunderstorm Generator's vacuum was used to treat liquid leachate, from the London Norlands land fill site, produced as a by-product from the normal operation of Landfill gas Methane drill hole extraction. This test was designed to determine whether the leachate, once treated within the MSAART EVO Creator component of the Thunderstorm Generator, would both supply and generate a fuel to the engine (by Ammonia and dissolved Methane boiling off into the vacuum created by the engines retreating piston), and remove toxins, including Cyanide, from the leachate. The test would also confirm or deny the MSAART EVO'S ability to remove or reduce the concentration of harmful mould, fungus, bacteria, viruses, amoeba, elements, salts and molecules from the leachate. The removal or reduction and addition of elements and molecules would prove or deny the presence, action and abilities of the MSAART EVO swarm once it reached its critical operational electron mass. Another aim of the test work described in the previous paragraph was to determine if the composition of the exhaust gas using the leachate would be free of Carbon Monoxide and Hydrocarbon pollution.

METHOD

The schematic and photo of the Atomic Protium Proton powered Thunderstorm Generator technology is shown in Appendix 1 on page 16. The methodology employed to treat, collect and assay the landfill leachate fluid, obtained from drill holes producing Methane tip gas, is documented here below. There is a detailed description of the Thunderstorm Generator Engine and component apparatus used are described below. The treatment times, collection methods and results of the treatment, within the MSAART EVO Creator component of the Thunderstorm Generator engine, of the leachate and their implications.

A one litre control Sample was taken of the untreated leachate (Sample A) and of samples treated for 3 minutes (sample C), 5 minutes (sample D) and 8 minutes (sample E) respectively. The samples were then sent to TERRA TEK for analysis. The results of this test work are included below in Appendix 2 and a full analysis of the results are included in Appendix 3.

DIAGRAM 37 –

APPLICATIONS FOR A MSAART EVO'S FORM AND FUNCTIONS

METHOD

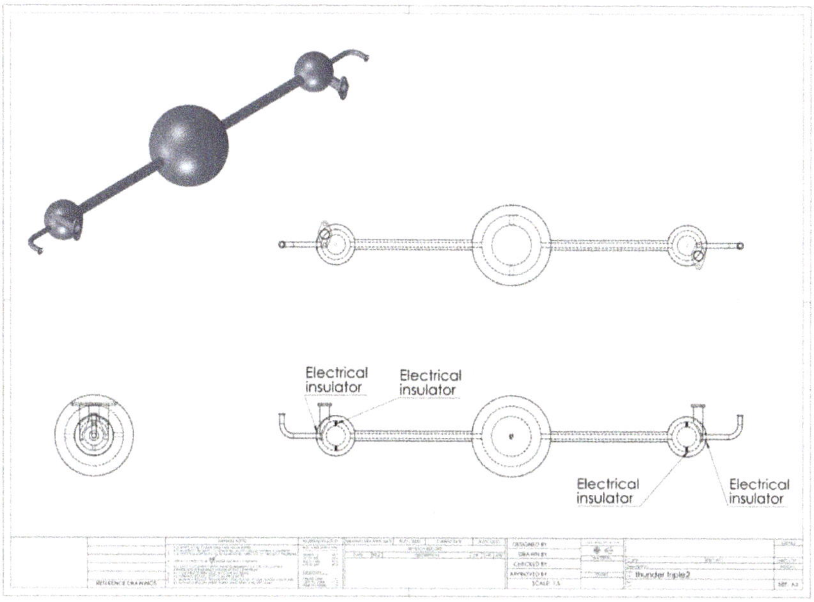

DIAGRAM 38 –

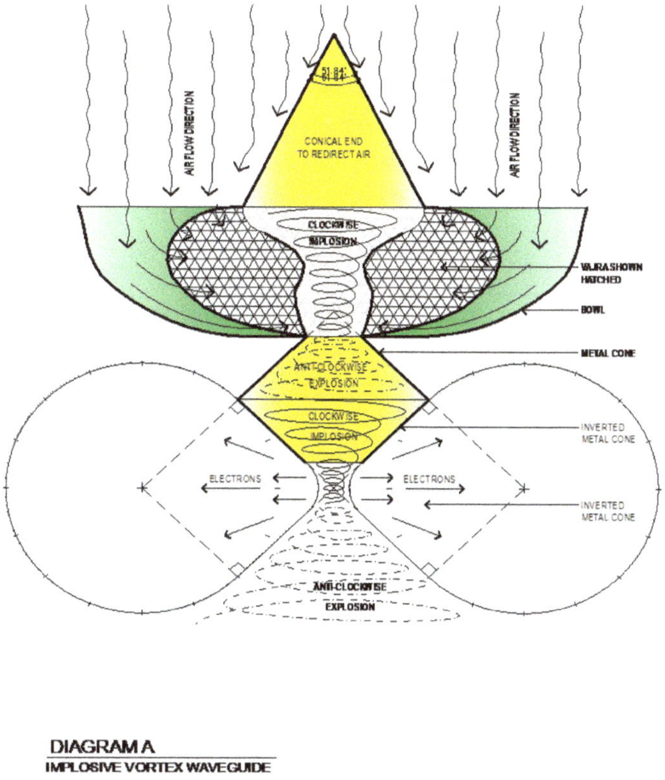

DIAGRAM A
IMPLOSIVE VORTEX WAVEGUIDE

DIAGRAM 39 –

METHOD

DIAGRAM 40 - THE MOLTEN SEA TORUS VAJRA IMPLOSIVE TURBINE SWIRL GUIDE & SPHERE

METHOD

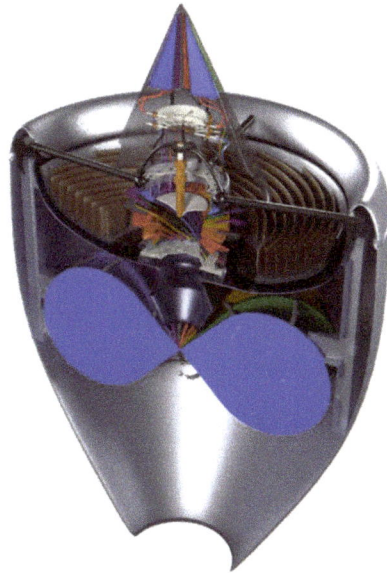

DIAGRAM 41 –

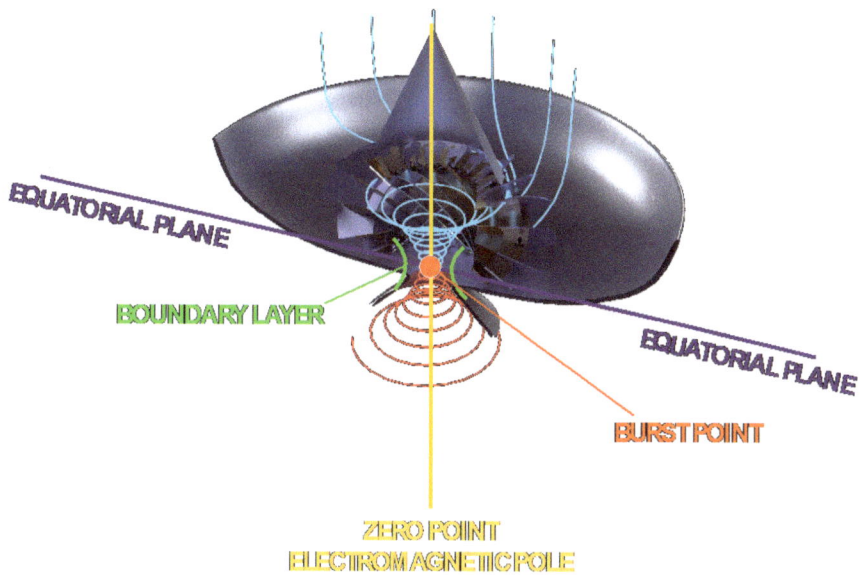

EQUATORIAL PLANE

BOUNDARY LAYER

EQUATORIAL PLANE

BURST POINT

ZERO POINT
ELECTROMAGNETIC POLE

DIAGRAM 42 –

METHOD

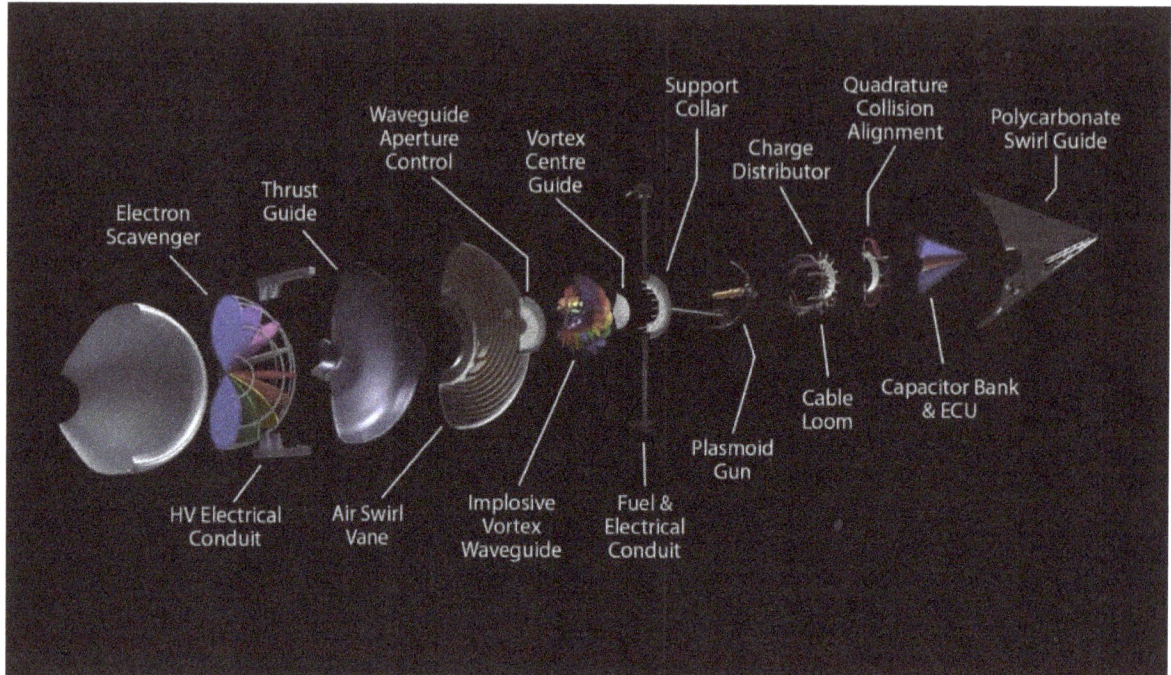

DIAGRAM 43 – MOLTEN SEA VAJRA SPACE IMPLOSIVE TURBINE.

The

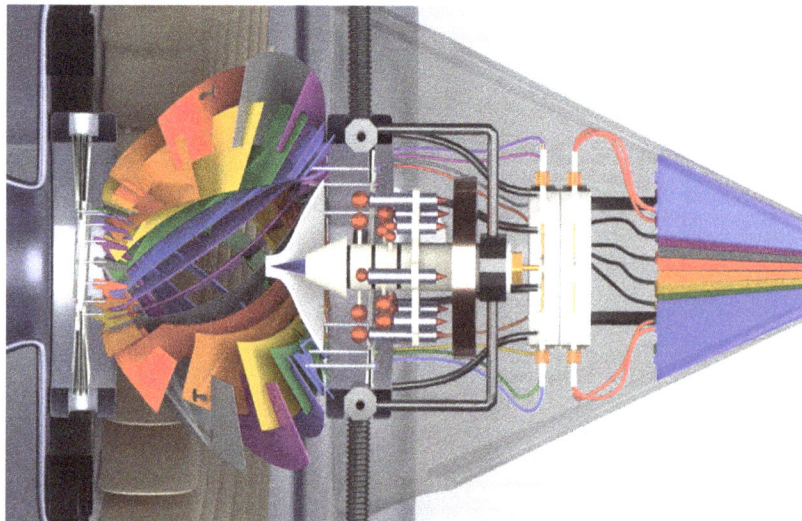

DIAGRAM 44– MOLTEN SEA VAJRA SPACE IMPLOSIVE TURBINE.

The

METHOD

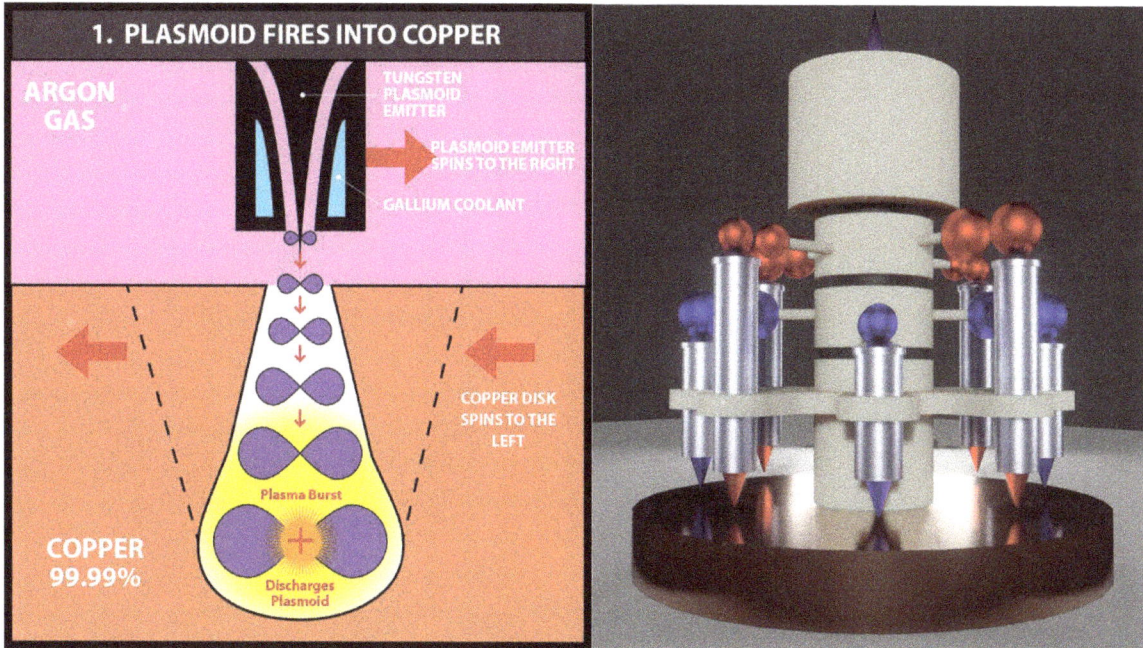

DIAGRAM 45 –

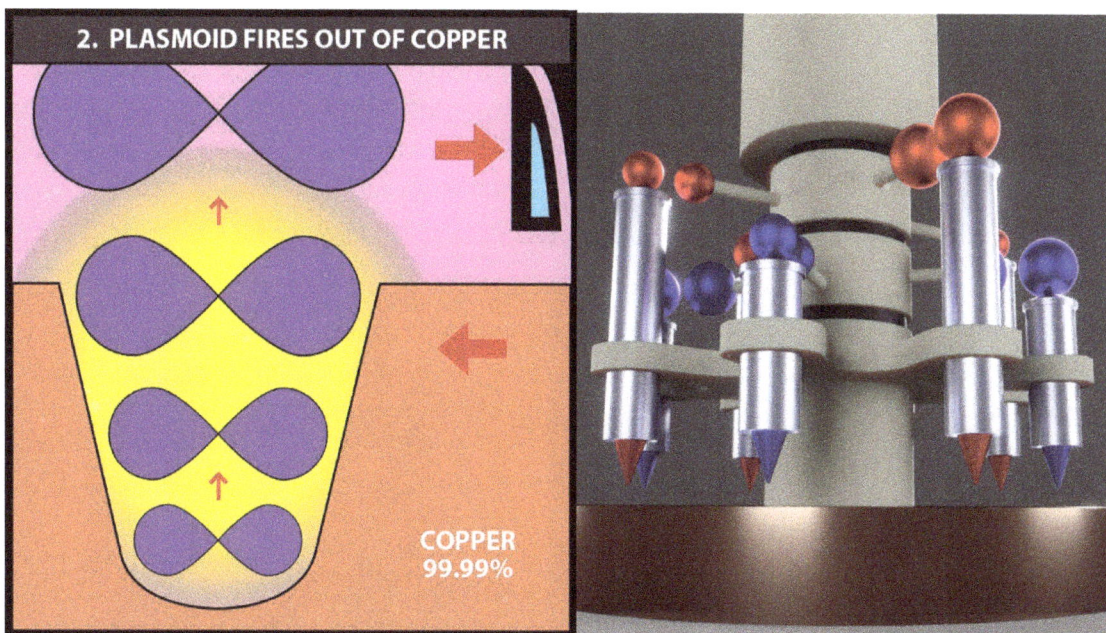

DIAGRAM 46 –

METHOD

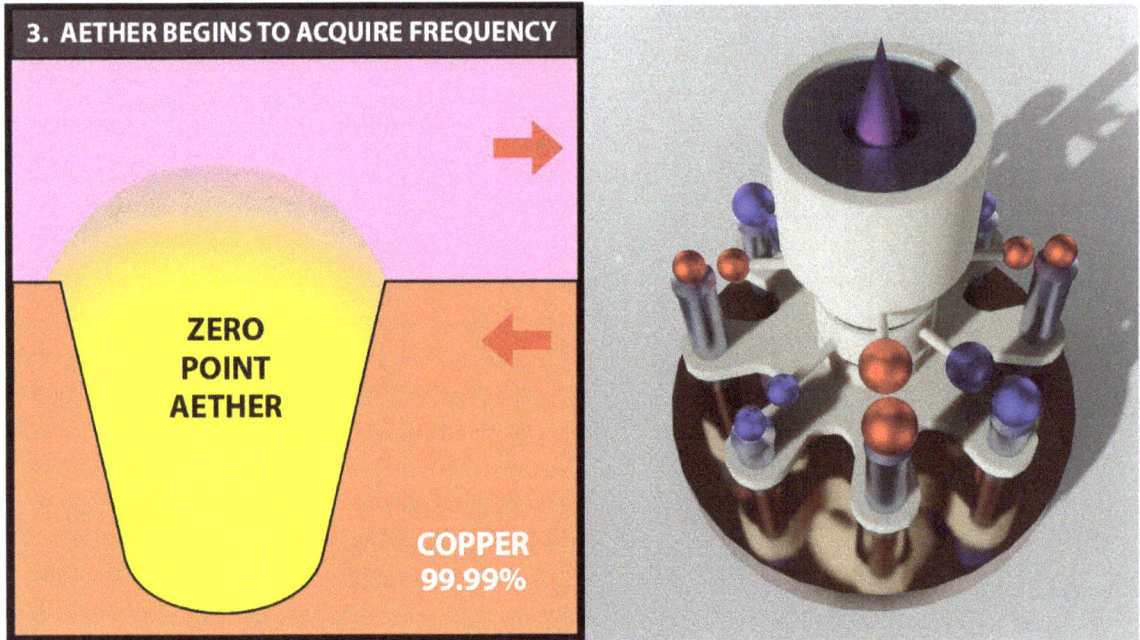

DIAGRAM 47 –

The

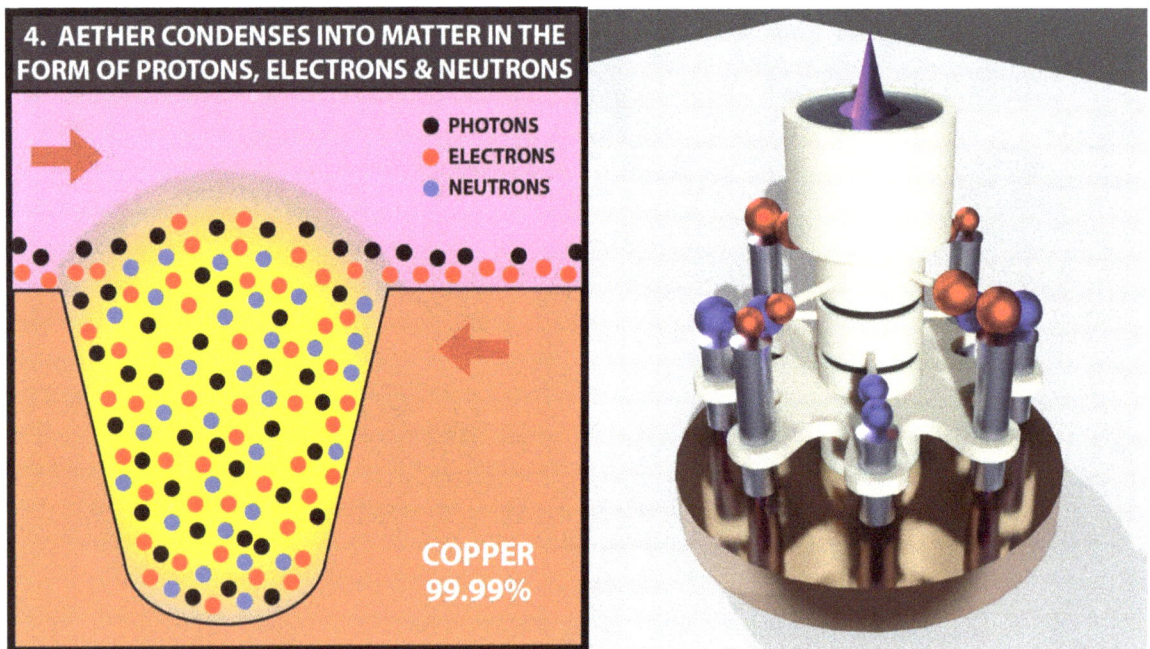

DIAGRAM 48 –

METHOD

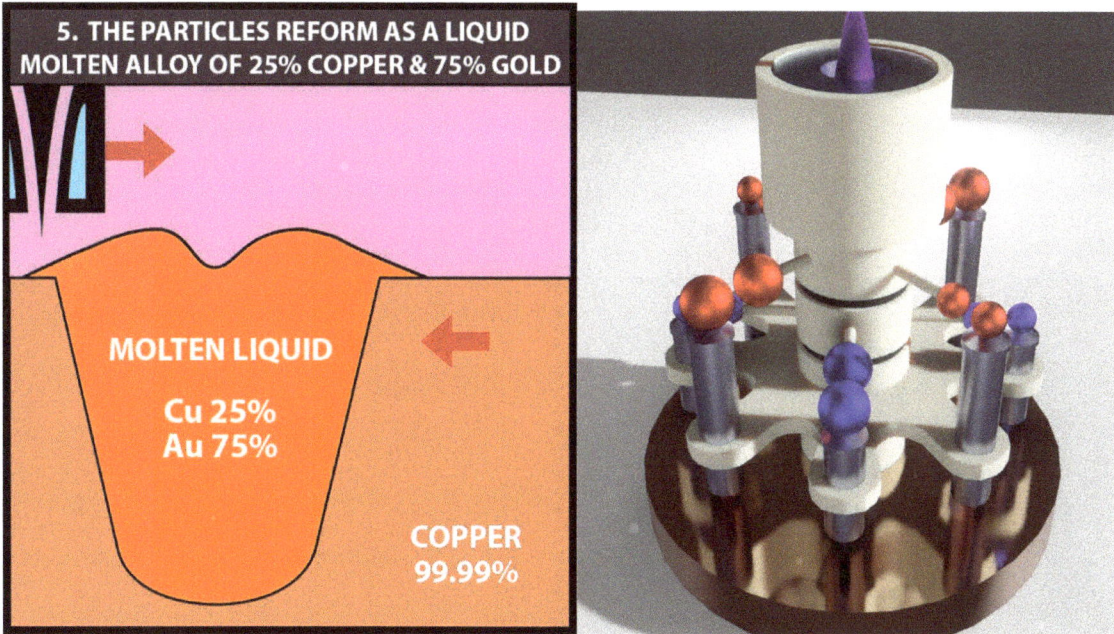

DIAGRAM 49 –

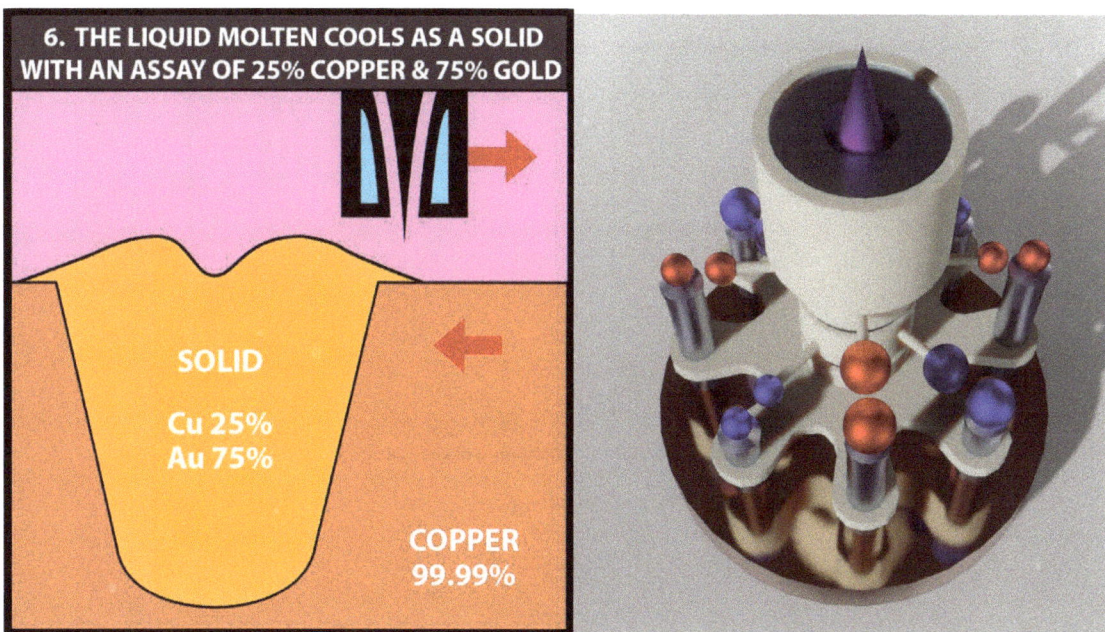

DIAGRAM 50 –

METHOD

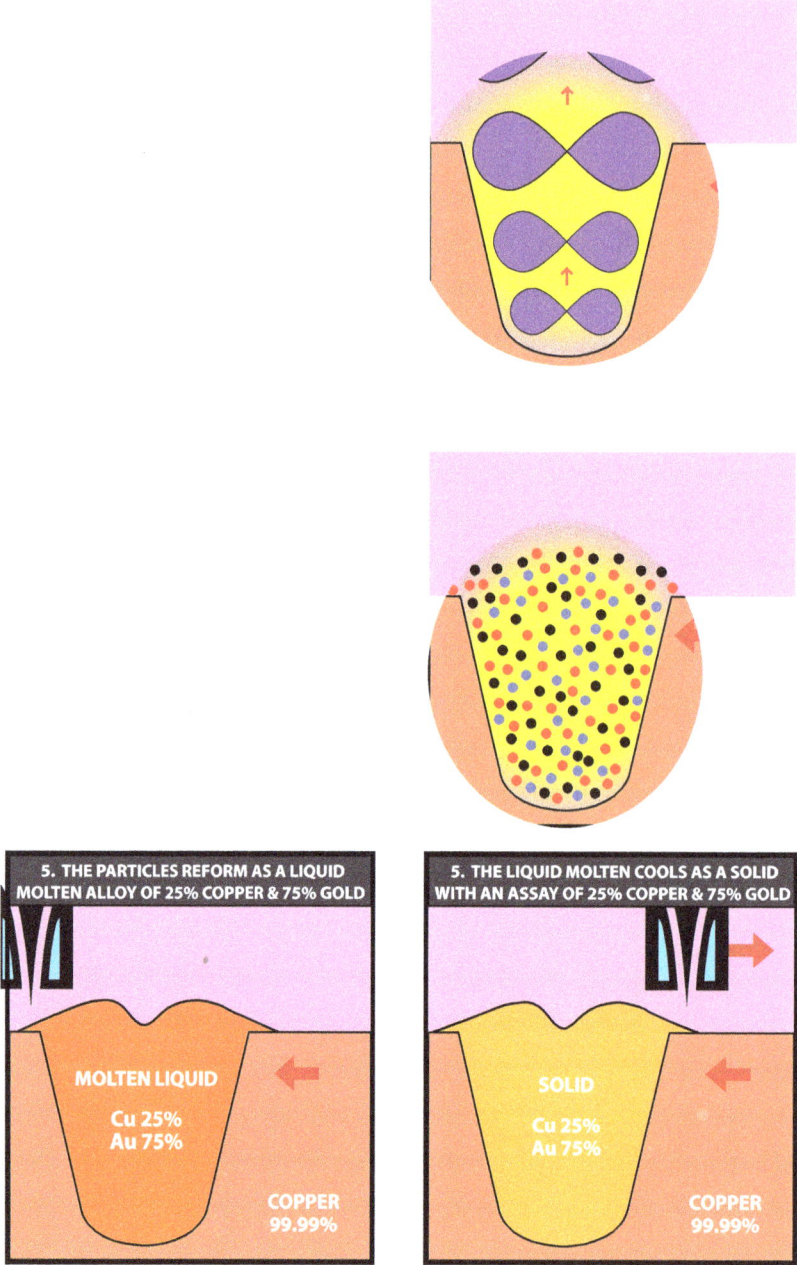

DIAGRAM 51 -

METHOD

DIAGRAM 52 – MSAART EVO GUN IMPLOSIVE THRUST ASSEMBLY.

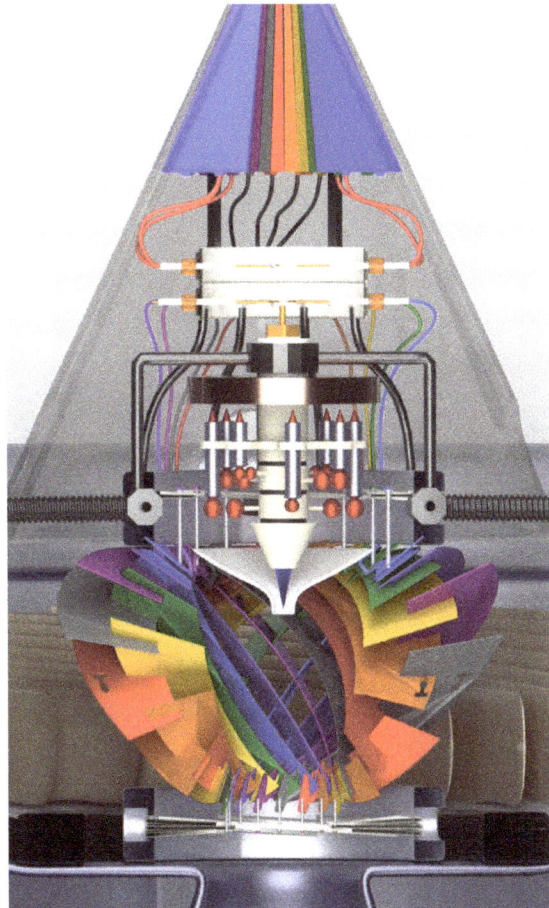

APPLICATIONS FOR A MSAART EVO'S FORM AND FUNCTIONS

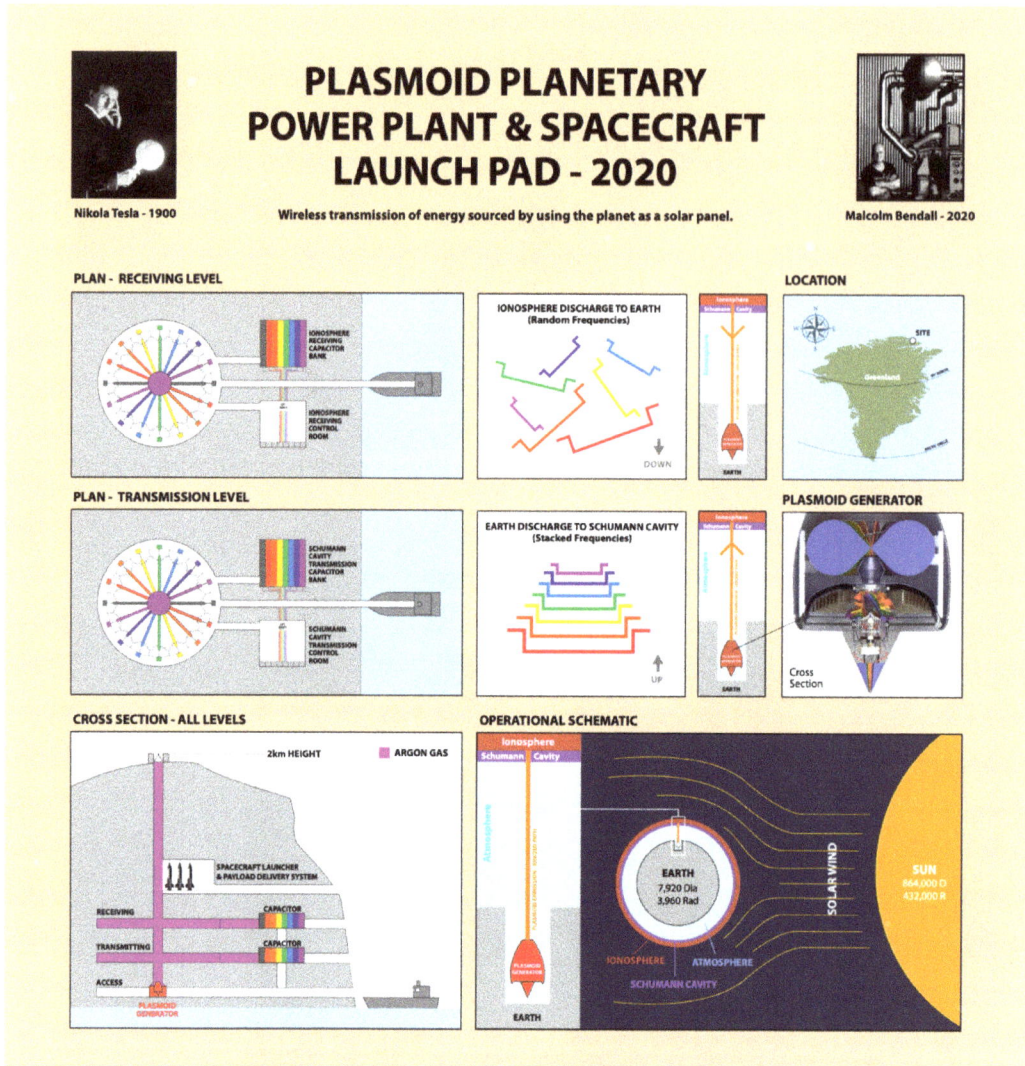

DIAGRAM 53 – MSAART EVO PLANETARY POWER PLANT & SPACECRAFT LAUNCHPAD.

RESULTS

The empirical results from the engine performance, exhaust gas analysis and TERRA TEK laboratory test work on the samples treated within the Thunderstorm Generator are described in Appendix 2. Their analysis and calculations of actual results and percentage changes are described in Appendix 3. The results documented in Appendix 1 were obtained from the operation for different periods of the Thunderstorm Generator technology. A one litre control Sample was taken of an untreated landfill leachate fluid (Sample A) and of samples treated for 3 minutes (Sample C), 5 minutes (Sample D) and 8 minutes (Sample E) respectively first within the MSAART EVO Generator then within the Catalytic Tornado Resonator component of the Thunderstorm Generator apparatus seen in the schematic and photo of Appendix 1. The control sample and treated leachate samples were then sent in one litre bottles to TERRA TEK for analysis. The results of this test work are included below in Appendix 2 and a full analysis of the results are in Appendix.

3.) The Thunderstorm Generator's vacuum and imploding symmetrical bubbles that generate MSAART EVO'S were used to treat liquid leachate, from the London Norland land fill site, within the MSAART EVO Creator bubbler component of the Thunderstorm Generator. The leachate was produced as a by-product from the normal operation of Landfill Methane gas extraction by drill holes. Both the TERRA TEK test results and engine performance determined that the treated leachate can supply a fuel to the engine, from Ammonia and dissolved Methane being boiled off from the leachate into the engines vacuum. The test results indicate that 2/3 of the Ammonia present in the leachate was boiled off within the first 3 minutes of operation. However, when the MSAART EVO'S did reach their critical mass, after 3 minutes, further Ammonia fuel was generated due to the infusion of Nitrogen under 14.7 psi pressure from the atmosphere and the release of free Hydrogen by the action of the UV light and MSAART EVO'S upon the water and certain Leachate component parts including the destroyed Cyanide. After 8 minutes the concentration of Ammonia had recovered to levels within the Leachate from the 3 minute sample C level of 576 mg/L to near those at the start of the test, from the zero time control sample A 1,788 mg/L back to the 8 minute test sample E of 1,698 mg/L. The engine after the 3 minute period increased in power and needed the idle setting to be wound back to prevent over revving of the motor. Throughout the test the throttle was held hard against the idle adjustment screw, as the Propane fuel from the bottle was only used to start and pre-heat the Catalytic Tornado Resonator

RESULTS

component, once the Thunderstorm Generator was engaged the full throttle was moved back to and held in the idle position.

These Ammonia and Methane fuels are in addition to the supply of bottled Methane, generated Hydrogen and charged MSAART EVO'S used or produced in the normal operation of the Thunderstorm Generator. The test also determined that the Thunderstorm Generator can remove or reduce both chemical and biological toxins, including Cyanide, from the leachate by the effects of the ionized air, imploding bubbles and MSAART EVO'S on the leachate in the normal operation of the Thunderstorm Generator. The Thunderstorm Generator has also proved capable of removing or reducing the concentration of harmful biological hazards such as mould, fungus, bacteria, viruses and amoeba by the production of High energy bursts including UV light from the North to South pole implosion of vacuum generated bubbles. These bubbles also create, at the centre Zero point of the implosion on their equatorial plane event horizon, the energy and geometry for the creation of MSAART EVO'S and the bursts of high energy particles and High energy UV light. The energy created at the bubbles implosive centre is amplified by the prior application of the UV light frequencies by the ioniser component of the Thunderstorm generator to the air infused by a diffuser prior to the bubbles collapse.

The results contained within Appendices 2 and 3 of this report prove a complex but proportional interaction of the MSAART EVO'S on elements, salts and molecules from the leachate. The orderly time exposure destruction of Cyanide by high energy UV light proves the existence of MSAART EVO'S and their pre-requisite imploded bubbles due to the atomic based Low Energy Nuclear Reaction [LEANR] created by the bubbles. These interactions when reduced to percentage changes of the afore mentioned groups display uniform effects and proportional changes that indicate the validity of the results and prove processes and forces at work which are not classical but quantum in nature. The results analysed in Appendices 2 and 3 can only be explained by complex atomic rearrangements of Matter and Energy resulting from forces and transformations not achievable without MSAART EVO influence, reaction interacting with Matter (mainly Protium) and Energy (mainly High Energy UV light and Gamma Rays).

RESULTS

The removal of all Carbon Monoxide and Hydrocarbons from the exhaust gas waste stream, as confirmed by our monitoring of the exhaust gas with our Bosche exhaust gas analyser, proving that the Catalytic Tornado Resonator and MSAART EVO components are effective in treating the gas phase constituents of the landfill leachate waste fluid.

DIAGRAM 54 –

CONCLUSIONS

The empirical results and analysis of them are documented in Appendices 2 and 3 the results were obtained from the operation for different time periods of the Thunderstorm Generator technology with one litre of leachate placed within the MSAART EVO Creator component of the Thunderstorm Generator Engine. A one litre control Sample was taken of the untreated leachate (sample A) and of samples treated for 3 minutes (sample C), 5 minutes (sample D) and 8 minutes (sample E) respectively within the Catalytic Tornado Resonator component of the Thunderstorm generator apparatus. The samples were then sent to TERRA TEK for analysis. The results of this test work are included below in Appendix 2 and a full analysis of the results are included in Appendix 3.

From the analysis of these results the following conclusions have been made:-

1.) The leachate generates its own Ammonia faster than the Thunderstorm. Generator Engine can use it due to the interactions between the leachate the MSAART EVO'S and diffused incoming ionised atmospheric gases at one atmosphere 14.7 psi pressure. The energy needed to create the Ammonia that is generated from the leachate comes from the release of Protons and Electrons when the MSAART EVO'S tear away the Proton and Electron that makes up Protium (part of the Hydrogen Group). In addition the MSAART EVO'S disassociates the water into Protium (part of the Hydrogen Group) and Oxygen releasing Protium (part of the Hydrogen Group) combined with ionized Nitrogen to create Ammonia. The results indicate that the Thunderstorm Generator Engine using the separation of the component parts of Protium (part of the Hydrogen Group) caused by the MSAART EVO'S could run for ever on Ammonia Hydroxide if supplied with water which is all it needs along with ionised atmospheric Nitrogen and Oxygen. The presence of every catalyst known to man is present as dissolved metals within the leachate as well as the Natron salts (Sodium, Potassium and Magnesium) so dear to Alchemists and those skilled in the art of mummification.

2.) The Cyanide component within the leachate has been effectively removed by the production from imploding symmetrical bubbles of high intensity bursts of UV light and HIGH ENERGY particles that disassemble the Cyanide into its component elemental parts of C, N, O and H. These elemental parts can be used to construct Ammonia Hydroxide NH4 (OH) and H both being fuels.

CONCLUSIONS

3.) The energy needed to change the concentrations of elements and molecules within the Leachate comes from the MSAART EVO'S destruction of Protium. There is no know effect in classical science to explain the anomalous changes both up and down in the concentrations of elements and molecules.

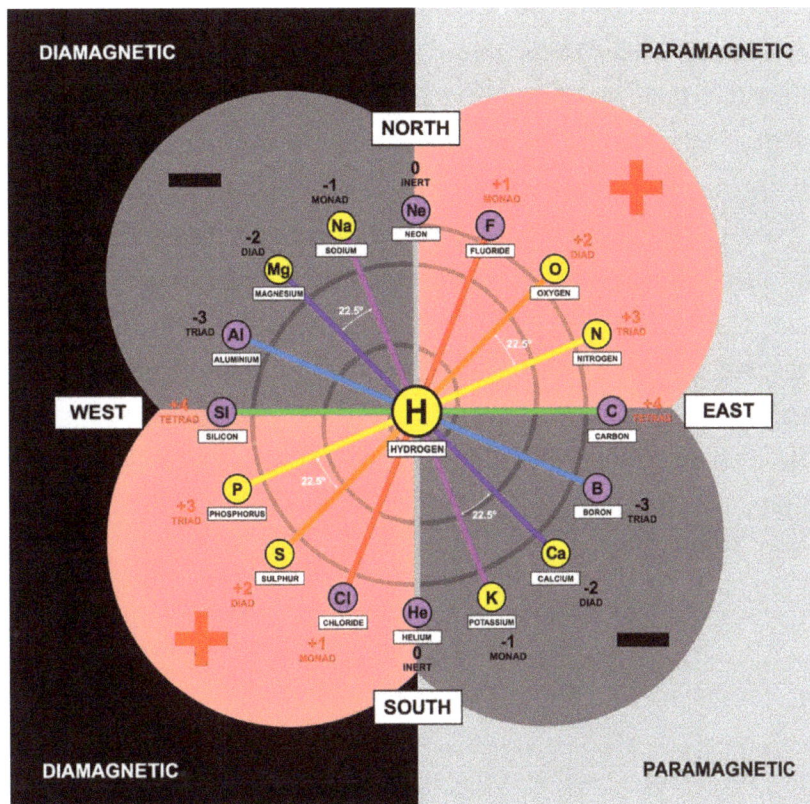

DIAGRAM 55 – MODEL OF THE ELEMENTS MAPPING RELATIONSHIPS IMPORTANT TO UNDERSTANDING THE DYNAMICS OF THE INTERACTION OF THE NORLANDS LANDFILL LEACHATE WITH THE THUNDERSTORM GENERATOR TECHNOLOGY ON AN ELEMENTAL CHARACTER AND PROPERTIES BASIS.

4.) The uniform changes of the concentration of SODIUM, MAGNESIUM and POTASSIUM in the results of 5 minutes (all down 15%) and 8 minutes (all down by 13%) indicate that they acted together as both catalysts and reagents and where consumed equally in the process that enabled Nitrogen to create Ammonia Hydroxide NH4 (OH) by adding Hydrogen and Oxygen available from the disassociation of the water molecules and other leachate components by MSAART EVO ACTIVITY.

CONCLUSIONS

5.) The suspended solids that disappeared were most likely organic compounds that were scavenged and therefore removed due to molecular disintegration caused by the MSAART EVO'S once the MSAART EVO'S reached their critical reactive level due to their substantial electron mass. This conclusion is based on the fact that the bulk PH (the negative log of free Hydrogen ions within the leachate) remained around 8.8 meaning dissolved elements and molecules that were held in solution could not have simply dropped out of solution due to PH change. Proof of the MSAART EVO action is further supported by the absence of any precipitate both observed at the time and later confirmed by empirical testing. The implication is also that despite all the changes within the leachate it has a PH balancing system not yet fully understood but is characteristic of a MSAART EVO Swarms homeostasis. The colour change of the test sample liquids demonstrated in the photo of the samples on pages 3 and 15 may be demonstrating the annihilation of the suspended solids into their component Electron and Proton parts and or the removal or addition of certain dissolved elements and molecules.

CONCLUSIONS

6.)

FIRST SOUND OCTAVES THEN THE ELEMENT OCTAVES

From the one ROD unit of 1.111 progressing in Octaves (simply doubling) first manifesting to us as Sound to the limit of human hearing of 20,000 vibrations per second then through the 144 known elements covering the ultrasonic range 20,000 to 60,000 vibrations per second to the high energy particles then through to First, Second and Third order light all governed by the 16 segments and ruled by the same spiral geometry using the molten sea as a standard. Aether and Time are ruled by 12, therefore the 7.5 degrees of separation, between 16 and 12 segments, also equals the second it takes light to circle the earth 7.5 times. As the Oxygen Isotopes progress 8, 9 and 10 so Music progresses as the table below demonstrates and the ratio of one Oxygen 88.999 to two Hydrogen 11.111 of living water.

TABLE OF ELEMENTAL OCTAVE PROGRESSION OF DAVID'S HARP MUSIC

C DOUBLE FLAT – 12   x 11.111   = 133.333
F DOUBLE FLAT - 16   x 11.111   = 177.777
A FLAT - 2 x 10 = 20   x 11.111   = 222.222
B FLAT - 2 x 11 = 22   x 11.111   = 244.444

C - 3 x 8 = 24   x 11.111 = 266.66 (C)
D - 3 x 9 = 27   x 11.111 = 300
E - 3 x 10 = 30   x 11.111 = 333.333
F - 4 x 8 = 32   x 11.111 = 355.555
G - 4 x 9 = 36   x 11.111 = 400
A - 4 x 10 = 40   x 11.111 = 444.444
B - 4 x 11 = 44   x 11.111 = 488.888
C - 4 x 12 = 48   x 11.111 = 533.333

533.333 (C#) – 266.666 C = 266.666 Interval

D SHARP - 6 x 9 = 54   x 11.111   = 600
E SHARP - 6 x 10 = 60   x 11.111   = 666.666
F SHARP - 6 x 11 = 66   x 11.111   = 711.11
G SHARP - 6 x 12 = 72   x 11.111   = 800
A SHARP - 8 x 10 = 80   x 11.111   = 888.888
B SHARP - 8 x 11 = 88   x 11.111   = 977.777
C SHARP - 8 x 12 = 96   x 11.111   = 1,066.64
D ##   - 9 x 12 = 108   x 11.111   = 1,200

CONCLUSIONS

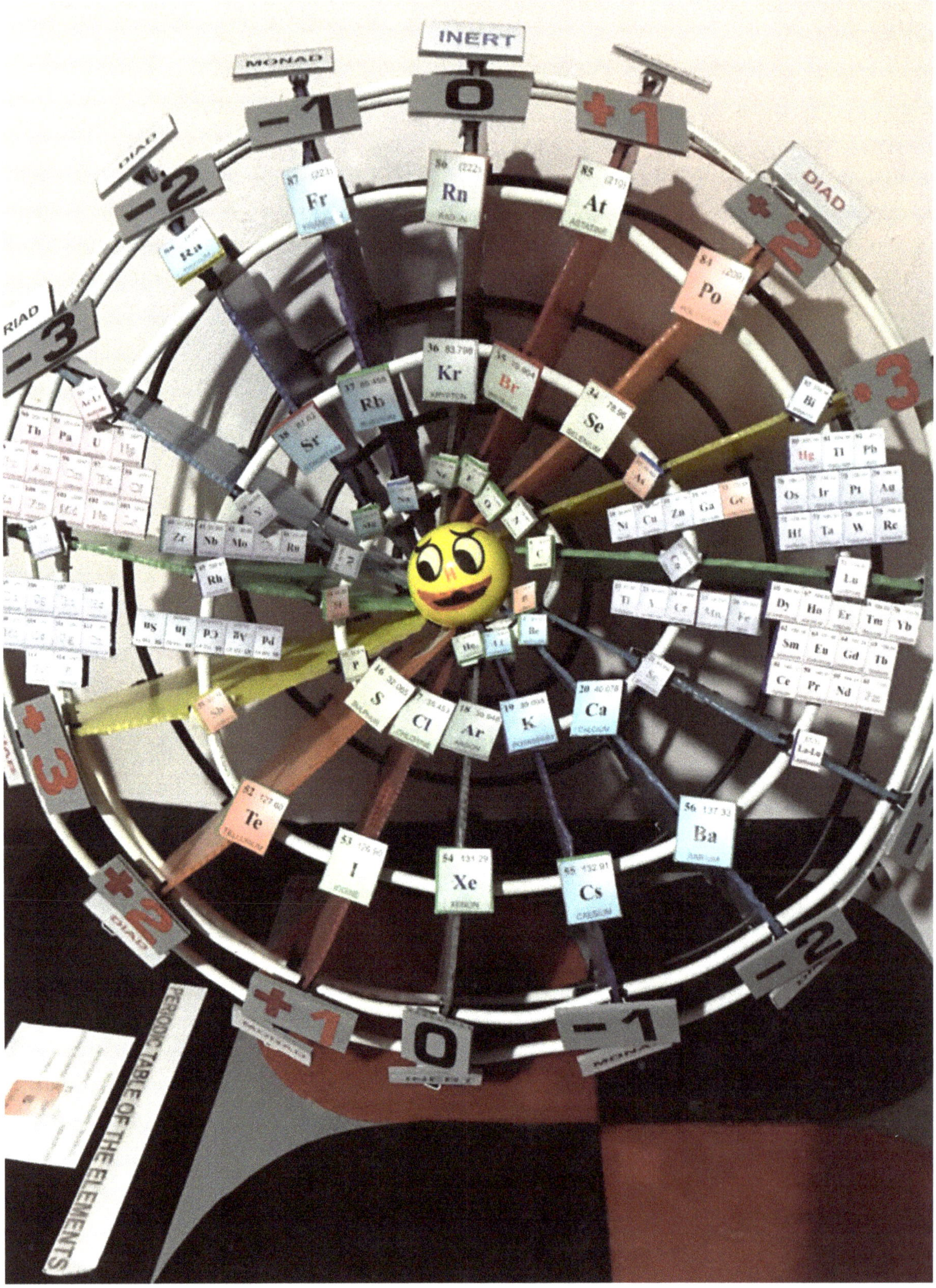

DIAGRAM 56 BELOW - TORUS PERIODIC TABLE OF ELEMENTS PLOTTED ON A VAJRA FIBANACCI CURVE WITH 8 PLANES AND 16 ELEMENTAL NODES PER TURN.

CONCLUSIONS

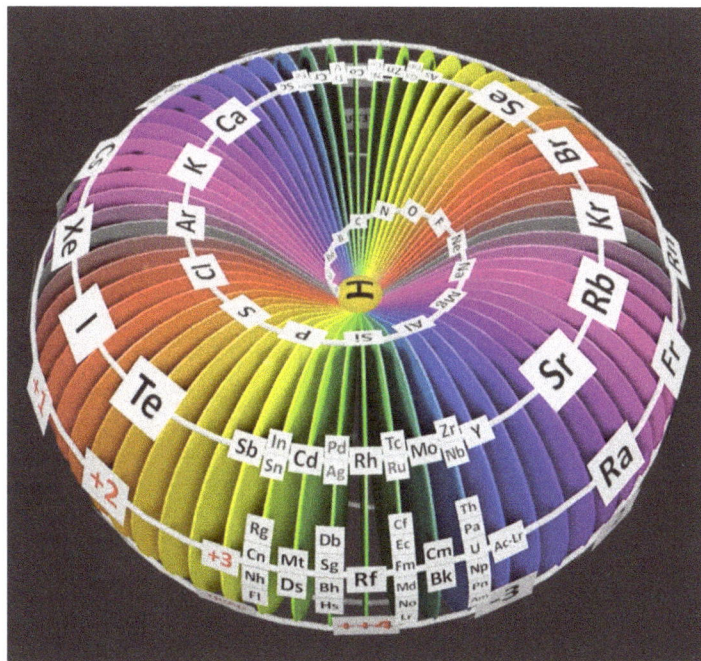

DIAGRAM 57 (ABOVE) - TORUS PERIODIC TABLE OF ELEMENTS PLOTTED ON A VAJRA FIBANACCI CURVE WITH 8 PLANES AND 16 ELEMENTAL NODES PER TURN.

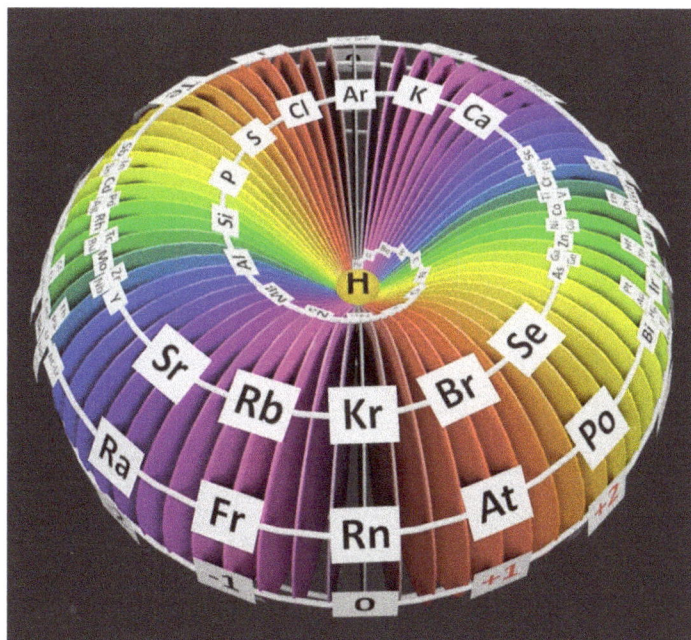

DIAGRAM 58 ABOVE –

CONCLUSIONS

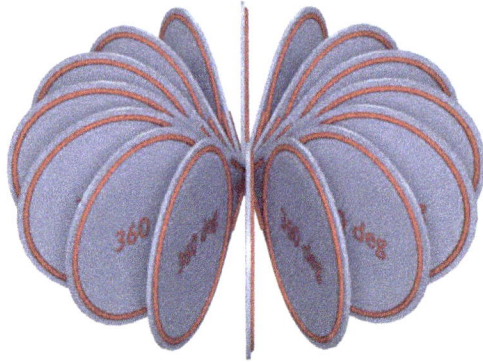

DIAGRAM 59 - TORUS IMPLODED 100 PI SPHERE SHOWING 360 DEGREE PLANES TIMES 16 EQUALS TOTAL OF 5,760 PLUS 16 TIMES 90 EQUALS 1,440 = 7,200 DEGREES.

The

DIAGRAM 60 – BASE MODEL 16 SECTOR IMPLODED SPHERE TORUS.

The

CONCLUSIONS

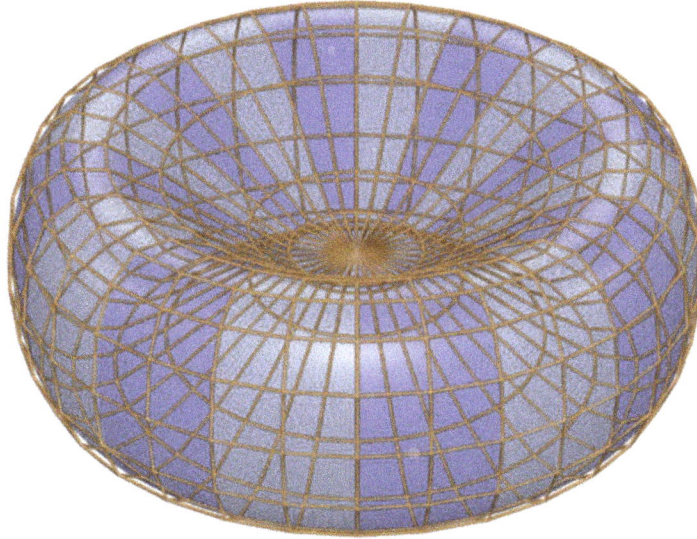

DIAGRAM 61 - ABOVE IMPLODED 100 PI SPHERE TORUS FABRIC PLOTTED ON A VAJRA FIBANACCI CURVE WITH 8 PLANES AND 16 ELEMENTAL NODES PER TURN.

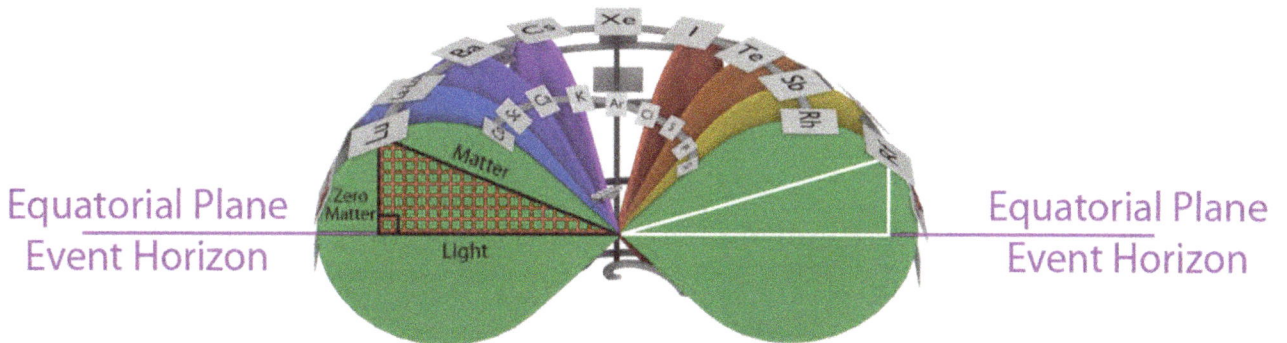

DIAGRAM 62 – MODEL OF THE ELEMENTS (MOE) 16 SECTOR IMPLODED 100 PI SPHERE TORUS VAJRA WITH THE ZERO MATTER, MATTER AND LIGHT RIGHT ANGLE TRIANGLE USED TO DEFINE THE AREA, TIME AND VOLUME (ATV) OF ANY ELEMENT AT THAT LOCATION ON THE FIBANACCI CURVE DEFINES ITS RESONANT FREQUENCY AND TIME BASE.

CONCLUSIONS

FG WILSON

www.FGWilson.com

P550-1

Image for illustration purposes only.

Output Ratings		
Generating Set Model	**Prime**	**Standby**
380-415V, 50Hz	500.0 kVA / 400.0 kW	550.0 kVA / 440.0 kW
	- / -	- / -

Ratings at 0.8 power factor.

Prime Rating
These ratings are applicable for supplying continuous electrical power (at variable load) in lieu of commercially purchased power. There is no limitation to the annual hours of operation and this model can supply 10% overload power for 1 hour in 12 hours.

Standby Rating
These ratings are applicable for supplying continuous electrical power (at variable load) in the event of a utility power failure. No overload is permitted on these ratings. The alternator on this model is peak continuous rated (as defined in ISO 8528-3).

Standard Reference Conditions
Note: Standard reference conditions 25°C (77°F) Air Inlet Temp, 100m (328 ft) A.S.L. 30% relative humidity.
Fuel consumption data at full load with diesel fuel with specific gravity of 0.85 and conforming to BS2869: 1998, Class A2, EN590.

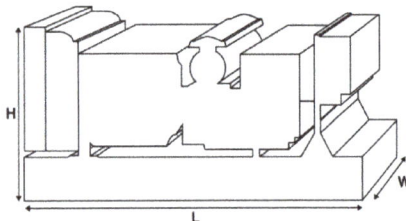

Ratings and Performance Data		
Engine Make & Model:	Perkins 2506A-E15TAG2	
Alternator manufactured for FG Wilson by:	Leroy Somer	
Alternator Model:	LL6114F	
Control Panel:	PowerWizard 1.1+	
Base Frame:	Heavy Duty Fabricated Steel	
Circuit Breaker Type:	3 Pole MCCB	
Frequency:	50 Hz	60 Hz
Engine Speed: RPM	1500	-
Fuel Tank Capacity: litres (US gal)	888 (234.6)	
Fuel Consumption: l/hr (US gal/hr)		
(100% Load) - Prime	98.5 (26.0)	-
- Standby	108.6 (28.7)	-

Available Options

FG Wilson offer a range of optional features to tailor our generating sets to meet your power needs. Options include:

- Upgrade to CE Certification
- A wide range of Sound Attenuated Enclosures
- A variety of generating set control and synchronising panels
- Additional alarms and shutdowns
- A selection of exhaust silencer noise levels

For further information on all of the standard and optional features accompanying this product please contact your local Dealer or visit: www.FGWilson.com

Dimensions and Weights				
Length (L) mm (in)	**Width (W)** mm (in)	**Height (H)** mm (in)	**Dry** kg (lb)	**Wet** kg (lb)
3800 (149.6)	1131 (44.5)	2215 (87.2)	3800 (8378)	3858 (8505)

Dry = With Lube Oil   Wet = With Lube Oil and Coolant

Ratings in accordance with ISO 8528, ISO 3046, IEC 60034, BS5000 and NEMA MG-1.22. Generating set pictured may include optional accessories.

FG Wilson has manufacturing facilities in the following locations:
Northern Ireland • Brazil • China • India • USA
With headquarters in Northern Ireland, FG Wilson operates through a Global Dealer Network.
To contact your local Sales Office please visit the FG Wilson website at www.FGWilson.com

DIAGRAM 63 – FJ WILSON GENERATOR, RE-BRANDED CATATERPILLAR.

CONCLUSIONS

9.)

DIAGRAM 64 - ABOVE -CAT 2506E FG WILSON 500KW GENSET.

DIAGRAM 65 - BELOW - PHOTO OF OUR CATERPILLAR G3508 E 670 Hp V8 GAS 8:1 COMPRESSION RATIO ENGINE WITH THE THUNDERSTORM GENERATOR TECHNOLOGY FITTED GENERATING MORE POWER AND ZERO CARBON MONOXIDE AND WASTE HYDROCARBONS IN THE EXHAUST STREAM. IN ADDITION THE EXHAUST GAS TEMPERATURE MAY BE REDUCED BY UP TO 20 TIMES DOWN TO AMBIENT LEVELS.

PART 2

PLASMOID SYSTEM FOR
CARS DESCRIPTION
SCHEMATIC AND PHOTOS

APPLICATIONS FOR A PLASMOIDS FORM AND FUNCTIONS

APPENDIX 1: - PLASMOID SYSTEM FOR CARS
DESCRIPTION SCHEMATIC AND PHOTOS

DRAFT 518,400 B KMV - PART TWO OF TWENTY

MALCOLM V of SCOTLAND | MALCOLM BENDALL
THURSDAY 22ND SEPTEMBER 2022

APPLICATIONS FOR A PLASMOIDS FORM AND FUNCTIONS

LIST OF DIAGRAMS IN PART 2

APPLICATIONS FOR A PLASMOIDS FORM AND FUNCTIONS THUNDERSTORM GENERATOR.

The Thunderstorm Generator is a system whereby the cold, vacuum and heat, pressure, shockwaves flow alternatively and sequentially from the exhaust and inlet ports of an internal combustion engine are utilized to retrieve and recycle that generated and stored potential energy. That energy is used to sequester, by the use of a Thunderstorm Tornado, generated free protons and electrons that are concentrated by a stream of Plasmoids (EVO's). The Plasmoids confine and store those free Electron and Protons by generating an imploded sphere torus geometry that manifests a homeostatic self-induced, self-structuring, self-sustained, fractal Toroidal electromagnetic confinement field that's captures and confines and isolates micro-plasma. That electromagnetic confinement field is effective and fractal once having been formed and energised by collapsing bubbles within a column of water. The column of water being subjected to alternating vacuum and pressure pulses sourced by the normal action of a piston within an internal combustion engine alternatively generate and collapse the bubbles. These are the same naturally occurring forces of nature that produces the enormous power of a Thunderstorm or Cyclone. Cool moist MSAART enriched air moving into the engine, structured using resonant spheres and cylinders of different diameters, interacts with hot dry air encapsulating it as it moves out in the opposite direction from the engine. This releases enough energy at an atomic level within the exhaust stream to fundamentally alter its composition eliminating toxic chemical wastes such as Carbon monoxide, nitrous oxide, Hydrocarbons and other toxic harmful compounds. The exhausts net positive ions which are also bad for life are replaced with net negative ions within the exhaust stream which support life. Simultaneously within the vacuum, imploding into the engine, together the MSAART'S and water vapor act to both disassociate the water into Hydrogen (Protium) and oxygen assisted by the catalytic and Tribone effects of the Catalytic Tornado Resonator's (CTR) 316 stainless steel spheres and cylinders. The MSAART Plasmoids alone, once reaching their effective charge density creates a viable Zero Singularity Zero Point, due to charging received by the Thunderstorm Tornado, dissociates the Hydrogen (Protium) into its component electron and Proton. This atomic and molecular fuel is fed back into the engine to add and enhance the burn and therefore explosive force of the normal Hydrocarbon fuel. Other elements that contain Neutrons within the imploding vacuum stream are unaffected by the forces applied by the MSAART Plasmoids as they are not powerful enough to act on the nucleus therefore producing no nuclear by products making the processes by-products non radio- active, toxin free and with a life enhancing Negative implosive charge.

APPLICATIONS FOR A PLASMOIDS FORM AND FUNCTIONS

THUNDERSTORM GENERATOR DESCRIPTION SCHEMATIC AND PHOTOS

SPHERICAL TORNADO ATOMIC THUNDERSTORM GENERATOR AND ELECTRON SEQUESTING PLASMOID TECHNOLOGY TEST RESULTS

INTRODUCTION

The invention, and now worldwide addiction and necessity, of the automobile has created the reliance on the hydrocarbon based fuel which powers them. This has in turn led to the first and second world wars and put the world at risk of a third world war fighting literally for power. Regrettably those who lust after this power mostly seek to empower themselves and in fact seek to disempower the individuals, communities and countries that rely on fuel.

This novel new Plasmoid-based device over the next decade will generate $4 Trillion USD for the global community, which will be noticeable on a grand scale at the grass roots level, that is, the petrol pump. In environmental terms, it means cleaner air, cheap water and heating, cheaper food, cheaper housing and higher oxygen levels in inner cities and around the globe. The device also empowers the individual. It dis-empowers the establishment, global corporations and banks that have enslaved the individual.

AIM

The aim is to implement the theories and facts outlined in Appendix 7 that are the result of 40 years of research. This includes 14 years away from Australia travelling around the world visiting ancient ruins and libraries and museums reading ancient Sanskrit, Sumerian, Greek, Egyptian and Hebrew texts. A further aim is to demonstrate and document the capture and use of the 66% of energy currently wasted by standard exhaust systems on all internal combustion engines. This will be shown by creating Plasmoids to act as thunderstorm harvesters. Plasmoids are a self-structuring closed system electron vacuum cleaners that capture and put into a toroidal orbit that generates an electromagnetic containment field that isolates the enormous amount of energy held within the Plasmoid. They can be a replacement for energy from hydrocarbon based fuels.

METHOD

A standard, off the shelf, single cylinder, 380 cc, four stroke, petrol, 5,500 Watt generator (Ref 2) has been retro - fitted with a waste energy recovery and simultaneous, electrical and Plasmoid based, fuel generation system. This novel system works by utilizing the waste 66% of exhaust energy, which escapes the internal combustion engine as heat and pressure, to create Plasmoids, which can be harvested within the combustion chamber.

This is achieved by using a plasma spark obtained by retrofitting the standard ignition system. The standard carburetor is used to supply fuel to the engine and therefore pre-heat the modified novel Spherical Tornado Atomic Thunderstorm Electron Sequestering Plasmoids

PLASMOID ANGEL ONCE AND FOR ALL DEFEATS MAXWELL'S DEMON

APPLICATIONS FOR A PLASMOIDS FORM AND FUNCTIONS

THUNDERSTORM GENERATOR DESCRIPTION SCHEMATIC AND PHOTOS

exhaust system to over 350 degrees Celsius. This pre-heating is critical to evoke the reverse Kelvin – Joule effect (Ref 3), the Kelvin thunderstorm effect (Ref 4), the Hilshe tube effect (Ref 5), the Bendall translates (Ref 1) and several other both defined and undefined but measurable observations.

These effects in combination, when timed and structured correctly, energise the Torus Plasmoids that in turn vacuum up all the free electrons generated by these multiple complementary effects. These Torus Plasmoids, being self - structuring and closed systems, do not emit heat themselves. But when free electrons are wrapped up and isolated, they have the power to extract all the free electrons, and therefore also heat from the device.
The Thunderstorm effect is induced by structuring the high pressure, hot, dry, exhaust gas flow anti-clockwise spin to move against a clockwise counter spin and flow of low pressure cold moist air. This generates an electric charge in balance. This cold air has been pre - seeded with Plasmoids created in water, by the implosive collapse of de-gassed bubbles.

These are created within a pulsed, structured vacuum chamber filled with pre-ionised air, tap water and stainless steel wool. These are designed on principles known to those skilled in the art of enhancing these effects.

RESULTS

We have measured the anomalous heating and cooling of this novel exhaust device with eight thermocouples and recorded the results on two separate data loggers, connected to two separate computers. We have done the following calculations based on our data. This demonstrates that energy, in excess of the waste heat and pressure, was being produced by this Spherical Tornado Atomic Thunderstorm Electron Sequestering Plasmoids device.

The energy (in Kwh) in a gas stream is the weight in Kilograms per second multiplied by the specific heat multiplied by the temperature difference in Kelvin = Kilo Joules per second, [1 Joule per second = 1 watt per second]. The motor is 380 cc say running at 4,000 RPM. This is 2,000 inspirations a minute, being a volume of approximately 760 Litres of air per minute. That is 0.760 Cu/M assuming it is normal air that has a specific gravity of 1.245 Kg/Cu M. (0.760 x 1.245 / 60 seconds = 0.0157Kg per second, multiplied by the 800 degrees Kelvin temperature (526 Degrees C). The difference is then multiplied by the specific heat of air (0.812Kj). This equals 10.918 Kj / per second = 10.918 Kwh.

These calculations have been done on temperatures taken from the outside of the one inch diameter stainless steel heat shield, not the central ¾ inch and ½ inch reactor pipes. Those reactor pipes have been proven to reach temperatures in excess of 1,200 Degrees C, as evidenced by the melting of brass fittings (950 Degrees C [1,223 Kelvin] melting temperature), and the deformation of the stainless steel reactor elements (forge temperature of 1,250 Degrees C [1,523 Kelvin]).

MSAART EVO ANGEL ONCE AND FOR ALL DEFEATS MAXWELL'S DEMON

APPLICATIONS FOR A PLASMOIDS FORM AND FUNCTIONS

DIAGRAM 66 – TEST RESULTS

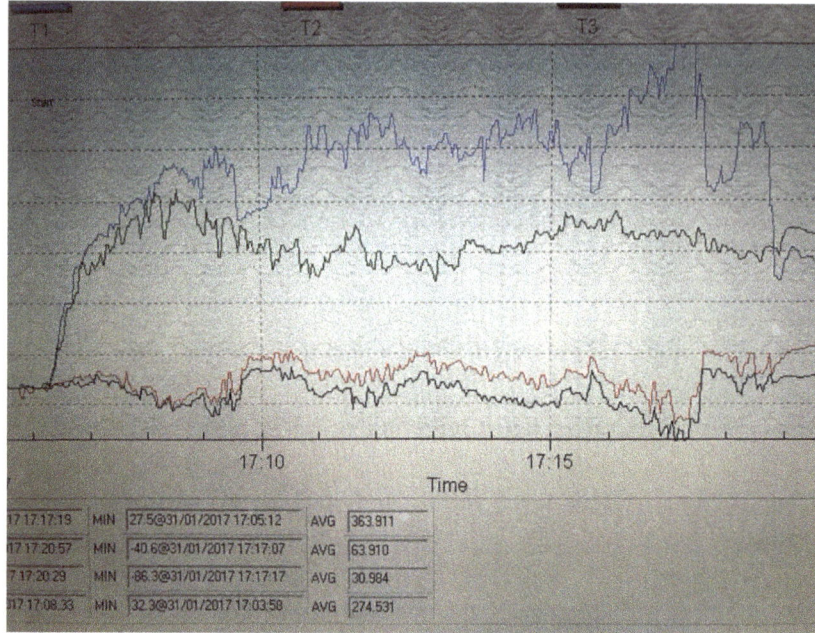

DIAGRAM 67 – TEST RESULTS

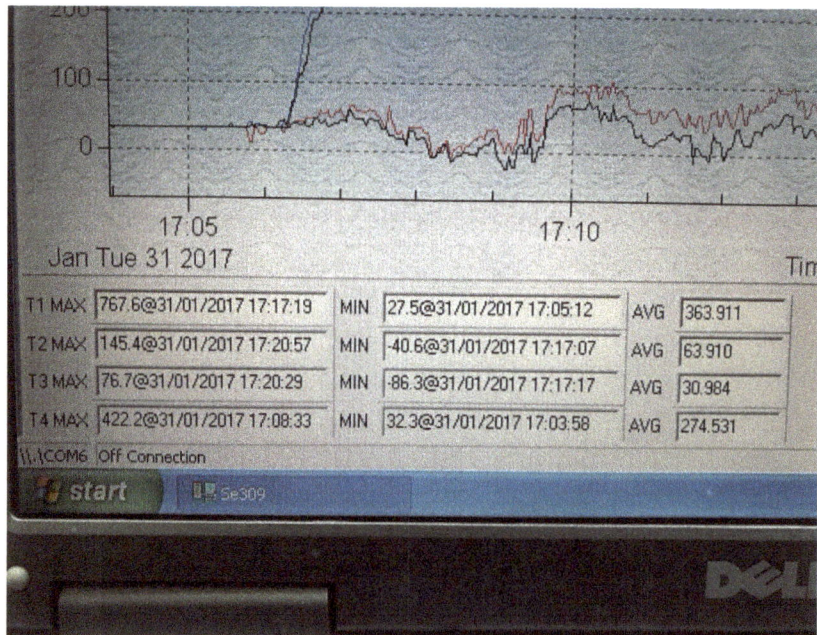

MSAART EVO ANGEL ONCE AND FOR ALL DEFEATS MAXWELL'S DEMON

APPLICATIONS FOR A PLASMOIDS FORM AND FUNCTIONS

THUNDERSTORM GENERATOR DESCRIPTION SCHEMATIC AND PHOTOS

THE THUNDERSTORM GENERATOR

Energy of over 22,000 Kwh, although not directly measured with thermocouples, has been generated as evidenced by examination of the early prototypes. The creation of volcano-like pits on the surface of the reactor elements, [as documented by a high magnification microscope linked to a computer], and visible light streaks escaping from the reactor chamber prove both the existence and charging of Plasmoids.

CONCLUSION

Plasmoids have been proven by our test results to be both an effective waste energy capturing mechanism and alternative fuel. They use and demonstrate quantum effects evoked by the creation of Plasmoids charged by a Thunderstorm. By simply imitating and copying the Thunderstorm in its most basic configuration we have tapped into one of natures most powerful electrical generators hitherto not achieved, let alone harvested by using the implosive Vortex principles of the Thunderstorm Generator Engine technology.

THE THUNDERSTORM GENERATOR

The devise is novel. The Plasmoids formed with pre-ionised air at a specific frequency are generated by collapsing cavitation bubbles, which are created by a vacuum and then imploded by applying pressure. The Plasmoids have proven to be effective heat sinks, electron storage devices, communication devices and safe energy discharge platforms.

The Plasmoids have demonstrated 'quantum swarm entanglement 'that was achieved by sharing the same frequency at the creation of the singularity zero point of the Toroidal structure. This common frequency blueprint enables the group to share equally the total electron input throughout the Plasmoid population, regardless of distance and time. This is because the singularity points in the centre of the Toroidal structure are not subject to either influence. By extracting and structuring the free electrons within the Plasmoid's internal Toroidal structure it does not emit any energy. But it is still capable of equalizing the electron distribution through the space - time continuum by mechanisms that are not completely understood. However, based on the data and the underlying knowledge base and Toroidal structure, one can confidently postulate what processes have been at work. Plasmoids have shown they have the ability to capture and release electrons to each other within the Plasmoid swarm. This statement is based in the fact that a swarm of Plasmoids must stay at the same frequency. Therefore the Plasmoids must expand in Area – Time – volume (ATV) at the same rate. Since they emit a lower or higher frequency, they will be absorbing or transmitting energy. In response they will create a group homeostasis that is immune to any direct influence applied to any one member of the group.

MSAART EVO ANGEL ONCE AND FOR ALL DEFEATS MAXWELL'S DEMON

APPLICATIONS FOR A PLASMOIDS FORM AND FUNCTIONS

THE THUNDERSTORM GENERATOR

This means that the Plasmoids are a base for humans to venture into space. They also represent a new clean, green global energy revolution.

Plasmoids are a fractal technology. They can be used in many ways other than as producing Atomic fuel from water and as Atomic batteries.

For instance:
1: Communication technologies at above light speed (at any scale);
2: Heat shields and cold shields;
3: Plasmoid weapons;
4: Plasmoid force field shields;
5: Plasmoid propulsion devices that are light weight;
6: Plasmoid energy storage and distribution systems. These can be applied to all of humankind's current and future needs and devices and can used on old technologies.

DIAGRAM

68 – PETROL TEST ENGINE

MSAART EVO ANGEL ONCE AND FOR ALL DEFEATS MAXWELL'S DEMON

APPLICATIONS FOR A PLASMOIDS FORM AND FUNCTIONS

THE THUNDERSTORM GENERATOR

DIAGRAM 69 - THE THUNDERSTORM GENERATOR TEST RESULTS DESCRIBED ABOVE ARE SHOWN BEING RECORDED

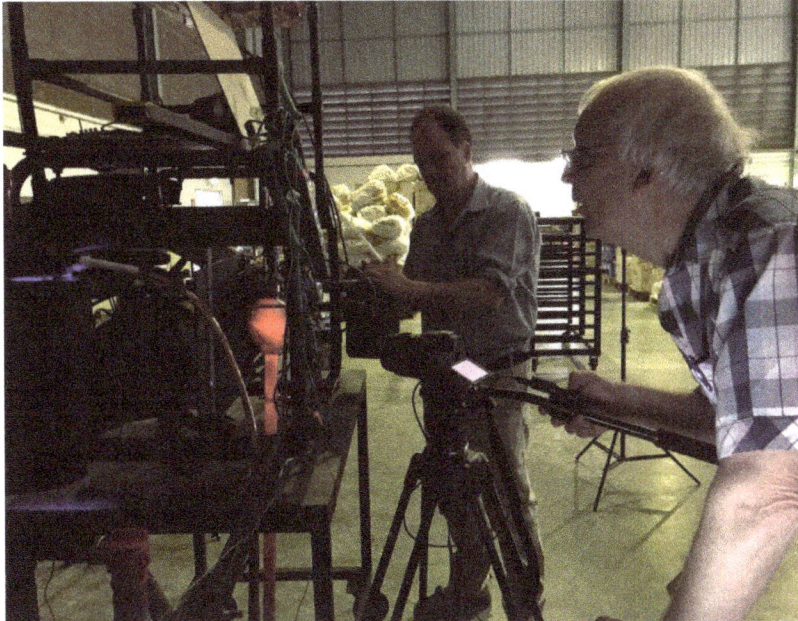

DIAGRAM 70 - THE PHOTOS BELOW CAPTURES AT 1,000 FRAMES PER SECOND AN OVERCHARGED PLASMOID ESCAPING FROM THE 4" 316 STAINLESS STEEL SPHERE THAT IS PART OF THE CATALYTIC TORNADO RESONATOR THE IMAGE PROVES IMPLODING LIGHT SEEN FOR THE FIRST TIME IN HISTORY ORIGINATED FROM THE ENERGY EMMITED AS A RESULT OF ATMOSPHERIC ARGONS IMPLODING ELECTRON ORBITS ENERGY RELEASE.

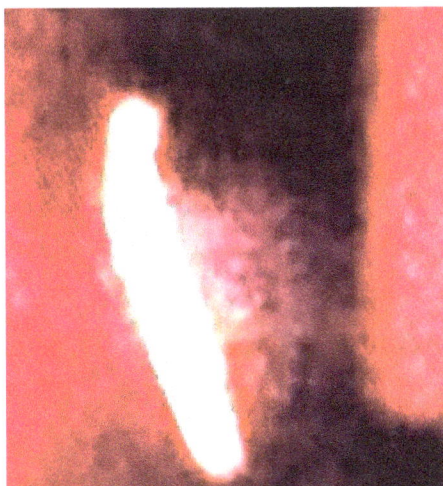

DIAGRAM 71 – IMPLODING LIGHT FROM TRAIL OF A PLASMOID

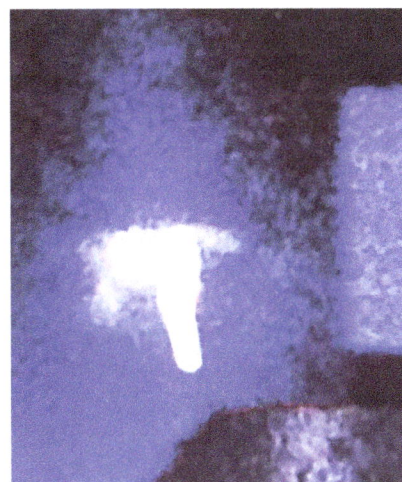

DIAGRAM 72 – IMPLODING LIGHT FROM TRAIL OF A PLASMOID

APPLICATIONS FOR A PLASMOIDS FORM AND FUNCTIONS

THE THUNDERSTORM GENERATOR

LIST OF REFERENCES

1.) BENDALL - DRAFT 266.666 SOTU

2.) WELCO - WEBSITE FOR GENERATOR

3.) KELVIN – JOULE PAPER

4.) LORD KELVIN 1867 THUNDERSTORM MACHINE

5.) RAUNGE - HILSCH TUBE PAPER

6.) W.H.BOSNICK - PLASMOID TERM FOUNDER BORN 1916 - DIED 1958

> **EXPERIMENTAL STUDY OF PLASMOIDS** REV. 106, 404 – PUBLISHED 1 MAY 1957; ERRATUM PHYS. REV 107, 1736 (1957)

7.) K.D.SINELNIKOV et al., Proceedings of the Second International conference on the peaceful uses of atomic energy (United Nations, Geneva,1958), Vol 31, p292. Google scholar

8.) KEPLER

9.) PYTHAGORAS

10.) NICOLA TESLA and SWAMI BAR

11.) WALTER RUSSELL

12.) KEN SHEPARD,

13.) FLYSHMAN AND PONDS

14.) STANLEY MEYER, PAUL PANTONE, GUY OBOLINSKY, JOE CELL

DEFINITIONS :- WIKIPEDIA PLASMOID DEFINITION

"A PLASMOID IS A COHERENT (TOROIDAL) STRUCTURE OF PLASMA AND MAGNETIC FIELDS. PLASMOIDS HAVE BEEN PROPOSED TO EXPLAIN NATURAL PHENOMENA SUCH AS BALL LIGHTNING, MAGNETIC BUBBLES IN THE MAGNETOSPHERE AND OBJECTS IN COMET TAILS, IN SOLAR WIND, IN THE SOLAR ATMOSPHERE AND IN THE HELIOSPHERIC SHEET."

MSAART EVO ANGEL ONCE AND FOR ALL DEFEATS MAXWELL'S DEMON

APPLICATIONS FOR A PLASMOIDS FORM AND FUNCTIONS

THE THUNDERSTORM GENERATOR

The Thunderstorm Generator System

DIAGRAM 73 -

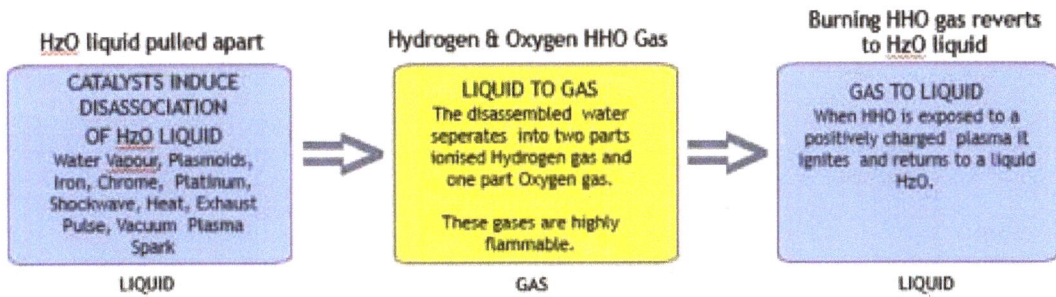

HzO liquid pulled apart	Hydrogen & Oxygen HHO Gas	Burning HHO gas reverts to HzO liquid
CATALYSTS INDUCE DISASSOCIATION OF HzO LIQUID Water Vapour, Plasmoids, Iron, Chrome, Platinum, Shockwave, Heat, Exhaust Pulse, Vacuum Plasma Spark	**LIQUID TO GAS** The disassembled water seperates into two parts ionised Hydrogen gas and one part Oxygen gas. These gases are highly flammable.	**GAS TO LIQUID** When HHO is exposed to a positively charged plasma it ignites and returns to a liquid HzO.
LIQUID	GAS	LIQUID

DIAGRAM 74 -

THE THUNDERSTORM GENERATOR

The Thunderstorm Generator System fully integrated schematic

APPLICATIONS FOR A MSAART EVO'S FORM AND FUNCTIONS

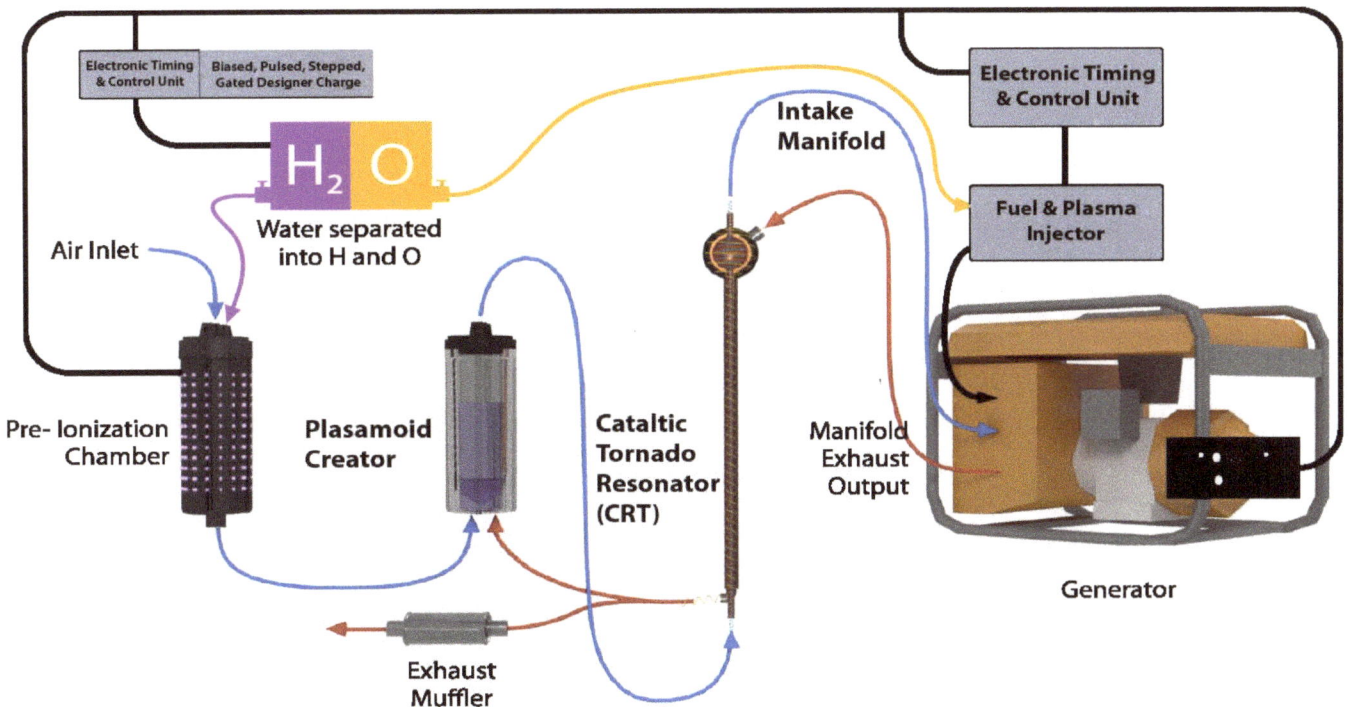

THE THUNDERSTORM GENERATOR

DIAGRAM 75 –

The Thunderstorm Generator System

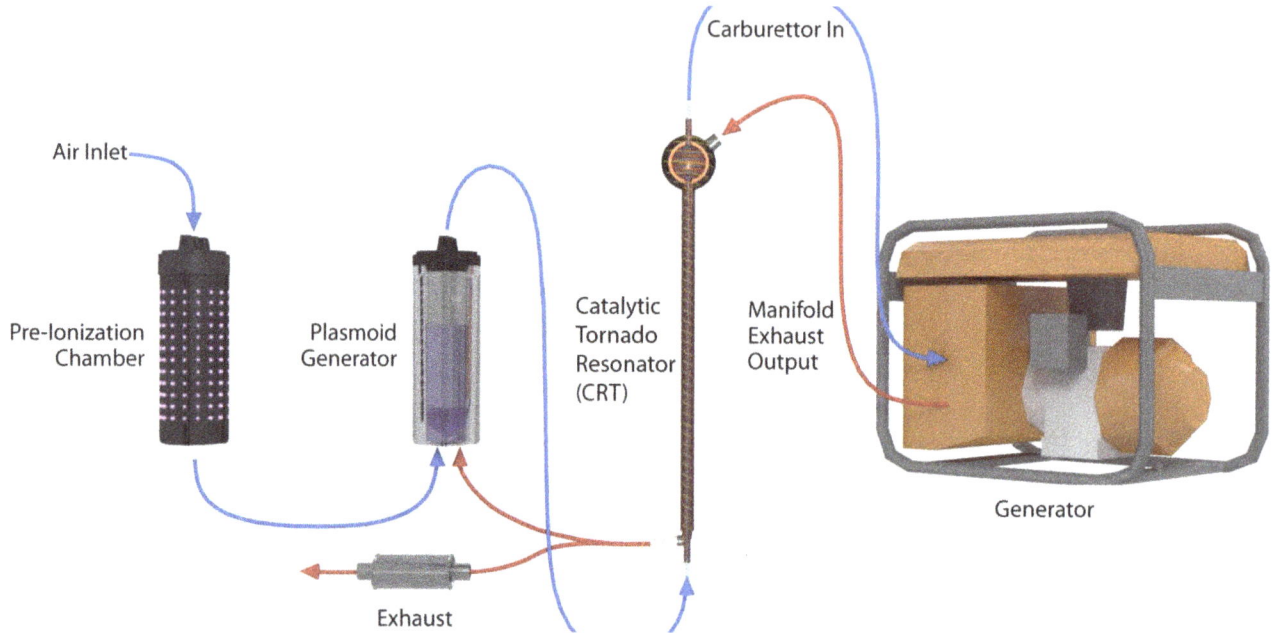

THE THUNDERSTORM GENERATOR SYSTEM BASIC BASE MODEL SCHEMATIC

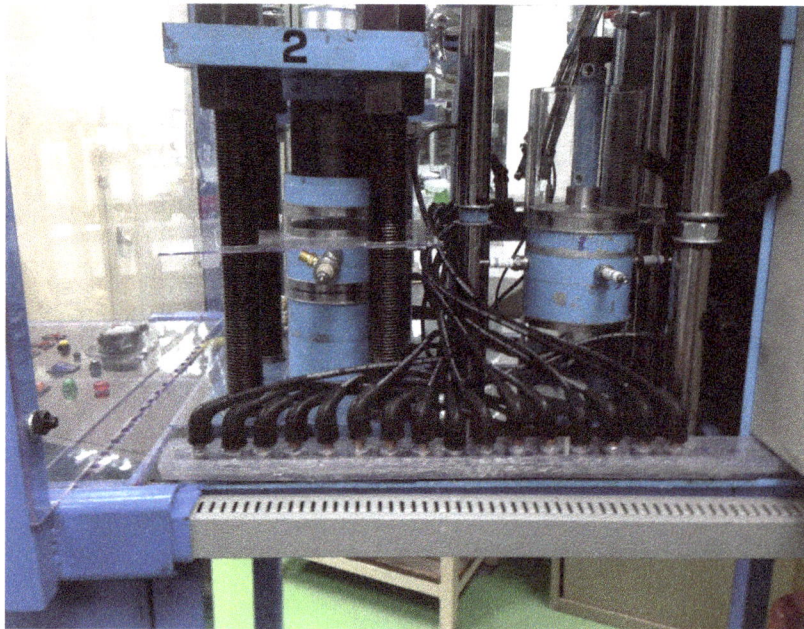

DIAGRAM 76 ABOVE – THUNDERSTORM GENERATOR TEST PAD

APPLICATIONS FOR A PLASMOIDS FORM AND FUNCTIONS

APPENDIX 2: - TERRA TEK RESULT SHEETS

DIAGRAM 64

APPLICATIONS FOR A PLASMOIDS FORM AND FUNCTIONS

DIAGRAM 77 -

TERRA TEK
SITE INVESTIGATION & LABORATORY SERVICES

Enitial
Enterprise Drive
Four Ashes Industrial Estate
Wolverhampton
WV10 7DE

For the attention of   Anne Morrison

Report No:	B24232
Issue No	01

LABORATORY TEST REPORT

Project Name	**NORLANDS**		
Project Number	B24232	Date samples received	19/02/2020
Your Ref		Date written instructions received	19/02/2020
Purchase Order		Date testing commenced	19/02/2020

Please find enclosed the results as summarised below

Figure / Table	Test Quantity	Description	ISO 17025 Accredited
1	5	Client Specified Suite - Water	See Report
App W1	~	Deviating Samples - Water	N/A
App W2	~	Summary of In-House Analytical Test Methods - Water	N/A

Remarks :

Issued by :   Stephen Langman   Date of Issue :   05/03/2020

Approved Signatories :   *S Langman* 05/03/2020

G Wilson (JMD/Laboratories Director), S Langman (Laboratory Coordinator)

Key to symbols used in this report
S/C : Testing was sub-contracted

Moor Lane, Witton, Birmingham, B6 7LG
Tel +44 (0)121 344 4838 Fax +44 (0)121 366 3209
birmingham@terratek.co.uk
www.terratek.co.uk
Terra Tek Ltd is registered in Scotland No. 124291
Offices in Airdrie, Birmingham, Belfast and Chesham

Head Office : 62 Rochsolloch Road, Airdrie, ML6 9BG

APPLICATIONS FOR A PLASMOIDS FORM AND FUNCTIONS

DIAGRAM 78 -

TERRA TEK — SITE INVESTIGATION AND LABORATORY SERVICES

4140 - Suite Maxi WATER - B24232 01.xls
Version 008 - 19/06/2007

Moor Lane, Witton, Birmingham, B6 7HG
Lab Project No B24232 : 05/03/2020 17:41:59

CHEMICAL ANALYSIS

Site: NORLANDS
Client:
Engineer:
Contract No **B24232**

Sample Identification				Arsenic	Cadmium	Chromium	Lead	Mercury	Copper	Nickel	Zinc	Iron	Manganese	Calcium	Magnesium	Sodium	Potassium	Total Cyanide	Nitrate	Nitrite	Alkalinity - Carbonate as CaCO3	Ammoniacal Nitrogen (as N)	Ammonia (as NH4)
Hole	Depth m	Sample Ref / Sample Type	Lab Sample ID	µg/l	µg/l	µg/l	µg/l	µg/l	µg/l	µg/l	µg/l	µg/l	µg/l	mg/l	mg/l	mg/l	mg/l	mg/l	mg/l	mg/l	mg/l	mg/l	mg/l
SAMPLE A		W1	723316	100	<0.08	53	8	<0.5	40	140	220	2.20	150	72	86	1,279	638	1.43	21.3	<0.01	7,906.8	1,385.8	1,787.7
SAMPLE C		W1	723319	52	<0.08	26	9	<0.5	1,800	64	270	2.00	120	32	24	462	221	0.53	26.1	0.59	2,824.7	446.7	576.3
SAMPLE D		W1	723320	85	<0.08	75	15	<0.5	5,400	130	260	1.20	74	47	73	1,080	532	0.03	23.4	1.28	6,730.1	1,166.4	1,504.6
SAMPLE E		W1	723321	98	<0.08	42	10	<0.5	7,500	110	320	2.10	120	51	75	1,109	549	0.07	28.6	<0.01	7,204.9	1,316.7	1,698.5
Limits of Detection				1	0.08	0.4	1	0.5	0.7	0.3	0.4	0.004	0.06	1	1	1	1	0.01	0.1	0.01	2.5	0.1	0.1
Terra Tek Analysis Method				S/C	S/C	S/C	S/C	S/C	S/C	S/C	S/C	S/C	S/C	TP117	TP117	TP117	TP117	TP062	TP184	TP184	TP184	TP184	TP184
Accreditation U=UKAS N=No accreditation				U	U	U	U	U	U	U	U	U	U	U	U	U	U	U	N	N	N	N	N

KEY

* - deviating result (refer to Appendix W1 for details)

Originator	Checked & Approved
	S. (signature)
DAB	05/03/2020

TiK

Figure 1
Sheet 1 of 2

APPLICATIONS FOR A PLASMOIDS FORM AND FUNCTIONS

DIAGRAM 80-

TERRA TEK — SITE INVESTIGATION AND LABORATORY SERVICES

4140 - Suite Maxi WATER - B24232 02.xls
Version 008 - 19/06/2007

Moor Lane, Witton, Birmingham, B6 7HG
Lab Project No B24232 : 05/03/2020 17:42:02

Site: NORLANDS
Client:
Engineer:
Contract No: **B24232**

CHEMICAL ANALYSIS

| Sample Identification | | | | Phosphate | Biochemical oxygen demand | Chemical oxygen demand | Total Organic Carbon | Total Oxidised Nitrogen | Suspended Solids | Chloride | Electrical conductivity | Sulphate (as SO4) | Sulphide | pH | Dissolved Methane | Visable Oil & Grease |
Hole	Depth m	Sample Ref	Sample Type	Lab Sample ID	mg/l	mg/l	mg/l	mg/l	mg/l	mg/l	mg/l	µS/cm	mg/l	mg/l		mg/l	
SAMPLE A			W1	723316	16.49	8.4	2,200	990.1	3.6	40	1,796.3	12,445	136	0.10	8.7	0.59	ND
SAMPLE C			W1	723319	5.95	8.0	770	355.6	2.6	108	503.6	5,375	45	0.03	8.8	0.53	ND
SAMPLE D			W1	723320	11.55	8.2	1,500	876.5	3.7	<4	1,593.9	11,545	123	0.11	8.8	0.53	ND
SAMPLE E			W1	723321	9.11	8.3	3,100	980.1	3.9	<4	1,737.8	13,160	109	0.12	8.8	0.50	ND
Limits of Detection					0.02	1	2	0.3	0.7	4	0.1	1	4	0.01	–	0.050	–
Terra Tek Analysis Method					TP184	S/C	S/C	TP162	S/C	TP081	TP184	TP108	TP170	TP066	TP020	S/C	N
Accreditation U=UKAS N=No accreditation					N	U	U	N	U	U	U	U	U	U	U	N	–

KEY
* - deviating result (refer to Appendix W1 for details)
ND - Not Detected

Originator: DAB — 05/03/2020
Checked & Approved: S. *[signature]* — 05/03/2020

Figure 1
Sheet 2 of 2

APPLICATIONS FOR A PLASMOIDS FORM AND FUNCTIONS

DIAGRAM 81 -

Version 017 - 22/01/2015
8051 - Deviating samples - WATER - B24232 01.xls

TERRA TEK
SITE INVESTIGATION AND LABORATORY SERVICES

								Deviating conditions					
Site	NORLANDS											Contract No	**B24232**
Client													
Engineer													

Sample Identification						Deviating conditions						
Exploratory Hole	Depth m	Sample Ref	Sample Type	Lab Sample ID	Date Sampled	Sampling date has not been provided	Exceeded maximum holding time for selected test(s)	Presence of headspace in sample vial	Poorly fitting cap or lid	Damaged container		Preservatives used
SAMPLE A			W1	723316	17/02/20							
SAMPLE B			W2	723318	17/02/20							
SAMPLE C			W1	723319	17/02/20							
SAMPLE D			W1	723320	17/02/20							
SAMPLE E			W1	723321	17/02/20							

Moor Lane, Witton, Birmingham, B6 7HG
Lab Project No B24232 - 05/03/2020 17:44:13

NOTES
1 Results reported for samples classified as deviating may be compromised. Deviation types are shown as "X" or "Yes" in the table above.
2 The absence of "X" or "Yes" in the table above indicates no reported deviations.
3 Deviations due to use of incorrect sample container are shown on result tables.
4 Deviating results are indicated within result tables.

Originator	Checked & Approved			
TGH	S. Langran 05/03/2020	**DEVIATING SAMPLES - WATER**	**TTk**	Appendix W1 Sheet 1 of 1

APPLICATIONS FOR A PLASMOIDS FORM AND FUNCTIONS

DIAGRAM 82 -

Version 009 - 24/06/2009
8200 - Test Methods Water - B24232 01.xls

TERRA TEK
SITE INVESTIGATION AND LABORATORY SERVICES

			Contract No	**B24232**
Site	NORLANDS			
Client				
Engineer				

Method Code	Reference	Description of Method	ISO17025 Accredited	
TP020	APHA/AWWA, 19th edition	Determination of pH using pH meter	Yes	
TP035	BS1377, Part 3, 1990: Soils for Civil Engineering Purposes.	Determination of dissolved solids by gravimetry	Yes	
TP054	MAFF Book 427: The Analysis of Agricultural Materials: Method 8	Determination of boron by colorimetry	Yes	
TP057	APHA/AWWA, 19th edition: Method 3500Cr-D	Determination of hexavalent chromium by colorimetry	Yes	
TP060	MEWAM method: Phenols in water and Effluents: 4-aminoantipyrine method	Determination of monohydric phenols by steam distillation/colorimetry	Yes	
TP061	MEWAM method: Cyanide in Waters etc	Determination of free cyanide by colorimetry	Yes	
TP062	MEWAM method: Cyanide in Waters etc	Determination of total cyanide by steam distillation/colorimetry	Yes	
TP063	MEWAM method: Cyanide in Waters etc	Determination of complex cyanide by calculation	Yes	
TP064	MEWAM method: Determination of Thiocyanate ,1985	Determination of thiocyanate by colorimetry	Yes	
TP066	MEWAM method: Sulphide in Waters and Effluents, Tentative Methods: 1983	Determination of sulphides by colorimetry	Yes	
TP068	APHA/AWWA, 19th edition: Method 4500-Cl-D	Determination of chlorides by titrimetry	Yes	
TP078	APHA/AWWA, 18th edition: Method 4500C	Determination of ammoniacal nitrogen by colorimetry		
TP079	In-house documented method	Determination of anionic detergent (MBAS) by colorimetry		
TP080	APHA/AWWA, 19th edition: Method 4500-F-C	Determination of fluoride by ion selective electrode	Yes	
TP081	APHA/AWWA, 19th edition: Method 2540D	Determination of suspended solids by gravimetry	Yes	
TP102	APHA/AWWA, 19th edition: Method 6640B USEPA Method 610	Determination of polyaromatic hydrocarbons extractable in dichloromethane, by GC/MS	Yes	
TP103	Texas Natural Resource Conservation Commission Method 1005 & USEPA Method 3510C	Determination of Extractable Petroleum Hydrocarbons (>C8 - C40) by GC/FID		
TP108	APHA/AWWA, 19th edition: Method 2510B	Determination of electrical conductivity by electrode	Yes	
TP112	USEPA Method 8100	Determination of polyaromatic hydrocarbons extractable in dichloromethane/hexane, by GC/MS		
TP113	APHA/AWWA, 19th edition: Method 6410 USEPA Method 2870D	Determination of phenol by GC/MS		

Notes

1. The laboratory records the date of analysis of each parameter. This information is available on request.
2. Where a parameter cannot be determined in house it is our policy to use a UKAS accredited laboratory wherever possible. Terra Tek will assume responsibility for the quality of subcontracted tests and the performance of the subcontractor chosen. Where there is no known UKAS laboratory for a particular parameter, a laboratory listed within the Terra Tek Approved Subcontractors list, which is subject to performance assessment, will be selected.

Moor Lane, Witton, Birmingham, B6 7HG
Lab Project No B24232 - 05/03/2020 17:42:06

Originator	Checked & Approved	SUMMARY OF IN-HOUSE ANALYTICAL TEST METHODS (WATER)	**TTk**	Appendix W2
N/A	N/A			Sheet 1 of 2

APPLICATIONS FOR A PLASMOIDS FORM AND FUNCTIONS

DIAGRAM 83 -

Version 009 - 24/06/2009
8200 - Test Methods Water - B24232 01.xls

Moor Lane, Witton, Birmingham, B6 7HG
Lab Project No B24232 : 05/03/2020 17:42:07

TERRA TEK SITE INVESTIGATION AND LABORATORY SERVICES

Site	NORLANDS	Contract No	B24232
Client			
Engineer			

Method Code	Reference	Description of Method	ISO17025 Accredited	
TP117	APHA/AWWA, 19th edition: Method 2340B	Determination of hardness of water (calculation)	Yes	
TP118	APHA/AWWA, 19th edition: Method 2320B	Determination of total alkalinity by titration	Yes	
TP128	APHA/AWWA, 19th edition: Method 6410 USEPA Method 2870D	Determination of Semi-Volatile Organic Compounds by GC/MS	Yes	
TP130	Texas Natural Resource Conservation Commission Method 1005 & 1006	Determination of Extractable Petroleum Hydrocarbons (EPH-CWG C8-C40) by GC/FID		
TP132	APHA/AWWA, 19th edition: Method 4500-NO2-B	Determination of nitrite by colorimetry	Yes	
TP133	In-house documented method	Determination of chemical oxygen demand by colorimetry		
TP146	USEPA Methods 8082A & 3665A	Determination of Total & Speciated 7 PCB Congeners by GC/MS SIM		
TP149	USEPA Methods 8082A & 3665A	Determination of Total & Speciated WHO 12 PCB Congeners by GC/MS SIM		
TP155	USEPA method 5021. Wisconsin DNR modified GRO method	Determination of volatiles in water by GC/MS headspace	Yes	
TP156	APHA/AWWA, 19th edition: Method 3030B (filtration)	Determination of dissolved metals by ICP-MS	Selected	
TP159	USEPA Method 1671	Determination of glycols in water by GC/FID DI		
TP160	USEPA Method 556	Determination of formaldeyde in water by GC/MS		
TP162	USEPA Method 9060A	Determination of TOC/DOC in water by HT Combustion/NDIR		
TP170	In-house documented method	Determination of sulphate by ICP-OES spectroscopy	Yes	
TP179	In-house documented method	Determination of nitrate by ion selective electrode		

Notes
1. The laboratory records the date of analysis of each parameter. This information is available on request.
2. Where a parameter cannot be determined in house it is our policy to use a UKAS accredited laboratory wherever possible. Terra Tek will assume responsibility for the quality of subcontracted tests and the performance of the subcontractor chosen. Where there is no known UKAS laboratory for a particular parameter, a laboratory listed within the Terra Tek Approved Subcontractors list, which is subject to performance assessment, will be selected.

Originator	Checked & Approved	SUMMARY OF IN-HOUSE ANALYTICAL TEST METHODS (WATER)	TTk	Appendix W2
N/A	N/A			Sheet 2 of 2

APPENDIX 3: - TERRA TEK RESULT SHEET ANALYSIS

APPLICATIONS FOR A PLASMOIDS FORM AND FUNCTIONS

ELEMENTS AND MOLECULES SUBTRACTED FROM THE NORLANDS TIP LEACHATE FLUID AFTER 3 MINUTES WITHIN THE PLASMOID CREATOR

ELEMENTS SUBTRACTED FROM THE LEACHATE AFTER 3 MINUTES OF OPERATION

MAGNESIUM (86 – 24 mg/L) ....DOWN 72%

SODIUM (1,279 – 462 mg/L) ....DOWN 64%

POTASSIUM (638 – 532 mg/L) ....DOWN 17%

CALCIUM (72 – 32 mg/L)... DOWN 44%

CHLORIDE (1796 -504)..DOWN 70%

ARSENIC (100 – 52ug/L)..DOWN 48%

CHROMIUM (53 – 26 ug/L)..DOWN 49%

NICKLE (140 – 64 ug/L)... DOWN 54%

IRON (2.20 – 2.00 ug/L)..DOWN 10%

MANGANESE (150 – 120 ug/L)..DOWN 20%

MOLECULES SUBTRACTED FROM THE LEACHATE AFTER 3 MINUTES OF OPERATION

DISSOLVED METHANE (0.59 – 0.53 mg/L)................................DOWN 10%

AMMONIA [AS NH4] (1,787 – 576)................................. DOWN 68%

AMMONIACAL NITROGEN [AS N] (1,386 – 446 mg/L) DOWN 68%

TOTAL OXIDISED NITROGEN (3.6 – 2.6 mg/L)................ DOWN 28%

NITRITE (0.01 – 0.59 mg/L)...UP 98%

NITRATE (21.3 – 26.1mg/L)..UP 18%

CHEMICAL OXYGEN DEMAND (2,200 – 770 mg/L)..................DOWN 65%

BIOCHEMICAL OXYGEN DEMAND (8.4 – 8.0 mg/L).................DOWN 5%

SULPHATE (136 – 45 mg/L)..DOWN 67%

SULPHIDE (0.10 – 0.03)..DOWN 70%

PHOSPHATE (16.49 – 5.95)..DOWN 65%

TOTAL ORGANIC CARBON (990 – 355 mg/L)............................DOWN 65%

ALKALINITY – CARBONATE as CaCO3 (7,906 – 2,825 mg/L).....DOWN 65%

TOTAL CYANIDE (1.43 – 0.53 mg/L)...DOWN 63%

ELECTRICAL CONDUCTIVITY (12,445 -5,375 uS/cm)..................DOWN 43%

ELEMENTS AND MOLECULES ADDED TO THE NORLANDS TIP LEACHATE FLUID AFTER 3 MINUTES WITHIN THE PLASMOID CREATOR

ELEMENTS ADDED TO THE LEACHATE AFTER 3 MINUTES OF OPERATION

COPPER (40 – 1,800 ug/L)..UP 4,500%

ZINC (220 – 270 ug/L)..UP 18%

LEAD (8 – 9 ug/L)..UP 11%

MOLECULES ADDED TO THE LEACHATE AFTER 3 MINUTES OF OPERATION

NITRITE (0.01 – 0.59 mg/L)..UP 98%

NITRATE (21.3 – 26.1mg/L)..UP 18%

SUSPENDED SOLIDS (40 – 108 mg/L)..UP 63%

PH (PH 8.7 - PH 8.8)...UP 1%

APPLICATIONS FOR A PLASMOIDS FORM AND FUNCTIONS

*ELEMENTS AND MOLECULES SUBTRACTED FROM THE NORLANDS TIP
LEACHATE FLUID AFTER 5 MINUTES WITHIN THE PLASMOID CREATOR*

ELEMENTS SUBTRACTED FROM THE LEACHATE AFTER 5 MINUTES OF OPERATION

MAGNESIUM (86 – 73 mg/L)..DOWN *15%*
SODIUM (1,279 – 1,080 mg/L)..DOWN *15%*
POTASSIUM (638 – 532 mg/L)..DOWN *15%*
CALCIUM (72 – 47 mg/L)..DOWN **35%**
CHLORIDE (1796 – 1594)...DOWN *11%*
IRON (2.20 – 1.20 ug/L)..DOWN 46%
MANGANESE (150 – 74 ug/L).. DOWN 50%
ARSENIC (100 – 85 ug/L)...DOWN 15%
NICKLE (140 – 130 ug/L)..DOWN 7%

MOLECULES SUBTRACTED FROM THE LEACHATE AFTER 5 MINUTES OF OPERATION

DISSOLVED METHANE (0.59 – 0.53 mg/L)................................DOWN *10%*
AMMONIA [AS NH4] (1,787 – 1,505)......................................DOWN *16%*
AMMONIACAL NITROGEN [AS N] (1,386 – 1,166 mg/L)   DOWN *16%*
TOTAL OXIDISED NITROGEN (3.6 – 3.7 mg/L)........................UP *3%*
TOTAL CYANIDE (1.43 – 0.03)...DOWN 98%
CHEMICAL OXYGEN DEMAND (2,200 – 1,500 mg/L)...............DOWN *32%*
BIOCHEMICAL OXYGEN DEMAND (8.4 – 8.2 mg/L).................DOWN **2%**
SUSPENDED SOLIDS (40 – >4 mg/L)..................................DOWN 90 - 100%
PHOSPHATE (16.49 – 11.55 mg/L)...DOWN 30%
ALKALINITY - CARBONATE as CaCO3 (7,907 – 6,730 mg/L).......DOWN 15%
TOTAL ORGANIC CARBON (990 – 876 mg/L).............................DOWN 12%
CHLORIDE (1,796 – 1,593mg/L)..DOWN 11%
SULPHATE (136 – 123 mg/L)...DOWN 10%
ELECTRICAL CONDUCTIVITY (12,445 – 11,545 uS/cm)...............DOWN 7%

*ELEMENTS AND MOLECULES ADDED TO THE NORLANDS TIP LEACHATE FLUID AFTER
5 MINUTES WITHIN THE PLASMOID CREATOR*

ELEMENTS ADDED TO THE LEACHATE *AFTER 5 MINUTES OF OPERATION*

CHROMIUM (53 – 75 ug/L)...UP   30%
LEAD (8 – 15 ug/L)..UP   47%
COPPER (40 – 5,400 ug/L) UP 135 X..UP 13,500%
ZINC (220 – 260 ug/L)...UP   15%

MOLECULES ADDED TO THE LEACHATE *AFTER 5 MINUTES OF OPERATION*

TOTAL OXIDISED NITROGEN (3.6 – 3.7 mg/L)...............................UP 3%
NITRATE (21.3 – 23.4 mg/L)...UP 9%
NITRITE (<0.01 – 1.28 mg/L)..UP 99 TO 100%
SULPHIDE (0.10 – 0.11 mg/L)...UP 9%
PH (8.7 – 8.8)...UP 1%

APPLICATIONS FOR A PLASMOIDS FORM AND FUNCTIONS

ELEMENTS *SUBTRACTED* FROM THE LEACHATE AFTER *8 MINUTES* OF OPERATION

MAGNESIUM DECREASE (86 – 75 mg/L)......................................DOWN *13%*
SODIUM DECREASE (1,279 – 1,109 mg/L)...............................DOWN *13%*
POTASSIUM DECREASE (638 – 549 mg/L)...............................DOWN *13%*
CALCIUM DECREASE (72 – 51 mg/L).......................................DOWN *31%*
CHLORIDE (1796 – 1738)..DOWN 3%
CHROMIUM DECREASE (53 - 42 ug/L)......................................DOWN 21%
NICKEL DECREASE (140 – 110 ug/L)...DOWN 21%
MANGANESE DECREASE (150 – 120 ug/L)...............................DOWN 21%
IRON DECREASE (2.20 – 2.10 ug/L)..DOWN 5%
ARSENIC DECREASE (100 - 98 ug/L)...DOWN 2%

MOLECULES *SUBTRACTED* FROM THE LEACHATE AFTER *8 MINUTES* OF OPERATION

DISSOLVED *METHANE* DECREASE (0.59 – 0.05mg/L)...................DOWN 8%
AMMONIA [AS NH4] (1,787 – 1,698 mg/L)...................................DOWN 5%
AMMONIACAL NITROGEN (1,385 – 1,316 mg/L)........................DOWN 5%
TOTAL OXIDISED NITROGEN (3.6 - 3.9 mg/L)...............................UP 8%
NITRITE (< 0.01 – < 0.01mg/L)..SAME 0%
NITRATE (21.3 – 28.6 mg/L)...UP 25%
CHEMICAL OXYGEN DEMAND (2,200 – 3,100 mg/L)........................UP 30%
BIOCHEMICAL OXYGEN DEMAND (8.4 – 8.3mg/L)......................DOWN 1%
TOTAL CYANIDE DECREASE (1.43 – 0.07 mg/L)............................DOWN 95%
PHOSPHATE DECREASE (16.49 – 9.11 mg/L)................................DOWN 55%
ALKALINITY - CARBONATE as CaCO3 (7,907 – 7204 mg/L)..........DOWN 9%
SULPHATE [AS SO4] (136 – 109)..DOWN 20%
CALCIUM CARBONATE (7,906 – 7,204 mg/L)................................DOWN 9%

ELEMENTS AND MOLECULES *ADDED* TO THE NORLANDS TIP LEACHATE FLUID

ELEMENTS *ADDED* TO THE LEACHATE AFTER *8 MINUTES* OF OPERATION

COPPER – 45 x INCREASE (40 – 7,500 ug/L)...................................UP 4,500%
ZINC – INCREASE (220 – 320 ug/L)...UP 30%
NITRATE – INCREASE (21.3 – 28.6 mg/L)..UP 25%
LEAD - INCREASE (8 – 10 ug/L)...UP 20%

MOLECULES *ADDED* TO THE LEACHATE AFTER *8 MINUTES* OF OPERATION

AMMONIA – NET INCREASE OF 10 LITRES RUNNING
TOTAL OXIDISED N (3.6 - 3.9 mg/L)..UP 8%
NITRITE (< 0.01 – < 0.01mg/L)...SAME 0%
NITRATE (21.3 – 28.6 mg/L)..UP 25%
SULPHIDE (0.10 – 0.12 mg/L)..UP 20%
ELECTRICAL CONDUCTIVITY (12,445 - 13,160 uS/cm)......................UP 5%
PH (Ph 8.7 to Ph 8.8)...UP 1%

APPLICATIONS FOR A PLASMOIDS FORM AND FUNCTIONS

ELEMENTS AND MOLECULES REMOVED FROM THE NORLANDS TIP LEACHATE FLUID AFTER 8 MINUTES WITHIN THE PLASMOID CREATOR

ELEMENTS ELIMINATED FROM THE LEACHATE AFTER 8 MIN   CYANIDE AND NITRITE

MOLECULES ELIMINATED FROM THE LEACHATE AFTER 8 MIN   SUSPENDED SOLIDS

APPLICATIONS FOR A PLASMOIDS FORM AND FUNCTIONS

ELEMENTS *SUBTRACTED* AND ADDED FROM LEACHATE AFTER *3 MINUTES* OF OPERATION

CHROMIUM (53 – 26 ug/L)..DOWN 49%
NICKLE (140 – 64 ug/L).. DOWN 46%
MANGANESE (150 – 120 ug/L)..DOWN 20%
IRON (2.20 – 2.00 ug/L)..DOWN 10%
ARSENIC (100 – 52ug/L)...DOWN 48%

ELEMENTS *SUBTRACTED* AND ADDED FROM LEACHATE AFTER *5 MINUTES* OF OPERATION

CHROMIUM (53 – 75 ug/L)..UP   30%
NICKLE (140 – 130 ug/L)...DOWN   7%
MANGANESE (150 – 74 ug/L).. DOWN 50%
IRON (2.20 – 1.20 ug/L)..DOWN 54%
ARSENIC (100 – 85 ug/L)...DOWN 15%

ELEMENTS *SUBTRACTED* AND *ADDED* FROM LEACHATE AFTER *8 MINUTES* OF OPERATION

CHROMIUM DECREASE (53 - 42 ug/L)...DOWN 21%
NICKEL DECREASE (140 – 110 ug/L)...DOWN 21%
MANGANESE DECREASE (150 – 120 ug/L).................................DOWN 21%
IRON DECREASE (2.20 – 2.10 ug/L)..DOWN 5%
ARSENIC DECREASE (100 - 98 ug/L)..DOWN 2%

DIAGRAM 84 –

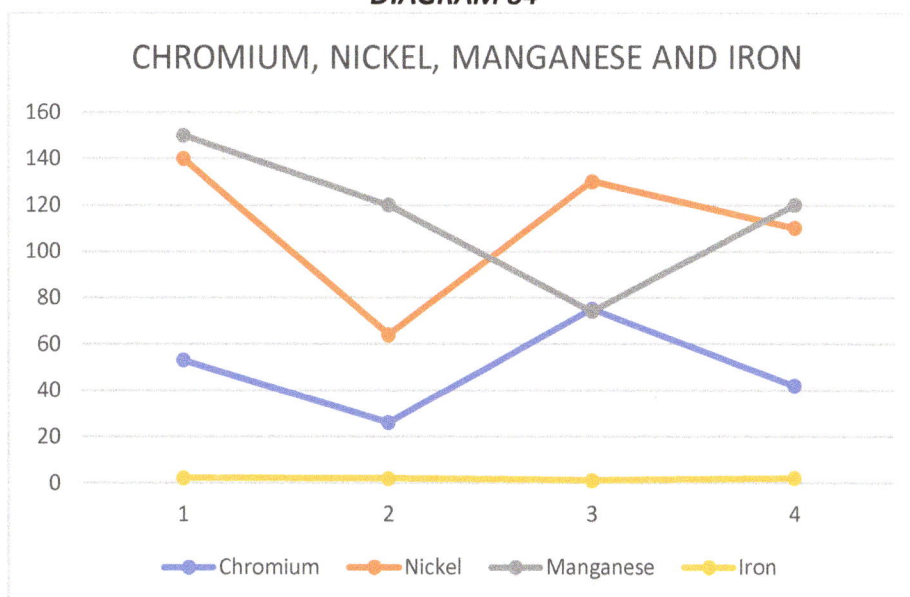

CHROMIUM, NICKEL, MANGANESE AND IRON

Chromium	Cr	53	26	75	42
Nickel	Ni	140	64	130	110
Manganese	Mn	150	120	74	120
Iron	Fe	2.2	2	1.2	2.1

APPLICATIONS FOR A PLASMOIDS FORM AND FUNCTIONS

DIAGRAM 85 –

MAGNESIUM, CALCIUM, ARSNIC AND IRON

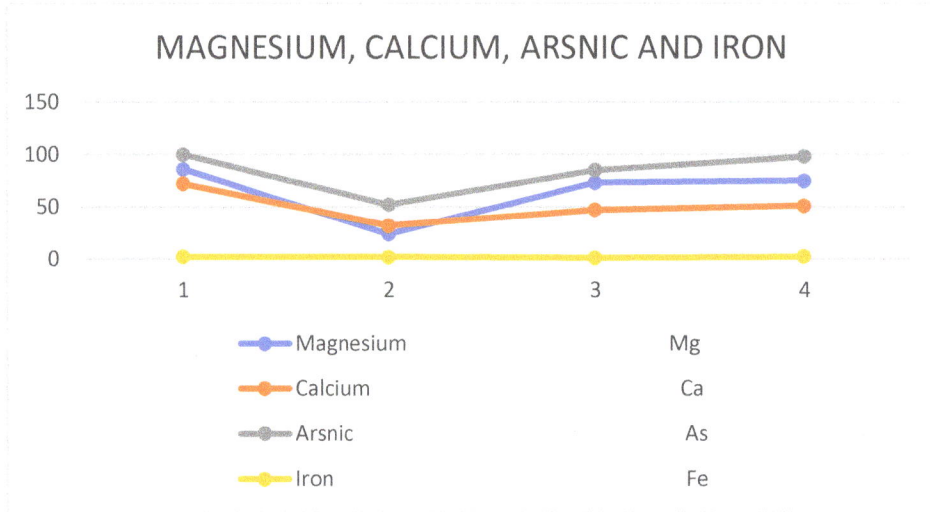

	ELEMENT	UNTREATED	3 MIN	5 MIN	8 MIN
Magnesium	Mg	86	24	73	75
Calcium	Ca	72	32	47	51
Arsenic	As	100	52	85	98
Iron	Fe	2.2	2	1.2	2.1

APPLICATIONS FOR A PLASMOIDS FORM AND FUNCTIONS

ELEMENTS** SUBTRACTED **FROM THE LEACHATE AFTER** 3 MINUTES **OF OPERATION

MAGNESIUM (86 – 24 mg/L) ....DOWN 72%
SODIUM (1,279 – 462 mg/L) ....DOWN 64%
POTASSIUM (638 – 532 mg/L) ....DOWN 17%

ELEMENTS** SUBTRACTED **FROM THE LEACHATE AFTER** 5 MINUTES **OF OPERATION

MAGNESIUM (86 – 73 mg/L)..DOWN 15%
SODIUM (1,279 – 1,080 mg/L)..DOWN 15%
POTASSIUM (638 – 532 mg/L)...DOWN 15%

ELEMENTS** SUBTRACTED **FROM THE LEACHATE AFTER** 8 MINUTES **OF OPERATION

MAGNESIUM DECREASE (86 – 75 mg/L).....................................DOWN 13%
SODIUM DECREASE (1,279 – 1,109 mg/L)..................................DOWN 13%
POTASSIUM DECREASE (638 – 549 mg/L)...................................DOWN 13%

DIAGRAM 86 -

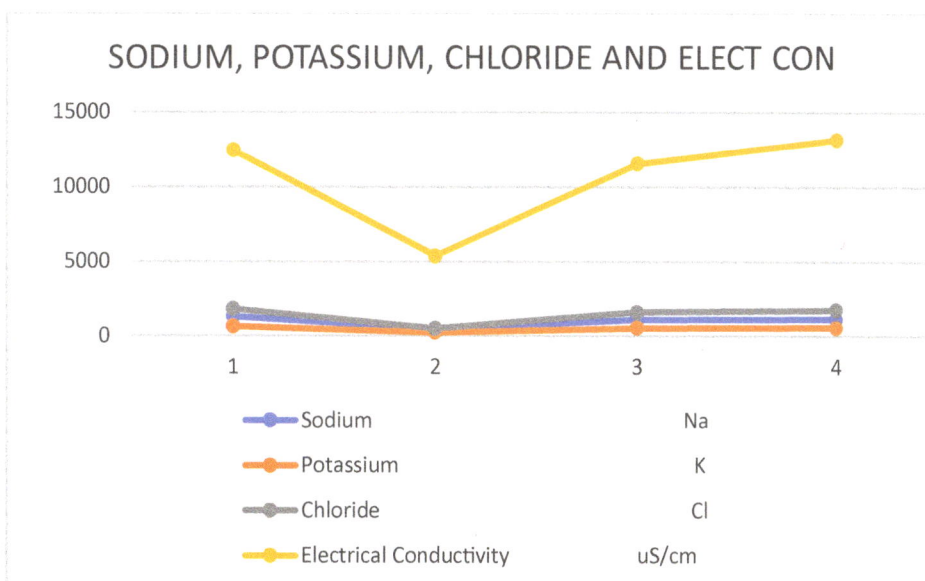

	ELEMENT	UNTREATED	3 MIN	5 MIN	8 MIN
Sodium	Na	1279	462	1080	1109
Potassium	K	638	221	532	549
Chloride	Cl	1796	504	1594	1738
Electrical Conductivity	uS/cm	12,445	5375	11545	13160

APPLICATIONS FOR A PLASMOIDS FORM AND FUNCTIONS

DIAGRAM 87 -

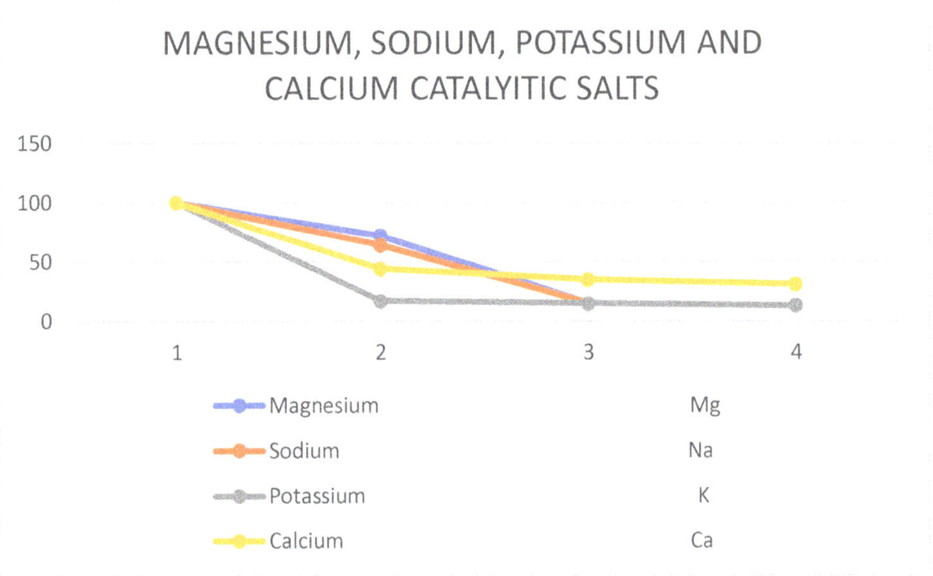

MAGNESIUM, SODIUM, POTASSIUM AND CALCIUM CATALYITIC SALTS

- Magnesium — Mg
- Sodium — Na
- Potassium — K
- Calcium — Ca

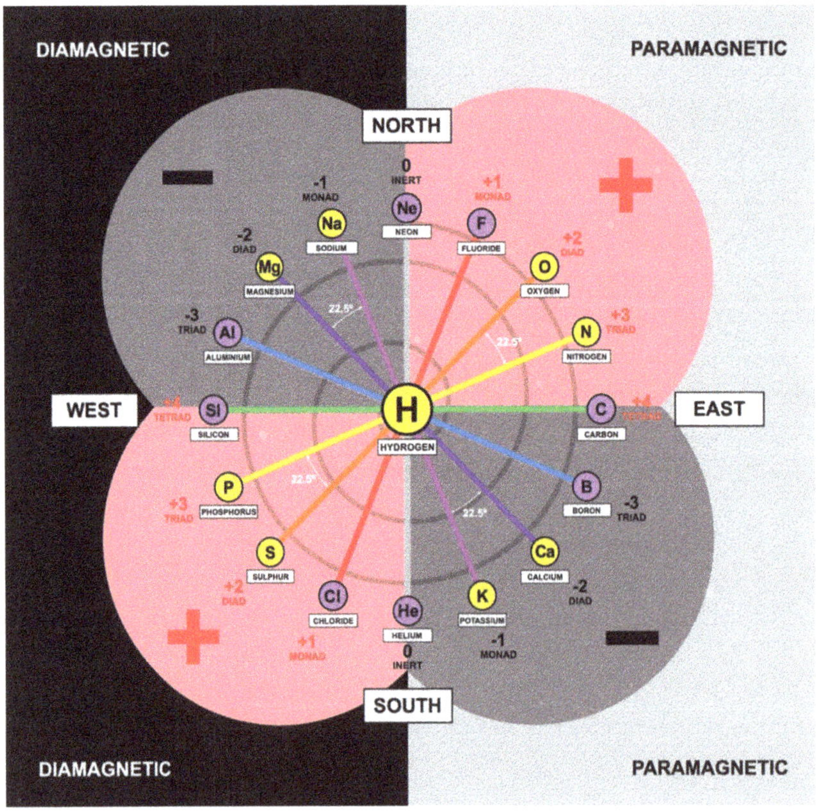

DIAGRAM 88 - Mg and Ca are opposites on the same purple elemental frequency plane Na and K share the next indigo elemental frequency plane

APPLICATIONS FOR A PLASMOIDS FORM AND FUNCTIONS

NITROGEN *SUBTRACTED* AND *ADDED* FROM LEACHATE AFTER 3 MINUTES OF OPERATION

AMMONIA [AS NH4] (1,787 – 576).................................... DOWN 68%
AMMONIACAL NITROGEN [AS N] (1,386 – 446 mg/L)   DOWN 68%
TOTAL OXIDISED NITROGEN (3.6 – 2.6 mg/L)................. DOWN 28%
NITRITE (0.01 – 0.59 mg/L)...UP 98%
NITRATE (21.3 – 26.1mg/L)....................................... UP 18%

NITROGEN *SUBTRACTED* AND *ADDED* FROM LEACHATE AFTER 5 MINUTES OF OPERATION

AMMONIA [AS NH4] (1,787 – 1,505)...........................DOWN 16%
AMMONIACAL NITROGEN [AS N] (1,386 – 1,166 mg/L)   DOWN 16%
TOTAL OXIDISED NITROGEN (3.6 – 3.7 mg/L)...................UP 3%
NITRATE (21.3 – 23.4 mg/L)......................................UP 9%
NITRITE (<0.01 – 1.28 mg/L).......................UP 99 TO 100%

NITROGEN *SUBTRACTED* AND *ADDED* FROM LEACHATE AFTER 8 MINUTES OF OPERATION

AMMONIA [AS NH4] (1,787 – 1,698 mg/L)..................DOWN 5%
AMMONIACAL NITROGEN (1,385 – 1,316 mg/L)........DOWN 5%
TOTAL OXIDISED NITROGEN (3.6 - 3.9 mg/L)...................UP 8%
NITRITE (0.01 – mg/L).......................................SAME 0%
NITRATE (21.3 – mg/L)..UP 25%

DIAGRAM 89 –

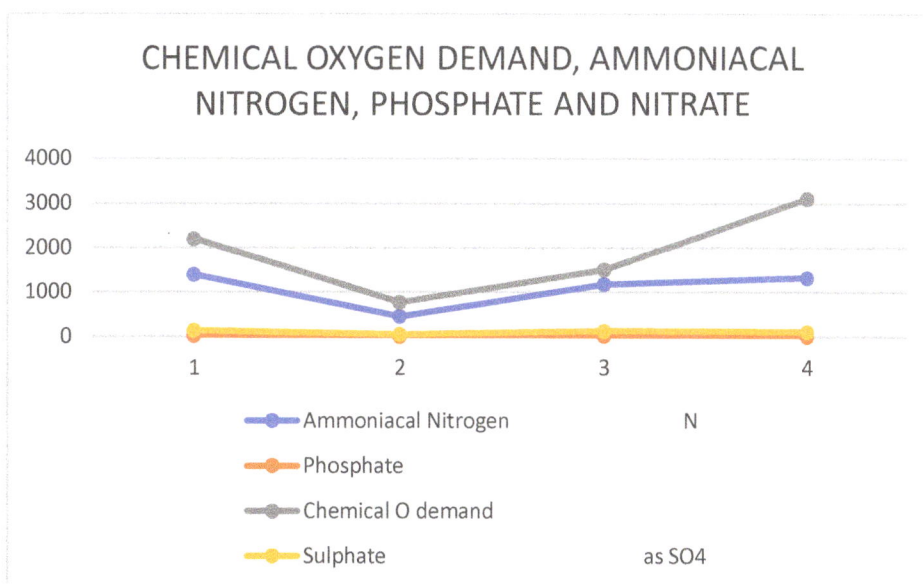

CHEMICAL OXYGEN DEMAND, AMMONIACAL NITROGEN, PHOSPHATE AND NITRATE

	ELEMENT	UNTREATED	3 MIN	5 MIN	8 MIN
Ammoniacal Nitrogen	N	1385	447	1166	1316
Phosphate	P	16.49	5.95	11.55	9.11
Chemical O demand	O	2200	770	500	3100
Sulphate as	SO4	136	45	123	109

APPLICATIONS FOR A PLASMOIDS FORM AND FUNCTIONS

DIAGRAM 90 –

ALKALINITY, NITROGEN, AMMONIA & TOTAL OX N

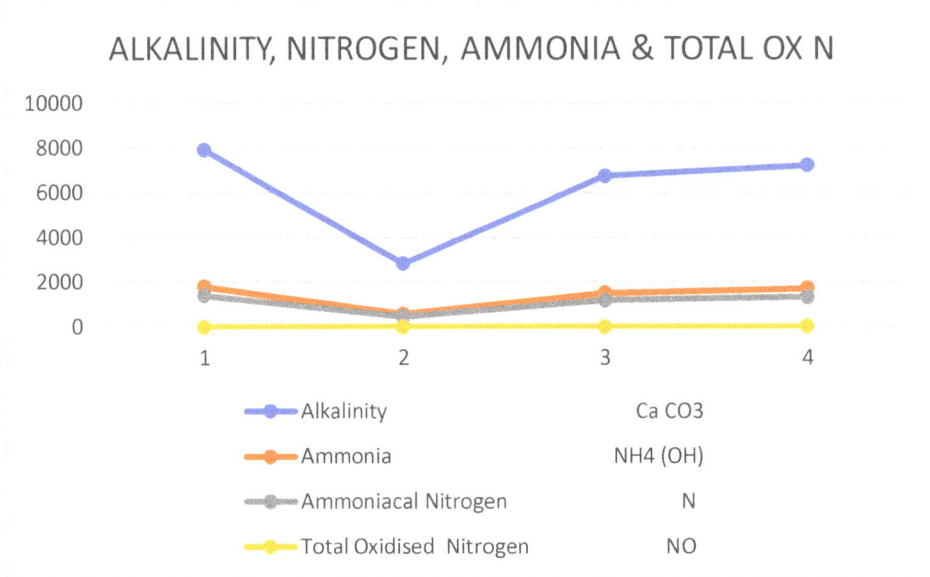

ELEMENTS	UNTREATED	3 MIN	5 MIN	8 MIN
Alkalinity	Ca CO3	7907	2825	6730	7205
Ammonia	NH4 (OH)	1788	576	1505	1699
Ammoniacal Nitrogen	N	1385	447	1166	1316
Total Oxidised Nitrogen	NO	3.6	2.6	3.7	3.9

APPLICATIONS FOR A PLASMOIDS FORM AND FUNCTIONS

DIAGRAM 91 –

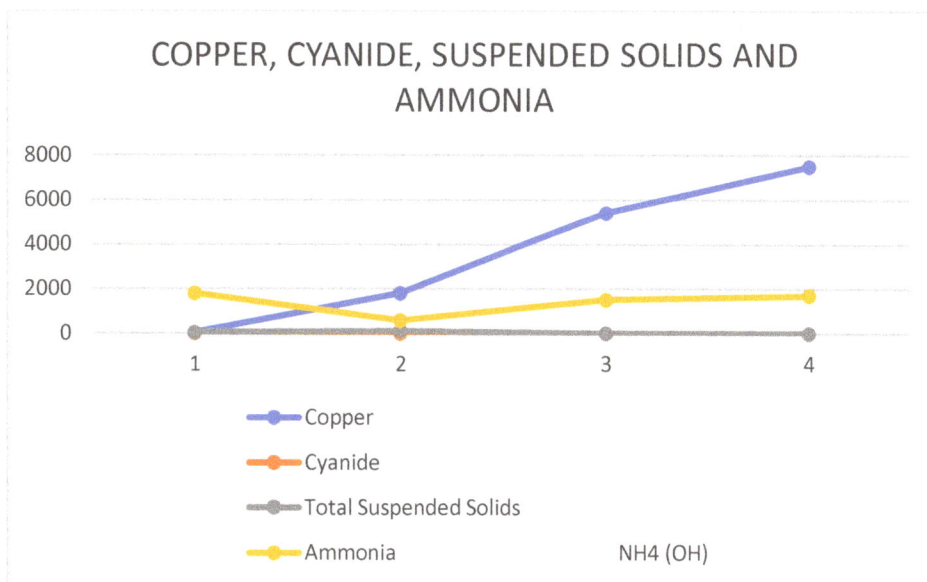

COPPER, CYANIDE, SUSPENDED SOLIDS AND AMMONIA

	ELEMENTS	UNTREATED	3 MIN	5 MIN	8 MIN
Copper	Cu	40	1,800	5400	7500
Cyanide	HCN	1.43	0.53	0.03	0.01
Total Suspended Solids		40	108	4	4
Ammonia	NH4 (OH)	1788	576	1505	1699

APPLICATIONS FOR A PLASMOIDS FORM AND FUNCTIONS

ELEMENTS AND MOLECULES REDUCED BY 50% FROM THE NORLANDS TIP LEACHATE FLUID AFTER 3 MINUTES WITHIN THE PLASMOID CREATOR

ELEMENTS REDUCED BY 50 % :-

ARSNIC (100 – 52 ug/L)..52%
CHROMIUM (53 – 26 ug/L)...49%
NICKLE (140 – 64 ug/L)..46%
CALCIUM (72 – 32 mg/L)..44%

MOLECULES REDUCED BY 50 % :-

ELECTRICAL CONDUCTIVITY (12,445 -5,375 uS/cm)...43%

ELEMENTS AND MOLECULES REDUCED BY 66 % (2/3) FROM THE NORLANDS TIP LEACHATE AFTER 3 MINUTES WITHIN THE PLASMOID CREATOR

ELEMENTS REDUCED BY 66 %:-

MAGNESIUM
SODIUM,
POTASSIUM,
CYANIDE,
CALCIUM
CARBONATE

MOLECULES REDUCED BY 66 %:-

PHOSPHATE
AMMONIACLE NITROGEN (AS N)
AMMONIA (AS NH4)
SULPHATE (AS SO4)
SULPHIDE
CHEMICAL OXYGEN DEMAND

APPLICATIONS FOR A PLASMOIDS FORM AND FUNCTIONS

ELEMENTS AND MOLECULES INCREASED BY 50% (1/2) FROM THE NORLANDS TIP LEACHATE FLUIDAFTER 5 MINUTES WITHIN THE PLASMOID CREATOR

ELEMENTS INCREASED BY 50 % :-

PHOSPHATE

MOLECULES INCREASED BY 50 %: -

CHEMICAL OXYGEN DEMAND
TOTAL ORGANIC CARBON

ELEMENTS AND MOLECULES REDUCED BY 66% (2/3) FROM THE NORLANDS TIP LEACHATE FLUIDAFTER 5 MINUTES WITHIN THE PLASMOID CREATOR

ELEMENTS REDUCED BY 66 %:-

MOLECULES REDUCED BY 66 %:-

APPLICATIONS FOR A PLASMOIDS FORM AND FUNCTIONS

ELEMENTS AND MOLECULES REDUCED BY 66% (2/3) FROM THE NORLANDS TIP LEACHATE FLUIDAFTER 8 MINUTES IN THE PLASMOID CREATOR

ELEMENTS REDUCED BY 66 %:-

MOLECULES REDUCED BY 66 %:-

ELEMENTS AND MOLECULES ELIMINATED BY 50% (1/2)FROM THE NORLANDS TIP LEACHATE FLUIDAFTER 8 MINUTES IN THE PLASMOID CREATOR

ELEMENTS ELIMINATED-

MOLECULES INCREASED BY 50 %:-

APPLICATIONS FOR A PLASMOIDS FORM AND FUNCTIONS

ELEMENTS AND MOLECULES REDUCED BY 66% (2/3)FROM THE NORLANDS TIP LEACHATE FLUIDAFTER 8 MINUTES IN THE PLASMOID CREATOR

ELEMENTS INCREASED BY 66 % :-

MOLECULES INCREASED BY 66 %: -

APPLICATIONS FOR A PLASMOIDS FORM AND FUNCTIONS

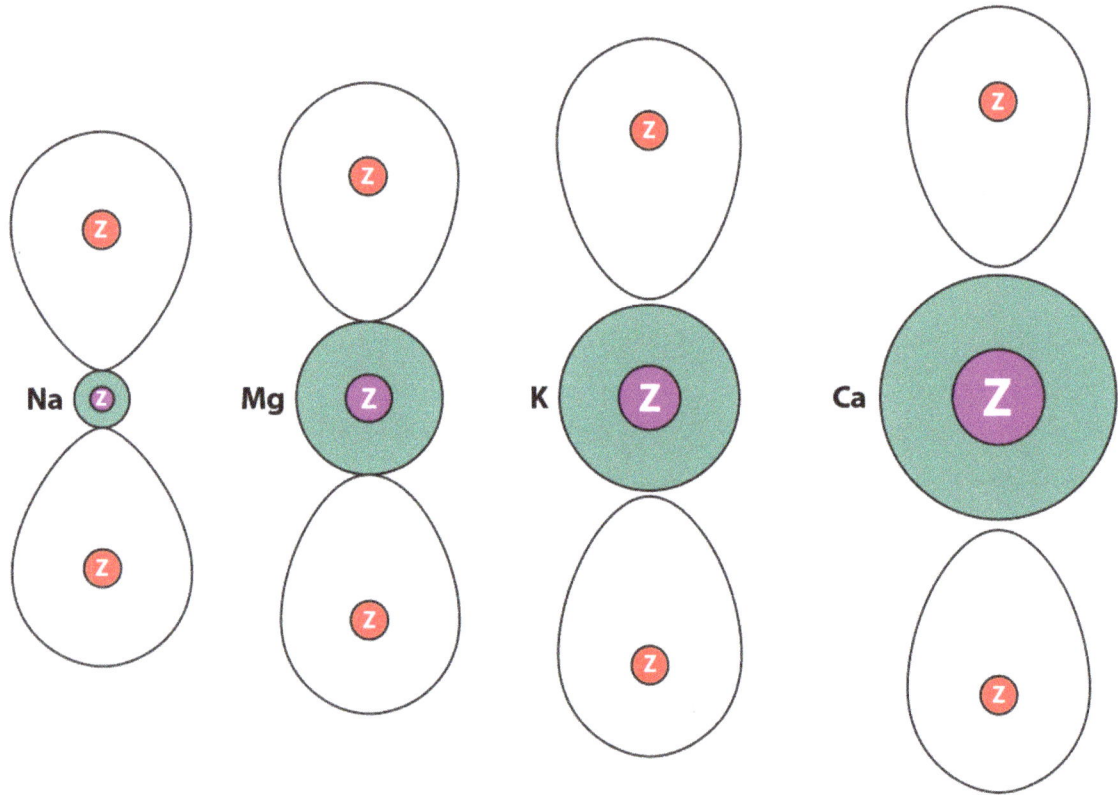

DIAGRAM 92 –

APPLICATIONS FOR A PLASMOIDS FORM AND FUNCTIONS

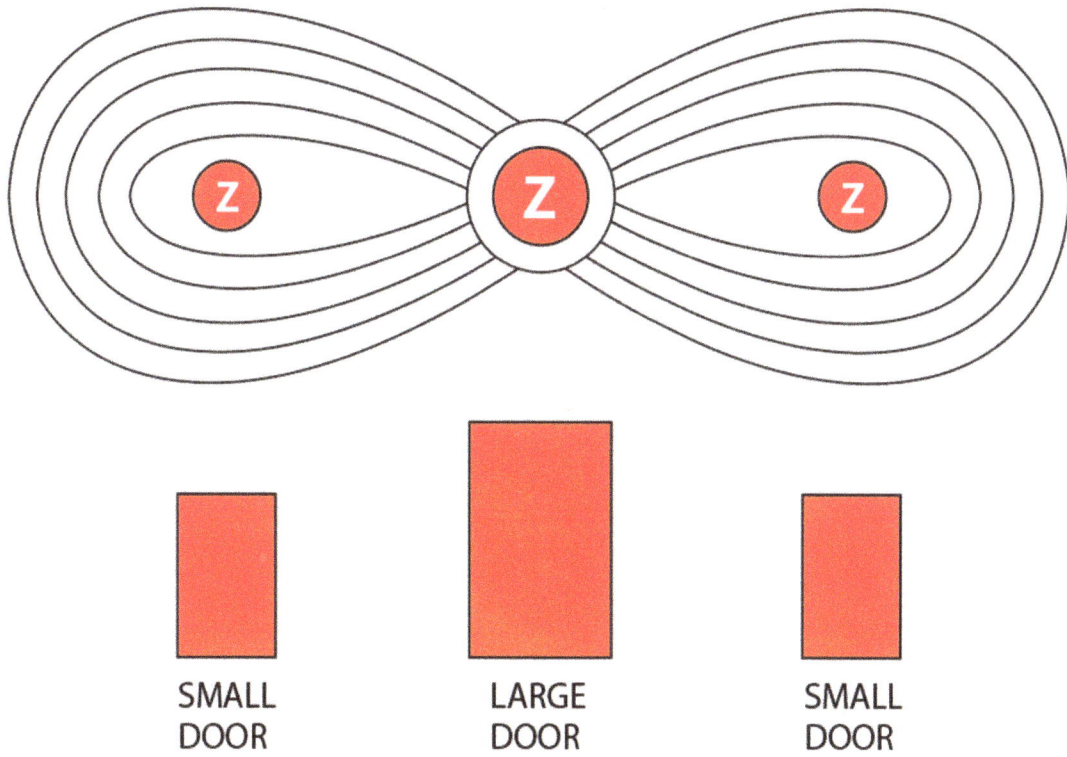

DIAGRAM 93 –

APPLICATIONS FOR A PLASMOIDS FORM AND FUNCTIONS

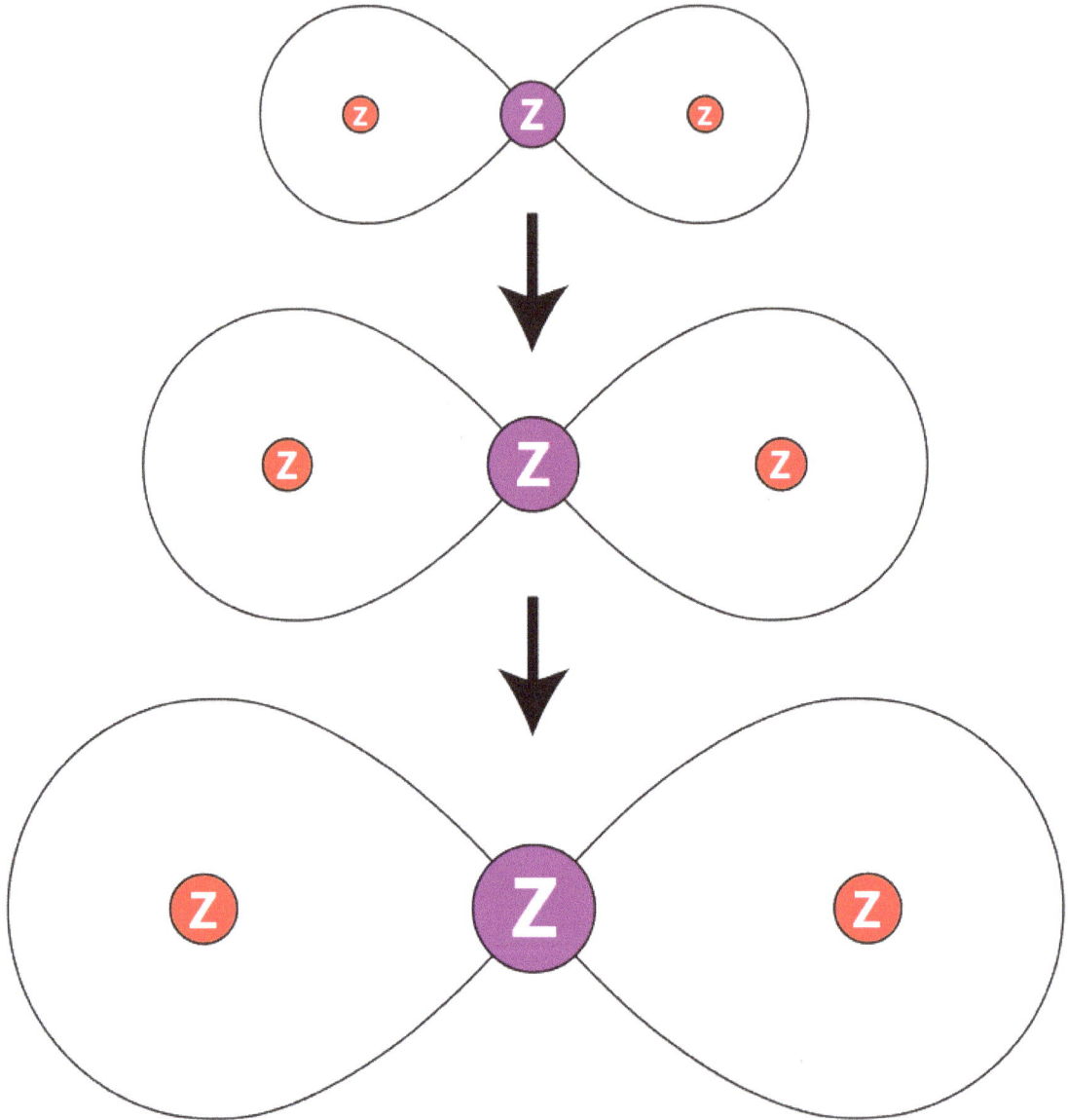

DIAGRAM 94 –

PART 3

PLASMOID SYSTEM
DESCRIPTION SCHEMATIC
AND PHOTOS

APPENDIX 5: - PLASMOID SYSTEM DESCRIPTION SCHEMATIC AND PHOTOS

DRAFT 518,400 B KMV - PART THREE OF TWENTY

MALCOLM V of SCOTLAND | MALCOLM BENDALL
THURSDAY 22ND SEPTEMBER 2022

APPENDIX 5 :- 4 / 3 / 2 RATIOS

APPLICATIONS FOR A PLASMOIDS FORM AND FUNCTIONS

**COMMENT ON 4,3,2 Base 50% = 1/2, 33% = 1/3, 25% = 1/4
RATIOS REFLECTED IN THE DATA**

APPLICATIONS FOR A PLASMOIDS FORM AND FUNCTIONS

APPENDIX 6: - MOLTEN SEA TORUS VAJRA

THE MOLTEN SEA TORUS VAJRA

ITS SCIENTIFIC PURPOSE

OPPOSITE CHARGED SPIRALS

TORUS 8 PLANE STRUCTURE

SINGULARITY ZERO POINT

PLASMOID SACRED GEOMETRY

ALIEN DIMENSIONAL QUANTUM

VORTEX MATHS CHEMISTRY

PHYSICS AND ATOMIC

UNIFICATION MODEL

DRAFT 266.666 22 – 01 – 2020

BY

MALCOLM ROY BENDALL

PHONE :- +66958345515

APPLICATIONS FOR A PLASMOIDS FORM AND FUNCTIONS

APPENDIX 7: - SPHERES, CONES, CYLINDERS, PYRAMIDS AND CUBES IMAGES AND CUT OUT PLANS

APPLICATIONS FOR A PLASMOIDS FORM AND FUNCTIONS

PATENT DESCRIPTION FOR THE BENDALL ENGINE

A system whereby the cold, vacuum and heat, pressure, shockwaves flow alternatively and sequentially from the exhaust and inlet ports of an internal combustion engine are utilized to retrieve and recycle that generated and stored potential energy. That energy is used to sequester, by the use of a Thunderstorm Tornado, generated free protons and electrons that are concentrated by a stream of Plasmoids (EV's). The Plasmoids confine and store those free Electron and Protons by generating an imploded sphere torus geometry that manifests a homeostatic self-induced, self-structuring, self-sustained, fractal Toroidal electromagnetic confinement field that's captures and confines and isolates micro-plasma. That electromagnetic confinement field is effective and fractal once having been formed and energised by collapsing bubbles within a column of water. The column of water being subjected to alternating vacuum and pressure pulses sourced by the normal action of a piston within an internal combustion engine alternatively generate and collapse the bubbles. These are the same naturally occurring forces of nature that produces the enormous power of a Thunderstorm or Cyclone.

Cool moist Plasmoid enriched air moving into the engine, structured using resonant spheres and cylinders of different diameters, interacts with hot dry air encapsulating it as it moves out in the opposite direction from the engine. This releases enough energy at an atomic level within the exhaust stream to fundamentally alter its composition eliminating toxic chemical wastes such as Carbon monoxide, nitrous oxide and hydrocarbons. The exhausts net positive ions which are also bad for life are replaced with net negative ions within the exhaust stream which support life. Simultaneously within the vacuum, imploding into the engine, together the Plasmoids and water vapor act to both disassociate the water into Hydrogen (Protium) and oxygen assisted by the catalytic and Tribone effects of the resonant 316 stainless steel spheres and cylinders. The Plasmoids alone, once reaching their effective charge density creating a viable Zero singularity point, due to charging received by the Thunderstorm Tornado, dissociate the Hydrogen (Protium) into its component electrons and Protons. This atomic and molecular fuel is fed back into the engine to add and enhance the explosive force of the normal hydrocarbon fuel.

Other elements that contain Neutrons within the imploding vacuum stream are unaffected by the forces applied by the Plasmoids as they are not powerful enough to act therefore producing no nuclear by products making the processes by-products non radio- active and life enhancing.

APPLICATIONS FOR A PLASMOIDS FORM AND FUNCTIONS

SPHERES

FIG 95: - The Catalytic Tornado Resonator

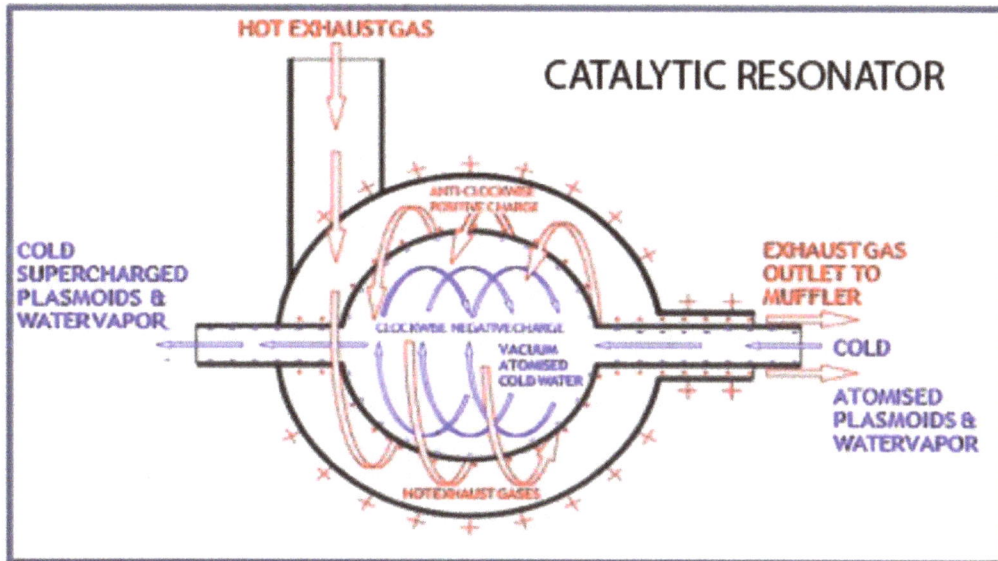

FIG 96 : - CATALYTIC TORNADO RESONATOR

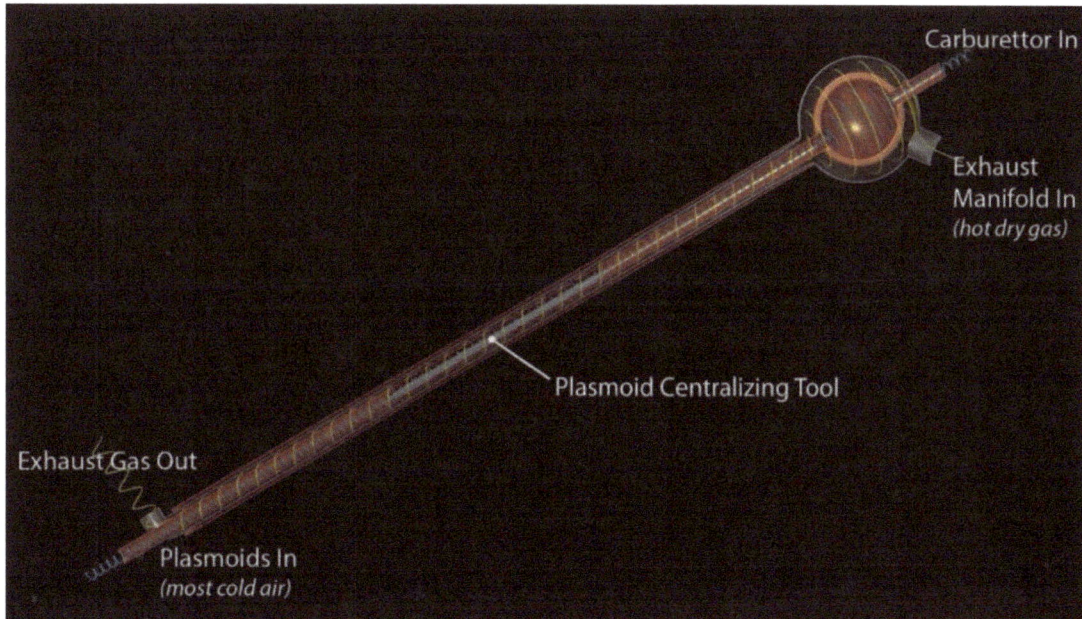

APPLICATIONS FOR A PLASMOIDS FORM AND FUNCTIONS

Fig 97 : - THREE SPHERES TWO 4-3-2 AND ONE CENTRAL 8-6-4

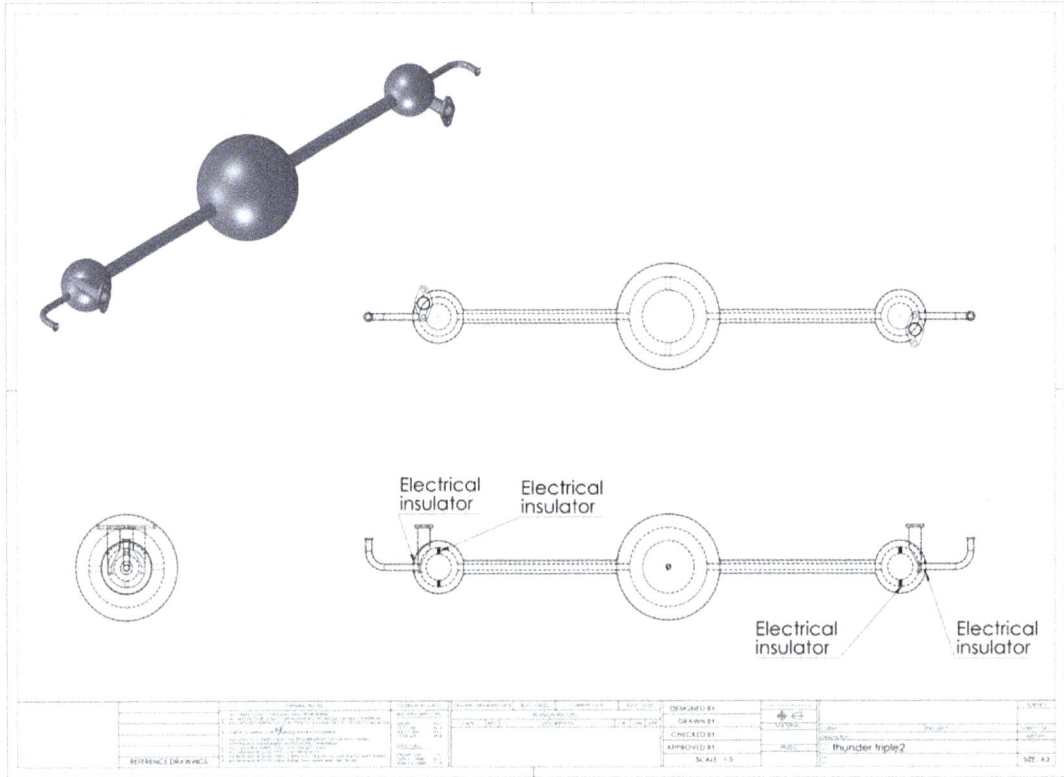

FIG 98 : - Catalytic Tornado Resonator Assembly – Double Spheres 4-3-2

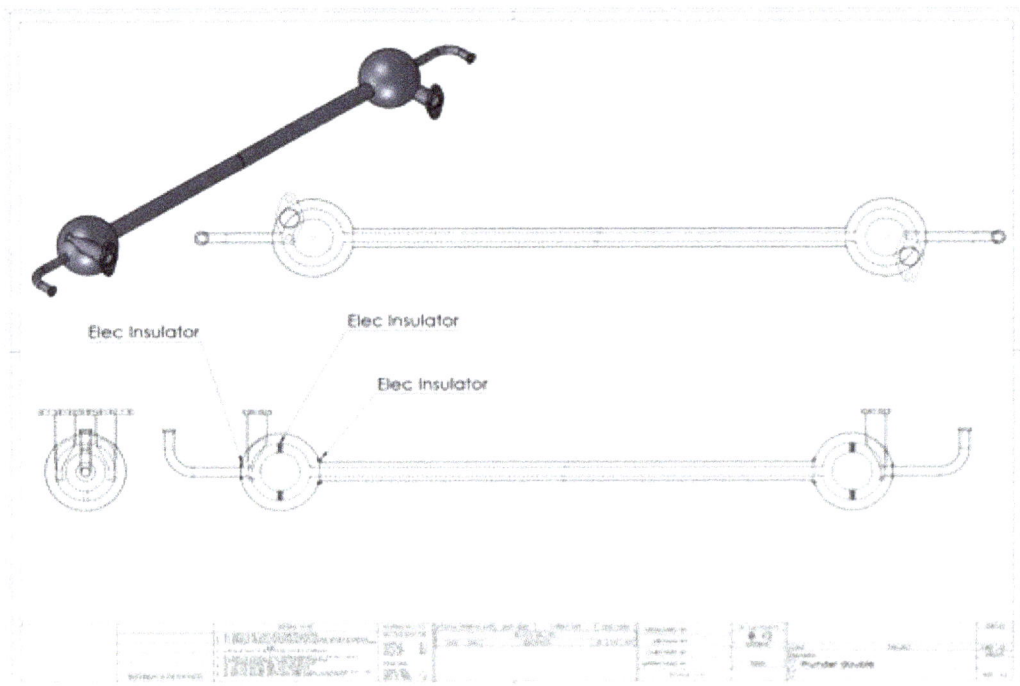

APPLICATIONS FOR A PLASMOIDS FORM AND FUNCTIONS

SPHERES

FIG 99 : - Catalytic Tornado Resonator Assembly – Insulated

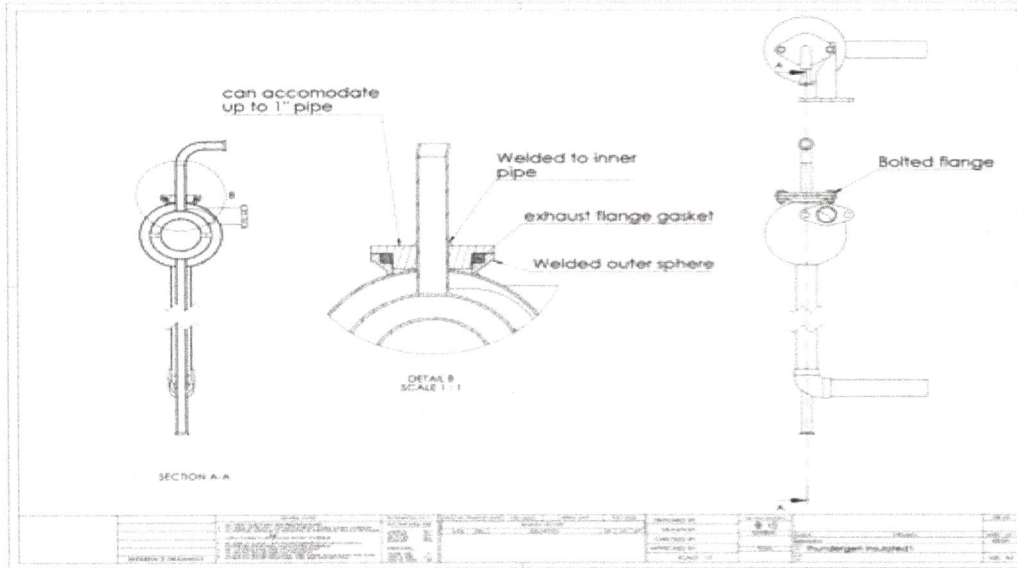

FIG 100 :- Catalytic Tornado Resonator Assembly – Single Spheres 4-3-2

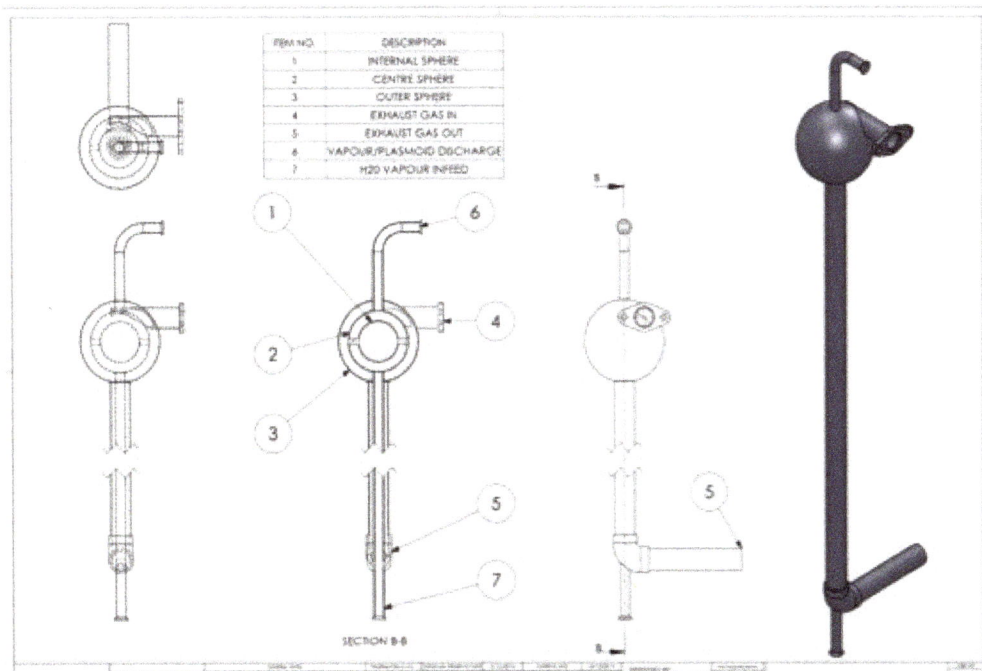

APPLICATIONS FOR A PLASMOIDS FORM AND FUNCTIONS

Fig 101 : - SOOTH END HALF SPHERES ON A SMOOTH CENTRAL CYLINDER COMBINATION

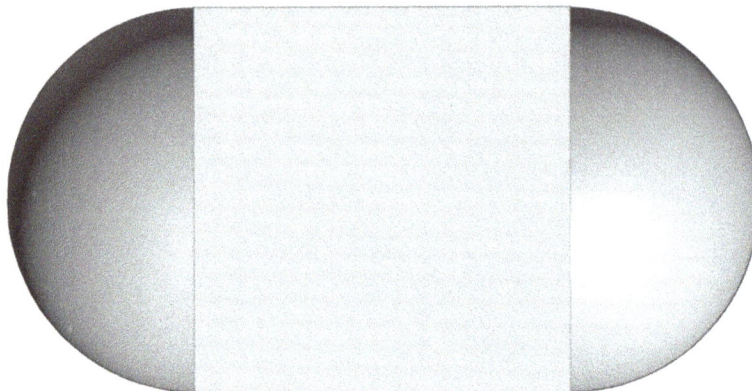

Fig 102 : - 16 SEGMENT HALF SPHERE END OBLIQUE END VIEW

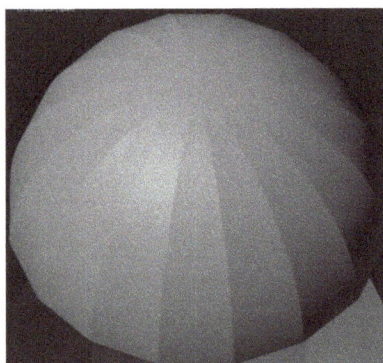

Fig 103 : - 16 SEGMENT END HALF SPHERES ON A 16 SEGMENT CENTRAL CYLINDER

APPLICATIONS FOR A PLASMOIDS FORM AND FUNCTIONS

FIG 104 : - SMOOTH END CONES AND SMOOTH CYLINDER COMBINATION

Fig 105 : - 16 SEGMENT END CONES AND SMOOTH CYLINDER COMBINATION

Fig 106 : - 16 SEGMENT END CONES AND 16 SEGMENT CYLINDER COMBINATION

APPLICATIONS FOR A PLASMOIDS FORM AND FUNCTIONS

FIG 107 : - 16 SIDED END CONES AND CYLINDERS

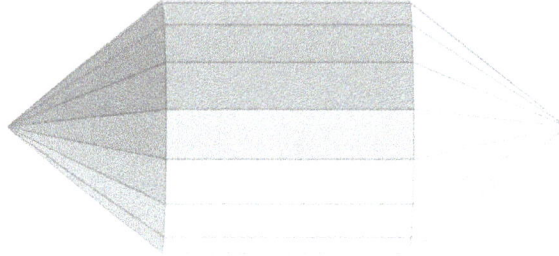

Fig 108 : - 16 SIDED END CONES ON CENTRAL CYLINDER

Fig 109 : - 16 SIDED END CONES ON A 16 SIDED CYLINDER CONSTRUCTION CUT-OUT

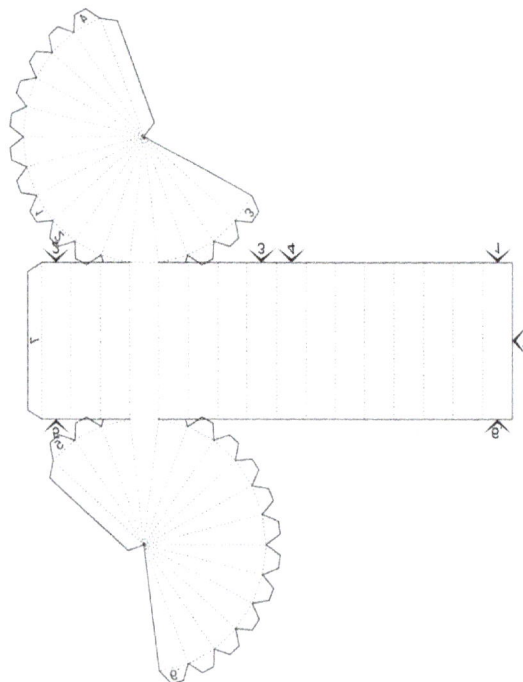

APPLICATIONS FOR A PLASMOIDS FORM AND FUNCTIONS

FIG 110: - 16 SIDED END CONES AND CYLINDERS CUT-AWAY IMAGE OF ITS SECTION

Fig 111: - 16 SIDED END CONES AND CYLINDERS CUT-AWAY IMAGE RIGHT ANGLE

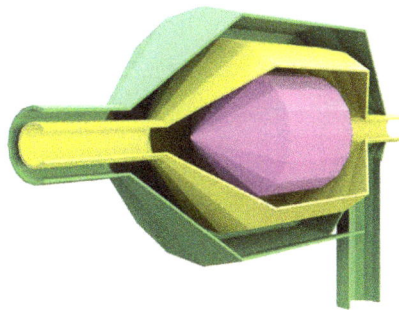

Fig 112 : - 16 SIDED END CONES AND CYLINDERS CUT-AWAY IMAGE LEFT ANGLE

Fig 113 : - 16 SIDED END CONES AND CYLINDERS SECTION FLOW DIAGRAM

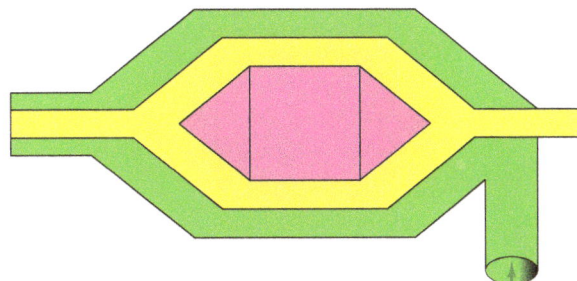

APPLICATIONS FOR A PLASMOIDS FORM AND FUNCTIONS

FIG 114 : - 8 SIDED END CONES AND CYLINDERS – WHITE AND SHADED

FIG 115 : - 8 SIDED END CONES AND CYLINDERS CONSTRUCTION CUT-OUT

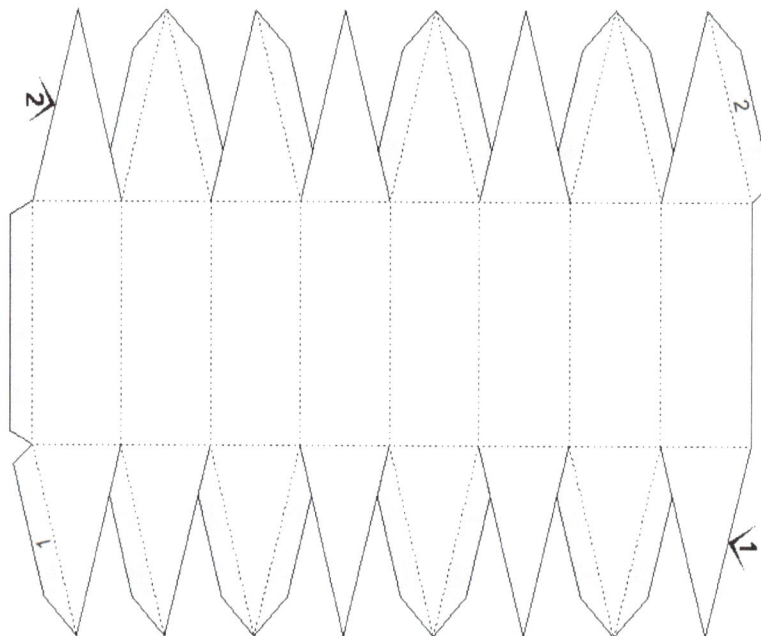

APPLICATIONS FOR A PLASMOIDS FORM AND FUNCTIONS

FIG 116 : - 8 SIDED END CONES AND CYLINDERS – CONSTRUCTION CUT-OUT

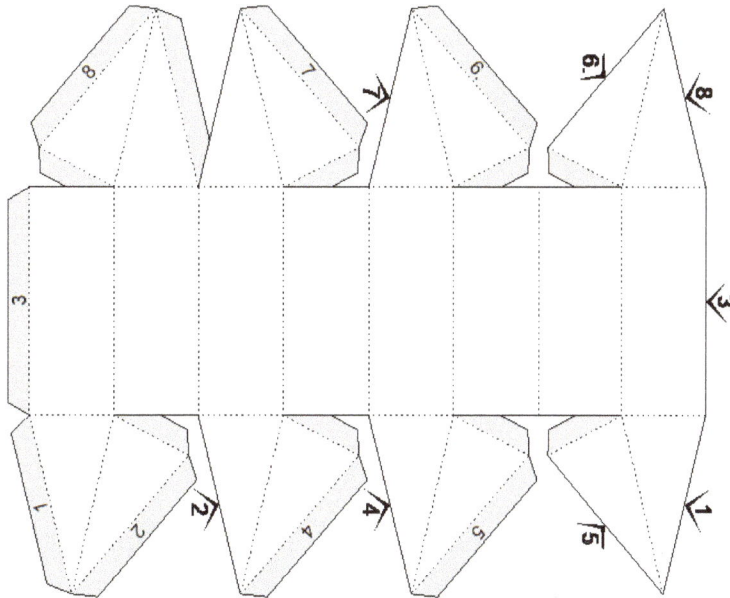

APPLICATIONS FOR A PLASMOIDS FORM AND FUNCTIONS

FIG 117 : -SMOOTH END CONES AND CUBE COLOURED AND WHITE

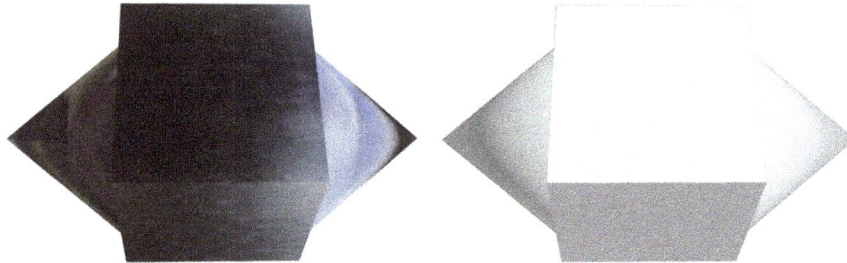

FIG 118 : -SMOOTH END CONES AND CUBE – CONSTRUCTION CUT-OUT

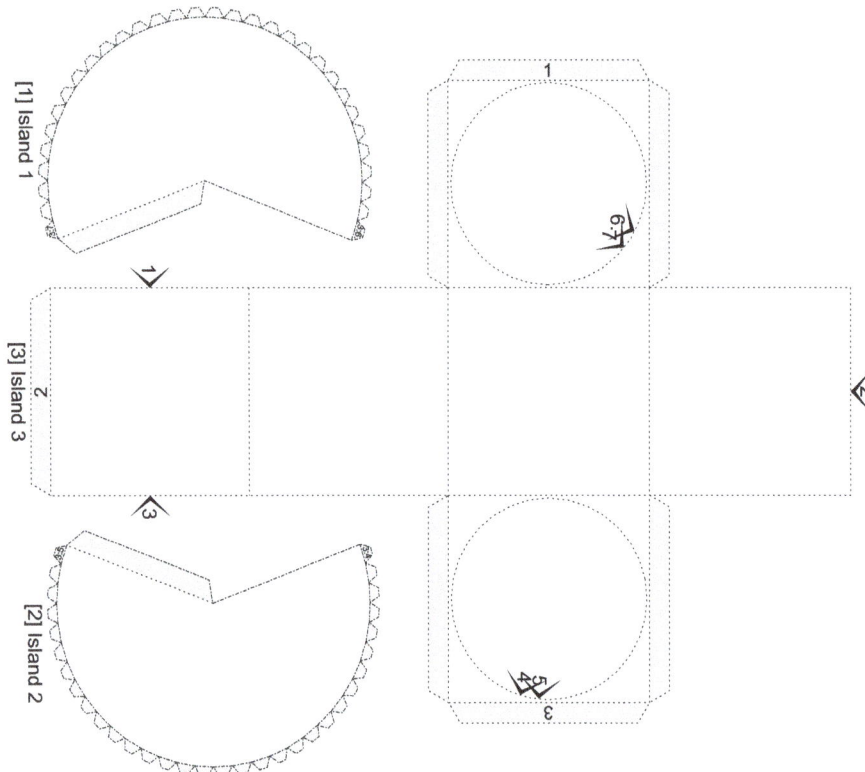

APPLICATIONS FOR A PLASMOIDS FORM AND FUNCTIONS

FIG 119 : - 16 SEGMENT END CONES AND CUBES – WHITE SIDE ANGLE

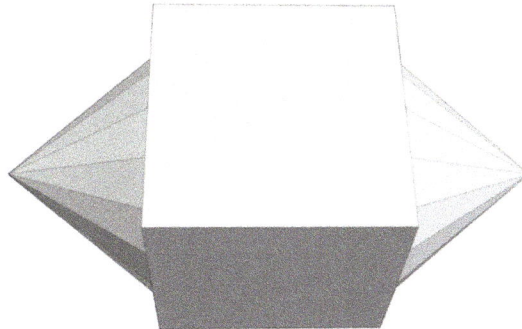

FIG 120 : - 16 SEGMENT END CONES AND CUBES – SIDE ANGLE COLOURED

FIG 121 : - 16 SEGMENT END CONES AND CUBE – CONSTRUCTION CUT-OUT

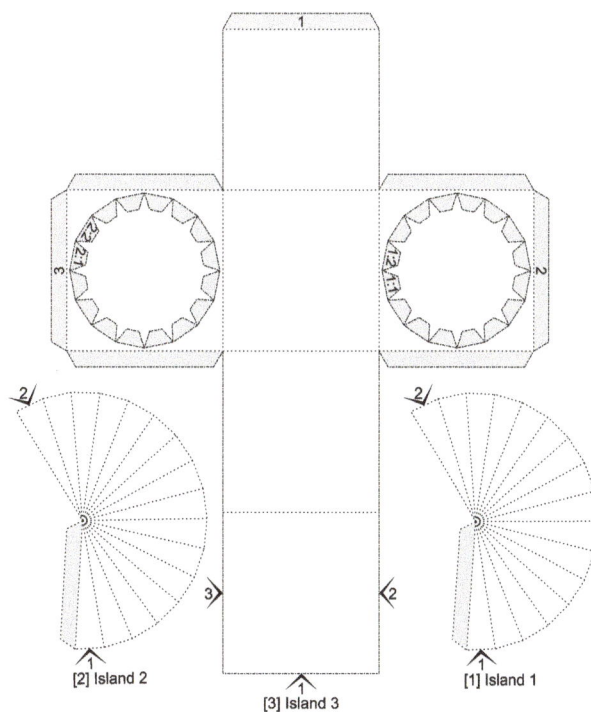

APPLICATIONS FOR A PLASMOIDS FORM AND FUNCTIONS

FIG 122 : - 51.84 END PYRAMIDS ON A CUBE – CONSTRUCTION CUT-OUT

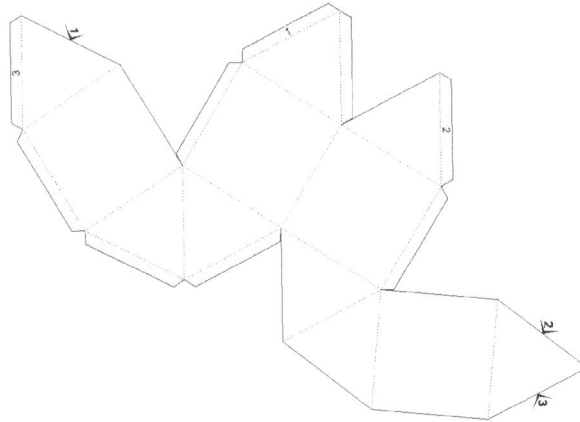

FIG 123 : - 51.84 END PYRAMIDS ON A CUBE – CONSTRUCTION CUT-OUT

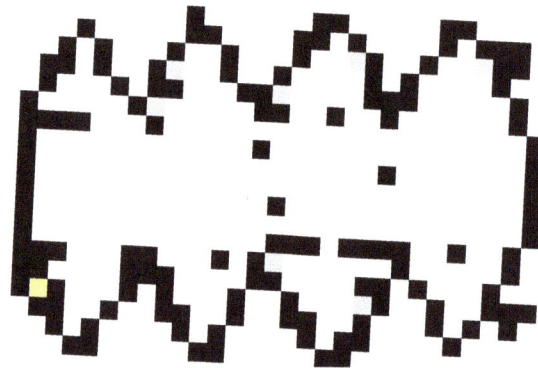

FIG 124 : - 51.84 END PYRAMIDS ON A CUBE – SIDE ANGLE

FIG 125 : - 51.84 END PYRAMIDS ON A CUBE – SIDE VIEW

APPLICATIONS FOR A PLASMOIDS FORM AND FUNCTIONS

FIG 126 : - FACETED 8-6-4 INTERNAL CUBES, STEPPED PYRAMIDS AND 8-6-4 CALCULATIONS

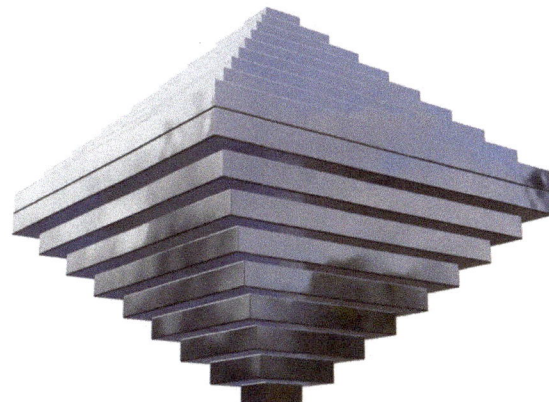

1 X 2 X 3 X 4 X 5 X 6 X 7 X 8 X 8 X 7 X 6 X 5 X 4 X 3 X 2 X 1 = 1,625,720,400 X 2 = 3,251,404,800 X 2 =   6,502,809,600

[6,502,809,600 / 24 = 270,950,400 / 60 = 4,515,840 / 60 = 75,264 / 60 = 1,254.4 ARC SECONDS]

1,625,720,400 / 518,400 = 3,136.0347222

1 X 2 X 3 X 4 X 5 X 6 X 6 X 5 X 4 X 3 X 2 X 1 = [720 X 720] = 518,400 X 2 = 1,036,800 X 2 = 2,073,600   2,073,600

[2,073,600 / 24 = 86,400 HOURS / 60 = 1,440 MINUTES / 60 = 24 SECONDS / 60 = 0.4 ARC SECONDS]

518,400 / 2,304 = 225

1 X 2 X 3 X 4 X 4 X 3 X 2 X 1 = [24 X 24] = 576 X 2 = 1,152 X 2 = 2,304   2,304

[2,304 / 24 = 96 / 60 = 1.6 / 60 = 0.0266.666 / 60 = 0.000444   GRAND TOTAL = 6,504,885,504

6,504,885,504 / 24 = 11,293,204 / 60 = 188,220.0666 / 60 = 317.00111 / 60 = 52.283351855185

6,504,885,504 / 2,160 = 3,011,521.0666   6,504,885,504 / 144 = 45,172,816

6,504,885,504 / 266.666 = 24,393,320.64   6,504,885,504 /4,924,800 = 1,320.842573099415

6,504,885,504 / 864,000 = 7,528.802666   6,504,885,504 / 108 = 60,230,421.333

6,504,885,504 / 25,920 = 250,960.0888

NUMBER OF CUBES WITHIN THE 8-6-4 STEPPED PYRAMID = 4,924,800

4,924,800 / 432,000 = 11.4   4,924,800 / 3,456,000 = 1.425   4,924,800 / 25,920 = 190

4,924,800 / 144 = 34,200   4,924,800 / 129,600 = 38   4,924,800 / 2,160 = 2,280

4,924,800 / 16 = 307,800   4,924,800 / 22.5 = 218,880   4,924,800 / 0.125 = 39,398,400

4,924,800 / = 4,924,800 / 720 = 6,840 x 4 = 25,920   Hydrogen   6,840 x 2 = 13,680

4,924,800 / 51.84 = 95,000   4,924,800 / 24 = 2052004924800 / 16 = 307800 / 22.5 = 13680 x 400 = 547,200 /

266.666 = 20,50   4,924,800 / 266.666 = 18,468   25 x 259.2 = 6,8400   24 x 259.2 = 6,220.8

16 x 259.2 = 4,147.2   51.84 x 259.2 = 13,436.928   Quartz crystal left and right hand spin

APPLICATIONS FOR A PLASMOIDS FORM AND FUNCTIONS

FIG 127 : - 45 DEGREE AND 51.84 DEGREE PYRAMID 12-8-6-4 WIDTHS

TOTAL WIDTH - 12 in
TOTAL HEIGHT - 7.6592 in
SIZE OF TOP SECTION - 1.6407 in

TOTAL WIDTH - 8 in
TOTAL HEIGHT - 5.0904 in
SIZE OF TOP SECTION - 1.0904 in

TOTAL WIDTH - 4 in
TOTAL HEIGHT - 2.5452 in
SIZE OF TOP SECTION - 0.5452 in

FIG 128 : -SMOOTH END HALF SPHERES ON A SMOOTH CUBE

APPLICATIONS FOR A PLASMOIDS FORM AND FUNCTIONS

FIG 129 : -SMOOTH END HALF SPHERES ON A SMOOTH CUBE

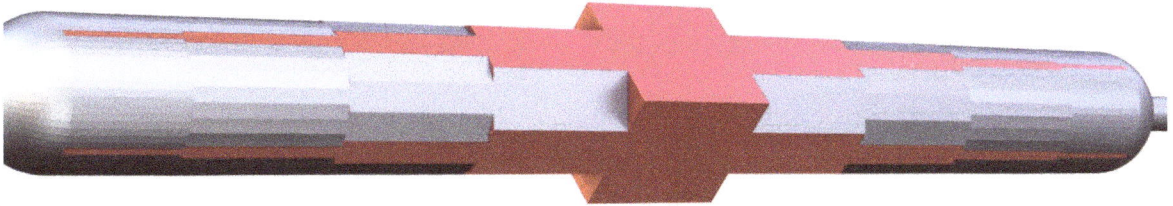

FIG 130 : -SMOOTH END HALF SPHERES ON A SMOOTH CUBE

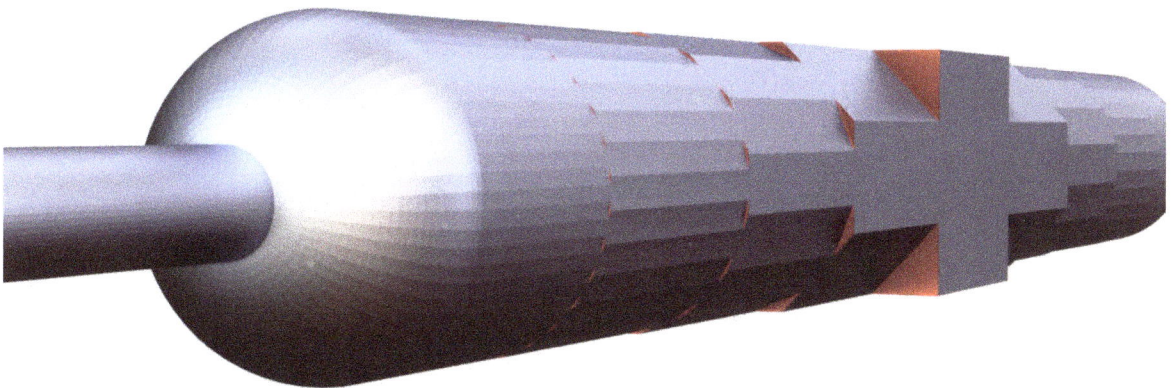

FIG 131 : -SMOOTH END HALF SPHERES ON A SMOOTH CUBE

APPLICATIONS FOR A PLASMOIDS FORM AND FUNCTIONS

FIG 132 : -SMOOTH END HALF SPHERES ON A SMOOTH CUBE

FIG 133 : -SMOOTH END HALF SPHERES ON A SMOOTH CUBE

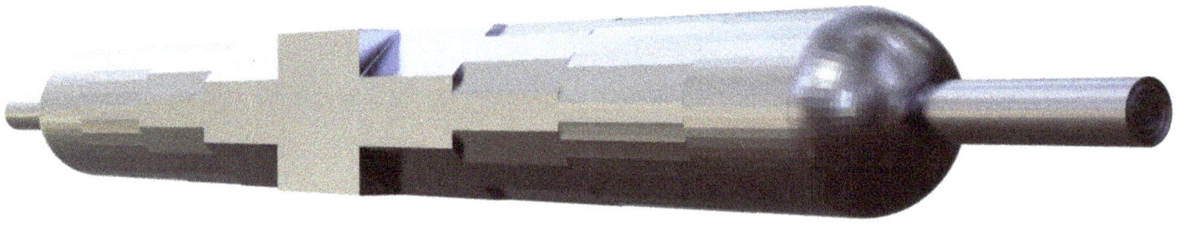

FIG 134 : -SMOOTH END HALF SPHERES ON A SMOOTH CUBE

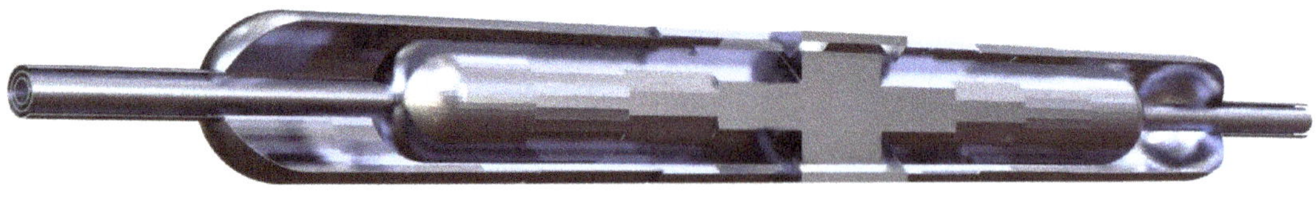

FIG 135 : -SMOOTH END HALF SPHERES ON A SMOOTH CUBE

PART 4

PLASMOID SYSTEM FOR
CARS DESCRIPTION
SCHEMATIC AND PHOTOS

APPLICATIONS FOR A PLASMOIDS FORM AND FUNCTIONS

APPENDIX 8:- VIRTUAL PLASMOID IMPLOSIVE TURBINE PLASMOID GUN AND QUADRITURE ASSEMBLY

DRAFT 518,400 B KMV – PART FOUR OF TWENTY

MALCOLM V of SCOTLAND | MALCOLM BENDALL
THURSDAY 22ND SEPTEMBER 2022

APPLICATIONS FOR A PLASMOIDS FORM AND FUNCTIONS

FIG 136 : - PLASMOID GUN ASSEMBLEY

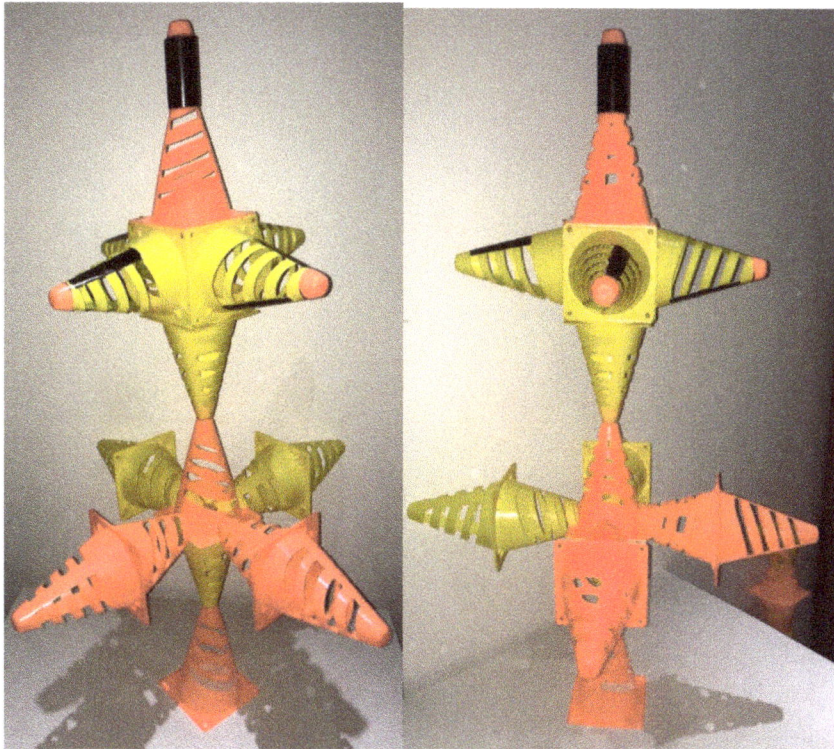

PART 4 OF 20 – MSAART PATENT APPLICATION NOTES DRAFT 518,400 – 22:22:22 THUR 22ND SEPT 2022   135
© STRIKE FOUNDATION GUARANTEE LIMITED | MALCOLM BENDALL 2022 | GRAPHICS - STEVE EARL

APPLICATIONS FOR A PLASMOIDS FORM AND FUNCTIONS

FIG 137 : - PLASMOID GUN ASSEMBLEY WITH THE BASE VAJRA QUADRITURE POSITIVE TO NEGATIVE

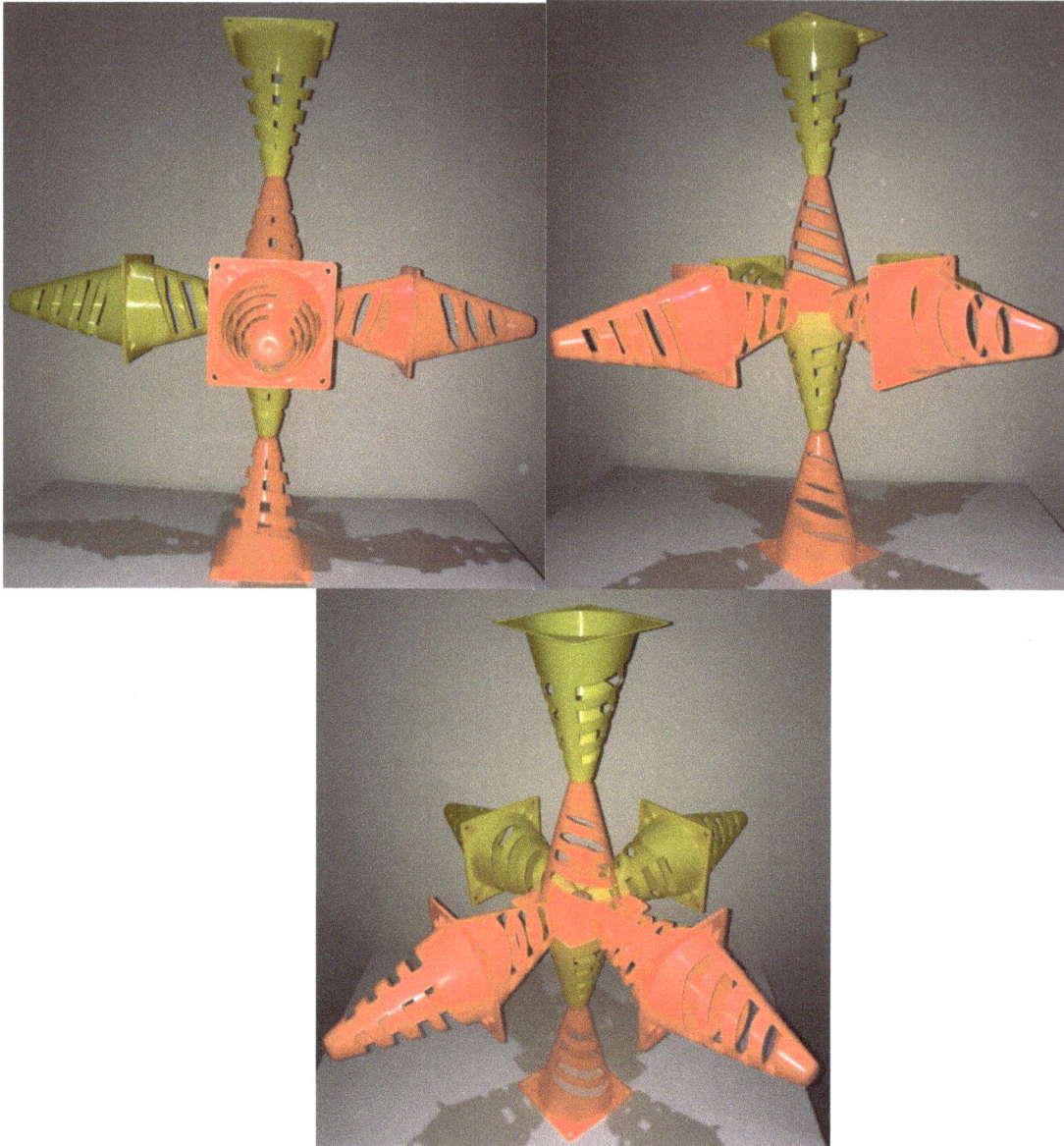

APPLICATIONS FOR A PLASMOIDS FORM AND FUNCTIONS

FIG 138 :- PLASMOID GUN ASSEMBLEY INTERLOCKING CONE ELECTRO-MAGNETIC CONFINEMENT FIELD ASSEMBLEY

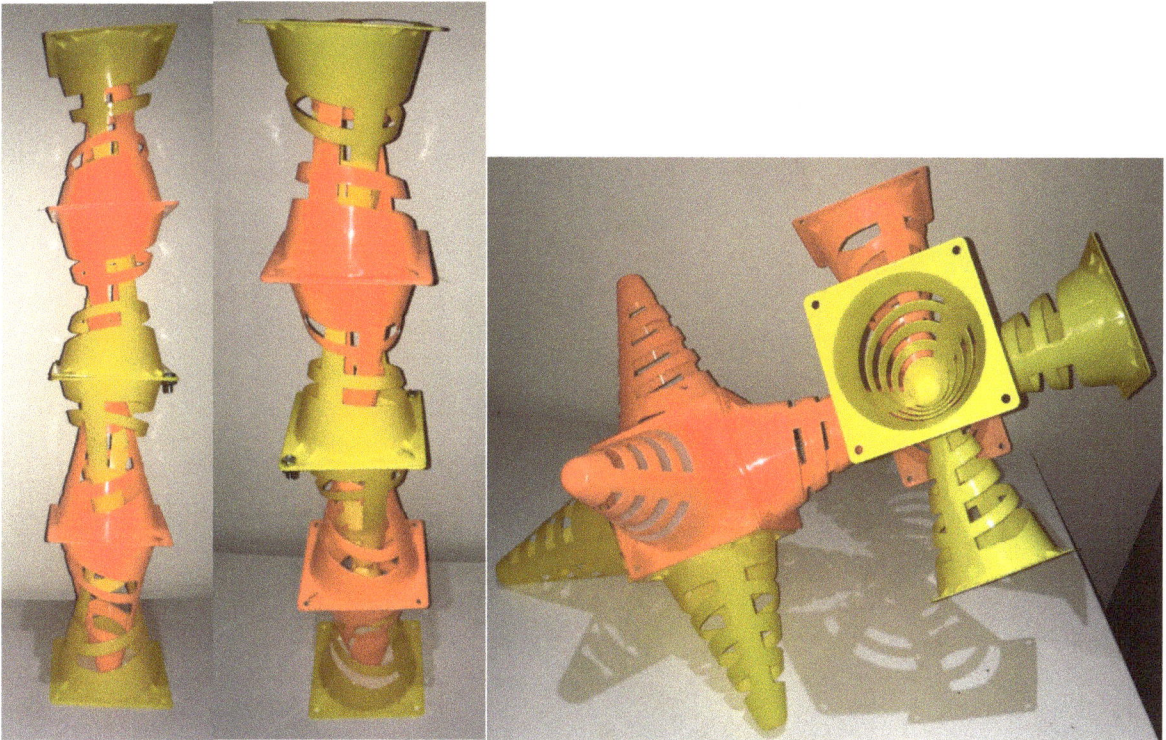

FIG 139 : - PLASMOID GUN ASSEMBLEY BASE VAJRA QUADRITURE

APPLICATIONS FOR A PLASMOIDS FORM AND FUNCTIONS

FIG 140 : - STANDARD CONVENTIONAL EXPLOSIVE JET ENGINE AS USED IN COMMERCIAL AND MILITARY JET PLANE OPERATIONS

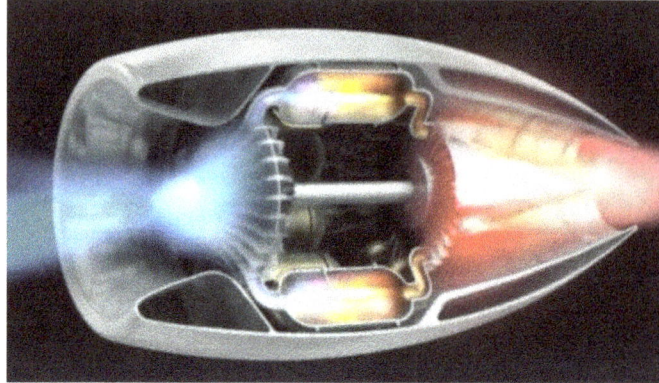

FIG 141 : - PLASMOID GUN AND VAJRA IMPLOSIVE JET TURBINE

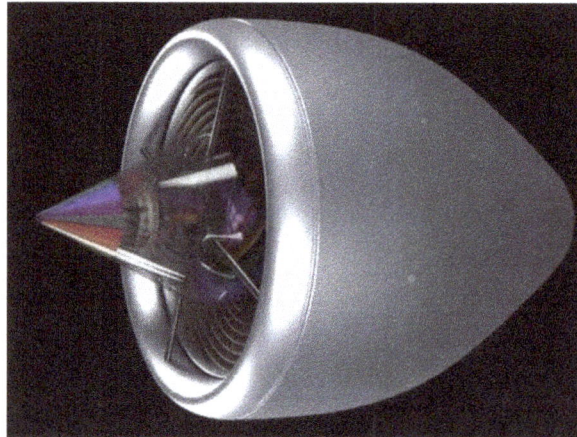

FIG 142 : - PLASMOID GUN AND VAJRA IMPLOSIVE JET TURBINE CROSS SECTION CUT-AWAY

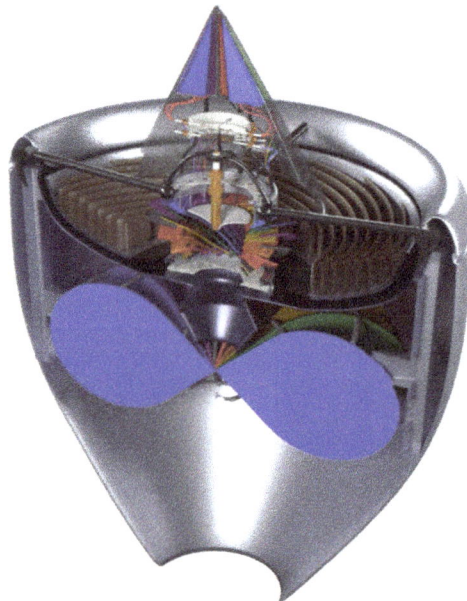

APPLICATIONS FOR A PLASMOIDS FORM AND FUNCTIONS

FIG 143 : - THE MOLTEN SEA TORUS VAJRA IMPLOSIVE TURBINE SWIRL GUIDE AND ENGINE CUT-AWAY

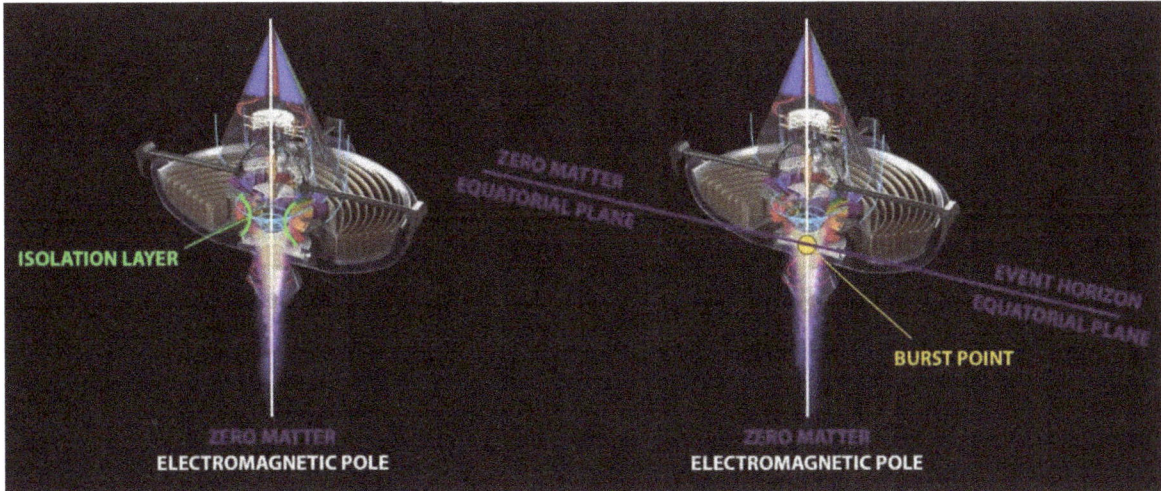

FIG 144 : - THE MOLTEN SEA TORUS VAJRA IMPLOSIVE TURBINE SWIRL GUIDE OPERATING POSITIONS OFF- HALF THRUST– FULL THROTTEL

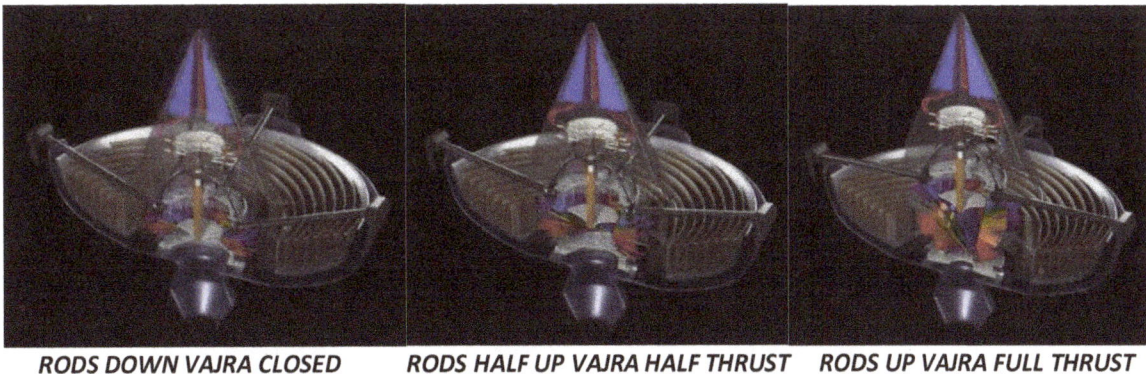

RODS DOWN VAJRA CLOSED   *RODS HALF UP VAJRA HALF THRUST*   *RODS UP VAJRA FULL THRUST*

APPLICATIONS FOR A PLASMOIDS FORM AND FUNCTIONS

FIG 145 : - THE MOLTEN SEA TORUS VAJRA IMPLOSIVE TURBINE SWIRL GUIDE

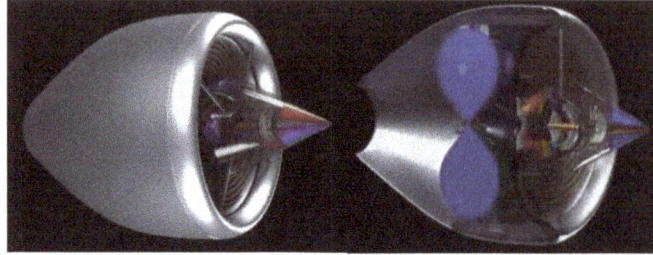

FIG 146 : - THE MOLTEN SEA TORUS VAJRA IMPLOSIVE TURBINE SWIRL GUIDE

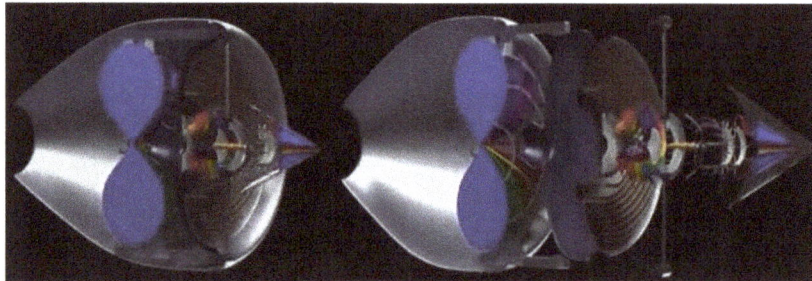

FIG 147 : - THE MOLTEN SEA TORUS VAJRA IMPLOSIVE TURBINE SWIRL GUIDE

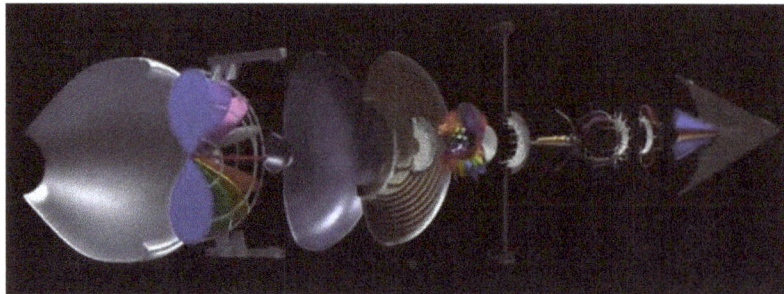

FIG 148 : - MOLTEN SEA TORUS VAJRA IMPLOSIVE TURBINE SWIRL GUIDE EXPLODED VIEW

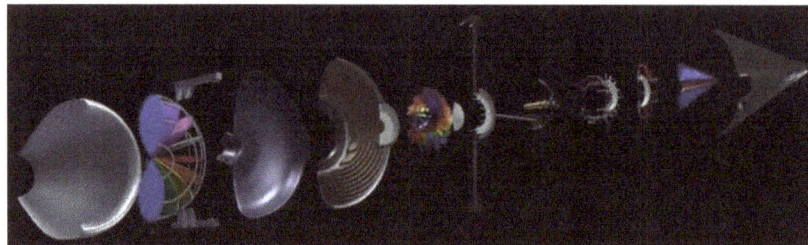

FIG 149 : - NAMED EXPLODED VIEW PARTS OF THE MOLTEN SEA TORUS VAJRA IMPLOSIVE TURBINE SWIRL GUIDE

APPLICATIONS FOR A PLASMOIDS FORM AND FUNCTIONS

FIG 150 : - THE MOLTEN SEA TORUS VAJRA IMPLOSIVE TURBINE DIAGRAM SHOWING, NOSE CONE, AIR COLLECTION BOWEL, SWIRL GUIDE, AIR FLOW (LIGHT BLUE AND RED), BOUNDARY LAYER (GREEN), EQUATORIAL PANE EVENT HORIZON (PURPLE) AND BURST POINT (ORANGE).

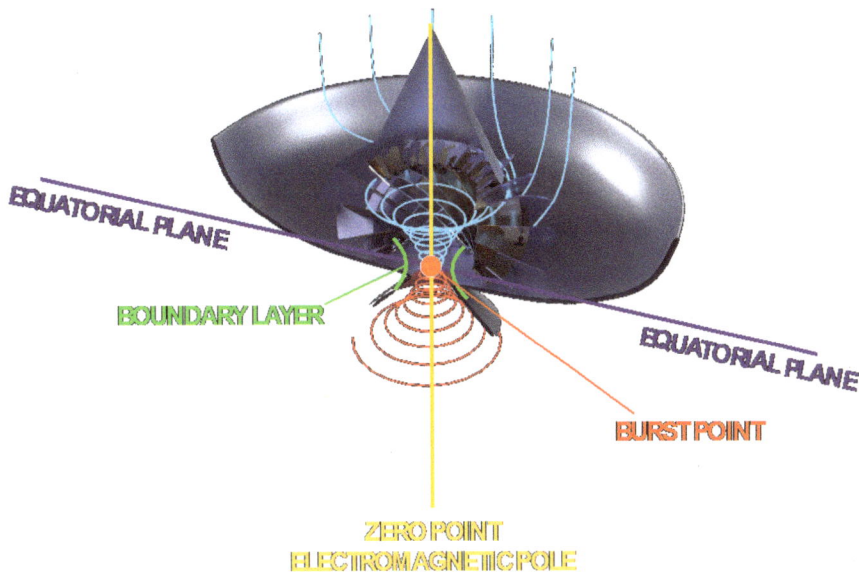

FIG 151 : - PLASMOID GUN ASSEMBLEY AND TORUS MOLTEN SEA VAJRA IMPLOSIVE TURBINE SWIRL GUIDE

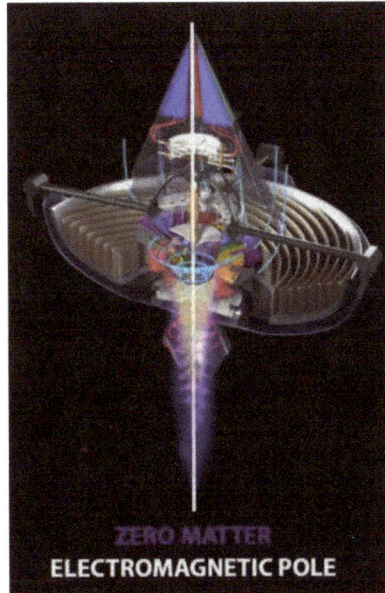

APPLICATIONS FOR A PLASMOIDS FORM AND FUNCTIONS

FIG 152 : - THE MOLTEN SEA TORUS VAJRA IMPLOSIVE TURBINE SWIRL GUIDE FLOW DIAGRAM

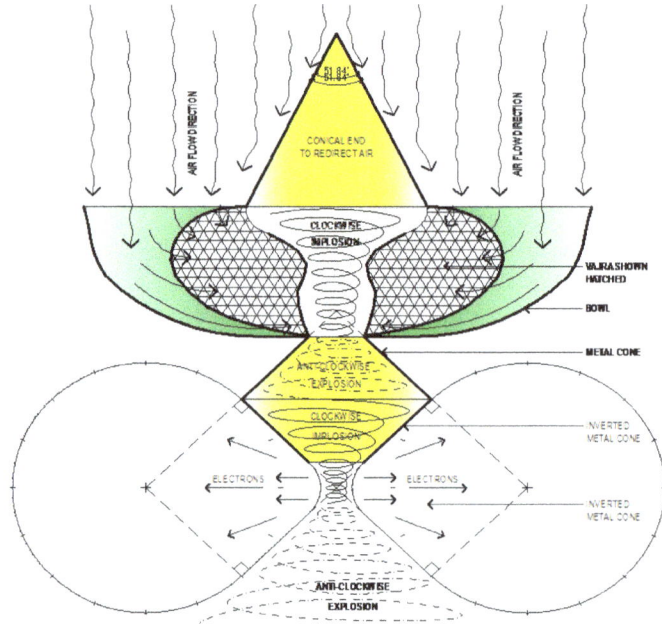

DIAGRAM A
IMPLOSIVE VORTEX WAVE GUIDE

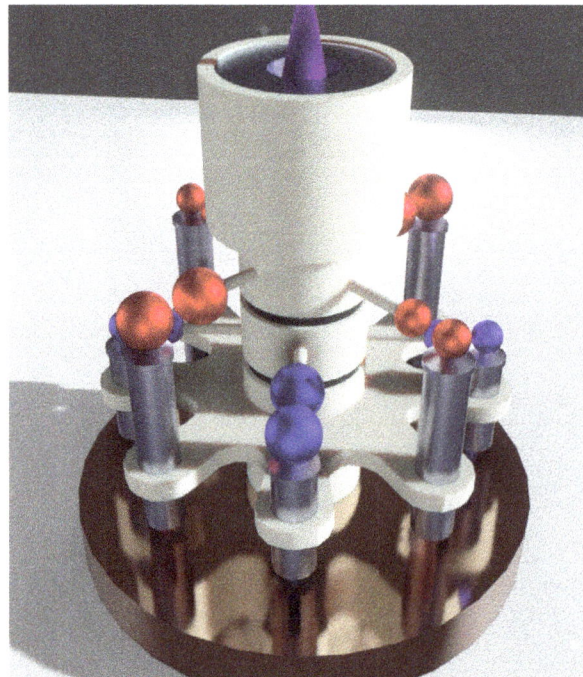

FIG 153 : - THE MOLTEN SEA TORUS VAJRA IMPLOSIVE TURBINE SWIRL GUIDE FLOW DIAGRAM

APPLICATIONS FOR A PLASMOIDS FORM AND FUNCTIONS

DIAGRAM 154 - TORUS MOLTEN SEA TORUS VAJRA IMPLOSIVE TURBINE SWIRL GUIDE & SPHERE

FIG 155 : - TORUS MOLTEN SEA VAJRA IMPLOSIVE TURBINE IN CLEAR POLYCARBONATE - BASE VIEW

FIG 156 : - TORUS MOLTEN SEA TORUS VAJRA IMPLOSIVE TURBINE IN CLEAR POLYCARBONATE – TOP VIEW

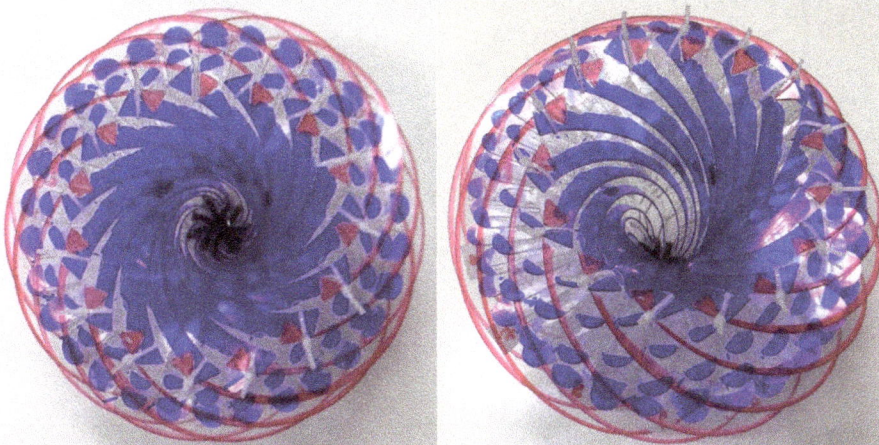

APPLICATIONS FOR A PLASMOIDS FORM AND FUNCTIONS

FIG 157 : - TORUS MOLTEN SEA TORUS VAJRA IMPLOSIVE TURBINE SWIRL GUIDE – TOP VIEW

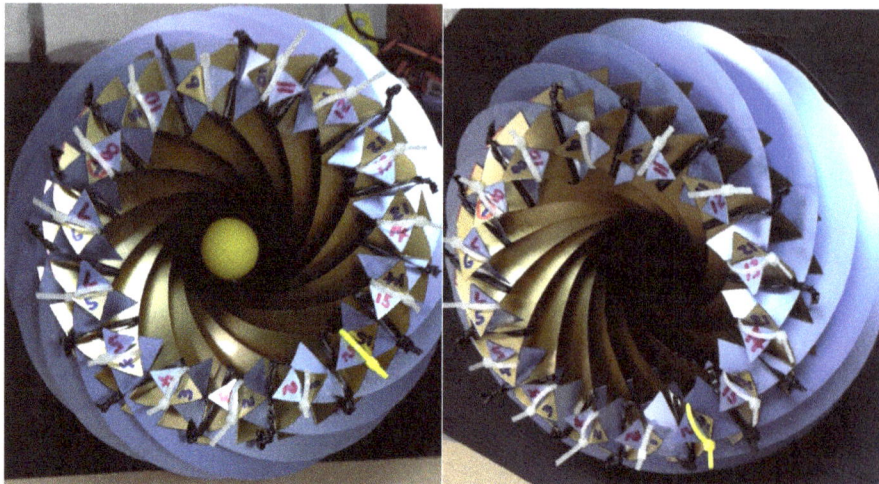

FIG 158 : - TORUS MOLTEN SEA VAJRA IMPLOSIVE TURBINE SWIRL GUIDE IN COPPER AND STAINLESS STEEL

FIG 159 : - TORUS MOLTEN SEA VAJRA IMPLOSIVE TURBINE SWIRL GUIDE IN COPPER AND STAINLESS STEEL

APPLICATIONS FOR A PLASMOIDS FORM AND FUNCTIONS

FIG 160 : - QUADRITURE FIRING SEQUENCE OF IMPLOSIVE TURBINE BLADES

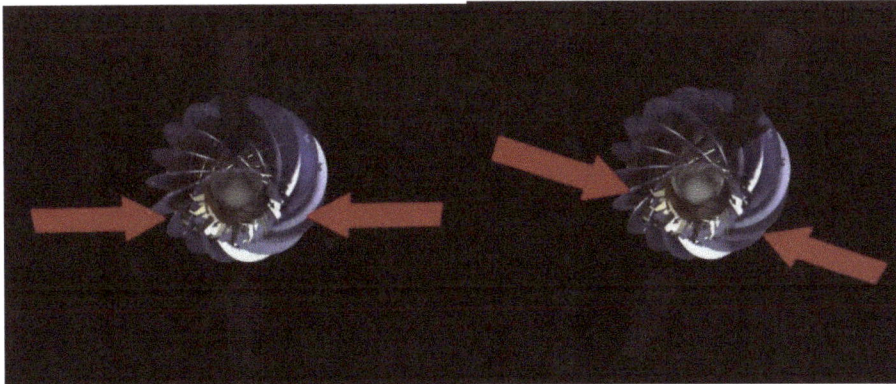

FIG 161 : - QUADRITURE FIRING SEQUENCE OF IMPLOSIVE TURBINE BLADES

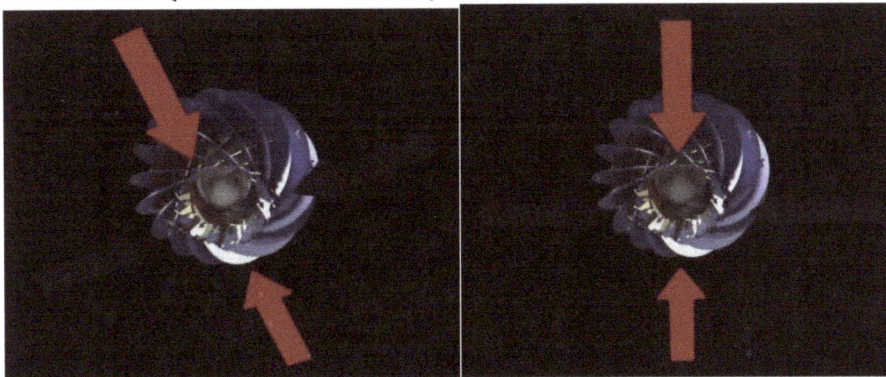

FIG 162 : - QUADRITURE FIRING SEQUENCE OF IMPLOSIVE TURBINE BLADES

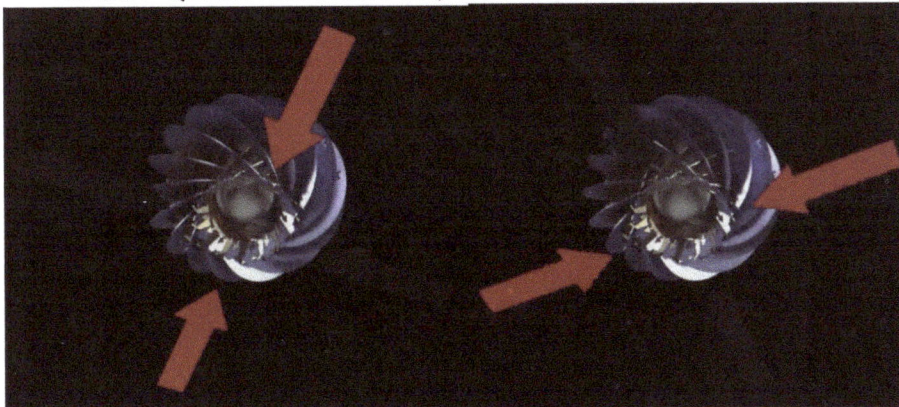

FIG 163 : - QUADRITURE FIRING SEQUENCE OF IMPLOSIVE TURBINE BLADES

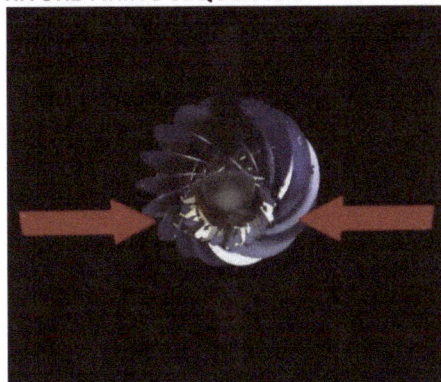

PART 5
SOLOMON'S MOLTEN SEA ARC

APPENDIX 11:- SOLOMON'S MOLTEN SEA LENSE

PLASMOID GENERATOR AND ARC CHARGER

DRAFT 518,400 B KMV – PART FIVE OF SIX

MALCOLM V of SCOTLAND | MALCOLM BENDALL
THURSDAY 22ND SEPTEMBER 2022

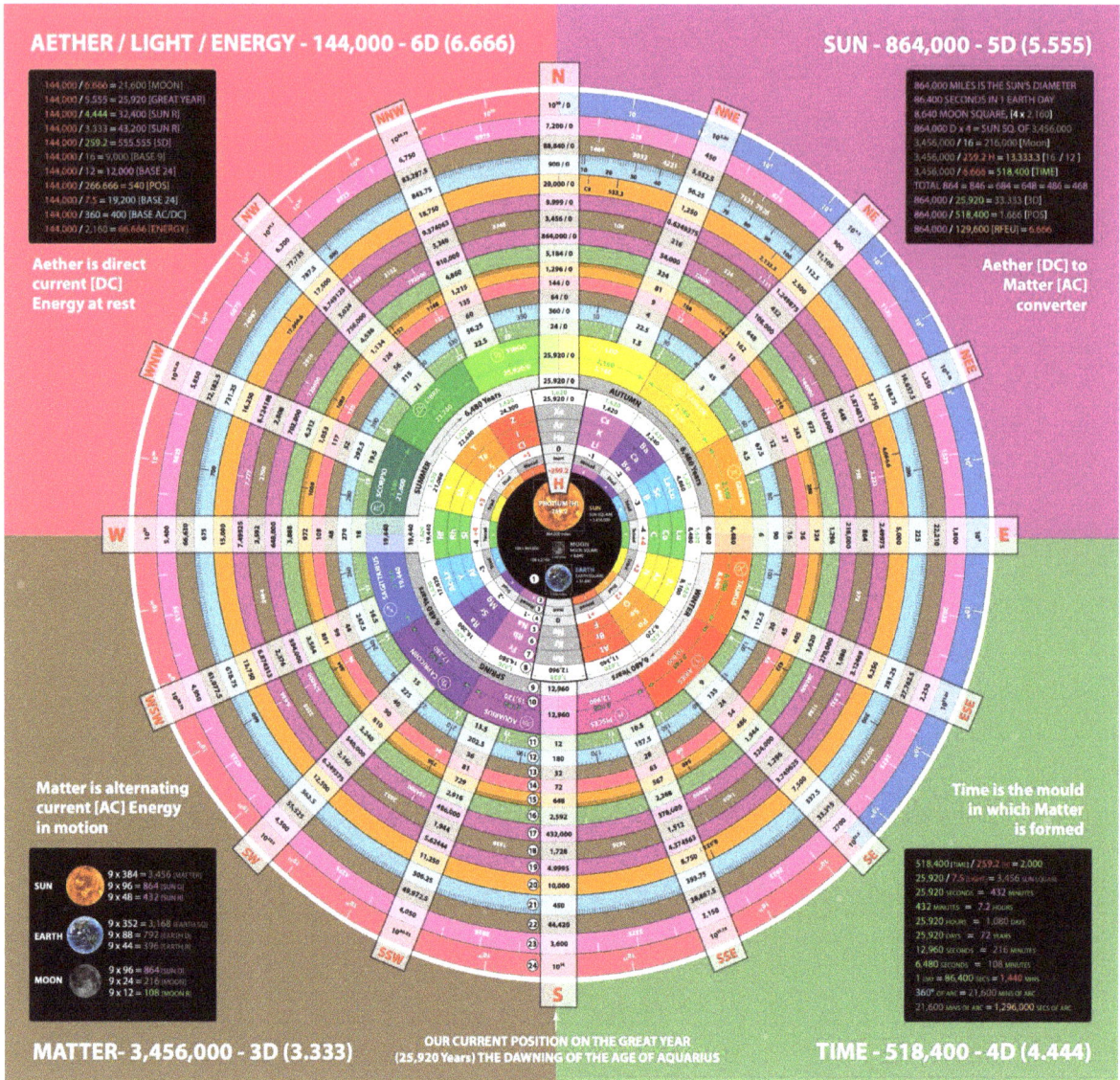

AETHER / LIGHT / ENERGY - 144,000 - 6D (6.666)

144,000 / 6.666 = 21,600 [MOON]
144,000 / 5.555 = 25,920 [GREAT YEAR]
144,000 / 4.444 = 32,400 [SUN R]
144,000 / 3.333 = 43,200 [SUN R]
144,000 / 259.2 = 555.555 [5D]
144,000 / 16 = 9,000 [BASE 9]
144,000 / 12 = 12,000 [BASE 24]
144,000 / 266.666 = 540 [POS]
144,000 / 7.5 = 19,200 [BASE 24]
144,000 / 360 = 400 [BASE AC/DC]
144,000 / 2,160 = 66.666 [ENERGY]

Aether is direct
current [DC]
Energy at rest

SUN - 864,000 - 5D (5.555)

864,000 MILES IS THE SUN'S DIAMETER
86,400 SECONDS IN 1 EARTH DAY
8,640 MOON SQUARE, [4 x 2,160]
864,000 D x 4 = SUN SQ. OF 3,456,000
3,456,000 / 16 = 216,000 [Moon]
3,456,000 / 259.2 H = 13,333.3 [16 / 12]
3,456,000 / 6.666 = 518,400 [TIME]
TOTAL 864 = 846 = 684 = 648 = 486 = 468
864,000 / 25,920 = 33.333 [30]
864,000 / 518,400 = 1.666 [POS]
864,000 / 129,600 = [RFEU] = 6.666

Aether [DC] to
Matter [AC]
converter

Matter is alternating
current [AC] Energy
in motion

SUN   9 x 384 = 3,456 [MATTER]
      9 x 96 = 864 [SUN D]
      9 x 48 = 432 [SUN R]

EARTH 9 x 352 = 3,168 [EARTH SQ]
      9 x 88 = 792 [EARTH D]
      9 x 44 = 396 [EARTH R]

MOON 9 x 96 = 864 [SUN D]
      9 x 24 = 216 [MOON D]
      9 x 12 = 108 [MOON R]

Time is the mould
in which Matter
is formed

518,400 [TIME] / 259.2 = 2,000
25,920 / 7.5 [SUN R] = 3,456 [SUN SQUARE]
25,920 SECONDS = 432 MINUTES
432 MINUTES = 7.2 HOURS
25,920 HOURS = 1,080 DAYS
25,920 DAYS = 72 YEARS
12,960 SECONDS = 216 MINUTES
6,480 SECONDS = 108 MINUTES
1 DAY = 86,400 SECS = 1,440 MINS
360° OF ARC = 21,600 MINS OF ARC
21,600 MINS OF ARC = 1,296,000 SECS OF ARC

MATTER - 3,456,000 - 3D (3.333)

OUR CURRENT POSITION ON THE GREAT YEAR
(25,920 Years) THE DAWNING OF THE AGE OF AQUARIUS

TIME - 518,400 - 4D (4.444)

PLASMOID UNIFICATION MODEL

Plasmoids are doughnut or toroidal shaped clusters of net Protons or net Electrons that once captured and placed into a Toroidal orbit are capable of absorbing, storing and releasing enormous amounts of energy present within their self-generated and structured electro-magnetic containment field. Plasmoids, in effect, function as an atomic battery that can be self-charging due to its ability to convert matter to available clean energy. Plasmoids by their unique geometry cause a consequential electro-magnetic containment field to generate a Zero point naturally and casually, without much effort, have the ability to convert the nuclear Mass of Protium (Atoms) into energy.

The Plasmoid Unification Model (PUM) posits that Plasmoids are epoch-making and that knowledge of them has been hidden in plain sight for centuries. This PUM 'slide rule' reveals the algorithmic relationships between life's elements critical to mankind's existence and development. It starts with Protium [H] which has a melting point of -259.2°C and is the most abundant element in our Solar System. Protium determines the 25,920 Great Year frequency of our Solar System. The resonant frequencies of all other elements can then be calculated when 25,920 years is reduced from years to days, hours and seconds.

The PUM is evidence that the Universe is an intelligent design. That design is in perfect octave harmonic resonance with itself. Therefore, all of creation from Galaxies to Planets to Elements all resonate in unison with a collective chord 'As Above So Below'. This is interconnected with an Energy 'web', the 24 components and laws of which are all based and governed on the same 16 sector Torus Plasmoid precepts shown. The concepts and ruling principles of the PUM can, and have, been applied to make Energy to Matter and Matter to Energy conversions. When applied to the modern hydrocarbon powered internal combustion engine, PUM technology removes exhaust toxic waste products and increases the engine power output by transforming waste energy back into fuel. Plasmoids employed in conjunction with the Plasmoid Toroidal Implosive Turbine provide a new novel Matter to Energy and Energy to Matter propulsion device for water, land, air and space travel.

LEGEND

Aether/Light   Sun   Time   Matter   Frequency   Degree

1. **Sun** (864,000) [DC], Earth (7,920) [AC], Moon (2,160) [AC]
2. **Music** [AC] (Do = C = 24 x 11.111 = 266.666)
3. **Elemental Crystal Forms** + / - Monad, Diad, Triad, Tetrad
4. **Elemental Valencies** (0, -1, -2, -3, -4 and 0, +1, +2, +3, +4)
5. **Elements 1-16** (He - Cl)
6. **Elements 17-32** (Ar - I)
7. **Elements 33-48** (Xe - 'Z')
8. **Elemental Frequencies** (1,620 x 16 = 25,920 light frequency of -259.2 C)
9. **Seasons of Great Year** (25,920 / 0 years)
10. **Zodiac Great Year** (25,920 / 0 years)
11. **Clock 24 Hour / 0 Hour** [AC] / 0 Hour [DC] Clock
12. **Compass 360° Degrees** (AC) / 0° Degrees [DC]
13. **Matter 64 / 0** 64 Points being 32 Resonant Planes / 0
14. **Light** 144 / 0
15. **Resonate Frequency Energy Unit (RFEU)** - 1,296 (129,600)
16. **All Time** 5,184 (518,400 secs)
17. **Aether - Sun** (864,000 miles diameter, 432,000 miles radius)
18. **Matter** 3,456 (3,456,000 miles - Sun Square)
19. **Dimensions** 3D Matter = 3.33, 4D Time = 4.44, 5D Aether = 5.55, 6D Light = 6.666
20. **Sound and Music** (0 - 20,000 Hz)
21. **Language** 1-9, 10-90, 100-900 (111,222,333,444,555,666,777,888,999) (45,450,4,500)
22. **Solar System** Sun & Planetary Diameters and Radii in miles
23. **Plasmoids** 7,200 Degrees / 0, 32 Planes 64 Radial Points, One Zero Point
24. **All Plasmoid Energy** = All Alternating Current [AC] Frequencies

(▬ = Non ionizing   ▬ = Ionizing)

© PLASMOID INTERNATIONAL

PART 6

MSAART SYSTEM APPLIED
TO A NEW CAR

APPLICATIONS FOR A MSAART'S FORM AND FUNCTIONS

PART SIX OF TWENTY

MSAART SYSTEM APPLIED TO A NEW CAR

MALCOLM V of SCOTLAND | MALCOLM BENDALL
THURSDAY 22ND SEPTEMBER 2022

DRAFT 518,400 B KMV

APPLICATIONS FOR A PLASMOIDS FORM AND FUNCTIONS

FIG 182 – Firing mechanism for the Atomic Plasmoid Protium Proton Power MATTER TO ENERGY Space, Air, Land and Sea Drive

FIG 183 – Firing mechanism for the Atomic Plasmoid Protium Proton Power MATTER TO ENERGY Space, Air, Land and Sea Drive

APPLICATIONS FOR A PLASMOIDS FORM AND FUNCTIONS

FIG 184 – Firing mechanism for the Atomic Plasmoid Protium Proton Power MATTER TO ENERGY Space, Air, Land and Sea Drive

FIG 185 – Firing mechanism for the Atomic Plasmoid Protium Proton Power MATTER TO ENERGY Space, Air, Land and Sea Drive

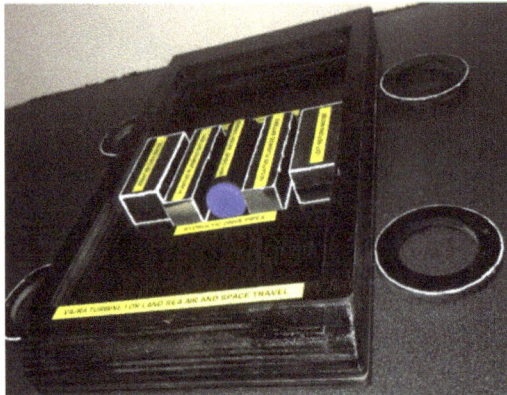

APPLICATIONS FOR A PLASMOIDS FORM AND FUNCTIONS

FIG 186 – Firing mechanism for the Atomic Plasmoid Protium Proton Power MATTER TO ENERGY Space, Air, Land and Sea Drive

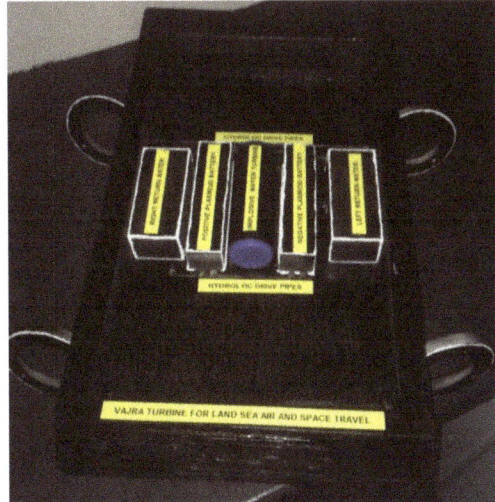

FIG 187 – Firing mechanism for the Atomic Plasmoid Protium Proton Power MATTER TO ENERGY Space, Air, Land and Sea Drive

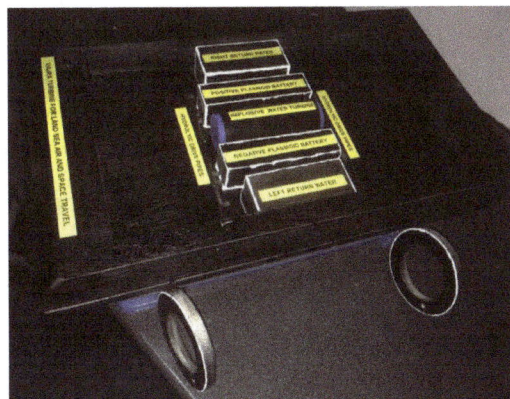

APPLICATIONS FOR A PLASMOIDS FORM AND FUNCTIONS

*FIG 188 – Firing mechanism for the Atomic Plasmoid Protium Proton Power
MATTER TO ENERGY Space, Air, Land and Sea Drive*

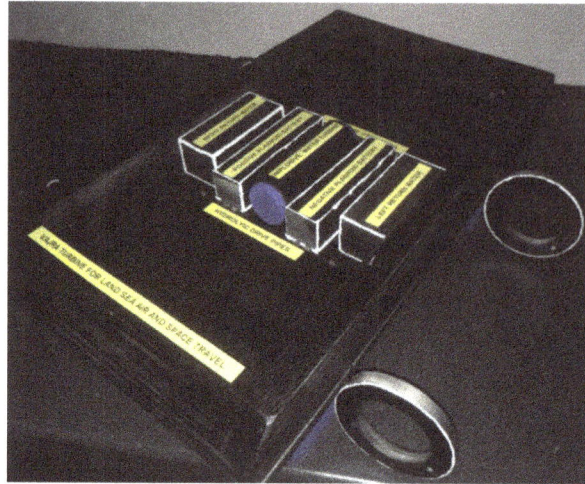

*FIG 189 – Firing mechanism for the Atomic Plasmoid Protium Proton Power
MATTER TO ENERGY Space, Air, Land and Sea Drive*

APPLICATIONS FOR A PLASMOIDS FORM AND FUNCTIONS

FIG 190 – Firing mechanism for the Atomic Plasmoid Protium Proton Power MATTER TO ENERGY Space, Air, Land and Sea Drive

FIG 191 – Firing mechanism for the Atomic Plasmoid Protium Proton Power MATTER TO ENERGY Space, Air, Land and Sea Drive

APPLICATIONS FOR A PLASMOIDS FORM AND FUNCTIONS

FIG 192 – Firing mechanism for the Atomic Plasmoid Protium Proton Power MATTER TO ENERGY Space, Air, Land and Sea Drive

APPLICATIONS FOR A PLASMOIDS FORM AND FUNCTIONS

APPENDIX 9: - PATENT CLAIMS

APPLICATIONS FOR A PLASMOIDS FORM AND FUNCTIONS

PATENT CLAIMS

1.)

2.)

3.)

4.)

5.)

6.)

7.)

8.)

9.)

10.)

APPLICATIONS FOR A PLASMOIDS FORM AND FUNCTIONS

PATENT CLAIMS

11.)

12.)

13.)

14.)

15.)

16.)

17.)

18.)

19.)

20.)

APPLICATIONS FOR A PLASMOIDS FORM AND FUNCTIONS

APPENDIX 10 : - REFERENCES

APPLICATIONS FOR A PLASMOIDS FORM AND FUNCTIONS

APPENDIX 10 : - REFERENCES

PART 7

MSAART ATOMIC OCTAVE
RESONANT CHAMBER
SHOCKWAVE GENERATOR
WITH IMPLODED SPHERE
SWIRL GUIDE

MSSART PATENT APPLICATION NOTES

APPENDIX 14:- MSAART ATOMIC OCTAVE

RESONANT CHAMBER SHOCKWAVE GENERATOR

WITH

IMPLODED SPHERE SWIRL GUIDE

DRAFT 518,400 B KMV – PART SEVEN OF TWENTY

MALCOLM V of SCOTLAND | MALCOLM BENDALL
THURSDAY 22ND SEPTEMBER 2022

APPLICATIONS FOR A PLASMOIDS FORM AND FUNCTIONS

FIG 193 : - PLASMOID RESONANT CAVITY

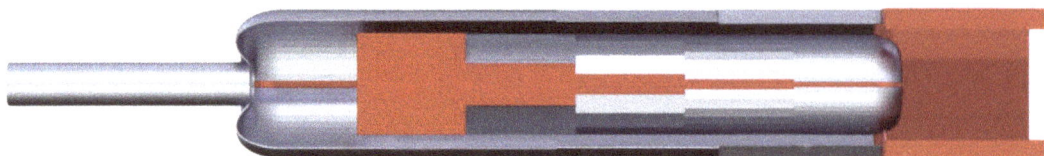

FIG 194 : - PLASMOID RESONANT CAVITY

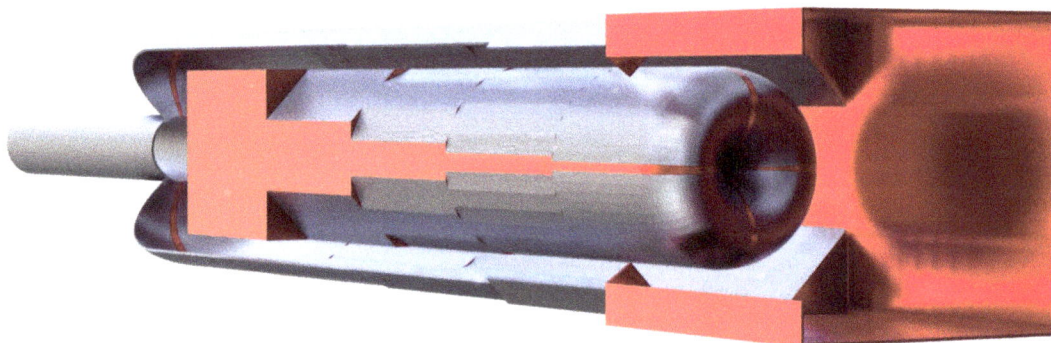

FIG 195 : - PLASMOID RESONANT CAVITY

APPLICATIONS FOR A PLASMOIDS FORM AND FUNCTIONS

FIG 196 : - PLASMOID RESONANT CAVITY

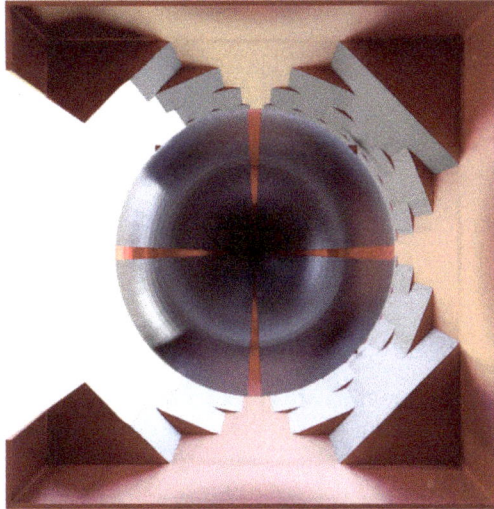

FIG 197 : - PLASMOID RESONANT CAVITY

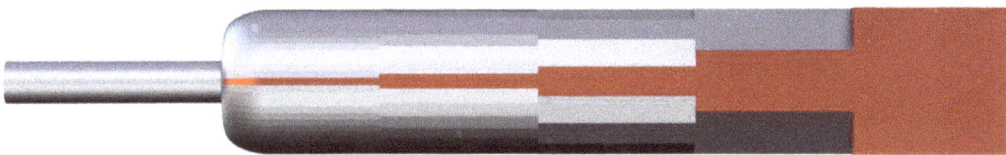

FIG 198 : - PLASMOID RESONANT CAVITY

APPLICATIONS FOR A PLASMOIDS FORM AND FUNCTIONS

FIG 199 : -SMOOTH END HALF SPHERES ON A SMOOTH CUBE

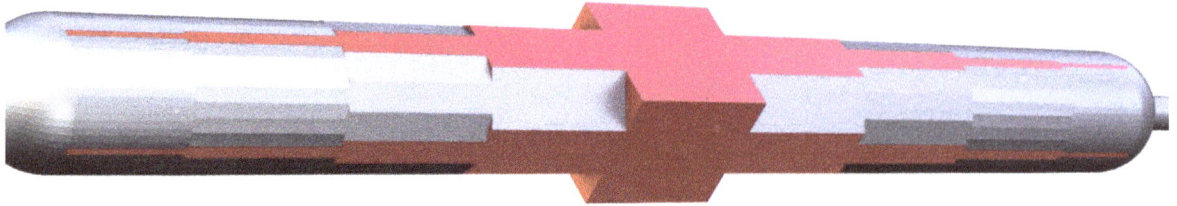

FIG 200 : -SMOOTH END HALF SPHERES ON A SMOOTH CUBE

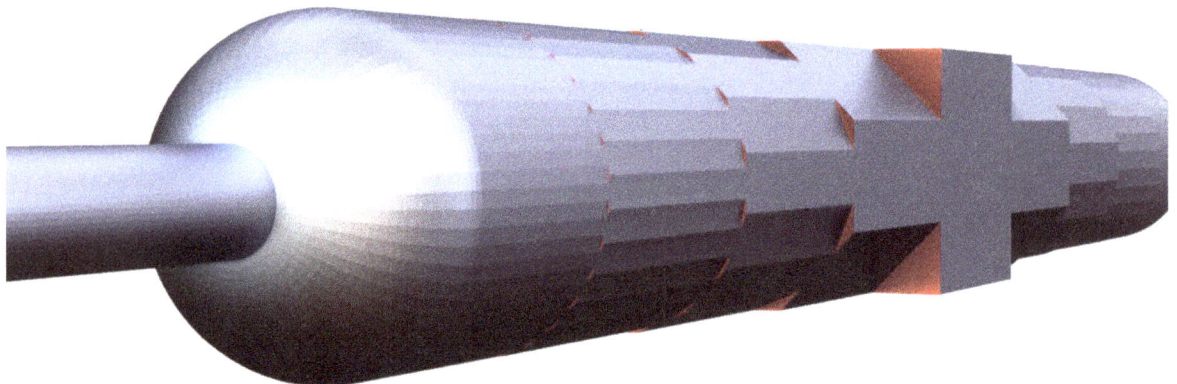

FIG 201 : -SMOOTH END HALF SPHERES ON A SMOOTH CUBE

APPLICATIONS FOR A PLASMOIDS FORM AND FUNCTIONS

FIG 202 : -SMOOTH END HALF SPHERES ON A SMOOTH CUBE

FIG 203 : -SMOOTH END HALF SPHERES ON A SMOOTH CUBE

FIG 204 : -SMOOTH END HALF SPHERES ON A SMOOTH CUBE

FIG 205 : -SMOOTH END HALF SPHERES ON A SMOOTH CUBE

PART 8

PLASMOID ENABLED BY ZP ASSIMULATION LOW ENERGY ATOMIC TRANSMUTATIONS (LEAT) AND OLD FUSION ELEMENTAL TRANSFORMATION

APPLICATIONS FOR A PLASMOIDS FORM AND FUNCTIONS

THE EXPLANATION AND DOCUMENTATION OF

PLASMOID ENABLED BY ZP ASSIMULATION

LOW ENERGY ATOMIC TRANSMUTATIONS (LEAT)

AND

COLD FUSION ELEMENTAL TRANSFORMATION

SUMMARY CONCLUSION
PART EIGHT OF TWENTY

DRAFT 518,400 B KMV – THURSDAY 22ND SEPTEMBER 2022

APPLICATIONS FOR A PLASMOIDS FORM AND FUNCTIONS

ABSTRACT

APPLICATIONS FOR A PLASMOIDS FORM AND FUNCTIONS

INTRODUCTION

APPLICATIONS FOR A PLASMOIDS FORM AND FUNCTIONS

GROUP ONE

PLASMOID ZP INDUCED LANDFILL LEACHATE LIQUID

LOW ENERGY ATOMIC TRANSMUTATIONS (LEAT) & COLD FUSION TRANSMUTATIONS

Elemental pairs Na – Mg and K – Ca melting points and AVT's.

Na (11) – Sodium's melting point = 97.79 C

9 x 7 x 7 x 9 = 3,969 x 9 x 7 x 7 x 9 = 15,752,961

9 + 7 + 7 + 9 = 32 + 9 + 7 + 7 + 9 = 64

AVT = 15,752,961 / 24 Hours / 60 Min / 60 sec / 60 Arc sec = 3.038765625

3.038765625 / 16 / 22.5 x 400 =

Mg (13) – Magnesium's melting point = 650 C

6 x 5 = 30 x 6 x 5 = 900

6 + 5 = 11

AVT = 900/24 = 37.5/60 = 0.625/60 = 0.01041667/60 = 0.0001736111   //////0.0111

K (19) – Potassium's melting point = 63.5 C

[Cl, Ca & Ni]

6 x 3 x 5 = x 6 x 3 x 5 = 8,100
Mirror Multiplied is 8,100

Mirror Added is 28

AVT =

Ca (20) – Calcium's melting point = 842 C

842 Calcium mirror multiplied is 4,096

842 Calcium mirror added is 28

AVT =

APPLICATIONS FOR A PLASMOIDS FORM AND FUNCTIONS

GROUP ONE

PLASMOID ZP INDUCED
LOW ENERGY ATOMIC TRANSMUTATIONS (LEAT) & COLD FUSION TRANSMUTATIONS

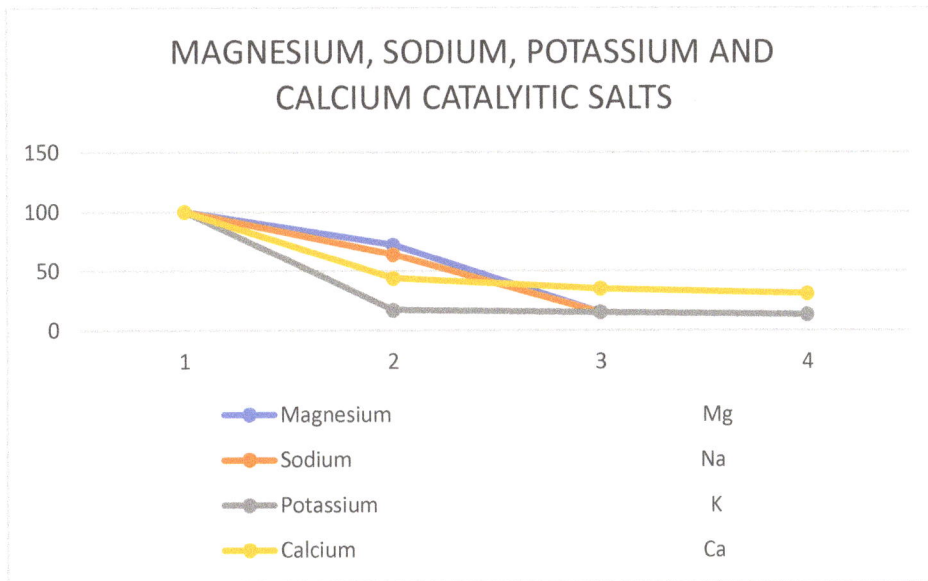

MAGNESIUM, SODIUM, POTASSIUM AND CALCIUM CATALYITIC SALTS

Magnesium		Mg
Sodium		Na
Potassium		K
Calcium		Ca

FIG 206 : - PLASMOID RESONANT CAVITY

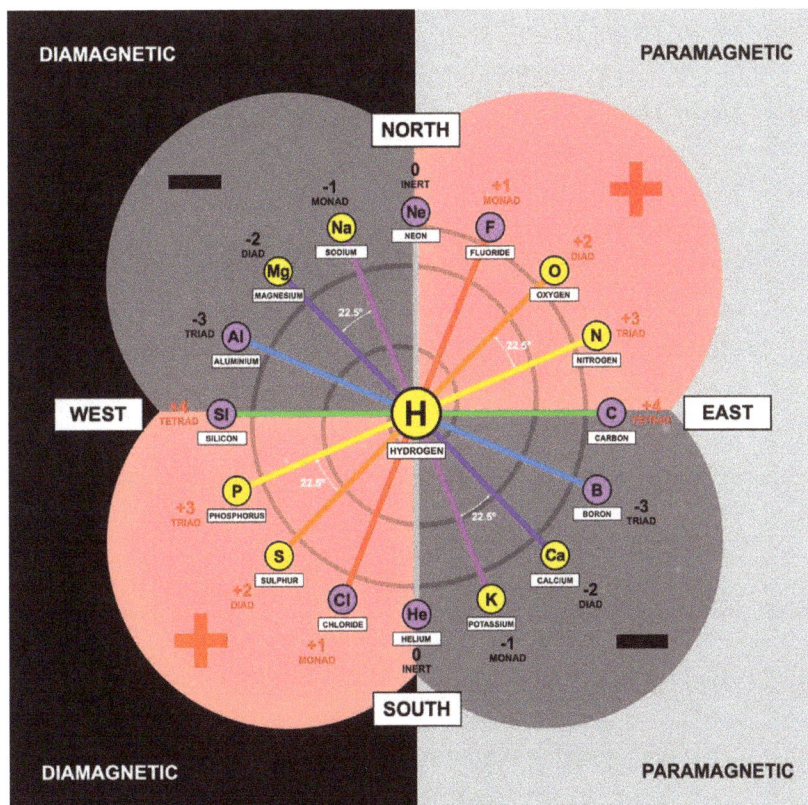

FIG 207 : - PLASMOID RESONANT CAVITY

APPLICATIONS FOR A PLASMOIDS FORM AND FUNCTIONS

ELEMENTS AND MOLECULES SUBTRACTED FROM THE NORLANDS TIP   LEACHATE FLUID AFTER 3 MINUTES WITHIN THE PLASMOID CREATOR

ELEMENTS SUBTRACTED FROM THE LEACHATE AFTER 3 MINUTES OF OPERATION

MAGNESIUM (86 – 24 mg/L) ....DOWN 72%
SODIUM (1,279 – 462 mg/L) ....DOWN 64%
POTASSIUM (638 – 532 mg/L) ....DOWN 17%
CALCIUM (72 – 32 mg/L)...   DOWN 44%

CHLORIDE (1796 -504)..DOWN 70%
ARSENIC (100 – 52ug/L)...DOWN 48%
CHROMIUM (53 – 26 ug/L)..DOWN 49%
NICKLE (140 – 64 ug/L)..   DOWN 54%
IRON (2.20 – 2.00 ug/L)..DOWN 10%
MANGANESE (150 – 120 ug/L)..DOWN 20%

MOLECULES SUBTRACTED FROM THE LEACHATE AFTER 3 MINUTES OF OPERATION
DISSOLVED METHANE (0.59 – 0.53 mg/L)...............................DOWN 10%
AMMONIA [AS NH4] (1,787 – 576)...............................   DOWN 68%
AMMONIACAL NITROGEN [AS N] (1,386 – 446 mg/L)   DOWN 68%
TOTAL OXIDISED NITROGEN (3.6 – 2.6 mg/L)................   DOWN 28%
NITRITE (0.01 – 0.59 mg/L)..UP 98%
NITRATE (21.3 – 26.1mg/L)..UP 18%
CHEMICAL OXYGEN DEMAND (2,200 – 770 mg/L)..................DOWN 65%
BIOCHEMICAL OXYGEN DEMAND (8.4 – 8.0 mg/L)................DOWN 5%
SULPHATE (136 – 45 mg/L)..DOWN 67%
SULPHIDE (0.10 – 0.03)..DOWN 70%
PHOSPHATE (16.49 – 5.95)..DOWN 65%
TOTAL ORGANIC CARBON (990 – 355 mg/L)...........................DOWN 65%
ALKALINITY – CARBONATE as CaCO3 (7,906 – 2,825 mg/L).....DOWN 65%
TOTAL CYANIDE (1.43 – 0.53 mg/L)...DOWN 63%
ELECTRICAL CONDUCTIVITY (12,445 -5,375 uS/cm)................DOWN 43%

ELEMENTS ADDED TO THE LEACHATE AFTER 3 MINUTES OF OPERATION

COPPER STANDARD (40 – 1,800 ug/L)....................................UP 4,500%
ZINC (220 – 270 ug/L)...UP   18%
LEAD (8 – 9 ug/L)...UP   11%

MOLECULES ADDED TO THE LEACHATE AFTER 3 MINUTES OF OPERATION

NITRITE (0.01 – 0.59 mg/L)...UP 98%
NITRATE (21.3 – 26.1mg/L)...UP 18%
SUSPENDED SOLIDS (40 – 108 mg/L)...UP 63%
PH (PH 8.7 - PH 8.8)..UP 1%

APPLICATIONS FOR A PLASMOIDS FORM AND FUNCTIONS

GROUP TWO

PLASMOID ZP INDUCED POTASSIUM
LOW ENERGY ATOMIC TRANSMUTATIONS (LEAT)
&

COLD FUSION TRANSMUTATIONS

***Ni (28)** – Nickel's melting point = 1,455 C

***Ca (20)** – Calcium's melting point = 842 C

*****K (19)** – Potassium Nucleus Potassium's melting point = 63.5 C

***Cl (17)** - Chlorine's melting point = 101.5 C

Nickel (Ni), Calcium (Ca) and **Chlorine Cl** are transmuted by the Plasmoid's capture of electrons, a Proton and Hydroxide (OH) using a Potassium (K) by mechanism of sharing the Plasmoid's Primary large central Zero Point (the large open door) with Potassium's Zero Point.

The Ring Structures Plasmoids often form are made from the capture of other Plasmoids or Elements into the outer Secondary peripheral Circular Zero point located on the Zero point Event Horizon Plane (The two small outer closed doors.) at the centre of the outer toroidal ring.

*****K (19)** – Potassium Nucleus Potassium's melting point = 63.5 C

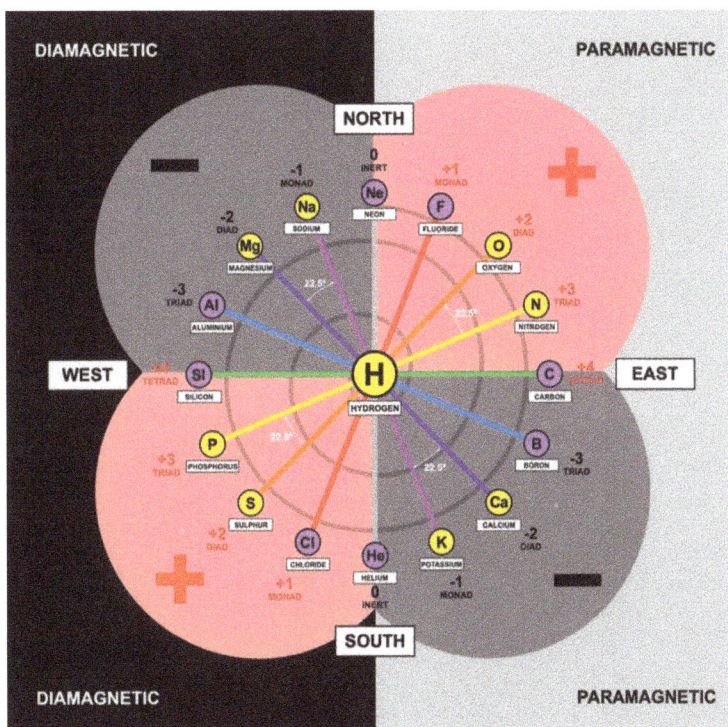

FIG 208 : - PLASMOID RESONANT CAVITY
APPLICATIONS FOR A PLASMOIDS FORM AND FUNCTIONS

ELEMENTS AND MOLECULES SUBTRACTED FROM THE NORLANDS TIP   LEACHATE FLUID AFTER *3 MINUTES* WITHIN THE PLASMOID CREATOR

ELEMENTS *SUBTRACTED* FROM THE LEACHATE AFTER *3 MINUTES* OF OPERATION

MAGNESIUM (86 – 24 mg/L) ....DOWN 72%
SODIUM (1,279 – 462 mg/L) ....DOWN 64%

POTASSIUM (638 – 532 mg/L) ....DOWN 17%
CALCIUM (72 – 32 mg/L)... DOWN 44%

CHLORIDE (1796 -504)...DOWN 70%
ARSENIC (100 – 52ug/L)..DOWN 48%
CHROMIUM (53 – 26 ug/L)..DOWN 49%
NICKLE (140 – 64 ug/L)... DOWN 54%
IRON (2.20 – 2.00 ug/L)..DOWN 10%
MANGANESE (150 – 120 ug/L)..DOWN 20%

MOLECULES SUBTRACTED FROM THE LEACHATE AFTER 3 MINUTES OF OPERATION

DISSOLVED METHANE (0.59 – 0.53 mg/L)....................................DOWN 10%
AMMONIA [AS NH4] (1,787 – 576)............................... DOWN 68%
AMMONIACAL NITROGEN [AS N] (1,386 – 446 mg/L) DOWN 68%
TOTAL OXIDISED NITROGEN (3.6 – 2.6 mg/L)................... DOWN 28%
NITRITE (0.01 – 0.59 mg/L)...…..UP 98%
NITRATE (21.3 – 26.1mg/L)..UP 18%
CHEMICAL OXYGEN DEMAND (2,200 – 770 mg/L)...................DOWN 65%
BIOCHEMICAL OXYGEN DEMAND (8.4 – 8.0 mg/L).................DOWN 5%
SULPHATE (136 – 45 mg/L)..DOWN 67%
SULPHIDE (0.10 – 0.03)..DOWN 70%
PHOSPHATE (16.49 – 5.95)...DOWN 65%
TOTAL ORGANIC CARBON (990 – 355 mg/L)...........................DOWN 65%
ALKALINITY – CARBONATE as CaCO3 (7,906 – 2,825 mg/L).....DOWN 65%
TOTAL CYANIDE (1.43 – 0.53 mg/L)...DOWN 63%
ELECTRICAL CONDUCTIVITY (12,445 -5,375 uS/cm)..................DOWN 43%

ELEMENTS ADDED TO THE LEACHATE AFTER 3 MINUTES OF OPERATION

COPPER STANDARD (40 – 1,800 ug/L)....................................UP 4,500%
ZINC (220 – 270 ug/L)…...UP 18%
LEAD (8 – 9 ug/L)…...UP 11%

MOLECULES ADDED TO THE LEACHATE AFTER 3 MINUTES OF OPERATION

NITRITE (0.01 – 0.59 mg/L)..…..UP 98%
NITRATE (21.3 – 26.1mg/L)..UP 18%
SUSPENDED SOLIDS (40 – 108 mg/L)..UP 63%
PH (PH 8.7 - PH 8.8)...UP 1%

GROUP THREE

PLASMOID ZP INDUCED PALLADIUM COLD FUSION
&
LOW ENERGY ATOMIC REACTIONS (LEAR) TRANSMUTATIONS

***Cd (48) - Cadmium's melting point = 321.1 C

*****Pd (46) - Palladium Nucleus. Palladium's melting point = 1,555 C

***I (53) – Iodine's melting point = 113.7 C

Cadmium (Cd) and **Iodine (I)** are transmuted by the Plasmoid's capture of a Proton and Oxygen (O) using a Palladium (Pd) by mechanism of sharing the Plasmoid's Primary large central Zero Point (the large open door) with Palladiums Zero Point.

The Ring Structures Plasmoids often form are made from the capture of other Plasmoids or elements into the outer Secondary ring peripheral Circular Zero point located on the Zero point Event Horizon Plane (The two small outer closed doors.) at the centre of the outer toroidal ring.

*****Pd (46) - Palladium Nucleus. Palladium's melting point = 1,555 C

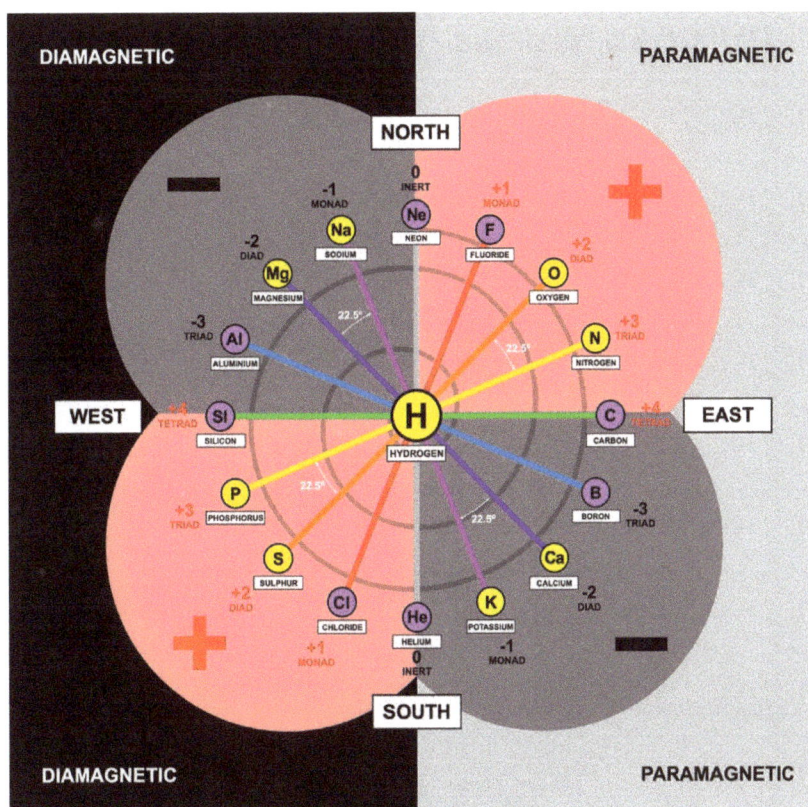

FIG 209 : - PLASMOID RESONANT CAVITY
GROUP THREE

PLASMOID ZP INDUCED PALLADIUM CENTRED COLD FUSION
&
LOW ENERGY ATOMIC REACTIONS (LEAR) TRANSMUTATIONS

***Ni (28) – Nickel's melting point = 1,455 C

***Ca **(20)** – Calcium's melting point = 842 C

***Cl **(17)** - Chlorine's melting point = 101.5 C

***K. (19) – Potassium Nucleus Potassium's melting point = 63.5 C

***Cd **(48)** - Cadmium's melting point = 321.1 C

***I **(53)** [126.9] – Iodine's melting point = 113.7 C

*****Pd (46) - Palladium Nucleus Palladium's melting point = 1,555 C

GROUP FOUR

THE PLASMOID ZP INDUCED 316 STAINLESS STEEL
LOW ENERGY ATOMIC TRANSMUTATIONS (LEAT)
&
COLD FUSION ELEMENTAL LIST

26   **Fe** 64% - Melting point = 1,538 C

28   ****Ni** 18% – Nickel's melting point = 1,455 C

24   **Cr** 14% – Chromium's Melting point = 1,907 C

42 Mo 3% – Molybdenum's Melting point = 2,623 C

25 Mn 2% – Manganese Melting point = 1,246 C

316 STAINLESS STEEL - Fe 63%, Cr 18%, Ni 14%, Mo 3% & Mn 2%.

4.1.) 26 *IRON (Fe) CALCULATIONS.*

IRON – Melting point – 1,538

IRON Melting point = 1,538 / 266.666 = 5.7675
and = 1,538 / 11.111 = 138.42
and = 1,538 / 1.333 = 1,153.5
and = 1,538 / 144 = 10.680555
and = 1,538 / 51.84 = 10.680555

1 + 5 + 3 + 8 = (17) + 8 + 3 + 5 + 1 = 34

1 x 5 x 3 x 8 = (120) x 8 x <u>3</u> x 5 x 1 = 14,400

(14,400) / 34 = 423.529412

(14,400) x 34 = 489,600

(14,400) / 266.666 = 54 // 108 // 216 // 432 // 864 // 1,728 // 3,456 // 6,912 //

(14,400) / 11.111 = 1,296 // 648 // 324 // 162 // 81 //   1,296 // 2,592 // 5,184 // 10,368

(14,400) / 1.333 = 10,800

(14,400) / 144 = 100

(14,400) / 51.84 = 277.777 // 555.555 // 1,111.111 // 2,222.222 // 4,444.444 // 8,888.888

4.2.) 24 *CHROMIUM (Cr) CALCULATIONS.*

CHROMIUM (Cr) – Melting point – 1,907 C

Melting point = 1,907 C / 266.666 = 7.15125   [1/x =]
and = 1,907 C / 11.111 = 171.63   [1/x =]
and = 1,907 C / 1.333 = 1,430.25   [1/x =]
and = 1,907 C / 144 = 13.2430556   [1/x =]
and = 1,907 C / 51.84 =

1 + 9 + 0 + 7 = (17) + 7 + 0 + 9 + 1 = 34

1 x 9 x 0 x 7 = (63) x 7 x 0 x 9 x 1 = 3,969

(3,969) / 34 = 116.735294

(3,969) x 34 = 134,946

(3,969) / 266.666 = 14.88375
(3,969) / 11.111 = 357.21
(3,969) / 1.333 = 2,976.75
(3,969) / 144 = 27.5625
(3,969) / 51.84 = 76.5625
(3,969) / 25.92 = 153.125

4.3.) 28 NICKLE (Ni) CALCULATIONS.

***NICKLE (Ni) – Melting point – 1,455 C**

Melting point = 1,455 C / 266.666 = 5.45625
              and =   1,455 C / 11.111 = 130.95
              and =   1,455 C / 1.333 = 1.091.25
              and =   1,455 C / 144 = 10.1041666
              and =   1,455 C / 51.84 = 28.0671296

1 + 4 + 5 + 5 = (15) + 5 + 5 + 4 + 1 = 30

1 x 4 x 5 x 5 = (100) x 5 x 5 x 4 x 1 = 10,000

 (10,000) / 30 = 333.333
 (10,000) x 30 = 300,000

 (10,000) / 266.666 = 37.5
 (10,000) / 11.111 = 900
 (10,000) / 1.333 = 7,500
 (10,000) / 144 = 69.444
(3,969) / 51.84 = 76.5625
(3,969) / 25.92 = 153.125
(5,184) / 12.96 = 400

4.4.) 42 MOLYBDENUM (Mo) CALCULATIONS.

MOLYBDENUM (Mo) – Melting point – 2,623 C

Melting point = 2,623 / 266.666 = 9.83625   [1 / x = 0.10166470452408]
              and =   2,623 / 11.111 = 236.07   [1 / x 0.004236031685517]
              and =   2,623 / 1.333 = 1,967.25   [1 / x = 0.000508323802262]
              and =   2,623 / 144 = 18.2152777   [1 / x = 0.0548989706443]
              and =   1,455 C / 51.84 = 28.0671296

2 + 6 + 2 + 3 = (13) + 3 + 2 + 6 + 2 = 26

2 x 6 x 2 x 3 = (72) x 3 x 2 x 6 x 2 = 5,184

(5,184) / 26 = 199.3846

(5,184) x 26 = 134,784

(5,184) / 266.666 = 19.44 // 9.72 // 4.86 // 2.43 //
(5,184) / 11.111 = 466.56
(5,184) / 1.333 = 3,888
(5,184) / 144 = 36
(5,184) / 51.84 = 100
(5,184) / 12.96 = 400

4.5.) 25 MANGANESE (Mn) CALCULATIONS.

MOLYBDENUM (Mo) – Melting point – 1,246 C

Melting point = 1,246 C / 266.666 = 4.6725
and = 1,246 C / 11.111 = 112.14
and = 1,246 C / 1.333 = 934.5
and = 1,246 C / 144 = 8.652777
and = 1,455 C / 51.84 = 28.0671296

1 + 2 + 4 + 6 = (13) + 6 + 4 + 2 + 1 = 26

1 x 2 x 4 x 6 = (48) x 6 x 2 x 4 x 6 = 2,304

(2,304) / 26 = 88.615384
(2,304) x 26 = 59,904

(2,304) / 266.666 = 8.64
(2,304) / 11.111 = 207.36 // 103.68 // 51.84 // 25.92 //12.96//6.48//3.24//1.62//0.81
(2,304) / 1.333 = 1,728 // 864 // 432 // 216 // 108 // 54 // 27 // 13.5 //
(2,304) / 144 = 16
(2,304) / 51.84 = 4.444

4.6.) SUMMARY

ELEMENTS SUBTRACTED AND ADDED FROM LEACHATE AFTER 3 MINUTES OF OPERATION

CHROMIUM (53 – 26 ug/L)..DOWN 49%
NICKLE (140 – 64 ug/L)...................................... DOWN 46%
MANGANESE (150 – 120 ug/L)......................................DOWN 20%
IRON (2.20 – 2.00 ug/L)...DOWN 10%
ARSENIC (100 – 52ug/L)..DOWN 48%

ELEMENTS SUBTRACTED AND ADDED FROM LEACHATE AFTER 5 MINUTES OF OPERATION

CHROMIUM (53 – 75 ug/L)...UP 30%
NICKLE (140 – 130 ug/L)..DOWN 7%
MANGANESE (150 – 74 ug/L)...................................... DOWN 50%

IRON (2.20 – 1.20 ug/L)..DOWN 54%
ARSENIC (100 – 85 ug/L)..DOWN 15%

ELEMENTS SUBTRACTED AND ADDED FROM LEACHATE AFTER 8 MINUTES OF OPERATION

CHROMIUM DECREASE (53 - 42 ug/L)...DOWN 21%
NICKEL DECREASE (140 – 110 ug/L)..DOWN 21%
MANGANESE DECREASE (150 – 120 ug/L)...................................DOWN 21%
IRON DECREASE (2.20 – 2.10 ug/L)..DOWN 5%
ARSENIC DECREASE (100 - 98 ug/L)...DOWN 2%

FIG 210 : - CHROMIUM, NICKEL, MANGANESE AND IRON

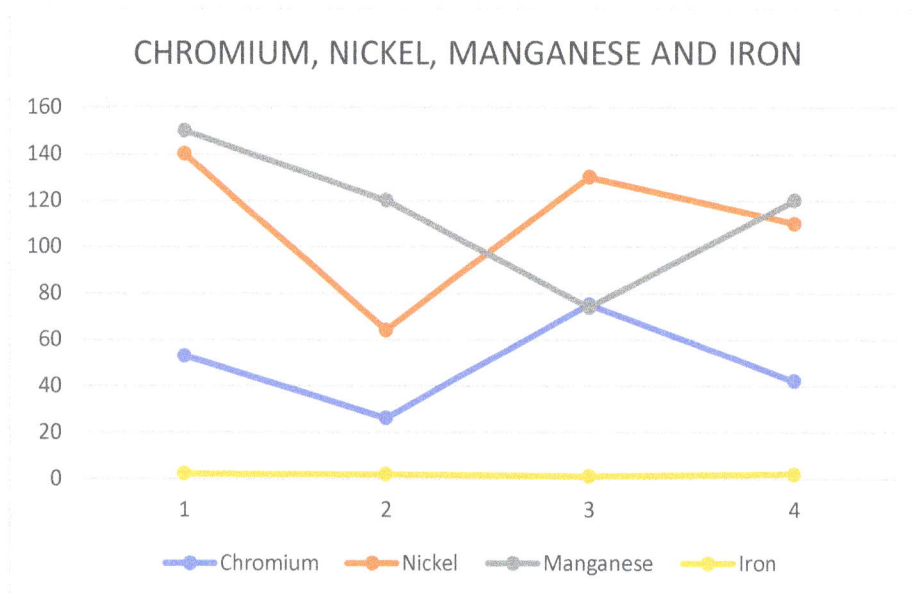

CHROMIUM, NICKEL, MANGANESE AND IRON

Chromium	Cr	53	26	75	42
Nickel	Ni	140	64	130	110
Manganese	Mn	150	120	74	120
Iron	Fe	2.2	2	1.2	2.1

GROUP FIVE

THE PLASMOID INDUCED COLD FUSION ELEMENTAL LIST

26 **Fe** 64% - Melting point = 1,538 C

28 ****Ni** 18% – Nickel's melting point = 1,455 C

24 **Cr** 14% – Chromium's Melting point = 1,907 C

42 **Mo** 3% – Molybdenum's Melting point = 2,623 C

25 Mn 2% – Manganese Melting point = 1,246 C

Ti (22) – Titanium's melting point = 1,668 C

Na (11) - Mirror melting point = 97.79 C

1 x 5 x 3 x 8 = (120) x 8 x 3 x 5 x 1 = 14,400
1 + 5 + 3 + 8 = (17) 8 + 3 + 5 + 1 = 34

Al (13) - Aluminum's melting point = 660.3 C

A Plasmoid turns 0.5 of a turn in one second x 60 Min = 30 x 60 min = 1,800 turns in one hour x 24 hours = 43,200 turns a day x 6 days equals creation time equivalent of 259,200.
Therefore the resonance of the Plasmoid determines the procession of the Earths Equinox and is the active action Radius of 5,184,000 Time Diameter it's-self.
Therefore also the resonance of the Plasmoid sets the seconds in a day.

259,200 Plasmoid / **14,400** Iron = 18   18 // 36 // 72 // 144 // 288 //

PART 9

THE MSAART SYSTEM FOR WASTE ENERGY RECOVERY FROM THE OPERATION OF THE INTERNAL COMBUSTION ENGINE

MSAART PATENT APPLICATION INFORMATION

THE MSAART SYSTEM FOR WASTE ENERGY RECOVERY FROM THE OPERATION OF THE INTERNAL COMBUSTION ENGINE

THIS SECTION DESCRIBES A METHODOLOGY AND PATHWAY TO CALCULATE AND INDUCE ATOMIC FUSION FROM BOTH ELEMENTAL AND MOLECULAR RESONANCE BY REFERENCING THE MSAART PLASMOID MODEL OF THE ELEMENTS (PMOE) AND MSAART PLASMOID UNIFICATION MODEL CALCULATOR (PUMC) TABLES TO ENGAGE THE NATURAL PROPERTIES OF THE TOROIDAL MSAART PLASMOID NATURE, GEOMETRY AND FUNCTION INCLUDING AN EXPLANATION OF THE MSAART PLASMOIDS
FORM AND STRUCTURE AS IT RELATES TO COLD FUSION AND LOW ENERGY ATOMIC RECONSTRUCTION (LEAR) INDUCED ATOMIC TRANSFORMATIONS AND TRANSMUTATIONS.

PART NINE OF TWENTY

DRAFT 518,400 B KMV – THURSDAY 22ND SEPTEMBER 2022

APPLICATIONS FOR A PLASMOIDS FORM AND FUNCTIONS

INDEX

ABSTRACT

INTRODUCTION

APPLICATIONS FOR A PLASMOIDS FORM AND FUNCTIONS

INDEX

APPLICATIONS FOR A PLASMOIDS FORM AND FUNCTIONS

INDEX

APPLICATIONS FOR A PLASMOIDS FORM AND FUNCTIONS

ABSTRACT

THIS SECTION DESCRIBES A METHODOLOGY AND PATHWAY TO CALCULATE AND INDUCE ATOMIC FUSION FROM BOTH ELEMENTAL AND MOLECULAR RESONANCE BY REFERENCING THE PLASMOID MODEL OF THE ELEMENTS (PMOE) AND PLASMOID UNIFICATION MODEL CALCULATOR (PUMC) TABLES TO ENGAGE THE NATURAL PROPERTIES OF THE TOROIDAL PLASMOID NATURE, GEOMETRY AND FUNCTION INCLUDING AN EXPLANATION OF THE PLASMOIDS FORM AND STRUCTURE AS IT RELATES TO COLD FUSION AND LOW ENERGY ATOMIC REACTIONS (LEAR) INDUCED ATOMIC TRANSFORMATIONS.

Time itself is constructed by equal and opposite mirror numbers. As for every force there is always an equal and opposite force, at all scales "As above so below". Time's number is 518,400 which manifests the product of 5 x 1 x 8 x 4 x 4 x 8 x 1 x 5 = 25,600 // 12,800 // 6,400 // 3,200 // 1,600 // 800.

Time's Base number is a product of 1 x 2 x 3 x 4 x 5 x 6 x 8 x 9 x 10 = 518,400. Seven is not in Time's Base number sequence as it represents Direct current (DC) not Matter which is constructed of Frequencies Alternating Current (AC) generated in the mould of Time. The equal and opposite mirror numbers for Time's 518,400 Base Number are as follows. 1 x 2 x 3 x 4 x 5 x 6 x 6 x 5 x 4 x 3 x 2 x 1 = *518,400* and 8 x 9 x 10 x 10 x 9 x 8 = *518,400* both are mirror numbers reflecting the basic symmetry of the Universal truth.

As there are no straight lines in the universe therefore every line is part of a curve.

Every curve is a part of a spiral.

Spirals are either imploding clockwise manifesting a Negative charge in its natural state (life force) or exploding anti-clockwise manifesting a Positive Charge (death and destruction force) in its natural state, reflecting Times opposed Numerical Base nature.

This is true as Time is the mould within which Matter is formed, therefore Elements moulded by Time must reflect both equal and opposite numbers, implicit and encoded in their phase change temperatures and equal and opposite positive and Negative charge and spiral spin components.

Therefore because and as a consequence of the above, as the Sun's diameter is 864,000 Miles it must factor out as 8 x 6 x 4 x 4 x 6 x 8 = *36,864*.

As Time is the mould in which Matter is formed, from 5th dimensional Aether (5.555), 6th dimensional Light Energy and Time the 4th dimension (4.444). Time therefore must logically and absolutely proceed Matter being the 3rd dimension (3.333).

So as the Sun is the mould in which the primary Matter Protium [Known as Hydrogen (H)] is formed, to be subsequently contained within the Sun's self-generated, self-structuring and self-regulated toroidal electromagnetic field, Protiums Base Mirror Number must again reflect the Sun's and Time's Base Numbers.

APPLICATIONS FOR A PLASMOIDS FORM AND FUNCTIONS

ABSTRACT

As Protiums Phase Change melting point, where applied external energy intensities reach the critical resonance of Protium its-self causing the implosive atomic focus, is minus 259.2 Degrees Celsius therefore Protiums Base Mirror Number calculation is as follows.
2 x 5 x 9 x 2 x 2 x 9 x 5 x 2 = *129,600* Resonant Frequency Energy Unit (RFEU)

129,600 is the Resonant Frequency Energy Unit (RFEU) number representing primal base Matter Protium (H) the first primary manifestation of the Third Dimension.

It is therefore also a proof that the Suns Diameter 864,000 divided by 129,600 equals 3,333.333 being 3.333 Matters dimensional number 3D Matter.

Another proof is that Protium's *129,600* / *36,864* Base Mirror Number = 3.515625 // 7.03125 // 14.0625 // 28.125 // 56.25 // 112.5 // 225 // 450 // 900 // 1,800 // 3,600 // 7,200 // 14,400 // 28,800 // 57,600 // 115,200 // 230,400 // 460,800 // 921,600 // 1,843,200 //3,686,400 [*36,864*] 20 Octaves up.

518,400 Time / *129,600* (RFEU) = 4 as the square represents Matter (AC) and the Circle represents (DC) Aether and *518,400* Time / *36,864* (Sun's Base Number) = 14.062 (PPOS).

THE PLASMOID SYSTEM FOR WASTE ENERGY RECOVERY FROM THE OPERATION OF THE INTERNAL COMBUSTION ENGINE, ELEMENTS AND MOLECULES INVOLVED

A Plasmoids merged orbits mechanism for the easy transmutation of Elements by the implosive merging of elemental Zero points at their common Toroidal centre. This process has been conclusively proven by the Plasmoid System leachate experiment using Copper as a standard and Frequency paired equal and opposite Na + K and Mg + Ca analytical results changing concentrations when exposed to Plasmoid action over an eight minute period.

APPLICATIONS FOR A PLASMOIDS FORM AND FUNCTIONS

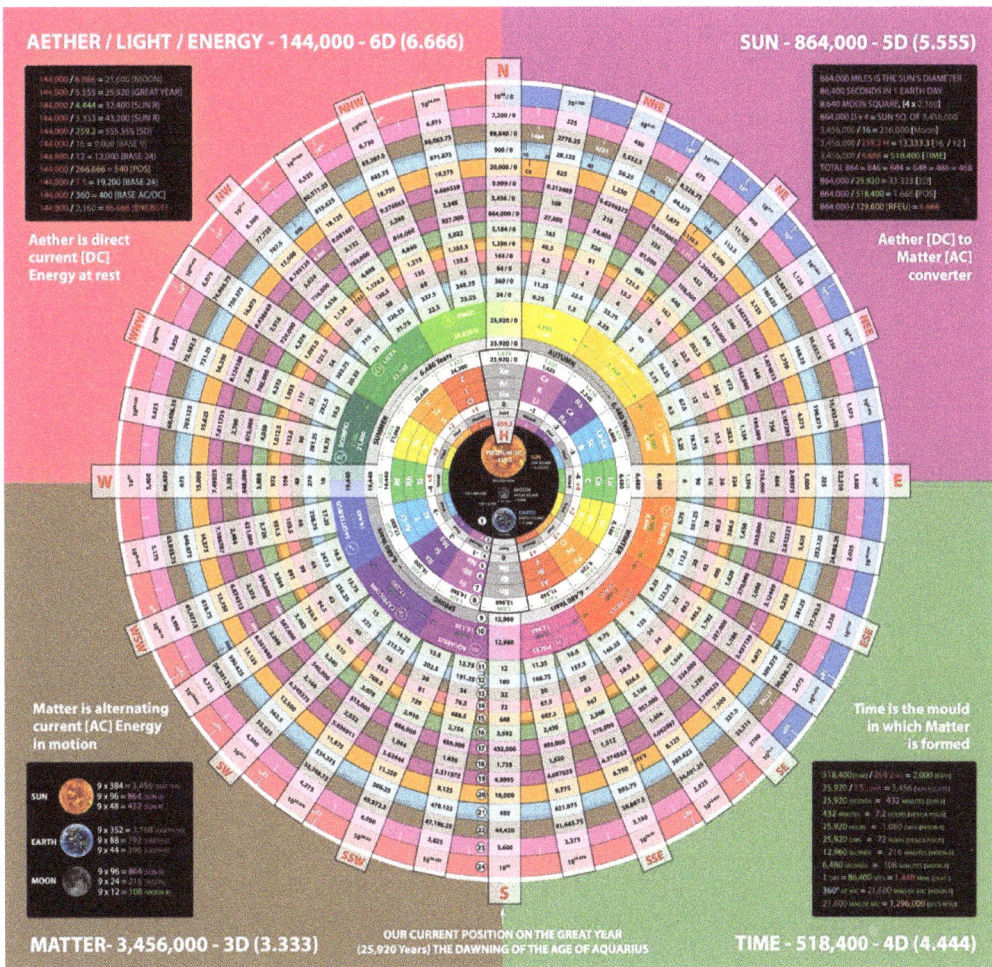

PLASMOID UNIFICATION MODEL

Plasmoids are doughnut or toroidal shaped clusters of net Protons or net Electrons that once captured and placed into a Toroidal orbit are capable of absorbing, storing and releasing enormous amounts of energy present within their self-generated and structured electro-magnetic containment field. Plasmoids, in effect, function as an atomic battery that can be self-charging due to its ability to convert matter to available clean energy. Plasmoids by their unique geometry cause a consequential electro-magnetic containment field to generate a Zero point naturally and casually, without much effort, have the ability to convert the nuclear Mass of Protium (Atoms) into energy.

The Plasmoid Unification Model (PUM) posits that Plasmoids are epoch-making and that knowledge of them has been hidden in plain sight for centuries. This PUM 'slide rule' reveals the algorithmic relationships between life's elements critical to mankind's existence and development. It starts with Protium (H) which has a melting point of -259.2°C and is the most abundant element in our Solar System. Protium determines the 25,920 Great Year frequency of our Solar System. The resonant frequencies of all other elements can then be calculated when 25,920 years is reduced from years to days, hours and seconds.

The PUM is evidence that the Universe is an intelligent design. That design is in perfect octave harmonic resonance with itself. Therefore, all of creation from Galaxies to Planets to Elements all resonate in unison with a collective chord 'As Above So Below'. This is interconnected with an Energy 'web', the 24 components and laws of which are all based and governed on the same 16 sector Torus Plasmoid precepts shown. The concepts and ruling principles of the PUM can, and have, been applied to make Energy to Matter and Matter to Energy conversions. When applied to the modern hydrocarbon powered internal combustion engine, PUM technology removes exhaust toxic waste products and increases the engine power output by transforming waste energy back into fuel. Plasmoids employed in conjunction with the Plasmoid Toroidal Implosive Turbine provide a new novel Matter to Energy and Energy to Matter propulsion device for water, land, air and space travel.

LEGEND

#		
	Aether/Light	Sun	Time	Matter	Frequency	Degrees
1	Sun (864,000) [DC], Earth (7,920) [AC], Moon (2,160) [AC]	
2	Music [AC] (Do = C = 24 x 11.111) = 266.666)	
3	Elemental Crystal Forms +/- Monad, Diad, Triad, Tetrad	
4	Elemental Valencies (0, - 1, - 2, - 3, - 4 and 0, +1, +2, +3, +4)	
5	Elements 1-16 (He - Cl)	
6	Elements 17-32 (Ar - I)	
7	Elements 33-48 (Xe - 'Z')	
8	Elemental Frequencies (1,620 x 16 = 25,920 light frequency of -259.2 C)	
9	Seasons of Great Year (25,920 / 0 years)	
10	Zodiac Great Year (25,920 / 0 years)	
11	Clock 24 Hour / 0 Hour [AC] / 0 Hour [DC] Clock	
12	Compass 360° Degrees [AC] / 0° Degrees [DC]	
13	Matter 64 / 0 64 Points being 32 Resonant Planes / 0	
14	Light 144 / 0	
15	Resonate Frequency Energy Unit (RFEU) - 1,296 (129,600)	
16	All Time 5,184 (518,400 secs)	
17	Aether - Sun (864,000 miles diameter, 432,000 miles radius)	
18	Matter 3,456 (3,456,000 miles - Sun Square)	
19	Dimensions 3D Matter = 3.333, 4D Time = 4.44, 5D Aether = 5.55, 6D Light = 6.666	
20	Sound and Music (0 - 20,000 Hz)	
21	Language 1-9, 10-90, 100-900 (111,222,333,444,555,666,777,888,999) (45,450,4,500)	
22	Solar System Sun & Planetary Diameters and Radii in miles	
23	Plasmoids 7,200 Degrees / 0, 32 Planes 64 Radial Points, One Zero Point	
24	All Plasmoid Energy = All Alternating Current [AC] Frequencies	
	(= Non-Ionizing = Ionizing)	

© PLASMOID INTERNATIONAL / MALCOLM BENDALL

FIG 211 : - CHROMIUM, NICKEL, MANGANESE AND IRON

APPLICATIONS FOR A PLASMOIDS FORM AND FUNCTIONS

ABSTRACT

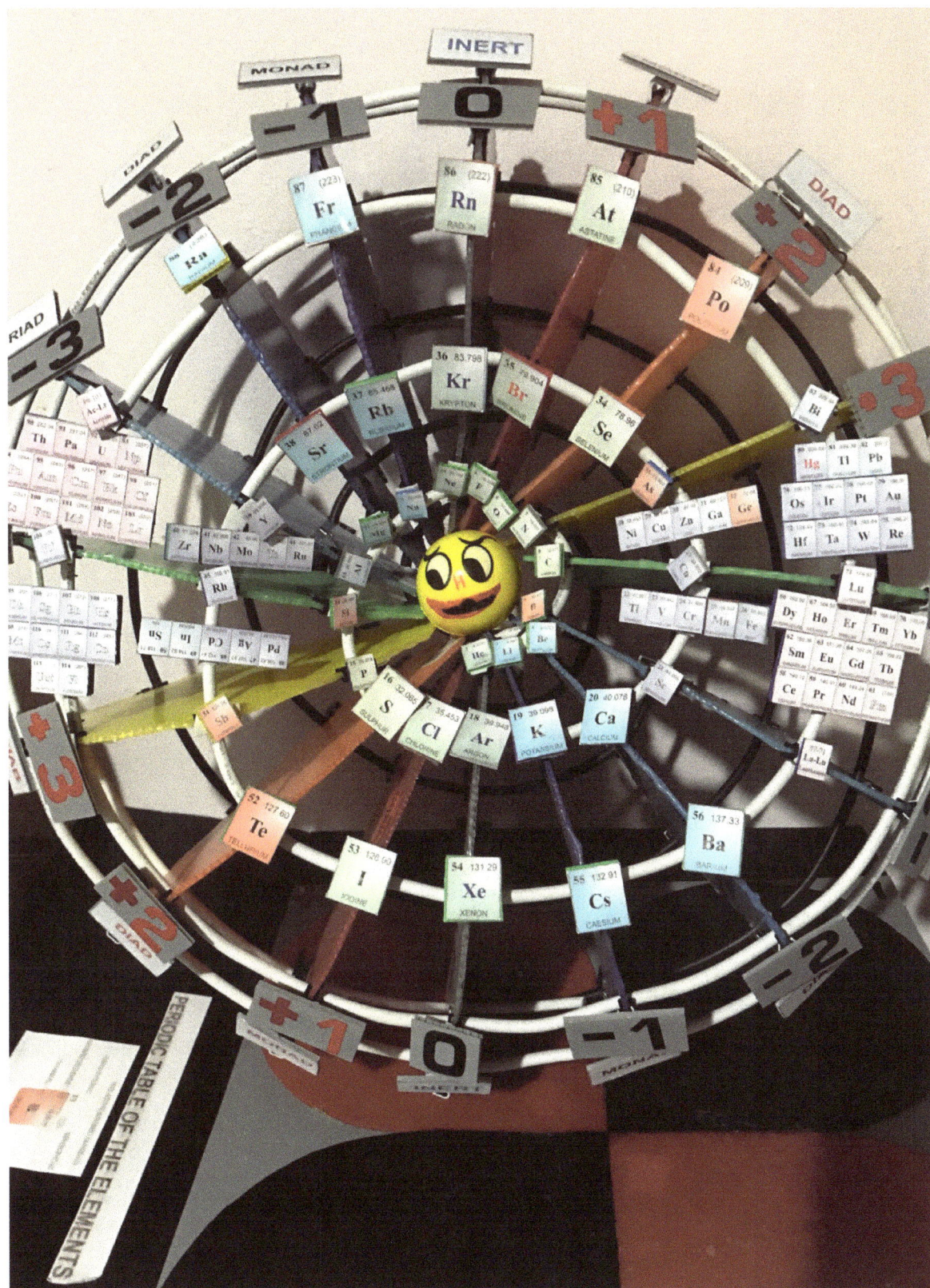

FIG 212 : - CHROMIUM, NICKEL, MANGANESE AND IRON

APPLICATIONS FOR A PLASMOIDS FORM AND FUNCTIONS

INTRODUCTION

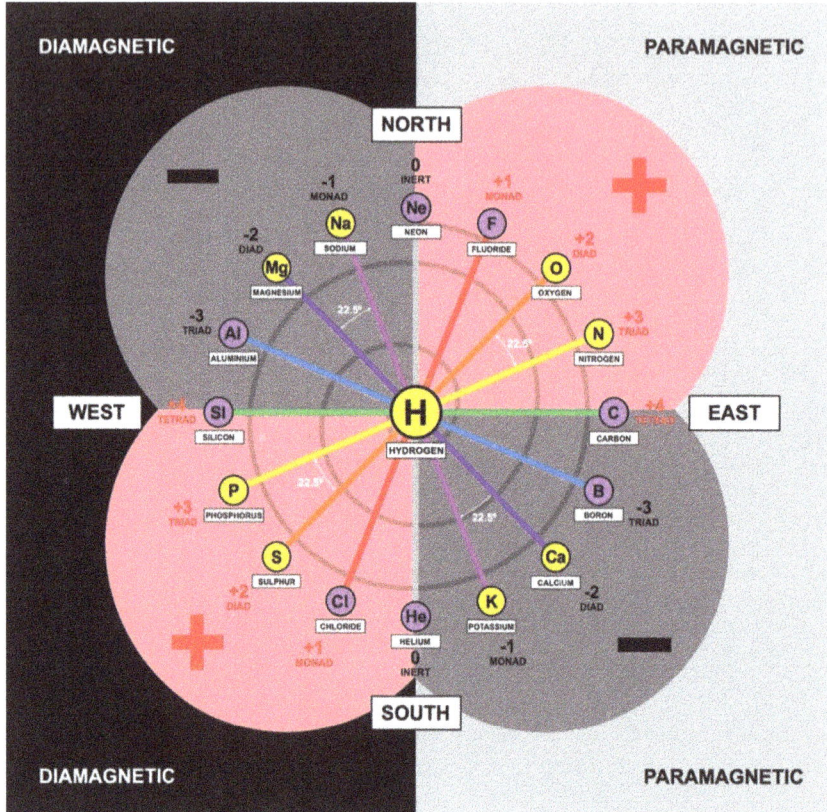

FIG 213 : - CHROMIUM, NICKEL, MANGANESE AND IRON

APPLICATIONS FOR A PLASMOIDS FORM AND FUNCTIONS

INTRODUCTION

THE PLASMOID SYSTEM FOR WASTE ENERGY RECOVERY FROM THE OPERATION OF THE INTERNAL COMBUSTION ENGINE

ELEMENTS AND MOLECULES INVOLVED WITHIN THE SYSTEM

FIG 214 : - AETHER AND MATTER ELEMENT AND DIMENSIONAL NUMBERS

1 TO 36 AETHER AND MATTER ELEMENT NUMBERS

36 x 1	2	3	4	5	6	7	8	9	10	11	12	13	14	15	16	17	18	19	20	21	22	23	24	25	26	27	28
= 36	72		144	180	216	252	288	324	360		432	468	504	540	576	612	648	702	756		828	900	936	972	1008		1080

= 37	74	111	148	185	222	259	296	333	370	407	444	481	518	555	592	629	666	703	740	777	814	851	888	925	962	999	1036
37 x 1	2	3	4	5	6	7	8	9	10	11	12	13	14	15	16	17	18	19	20	21	22	23	24	25	26	27	28

36 x	29	30	31	32	33	34	35	36	37	38	39	40	41	42	43	44	45	46	47	48
= 1044	1081	1118	1155	1192	1229	1266	1303	1340	1377	1414	1451	1488	1525	1562	1599	1636	1673	1710	1747	

= 1073	1110	1147	1184	1221	1258	1295	1332	1369	1406	1443	1480	1517	1554	1591	1628	1665	1702	1739	1776
37 x	29	30	31	32	33	34	35	36	37	38	39	40	41	42	43	44	45	46	47	48

36x	49	50	51	52	53	54	55	56	57	58	59	60	61	62	63	64	65	66	67	68	69	70
= 1764	1800	1836	1872	1908	1944	1980	2016	2052	2088	2124	2160	2196	2232	2268	2304	2340	2376	2412	2448	2484	2520	

= 1813	1850	1887	1924	1961	1998	2035	2072	2109	2146	2183	2220	2257	2294	2331	2368	2405	2442	2479	2516	2553	2516
37x	49	50	51	52	53	54	55	56	57	58	59	60	61	62	63	64	65	66	67	68	69	70
	80	90	100	110	120	160	180	225	360													

37 AETHER, TIME AND DIMENSIONAL NUMBERS

APPLICATIONS FOR A PLASMOIDS FORM AND FUNCTIONS

INTRODUCTION

PART 1 - *DESCRIPTION OF PLASMOID TORUS FORM AND FUNCTION*

The Plasmoid Clockwise Implosive Vortex Torus is an octave, Pi and Phi based doughnut shape, where Pi defines the surface the Phi curve intersects the 8 cardinal planes octave nodes (insert Torus formula) centred on a singularity point G.

At G, time does not exist, as on an atomic scale the slowing of time to D max and the later acceleration of time returning to D min G must have a Zero net effect on time being that they are acted upon by equal but opposite forces. I am therefore defining G as a Zero-time and zero entropy point, with V the node size as a constant and the torque induced changes in charge density as variable along with time, speed of light, node spin direction of both spin charge and entropy. As entropy is the fabric of Time both are by definition clearly subject to varied states and are not constants as the previous observers have declared. The fabric of Time entropy and at 90 degrees centropy, clearly and demonstrably acts on different states at variable rates and requires different energy inputs to resist falling apart and to maintain their operational designed function, but with time all creation is demonstrably all affected by death and destruction and ultimately, in time, fail. The connection of clockwise Female imploding centripetal negative entropy, to a slowing and or reversal of manifest male counter clockwise destructive exploding centrifugal vortexes can be demonstrated when counteracted by applying in force centripetal imploding forces and by maintaining stasis the reversal of time is also manifest when reversal of Entropy is achieved that is negative Entropy is applied with effect.

The node has a volumetric constant but its singularity point G is the focus of its own inward implosive forces interacting with the weakening counterclockwise explosive outer shell positive charges. The weakening outer positive charge allows the clockwise negative charge previously attracted towards the outer shell to congregate back toward the center G creating an internally stronger net negative charge density precipitating the zero point quadrature phase change seen at point D max. The concentration back to the center of the node effectively switches the polarity at the outer shells potential difference with the SN decreases. Conversely as the outer positive charge increases the internal node negative charge is attracted out, charge moving out decreases the charge density, as the torque induced counter spin slows the spin the moving out of the charge also slows the spin in the same way that the increasing negative outer charge increases the charge density at the center and therefore naturally increases the rate of spin as the outer shells net positive charge is overcome with the infusing negative charge.

The Torus has eight inner layers, four with a decreasing angle and effect of female spiral and four with a decreasing angle and effect of male spiral. It is divided on the outer skin into 16 primary points of origin with 32 secondary and 64 tertiary origins. These describe both a positively charged male spiral around the Torus, crossing at predetermined and geometrically defined nodes manifesting divergent centrifugal forces following both the long Fibonacci curves moving away from the singularity point but with a latitudinal straight series of discharge points perpendicular to and intersecting all the nodes, with the charge directed over the curve of the Torus to the singularity point S. until Dmax central

APPLICATIONS FOR A PLASMOIDS FORM AND FUNCTIONS

PART 1 - *DESCRIPTION OF PLASMOID TORUS FORM AND FUNCTION*

torus longitudinal outside equator Dmax point is reached. There at Dmax the minimum positive charge density within the constant volume V is obtained with the corresponding proportional response of the maximum mass and also, and directly proportional to it, the slowest spin and slowest progression of time are obtained. The outward velocity of the node being overcome by the pull of the singularity point G. it first neutralizes its susceptibility by rotating through 90 degrees therefore initiating the first step of reverses its poles after it has briefly spun at 90 degrees to both forces then after progressing past the point of quadrature at D max with no pole or interference then flips poles by reversing its apparent spin by inverting the node and starts its arc back to the line of least resistance and maximum infusing torsional negative charge route along the long Fibonacci curve to G.

PLASMOID TORUS MODEL (INSERT HERE)

Beyond the inflection point at Dmax the charge density is increased thru the infusion of negatively charged male centripetal, imploding, clockwise spiral energy flowing out through the latitudinal straight lines, torque on the node is also applied by the centripetal converging forces returning along the longitudinal Fibonacci curves to G where time is at its zero and inflection point.

The Plasmoid Torus model also describes through the ROD and Bendall Translate every manifest measurable characteristic of gods creation a person is capable of perceiving, and some not, all moving up or down in circulate left and right spiral octave steps from knop to knop, to and from G identifiable to us as elemental crystal forms, as sounds, as elements, as high energy states and color. This progression continues right up to the second and third order colors of the light spectrum three visible, first second and third order colors we see, and four we do not.

The progression of the entropy male counterclockwise spirals, emitting their moving out and away, dissipating positive charge effusions in exploding destructive phase is countered by centropy, clockwise, moving in, imploding, Female aggregating negative infusion in constructive imploding stage. Paramagnetic and Diamagnetic character, Monad, Diad, Triad and Tetrad dispositions are predicted, along with the octave scaling of the charge density, centrifugal energy, Osmotic charge attraction manifest as Mass and their relationships to Aether, Light, Dimensions, Music, Matter, Energy, color, Area / Time / Space and Motion.

APPLICATIONS FOR A PLASMOIDS FORM AND FUNCTIONS

PLASMOID TORUS RESONANT FORM WITH IMPLOSIVE VORTEX AND ZP

The Implosive vortex can have 14 clockwise oriented semicircles of outer 240 + inner 120 degrees equaling 360 degrees times 14 = 5,040.

50,040 / 16 (Segments)= 315
5,040 / 22.5 (Degrees per segment) = 224
5,040 / 266.666 = 18.9 / 16 = 1.18125 1/x = 0.846560846560 / .125 = 6.77248677248677
5,040/ 11.111 = 453.6 x 22.5 = 10,206
7,776 /2/2/2/ = 972 /2 = 486 / 2 = 243

5,040 times two for the anticlockwise equals 10,080.
10,080 / 16 = 630
10,080 / 22.5 = 488
10,080 / 266.666 = 37.8
10,080 / 11.111 = 907.2 x 7.5 = 6,804

The Implosive vortex can also have 15 clockwise oriented semicircles of 240 degrees + inner 120 Degrees equaling 360 degrees times 15 = 5,400.
5,400 / 216 = 25
5,400 / 22.5 = 240
5,400 / 266.666 = 20.2500506251562816 / 16 = 1.2656281 1/x = 0.790121481 / .125 = 6.3
5,400 / 11.111 = 486 / 1.333 = 364.5
5,400 / 7.5 = 720
5,400 x 7.5 = 3,888,000
3,888,000 /2/2/2/ = 486,000 / 2 = 243

5,400 Degrees times two for the anticlockwise equals 10,800
10,800 / 2,160 = 5
10,800 / 16 = 675
10,800 / 22.5 = 480
10,800 / 266.666 = 40.50
10,800 / 11.111 = 972   972 x 7.5 = 129.600
10,800 / 7.5 = 1,440
10,800 x 7.5 = 81,000

The Implosive vortex can also have 16 clockwise oriented semicircles of 360 degrees equaling 5,760.
5,760 / 216 = 26.666
5,760 / 432 = 13.333
5,760 / 266.666 = 21.6
5,760 / 1.333 = 4,320
5,760 / 22.5 = 256
5,760 / 266.666 = 21.6 / 16 = 1.35 1/x =   / .125 =
5,760 / 11.111 = 518.4 x 22.5 = 11,664

APPLICATIONS FOR A PLASMOIDS FORM AND FUNCTIONS

5,760 / 2 = 2,880 / 2 = 1,440 / 2 = 720 / 2 = 360 / 2 = 180 / 2 = 90 / 2 = 45 / 2 = 22.5 / 2 = 11.25 / = 5.625
5.625 / 2 = 2.8125 C / 2 = 1.4625 C / 2 = 0.703125 C / 2 = 0.3515625 / 2 = 0.17578125 / 2 = 0.08789062

5,760 times two for the 16 anticlockwise equals 11,520.
11,520 / 2160 = 5.333
11,520 / 16 = 720
11,520 / 22.5 = 512
11,520 / 266.666 = 43.2
11,520 / 11.111 = 1,036.8 x 22.5 = 23,328
.125 or one eighth is half the 2.5 of the torus

The Implosive vortex can also have 16 clockwise oriented semicircles of 240 degrees equaling.
3,840 / 720 = 5.333
3,840 / 16 = 240
3,840 / 22.5 = 170.6666666666667 octave of 266.666
3,840 / 266.666 = 14.4 / 16 = .9 1/x = 1.111 14.4 / .125 =
3,840 / 11.111 = 345.6 x 22.5 = 7,776 x 2 x 2 = 31,104
7,776 /2/2/2/ = 972/2 = 486 / 2 = 243

3,840 times two for the 16 anticlockwise equals 7,680.
7,680 / 22.5 = 341.333 octave of 266.666 and .666
7,680 / 266.666 = 28.800000072
7,680 / 11.111 = 691.2 x 22.5 = 15,552

.125 or one eighth is half the 2.5 of the torus
Torus has the sum of internal angles, Pythagorean resonance being the sum of the internal angles, on one of its eight planes of 360 plus joiners 90 + 90 = 180 plus 360 = 900 degrees times eight = 7,200 / 22.5 = 320.

7,200 / 16 = 450 (molten sea 45 hands)
7,200 / .125 = 57,600
7,200 / 11.111 = 648.0000000000001
7,200 / 266.666 = 27
27 / 16 = 1.6875
1.6875 the up octaves of which are all the planetary diameters
3.375000000000001 // 6.75 // 13.5 // 27 // 54 // 108 // 216 Moon // 432 Sun radius // 864 Suns diameter // 1,728 // 3,456 SUN SQUARE // 6,912 RFEU // 13,824 // 27,648

900 vertical plus 360 horizontal =

1,260 / 11.1111 = 113.4
1,260 / 266.666 = 4.725

1,260 // 2,520 // 5,040 // 10,080 // 20,160 // 40,320 // 80,640 // 161,280 // 322,560

518,400 x 2.5 = 1,296,000 RFEU

Tests for Light / 6.666 for Aether / 5.555 for Time / 4.444 for Matter 3.333 and 266.666 Divide by 37 AND 5.555 for ENERGY and Divide by 36 AND 3.333 for MATTER

APPLICATIONS FOR A PLASMOIDS FORM AND FUNCTIONS

2.) *DESCRIPTION OF ALIEN MATHS AND CHEMISTRY*

HYDROGEN, NOBLE GASES, METHANE AND CARBON MONOXIDE

The melting point and boiling point of any element are the points at which enough energy has been put into the dynamic system for the resonant frequency intensity to be met. This is determined by a combination of dynamic factors. Primary of these factors is the sum of the internal angles of the elements or compounds natural crystal form manifesting in its solid state. These are adjusted to their atomic charge density and the Area / time / space that it occupies which are the secondary forces. The Area / Time / Space its self introduces the additional variable of the external positive and negative atmospheric and atomic pressure. This is in addition to the tertiary effect of implosive and explosive environmental and atomic forces. Contamination of other matter further varies the frequency /area / time / space equation and therefore must be measured and accounted for to determine a "group " mass resonance.

APPLICATIONS FOR A PLASMOIDS FORM AND FUNCTIONS

3.) WATER H2O – TWO PROTIUM (HYDROGEN) AND ONE OXYGEN.

3.1.) PROTIUM = HYDROGEN CALCULATIONS (H).

HYDROGEN MELTING POINT = **− 259.2** so - 259.2 / 266.666 = 0.972

- 259.2 / 11.111 = 23.328

- 259.2 / 1.333 = 194.4 // 97.2//48.6//24.3

- 259.2 added sum = 2 + 5 + 9 + 2 [18] + 2 + 5 + 9 + 2 = 36

- 259.2 multiplied product = 2 x 5 x 9 x 2 (180) x 2 x 9 x 5 x 2 = 32,400

32,400 / 720 = 45   32,400 / 16 = 2,025   32,400 / 22.5 = 1,440   32,400 / 0.125 = 259,200

32,400 x 16 = 518,400   32,400 x 22.5 = 729,000   32,400 x .125 = 4,050

32,400 x 400 = 12,960,000   32,400 x 720 = 23,328,000   32,400 x 900 = 29,160,000

518,400 / 16 = 32,400   518,400 / 22.5 = 23,040   518,400 / 720 = 720

32,400 / 144 = 225   32,400 / 36 = 900   36 / 32,400 = 0.00111

144,000 / 32,400 = 4.4444   32,400 / 7.5 = 4,320 / 7.5 = 576 / 2 = 288 / 2 = 144

144 x 32,400 = 4,665,600 / 432 = 10,800 / 2,160 = 5

32,400 / 259.2 = 125
32,400 / 6.666 = 4,860   32,400 / 5.555 = 5,850   32,400 / 4.444 = 7,290
32,400 / 3.333 = 9,720   32,400 / 2.222 = 14,580   32,400 / 11.111 = 2,916

HYDROGEN RATIOS TO ALL OTHER NOBLE GASES

32,400 (Hydrogen -259.2) / 3,136 (Helium -272.2) = 10.33163265306122
3,136 (Helium -272.2) / 32,400 (Hydrogen -259.2) = 0.09679012345679

32,400 (Hydrogen -259.2) / 147,456 (Neon -248.6) = 0.2197265625
147,456 (Neon -248.6) / 32,400 (Hydrogen -259.2) = 4.55111

32,400 (Hydrogen -259.2) / 82,944 (Argon -189.4) = 0.390625
82,944 (Argon -189.4) / 32,400 (Hydrogen -259.2) = 2.56

32,400 (Hydrogen -259.2) / 19,600 (Krypton 157.4) = 1.653061224489796
19,600 (Krypton 157.4) / 32,400 (Hydrogen -259.2) = 0.604938271604938

32,400 (Hydrogen -259.2) / 64 (Xeon -111.8) = 506.25
64 (Xeon -111.8) / 32,400 (Hydrogen -259.2) = 0.00195308641975

32,400 (Hydrogen -259.2) / 1,225 (Radon -71.15) = 26.44897959183673
1,225 (Radon -71.15)/ 32,400 (Hydrogen -259.2) = 0.037808641975309 Argon/Helium

ALL SIX 864,000 COMBINATIONS

864,000	846,000	684,000	648,000	486,000	468,000
56,623,104*	55,443,456*	44,826,624*	42,467,328*	31,850,496*	30,670,848*
28,311,552*	27,721,728*	22,413,312*	21,233,664*	15,925,248*	15,335,424*
14,155,776*	13,860,864*	11,206,656*	10,616,832*	7,962,624*	7,667,712*
7,077,888*	6,930,432*	5,603,328*	5,308,416*	3,981,312*	3,833,856*
3,538,944*	3,465,216*	2,801,664*	2,654,208*	1,990,656*	1,916,928*
1,769,472*	1,732,608*	1,400,832*	1,327,104*	995,328*	958,464*
884,736*	866,304*	700,416*	663,552*	497,664*	479,232*
442,368*	433,152*	350,208*	331,776*	248,832*	239,616*
221,184*	216,576*	175,104*	165,888*	124,416	119,808*
110,592*	108,288*	87,552*	82,944*	62,208*	59,904*
55,296*	54,144*	43,776*	41,472*	31,104*	29,952*
27,684*	27,072*	21,888*	20,736*	15,552*	14,976*
13,824*	13,536*	10,944*	10,368*	7,776*	7,488*
6,912,000	6,768,000	5,472,000	5,184,000	3,888,000	3,744,000
3,456,000	3,384,000	2,736,000	2,592,000	1,944,000	1,872,000
1,728,000	1,692,000	1,368,000	1,296,000	972,000	936,000

864,000	846,000	684,000	648,000	486,000	468,000
432,000	423,000	342,000	H-324,000	243,000	234,000
216,000	211,500	171,000	162,000	121,500	117,000
108,000	105,000	85,500	81,000	60,750	58,500
54,000	52,875	42,750	40,500	30,375	29,250
27,000	26,437.5	21,375	20,250	15,187.5	14,625
13,500	13,218.75	10,687.5	10,125	7,593.75	7,312.5
6,750	6,609.375	5,343.75	5,062.5	3,796.875	
3,656.25					
3,375	3,304.6875	2,671.875	2,531.25	1,898.43	1,828.1
1,687.5	1,652.34375	1,335.9375	1,265.62	949.2187	914.06
843.75	826.171875	667.96875	632.812	474.6093	457.03
421.875	413.0859375	333.984375	316.406	237.3046	228.51
210.9375	206.54296875	166.99218	158.2031	118.6523	114.25
105.46875	103.271484375	83.496093	79.10156	59.32617	57.128
52.734375	51.6357421875	41.7480468	39.55078	29,66308	28.564
26.3671875	25.8187109375	20.8740234	19.77539	14.83154	14.282

864,000 846,000   684,000   648,000 486,000 468,000

HYDROGEN RATIOS TO ALL OTHER NOBLE GASES

32,400 (Hydrogen -259.2) / 3,136 (Helium -272.2) = 10.33163265306122
3,136 (Helium -272.2) / 32,400 (Hydrogen -259.2) = 0.09679012345679

32,400 (Hydrogen -259.2) / 147,456 (Neon -248.6) = 0.2197265625
147,456 (Neon -248.6) / 32,400 (Hydrogen -259.2) = 4.55111

32,400 (Hydrogen -259.2) / 82,944 (Argon -189.4) = 0.390625
82,944 (Argon -189.4) / 32,400 (Hydrogen -259.2) = 2.56

32,400 (Hydrogen -259.2) / 19,600 (Krypton 157.4) = 1.653061224489796
19,600 (Krypton 157.4) / 32,400 (Hydrogen -259.2) = 0.604938271604938

32,400 (Hydrogen -259.2) / 64 (Xeon -111.8) = 506.25
64 (Xeon -111.8) / 32,400 (Hydrogen -259.2) = 0.00195308641975

32,400 (Hydrogen -259.2) / 1,225 (Radon -71.15) = 26.44897959183673
1,225 (Radon -71.15)/ 32,400 (Hydrogen -259.2) = 0.037808641975309 Argon/Helium

HYDROGEN RIGHT TRIANGLE MELTING POINT = - 259.2 C (Earths Equinox 25,920)

864,000 / 259.2 = 3,333.3   864,000 / 345.6 = 2,500   864,000/ 432 = 2,000
2,160 / 259.2 = 8.333   259.2 / 2,160 = 0.972

Triangles Area / Time / Space calculation

259.2 x -345.6 x 432 = 38,698,352.6
259.2 x -345.6 = - 89,579.5 (Area of the square) / 2 = 44,789.76 (Area of the Right Triangle)

44,789.76 / 24 Hrs = 1,866.24   1,866.24 / 60 Min = 31.104   31.104 / 60 Sec = 0.5184
0.5184 / 60 Arc = 0.00864   0.00864 / 60 = 0.000144   0.000144 / 60 = 0.0000024
0.0000024 / 60 = 0.00000004   0.00000004 / 24 = 0.000000001666
0.00000004 / 60 = 0.0000000000666   0.0000000000666 / 60 = 0.00000000001111

0.5184 / 7.5 = 0.06912 / 0.266666 = 0.2592 x 400 = 103.68 / 16 = 6.48 / 22.5 = 0.288
0.288/ 0.125 = 2.304 x 400 = 921.9 / 1.333 = 691.2 / 1.333 = 518.4 /1.333 = 388.8 / 1.333 =
291.6

-259.2 / 3 = - 86.4

86.4 x 5 = 432

- 345.6 = - 86.4 x 4

HYDROGEN SQUARE MELTING POINT -259.2 (Earths Equinox) x 4 = 1,036.8

1,036.8/3,456 = 0.3
259.2 x 259.2 = 67,184.64 / 11.111 = 6,046.6   67,184.64 / 266.666 = 251.9424
67,184.64 / 144 = 466.56
67.184.64 / 7.5 = 8,957.952
67,184.64 / 2 = 33,592.32
33,592.32 / 3 = 11,197.44   11,197.44 x 4 = 44,789.76   11,197.44 x 5 = 55,987.2

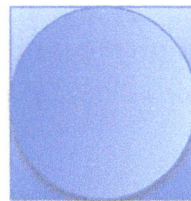

HYDROGEN CUBE MELTING POINT CUBED = 259.2 x 259.2 x 259.2 = 17,414,258.688
1 year, 360 day Time in seconds cube 17,414,258.688 / 360 = 48,372.9408
48,372.9408/24 = 2,016/60 = 33.6/60 = 0.56/ 720 = 0.000777/ 16 = 0.0000486111/ 22.5 =
0000002160493827

17,414,258.688 / 144,000 = 120.932352 / 7.5 = 16.1243136 / 16 = 1.0077696
 = 1.0077696 / 22.5 = 0.04478976
17,414,258.688 / 1.333 = 13,060,694.016
17,414,258.688 / 266.666 = 65,303.47008000002
17,414,258.688 / 6.666 = 2,612,138.8032
17,414,258.688 / 5.555 = 3,134,566.56384
17,414,258.688 / 4.444 = 3,918,208.2048

17,414,258.688 / 3.333 = 5,224,277.606400001
17,414,258.688 / 2.222 = 7,836,416.409600001
17,414,258.688 / 1.111 = 15,672,832.8192 / 9 = 1,741,425.8688

HYDROGEN RIGHT ANGLE TRIANGLE BOILING POINT = - 252.8 C

HYDROGEN Boiling point = - 252.87 so -252.87 / 266.666 = .9482625
- 252.87 so -252.87 / 11.111 = 22.7583
- 252.87 so -252.87 / 1.333 = 189.699925

252.87 = 2 + 5 + 2 + 8 + 7 (24) +7 + 8 + 2 + 5 + 2 = 48

252.87 = 2 x 5 x 2 x 8 x 7 (1,120) x 7 x 8 x 2 x 5 x 2 = 1,254,400 / 5.555 = 225,792

252.8 x 337.0666 = 85,210.4365 therefore triangle area 85,210.4365 / 2 = 42,605.2182
42,605.2182 / 24 Hours = 1,775.21743   1,775.21743 / 60 Minutes = 29.5869571
29.5869571 / 60 Seconds = 0.493115951
0.493115951 / 60 Arc seconds = 0.0821859918333

- 252.8 / 3 = 84.2666

421.333 = 84.2666 x 5

337.0666 = 84.2666 x 4

3.2.) OXYGEN CALCULATIONS (O2).

OXYGEN

OXYGEN Melting point = - 218.8 / 266.666 = 0.8205   [1/x = 1.218769043266301]

and = -218.8 / 11.111 = 19.692   [1/x = 0.050782043469429]

and = -218.8 / 1.333 = 164.1   [1/x = 0.006093845216332]

and = -218.8 / 144 = 1.519444   [1/x = 0.658135283363803]

2 + 1 + 8 + 8 = (19) + 8 + 8 + 1 + 2 = 38

2 x 1 x 8 x 8 = (128) x 8 x 8 x 1 x 2 = 16,384

16,384 (Oxygen 218.8) / 22 = 2.909090
16,384 (Oxygen 218.8) x 22 = 1,408

16,384 (Oxygen 218.8) / 266.666 = 61.44   [1/x = 0.16276041666]
16,384 (Oxygen 218.8) / 11.111 = 1,474.56   [1/x = 0.000678168402778]
16,384 (Oxygen 218.8) / 1.333 = 122,288   [1/x = 0.000081380208333]
16,384 (Oxygen 218.8) / 144 = 113.777   [1/x = 0.0087890625]

OXYGEN RATIOS TO ALL OTHER NOBLE GASES INCLUDING HYDROGEN

OXYGEN RIGHT ANGLE TRIANGLE MELTING POINT – 218.8

-218.8 x 291.7333 = 63,831.25333
63,831.25333 / 2 = 31,915.62666
31,915.62666 / 24 Hours = 1,329.81 / 60 Minutes = 22.1636296296296
22.1636296296296 / 60 Seconds = 0.369393827160494
0.369393827160494 / 60 Arc Seconds = 0.006156563786008

-218.8 / 3 = 72.9333

364.666 = 5 x 72.9333

291.7333 = 4 x 72.9333

OXYGEN SQUARE MELTING POINT -218.8 x 4 = 875.2 / 3,456 = 0.25324074074074

-218.8

OXYGEN

OXYGEN CUBE MELTING POINT -218.8 CUBED = -10,474,708.672

1 year, 360 day Time in seconds cube 10,474,708.672 / 360 = 29,096.4129777
29,096.4129777 / 24 Hours = 1212.35054074074 / 60 Minutes = 20.20584234567901
20.20584234567901 / 60 Seconds = 0.33676403909465
0.33676403909465 / 60 Arc Seconds = 0.005612733984911
0.005612733984911 / 16 = 0.0003507958740570 / 22.5 = 0.000015509027736

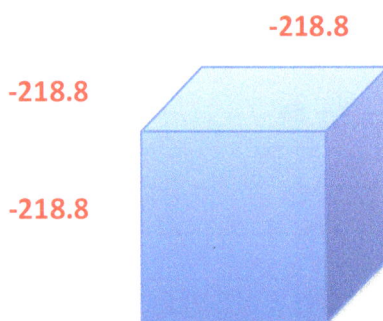

10,474,708.672 / 360 (1 year) Time in Seconds cube = 29,096.4129777
10,474,708.672 / 24 = 436,446.194666 /60 = 7,274.1-32444 /60= 121.235 /60= 2.00205842345679
10,474,708.672 / 144,000 = 72.7410324 / 24 = 3.03087635185 / 60 = / 60 = / 60 =
10,474,708.672 / 266.666 = 39,280.15 / 24 = / 60 = / 60 = / 60 =
10,474,708.672 / 6.666 = 1,571,206.3008 / 24 = / 60 = / 60 = / 60 =
10,474,708.672 / 5.555 = 1,885,447.56096 / 24 = / 60 = / 60 = / 60 =
10,474,708.672 / 4.444 = 2,356,8094512 / 24 = / 60 = / 60 = / 60 =
10,474,708.672 / 3.333 = 3,142,412.6016 / 24 = / 60 = / 60 = / 60 =
10,474,708.672 / 2.222 = 4,713,618.9024 / 24 = / 60 = / 60 = / 60 =
10,474,708.672 / 1.111 = 9,427,237.8048 / 24 = / 60 = / 60 = / 60 =

4.) *HYDROCARBONS H(x)C(y) – HYDROGEN AND CARBON.*

4.1.) *HYDROGEN CALCULATIONS.*
HYDROGEN

HYDROGEN MELTING POINT = − 259.2 so - 259.2 / 266.666 = 0.972

- 259.2 / 11.111 = 23.328

- 259.2 / 1.333 = 194.4 // 97.2//48.6//24.3

- 259.2 added sum = 2 + 5 + 9 + 2 [18] + 2 + 5 + 9 + 2 = 36

- 259.2 multiplied product = 2 x 5 x 9 x 2 (180) x 2 x 9 x 5 x 2 = 32,400

32,400 / 720 = 45   32,400 / 16 = 2,025   32,400 / 22.5 = 1,440   32,400 / 0.125 = 259,200

32,400 x 16 = 518,400   32,400 x 22.5 = 729,000   32,400 x .125 = 4,050

32,400 x 400 = 12,960,000   32,400 x 720 = 23,328,000   32,400 x 900 = 29,160,000

518,400 / 16 = 32,400   518,400 / 22.5 = 23,040   518,400 / 720 = 720

32,400 / 144 = 225   32,400 / 36 = 900   36 / 32,400 = 0.00111

144,000 / 32,400 = 4.4444   32,400 / 7.5 = 4,320 / 7.5 = 576 / 2 = 288 / 2 = 144

144 x 32,400 = 4,665,600 / 432 = 10,800 / 2,160 = 5

32,400 / 259.2 = 125
32,400 / 6.666 = 4,860   32,400 / 5.555 = 5,850   32,400 / 4.444 = 7,290
32,400 / 3.333 = 9,720   32,400 / 2.222 = 14,580   32,400 / 11.111 = 2,916

HYDROGEN RATIOS TO ALL OTHER NOBLE GASES

32,400 (Hydrogen -259.2) / 3,136 (Helium -272.2) = 10.33163265306122
3,136 (Helium -272.2) / 32,400 (Hydrogen -259.2) = 0.09679012345679

32,400 (Hydrogen -259.2) / 147,456 (Neon -248.6) = 0.2197265625
147,456 (Neon -248.6) / 32,400 (Hydrogen -259.2) = 4.55111

32,400 (Hydrogen -259.2) / 82,944 (Argon -189.4) = 0.390625
82,944 (Argon -189.4) / 32,400 (Hydrogen -259.2) = 2.56

32,400 (Hydrogen -259.2) / 19,600 (Krypton 157.4) = 1.653061224489796
19,600 (Krypton 157.4) / 32,400 (Hydrogen -259.2) = 0.604938271604938

32,400 (Hydrogen -259.2) / 64 (Xeon -111.8) = 506.25
64 (Xeon -111.8) / 32,400 (Hydrogen -259.2) = 0.00195308641975

32,400 (Hydrogen -259.2) / 1,225 (Radon -71.15) = 26.44897959183673
1,225 (Radon -71.15)/ 32,400 (Hydrogen -259.2) = 0.037808641975309 Argon/Helium

ALL SIX 864,000 COMBINATIONS

864,000	846,000	684,000	648,000	486,000	468,000
56,623,104*	55,443,456*	44,826,624*	42,467,328*	31,850,496*	30,670,848*
28,311,552*	27,721,728*	22,413,312*	21,233,664*	15,925,248*	15,335,424*
14,155,776*	13,860,864*	11,206,656*	10,616,832*	7,962,624*	7,667,712*
7,077,888*	6,930,432*	5,603,328*	5,308,416*	3,981,312*	3,833,856*
3,538,944*	3,465,216*	2,801,664*	2,654,208*	1,990,656*	1,916,928*
1,769,472*	1,732,608*	1,400,832*	1,327,104*	995,328*	958,464*
884,736*	866,304*	700,416*	663,552*	497,664*	479,232*
442,368*	433,152*	350,208*	331,776*	248,832*	239,616*
221,184*	216,576*	175,104*	165,888*	124,416	119,808*
110,592*	108,288*	87,552*	82,944*	62,208*	59,904*
55,296*	54,144*	43,776*	41,472*	31,104*	29,952*
27,684*	27,072*	21,888*	20,736*	15,552*	14,976*
13,824*	13,536*	10,944*	10,368*	7,776*	7,488*
6,912,000	6,768,000	5,472,000	5,184,000	3,888,000	3,744,000
3,456,000	3,384,000	2,736,000	2,592,000	1,944,000	1,872,000
1,728,000	1,692,000	1,368,000	1,296,000	972,000	936,000

864,000	846,000	684,000	648,000	486,000	468,000
432,000	423,000	342,000	H-324,000	243,000	234,000
216,000	211,500	171,000	162,000	121,500	117,000
108,000	105,000	85,500	81,000	60,750	58,500
54,000	52,875	42,750	40,500	30,375	29,250
27,000	26,437.5	21,375	20,250	15,187.5	14,625
13,500	13,218.75	10,687.5	10,125	7,593.75	7,312.5
6,750	6,609.375	5,343.75	5,062.5	3,796.875	
3,656.25					
3,375	3,304.6875	2,671.875	2,531.25	1,898.43	1,828.1
1,687.5	1,652.34375	1,335.9375	1,265.62	949.2187	914.06
843.75	826.171875	667.96875	632.812	474.6093	457.03
421.875	413.0859375	333.984375	316.406	237.3046	228.51
210.9375	206.54296875	166.99218	158.2031	118.6523	114.25

105.46875	103.271484375	83.496093	79.10156	59.32617	57.128
52.734375	51.6357421875	41.7480468	39.55078	29,66308	28.564
26.3671875	25.8187109375	20.8740234	19.77539	14.83154	14.282

| 864,000 | 846,000 | 684,000 | 648,000 | 486,000 | 468,000 |

HYDROGEN RATIOS TO ALL OTHER NOBLE GASES

32,400 (Hydrogen -259.2) / 3,136 (Helium -272.2) = 10.33163265306122
3,136 (Helium -272.2) / 32,400 (Hydrogen -259.2) = 0.09679012345679

32,400 (Hydrogen -259.2) / 147,456 (Neon -248.6) = 0.2197265625
147,456 (Neon -248.6) / 32,400 (Hydrogen -259.2) = 4.55111

32,400 (Hydrogen -259.2) / 82,944 (Argon -189.4) = 0.390625
82,944 (Argon -189.4) / 32,400 (Hydrogen -259.2) = 2.56

32,400 (Hydrogen -259.2) / 19,600 (Krypton 157.4) = 1.653061224489796
19,600 (Krypton 157.4) / 32,400 (Hydrogen -259.2) = 0.604938271604938

32,400 (Hydrogen -259.2) / 64 (Xeon -111.8) = 506.25
64 (Xeon -111.8) / 32,400 (Hydrogen -259.2) = 0.00195308641975

32,400 (Hydrogen -259.2) / 1,225 (Radon -71.15) = 26.44897959183673
1,225 (Radon -71.15)/ 32,400 (Hydrogen -259.2) = 0.037808641975309 Argon/Helium

HYDROGEN RIGHT TRIANGLE MELTING POINT = - 259.2 C (Earths Equinox 25,920)

864,000 / 259.2 = 3,333.3   864,000 / 345.6 = 2,500   864,000/ 432 = 2,000
2,160 / 259.2 = 8.333   259.2 / 2,160 = 0.972

Triangles Area / Time / Space calculation

259.2 x -345.6 x 432 = 38,698,352.6
259.2 x -345.6 = - 89,579.5 (Area of the square) / 2 = 44,789.76 (Area of the Right Triangle)

44,789.76 / 24 Hrs = 1,866.24   1,866.24 / 60 Min = 31.104   31.104 / 60 Sec = 0.5184
0.5184 / 60 Arc = 0.00864   0.00864 / 60 = 0.000144   0.000144 / 60 = 0.0000024
0.0000024 / 60 = 0.00000004   0.00000004 / 24 = 0.000000001666
0.00000004 / 60 = 0.0000000000666   0.0000000000666 / 60 = 0.00000000001111

0.5184 / 7.5 = 0.06912 / 0.266666 = 0.2592 x 400 = 103.68 / 16 = 6.48 / 22.5 = 0.288
0.288/ 0.125 = 2.304 x 400 = 921.9 / 1.333 = 691.2 / 1.333 = 518.4 /1.333 = 388.8 / 1.333 =
291.6

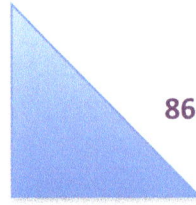

-259.2 / 3 = - 86.4

86.4 x 5 = 432

- 345.6 = - 86.4 x 4

HYDROGEN SQUARE MELTING POINT -259.2 (Earths Equinox) x 4 = 1,036.8

1,036.8/3,456 = 0.3
259.2 x 259.2 = 67,184.64 / 11.111 = 6,046.6   67,184.64 / 266.666 = 251.9424
67,184.64 / 144 = 466.56
67.184.64 / 7.5 = 8,957.952
67,184.64 / 2 = 33,592.32
33,592.32 / 3 = 11,197.44   11,197.44 x 4 = 44,789.76   11,197.44 x 5 = 55,987.2

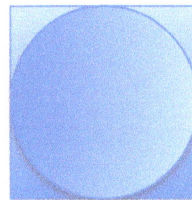

HYDROGEN CUBE MELTING POINT CUBED = 259.2 x 259.2 x 259.2 = 17,414,258.688
1 year, 360 day Time in seconds cube 17,414,258.688 / 360 = 48,372.9408
48,372.9408/24 = 2,016/60 = 33.6/60 = 0.56/ 720 = 0.000777/ 16 = 0.0000486111/ 22.5 =
0000002160493827

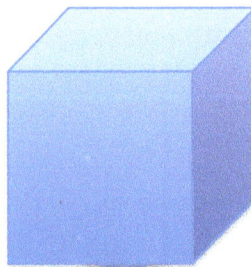

17,414,258.688 / 144,000 = 120.932352 / 7.5 = 16.1243136 / 16 = 1.0077696
 = 1.0077696 / 22.5 = 0.04478976
17,414,258.688 / 1.333 = 13,060,694.016
17,414,258.688 / 266.666 = 65,303.47008000002
17,414,258.688 / 6.666 = 2,612,138.8032
17,414,258.688 / 5.555 = 3,134,566.56384
17,414,258.688 / 4.444 = 3,918,208.2048
17,414,258.688 / 3.333 = 5,224,277.606400001
17,414,258.688 / 2.222 = 7,836,416.409600001
17,414,258.688 / 1.111 = 15,672,832.8192 / 9 = 1,741,425.8688

HYDROGEN RIGHT ANGLE TRIANGLE BOILING POINT = - 252.8 C

HYDROGEN Boiling point = - 252.87 so -252.87 / 266.666 = .9482625
                         - 252.87 so -252.87 / 11.111 = 22.7583
                         - 252.87 so -252.87 / 1.333 = 189.699925

252.87 = 2 + 5 + 2 + 8 + 7 (24) +7 + 8 + 2 + 5 + 2 = 48

252.87 = 2 x 5 x 2 x 8 x 7 (1,120) x 7 x 8 x 2 x 5 x 2 = 1,254,400 / 5.555 = 225,792

252.8 x 337.0666 = 85,210.4365 therefore triangle area 85,210.4365 / 2 = 42,605.2182
42,605.2182 / 24 Hours = 1,775.21743   1,775.21743 / 60 Minutes = 29.5869571
29.5869571 / 60 Seconds = 0.493115951
0.493115951 / 60 Arc seconds = 0.0821859918333

- 252.8 / 3 = 84.2666

421.333 = 84.2666 x 5

337.0666 = 84.2666 x 4

4.2.) CARBON CALCULATIONS - CARBON'S RESONANT FORMS

Carbon has 5 states, each state has a unique melting point or phase change

1) - DIAMOND

MELTING POINT = 3,700K or 3,426.85C   The Sum = 56 and Product = 33,177,600 (5,760 Sq)
33,177,600 / 144,000 = 230.4   33,177,600 / 518,400 = 64   33,177,600 / 31,104,000 = 1.0666 /// 0.2666
33,177,600 / 24 = 1,382,400 / 60 = 23,040 / 60 = 384 / 60 = 6.4 / 7.5 = 0.85333 \\\\\ 0.02666
33,177,600 / 864 = 38.4 / 144 = 0.2666   33,177,600 / 2,160 = 15,360   33,177,600 / 129,600 = 256\\\\16

MELTING POINT = 5,000K or 4,726.85C The Sum = 32 and Product = 180,633,600 (13,440 Sq)
180,633,600 / 144,000 = 1,254.4   180,633,600/518,400 = 384.444   180,633,600/31,104,000 = 5.80740
180,633,600 / 7.5 = 24,084,480 / 24 = 1,003,520 / 60 = 16,725.333 / 60 = 278.7555 / 60 = 4.64592592
180,633,600 / 25,920 = 6,968.888   180,633,600 /864,000 = 209.0666 180,633,600 / 266.666 = 677,376

2) - GRAPHENE

MELTING POINT = 4,510K or 4,236.85C The Sum= 28 and Product= 33,177,600 SAME AS DIAMOND
33,177,600 / 144,000 = 230.4   33,177,600 / 518,400 = 64   33,177,600 / 31,104,000 = 1.0666 \\\0.2666

MELTING POINT = 4,900K or 4,626.85C The Sum = 31 and Product = 132,710,400 (11,520 Sq)
132,710,400 / 24 = 5,529,600 / 60 = 92,160 / 60 = 1,536 / 60 = 25.6 / 7.5 = 3.41333\\\\\\\0.02666
132,710,400 / 144,000 = 921.6 / 266.666 = 3.456 / 11.111 = 0.31104000031104

3) - GRAPHITE

MELTING POINT = 3,800K or 3,526.85C Sum = 58 and Product = 51,840,000 (7,200 Squared)
51,840,000 / 24 Hours = 2,160,000 / 60 min = 36,000 / 60 Sec = 600 / 60 Arc = 10 / 7.5 = 1.333 \266.666
51,840,000/864,000=60 51,840,000/144,000=360 51,840,000/129,600=400 51,840,000/266.666=1944
51,840,000 / 25,920 (EQUINOX 25,920 Years) = 2,000 (2,000 BATHS IN THE 2/3 FULLMOLTEN SEA)

4) - BUCKEY BALLS CARBON FULLERENE C60

MELTING POINT = 1523K or 1,249.85C The Sum = 58 and Product = 8,294,400 (2,880 Sq)
8,294,400/ 25,920= 320 8,294,400/864,000= 9.6 8,294,400/144= 57,600 8,294,400/21,600=384\\\\24
8,294,400 / 24 = 345,600 / 60 = 5,760 / 60 = 96 / 60 = 1.6 / 7.5 = 0.21333\\\0.02666
MELTING POINT = 278.15K or 5C The Sum = 10 and Product = 25

MELTING POINT = 800K or 526.85C The Sum = 52 and Product = 5,760,000 (2,400 Sq)
5,760,000 / 25,920= 222.222 5,760,000/864,000= 6.666 5,760,000/144= 40 5,760,000/21,600= 266.666
5,760,000 / 24 = 240,000 / 60 = 4,000 / 60 = 6.666 / 60 = 1.111 / 7.5 = 0.148148148

MELTING POINT = 707K or 433.85C The Sum = 46 and Product = 2,073,600 (1,440 Sq)
2,073,600 / 24 = 86,400 / 60 = 1,440 / 60 = 24 / 60 = 0.4 / 0.00075 = 266.666 (C music elemental Octave)
MELTING POINT = 739K or 465.85C The Sum = 56 and Product = 23,040,000 (4,800)
23,040,000 / 24 = 960,000 / 60 = 16,000 / 60 = 266.666 / 60 = 4.444   23,040,000 / 25,920 = 888.888

5) - NANOTUBULES

MELTING POINT = 5,200K or 4,926.85C The Sum = 68 and Product = 298,598,400 (17,280 Sq)
298,598,400 /24 = 12,441,600 / 60 = 207,360 / 60 = 3,456 / 60 = 57.6 (14.4 x 3)/ 7.5 = 7.68 \\\\\0.24
298,598,400 /144 = 2,073.6\\\259.2(H)   298,598,400/25,920 = 11,520   298,598,400/= 129,600 = 2,304
MELTING POINT VACUME = 3,073 K or 2,800C The Sum = 20 and Product = 256 (16 Sq)
MELTING POINT AIR = 1,023K or 750C The Sum = 13 and Product = 1,225 (35 Sq)

ALL SIX 864,000 COMBINATIONS

864,000	846,000	684,000	648,000	486,000	468,000
56,623,104*	55,443,456*	44,826,624*	42,467,328*	31,850,496*	30,670,848*
28,311,552*	27,721,728*	22,413,312*	21,233,664*	15,925,248*	15,335,424*
14,155,776*	13,860,864*	11,206,656*	10,616,832*	7,962,624*	7,667,712*
7,077,888*	6,930,432*	5,603,328*	5,308,416*	3,981,312*	3,833,856*
3,538,944*	3,465,216*	2,801,664*	2,654,208*	1,990,656*	1,916,928*
1,769,472*	1,732,608*	1,400,832*	1,327,104*	995,328*	958,464*
884,736*	866,304*	700,416*	663,552*	497,664*	479,232*
442,368*	433,152*	350,208*	331,776*	248,832*	239,616*
221,184*	216,576*	175,104*	165,888*	124,416	119,808*
110,592*	108,288*	87,552*	82,944*	62,208*	59,904*
55,296*	54,144*	43,776*	41,472*	31,104*	29,952*
27,684*	27,072*	21,888*	20,736*	15,552*	14,976*
13,824*	13,536*	10,944*	10,368*	7,776*	7,488*
6,912,000	6,768,000	5,472,000	5184,000	3,888,000	3,744,000
3,456,000	3,384,000	2,736,000	2,592,000	1,944,000	1,872,000
1,728,000	1,692,000	1,368,000	1,296,000	972,000	936,000

864,000	846,000	684,000	648,000	486,000	468,000
432,000	423,000	342,000	324,000	243,000	234,000
216,00	211,500	171,000	162,000	121,500	117,000
108,000	105,000	85,500	81,000	60,750	58,500
54,000	52,875	42,750	40,500	30,375	29,250
27,000	26,437.5	21,375	20,250	15,187.5	14,625
13,500	13,218.75	10,687.5	10,125	7,593.75	7,312.5

6,750	6,609.375	5,343.75	5,062.5	3,796.875	
3,656.25					
3,375	3,304.6875	2,671.875	2,531.25	1,898.43	1,828.1
1,687.5	1,652.34375	1,335.9375	1,265.62	949.2187	914.06
843.75	826.171875	667.96875	632.812	474.6093	457.03
421.875	413.0859375	333.984375	316.406	237.3046	228.51
210.9375	206.54296875	166.99218	158.2031	118.6523	114.25
105.46875	103.271484375	83.496093	79.10156	59.32617	57.128
52.734375	51.6357421875	41.7480468	39.55078	29,66308	28.564
26.3671875	25.8187109375	20.8740234	19.77539	14.83154	14.282
864,000	846,000	684,000	648,000	486,000	468,000

MATTER-MUSIC 266.666 OCTAVE SEQUENCE

1,118,481,066.666

559,240,533.333

279,620,266.666

139,810,133.333

69,905,066.666

34,952,533.333

17,476,266.666

8,738,133.333

4,369,066.666

2,184,533.333

1,092,266.666

546,133.333

273,066.666

136,533.333

68,266.666

34,133.333

17,066.666

8,533.333

4,266.666

2,133.333 BUCKEY BALLS FULLERINE

1,066.666 DIAMOND AND GRAPHENE

533.333

266.666 MATTER ELEMENTAL OCTAVES

3,456,000   1,728,000   864,000   432,000

31,104,000,000   518,400

144,000   7.5

266.666 x 7.5 = 2,000 x 7.5 = 15,000

133.333 GRAPHITE

66.666

33.333

16.666
8.333
4.1666
2.08333
1.041666
0.5208333
0.26041666
0.130208333
0.0651041666
0.03255208333
0.016276041666
0.0081380208333
0.00406901041666
0.002034505208333

CARBON 3,500 C + OXYGEN -183 C
HYDROGEN -259.2

This table below needs editing for the combined numbers
HYDROGEN PLUS PALLADIUM COLD FUSION INTERACTION STRIPPING THE ELECTRON FROM THE PROTON NON NUCLEAR REACTION, NO NUCLEUS, NO POLLUTION

2,963 (Palladium) x 252.87 (Hydrogen) = 749,253.81 x 1.333 = 999,005.078 / 11.111 / 7.5 / 16 / 22.5 =
2,963 (Palladium) / 252.87 = 11.71748329181   252.87 / 2,963 = 0.085342

252.87 + 2,963 (Palladium) = 3,215.87 is [3 + 2 + 1 + 5 + 8 + 7 (26) + 7 + 8 +5 + 1 + 2 + 3] = 52
3,215.87 = 3 x 2 x 1 x 5 x 8 x 7 (1,680) x 7 x 8 x 5 x 1 x 2 x 3 = 2,822,400 / 3.333 = 864,720

2,822,400 / 5.555 = 508,032   2,822,400 / 6.666 = 423,360
2,822,400/ 144 = 19,600   3,215.87/ 144,000 = / 7.5 = 2.61

CARBON RIGHT ANGLE TRIANGLE MELTING POINT 4,926.85 C

4,926.85 (3 side) x 6,569.1333 (4 side) = 32,365,134.56333 (Area of square)

32,365,134.56333 / 2 = 16,182,567.281666 (Area of the Right Triangle)

16,182,567.281666 / 24 Hours = 674,273.636736111
674,273.636736111 / 60 Minutes = 11,237.89394560185
11,237.89394560185 / 60 Seconds = 187.2982324266975
187.2982324266975 / 60 Arc Seconds = 3.121637207111625

4,926.85 / 3 = 1,642.28333

1,642.28333

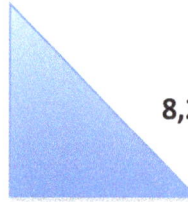

8,211.41666 = 5 x

6,569.1333 = 4 x 1,642.28333

CARBON SQUARE MELTING POINT 4,926.85 x 4,926.85 = 24,273,850.9225

24,273,850.9225 / 518,400 = 46.82455810667438
24,273,850.9225 / 311,040,000 = 78.4093017779064
24,273,850.9225 / 266.666 = 91,026.94095937502
24,273,850.9225 / 11.111 = 2,184,646.583025
24,273,850.9225 / 1.333 = 18,205,388.191875
24,273,850.9225 / 144 = 168.568.4091840278
24,273,850.9225 / 7.5 = 3,236,513.456333

4,926.85 x 4,926.85 (Square) = 24,273,850.9225

24,273,850.9225 / 2 = 12,136,9256.46125
12,136,9256.46125 / 24 Hours = 505,705.2275520833
505,705.2275520833 / 60 Minutes = 8,428.420459201389
8,428.420459201389 / 60 Seconds = 140.4736743200231
140.4736743200231 / 60 Arc Sec = 0.009528400205761

CARBON NANOTUBULE MELTING POINT 4,926.85 CUBED = 119,593,622,417.5191

119,593,622,417.5191 / 266.666 =
119,593,622,417.5191 / 11.111 =
119,593,622,417.5191 / 1.333 =
119,593,622,417.5191 / 144 =
119,593,622,417.5191 / 7.5 =
119,593,622,417.5191 / 24 =
119,593,622,417.5191 / 864 =
119,593,622,417.5191 / 432
119,593,622,417.5191 / 216
119,593,622,417.5191 / 108

1 year, 360 day Time in seconds cube 119,593,622,417.5191/ =

/ 24 = / 60 = / 60 = / 60 =
/ 3,456,000 = / 720 = / 16 = / 22.5 =

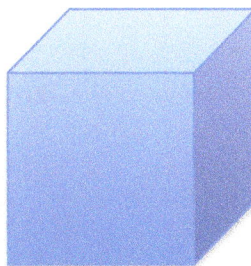

CARBON NANOTUBULES RATIOS TO HYDROGEN

CARBON NANOTUBULES 4,926.85 / HYDROGEN -259.2 = 19.007909
CARBON NANOTUBULES 4,926.85 x HYDROGEN -259.2 = 1,277,039.52
CARBON NANOTUBULES 4,926.85 + HYDROGEN -259.2 = 5,186.05
5 + 1 + 8 + 6 + 5 (25) 5 + 1 + 8 + 6 + 5 = 50
5 x 1 x 8 x 6 x 5 (1,200) x 5 x 1 x 8 x 6 x 5 = 1,440,000
1,440,000 / 50 = 28,800   1,440,000 x 50 = 72,000,000 / 2,160 = 33,333.333
72,000 / 7.5 = 9,600 / 24 = 400 / 60 = 6.666 /60 = 0.111 / 60 = 0.00185185185
72,000 / 24 = 3,000 / 60 = 50 / 60 = 0.8333 / 60 = 0.013888 x 400 = 5.555 / 0.125 = 44.444
72,000 x 720 = 51,840,000 / 864 = 60,000 / 144 = 416.666 / 720 = 0.57870370 / 16 = 0.036 / 22.5 = x
400 = /0.125 =

PALLADIUM MELTING POINT

MELTING POINT = 1,555.2 x 2 = 3,110.4 (Sanskrit long year TIME) 1,555.2 / 11.111 = 139.68
1 x 5 x 5 x 5 x 2 (250) x 2 x 5 x 5 x 5 x 1 = 62,500 (6.25 Pi square cubits 1/16 th sphere x 10,000)
62,500 + 250 + 250 = 63,000
The product of Palladium's BOILING POINT (2,963) is 46,656 divided by the melting point (1,555.2) is 62,500
equals 0.746496 and 62,500 / 46,656 = 1.333 x 2 = 2.666
1 + 5 + 5 + 5 + 2 (18) + 2 + 5 + 5 + 5 + 1 = 36

PALLADIUM RIGHT ANGLE TRIANGLE BOILING POINT 1,555.2

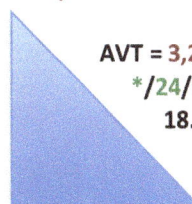

AVT = 3,224,862.72 /2 = 1,612,431.36*
*/24/60/60/60 = 18.6624
18.6624 /16 / 22.5 = 0.05184
*/720/16/22.5 = 6.2208 (648)

1,555.2 / 3 = 518.4   518.4 x 5 = 2,592

2,073.6 = 518.4 x 4

PALLADIUM SQUARE MELTING POINT 1,555.2 x 4 = 6,220.8 / 3,456 = 1.8 (1/x = 0.555)

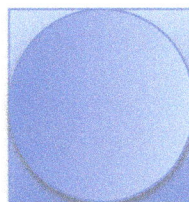

PALLADIUM CUBE MELTING POINT 1,555.2 CUBED = 3,761,479,876.608
1 yr = 360 days in Seconds = 31,104,000 Sec cubed = 30,091,839,012,860,000,000,000

30,091,839,012,860,000,000,000 / 3,761,479,876.6.608 = 8,000,000,000,000.
3,761,479,876.608 / 24 Hrs / 60 Min / 60 Sec / 60 Arc Sec / 7.5 = 96.7458816
3,761,479,876.608 / 3,456 = 1,088,391.168
8,000,000,000,000 / 7,200,000,000 = 1,111.111 / 16 = 69.444 / 22.5 =
3.08641975308641975 / 0.125 = 24.69135802469136 x 400 = 987.654320987654320

1 year Time in Seconds cubed /   3,761,479,876.608 =
8,000,000,000,000/24 = 333,333,333,333.333 /60 =   5,555,555,555.555 /60 =
92,592,592.592 /60 = 154,320.987654321   /   3.333 = 46,296.296296296

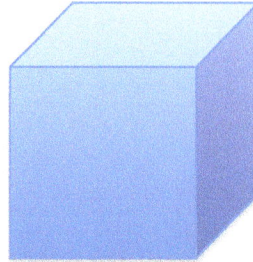

8,000,000,000,000 / 144,000 = 5,555,555.555/ 7.5 = 740,740.740 (1/x = 0.00000135)
740,740.740 / 720 = 1,028.80658436214 /16 = 64.3004115226337448559 /22.5 = 2.857796067672611
8,000,000,000,000 /129,600= 6,172,839,50617284/1.333/144,000=32.150205761316881/x=0.031104
8,000,000,000,000 / 266.666 = 3,000,000,000.
8,000,000,000,000 / 1.333 = 600,000,000,000
8,000,000,000,000 / 6.666 = 120,000,000,000
8,000,000,000,000 / 5.555 = 144,000,000,000
8,000,000,000,000 / 4.444 = 180,000,000,000
8,000,000,000,000 / 3.333 = 240,000,000,000
8,000,000,000,000 / 2.222 = 360,000,000,000
8,000,000,000,000 / 1.111 = 720,000,000,000 / 9 = 80,000,000,000

PALLADIUM BOILING POINT

BOILING POINT = 2,963 / 266.666 = 11.11125

2,963 (Palladium) so 2 + 9 + 6 + 3 (20) + 3 + 6 + 9 + 2 = 40
2,963 (Palladium) so 2 x 9 x 6 x 3 (216) 3 x 6 x 9 x 2 = 46,656 / 5.555 = 8,398.080

2,963 (Palladium) / 11.111 = 266.666 / 24 = 11.111
2,963 (Palladium) / 5.555 = 533.34   2,963 x 1.333 = 2,222.25   2,963 / 3.333 = 888.88

PALLADIUM RIGHT ANGLE TRIANGLE BOILING POINT 2,963

AT = 5,852,911.679*
*/24/60/60 = 67.74 / 16 = 4.23
*/720/16/22.5/ .125 = 180.064

2,963 / 3 = 987.666   987.666 x 5 = 4,938.333

3,950.666 = 987.666 x 4

PALLADIUM SQUARE BOILING POINT 2,963 x 4 = 11,852 / 3,456 = 3.429398148148

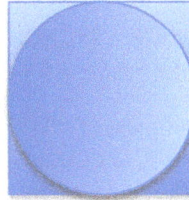

/
PALLADIUM CUBE BOILING POINT 2,963 CUBED = 26,013,270,347
1 year, 360 days in seconds 31,104,000 cubed 30,091,839,012,860,000,000,000
30,091,839,012,860,000,000,000 / 26,013,270,347 = 1,156,788,001,333.879
1,156,788,001,333.879/24/60/60 = 13,338,750.0154384
1,156,788,001,333.879/3,456,000 = 334,718.7503859605 /720/16/22.5 /.125 = 10.33

1 year Time in Seconds cube /   26,013,270,347 =
1,156,788,001,333.879
1,156,788,001,333.879 /24/60/60/60 =   223,145.8335906403
 223,145.8335906403 / 3.333 =   66,943.7500771921
1,156,788,001,333.879 / 144,000 / 7.5   =   1,071,100.001235073
/ 720 / 16 / 22.5 = 148.7638890604269
1,156,788,001,333.879 / 864,000 =   1,338,875.001543842
1,156,788,001,333.879 / 129,600 = 8,925,833.343625612
1,156,788,001,333.879 / 266.666 = 4,337,955,005.002049
1,156,788,001,333.879 / 6.666 = 173,518,200200.0819
1,156,788,001,333.879 / 5.555 = 208,221,840,240.0983
1,156,788,001,333.879 / 4.444 = 260,277,300,300.1229
1,156,788,001,333.879 / 3.333 = 347,036,400,400.1638
1,156,788,001,333.879 / 2.222 = 520,554,600,600.2458
1,156,788,001,333.879 / 1.111 = 1,041,109,201,200.492 / 9 = 115,678,800,133.3879
PLATINIUM - Boiling point = - 3,825

OXYGEN – Boiling point = -183
            - Melting point =

HYDROGEN – Boiling point = - 252.87
                - Melting point = -259.20

CARBON – MELTING POINT - CARBON SHEETS DIAMOND = 3,426.85
               - MELTING POINT - CARBON SHEETS GRAPHENE = 4,236.85
               - MELTING POINT - CARBON MATRIX GRAPHITE = 3,526.85
               - MELTING POINT CARBON FORMED SPHERES   = 1,249.85
               - MELTING POINT CARBON FORM NANOTUBULE = 4,926.85

HELIUM – Boiling point = -268.93
              - Melting point = -272.2

DEUTERIUM – Boiling point = 101.4

TUNGSTEN – Boiling point 5,555 (Sun Aether) / 11.111 = 500
                  5,555 / 266.666 = 20.83125

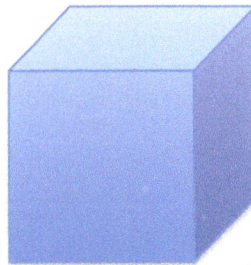

5 x 5 x 5 x 5 (625) x 5 x 5 x 5 x 5 = 390,625
390,625 + 625 + 625 = 391,875
5 + 5 + 5+ 5 (20) + 5 + 5 + 5 + 5 = 40

The product of Tungsten's boiling point (5,555) product is 390,625 divided by the MELTING POINT (3,420) product 576 is 678.168
MELTING POINT 3,420 (Sun's radius)

3 x 4 x 2 (24) 2 x 4 x 3 = 576 / 24 = 24
576 + 24 + 24 = 624 / 24 = 26
3 + 4 + 2 (9) + 2 + 4 + 3 = 18

GOLD – **Boiling point** 2,970 (Earth)
          - **Melting point 1,063**

SILVER – **Boiling point** 2,162 (Moon)

COPPER – **Boiling point** 2,562 (Earths Equinox)

BRASS - **Melting point** 900

RED BRASS – **Melting point 1,000**

IRON – **Boiling point** – **2,861**
          Melting point – **1,538**

CARBON – **Boiling point** - **4,827**
            – **Melting point** - **3,550**

NITROGEN – **Boiling point** – **195.8**

4.3.) OXYGEN CALCULATIONS – RESONANCE FOR THE OXIDATION OF HYDROCARBONS.

OXYGEN

OXYGEN Melting point = - 218.8 / 266.666 = **0.8205** [1/x = 1.218769043266301]
            and = -218.8 / 11.111 = **19.692** [1/x = 0.050782043469429]
            and = -218.8 / 1.333 = **164.1** [1/x = 0.006093845216332]
            and = -218.8 / 144 = **1.519444** [1/x = 0.658135283363803]

2 + 1 + 8 + 8 = (19) + 8 + 8 + 1 + 2 = 38

2 x 1 x 8 x 8 = (128) x 8 x 8 x 1 x 2 = 16,384

16,384 (Oxygen 218.8) / 22 = **2.909090**
16,384 (Oxygen 218.8) x 22 = **1,408**

16,384 (Oxygen 218.8) / 266.666 = **61.44** [1/x = 0.16276041666]
16,384 (Oxygen 218.8) / 11.111 = **1,474.56** [1/x = 0.000678168402778]

16,384 (Oxygen 218.8) / 1.333 = 122,288   [1/x = 0.000081380208333]
16,384 (Oxygen 218.8) / 144 = 113.777   [1/x = 0.0087890625]

OXYGEN RATIOS TO ALL OTHER NOBLE GASES INCLUDING HYDROGEN

OXYGEN RIGHT ANGLE TRIANGLE MELTING POINT – 218.8

-218.8 x 291.7333 = 63,831.25333
63,831.25333 / 2 = 31,915.62666
31,915.62666 / 24 Hours = 1,329.81 / 60 Minutes = 22.1636296296296
22.1636296296296 / 60 Seconds = 0.369393827160494
0.369393827160494 / 60 Arc Seconds = 0.006156563786008

-218.8 / 3 = 72.9333

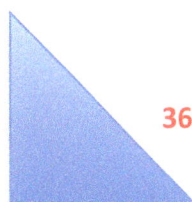

364.666 = 5 x 72.9333

291.7333 = 4 x 72.9333

4.3.) OXYGEN CALCULATIONS – RESONANCE FOR THE OXIDATION OF HYDROCARBONS.

OXYGEN

OXYGEN SQUARE MELTING POINT -218.8 x 4 = 875.2 / 3,456 = 0.25324074074074

-218.8

218.8

-

OXYGEN

OXYGEN CUBE MELTING POINT -218.8 CUBED = -10,474,708.672

1 year, 360 day Time in seconds cube 10,474,708.672 / 360 = 29,096.4129777
29,096.4129777 / 24 Hours = 1212.35054074074 / 60 Minutes = 20.20584234567901
20.20584234567901 / 60 Seconds = 0.33676403909465
0.33676403909465 / 60 Arc Seconds = 0.005612733984911
0.005612733984911 / 16 = 0.0003507958740570 / 22.5 = 0.000015509027736

-218.8

-218.8

-218.8

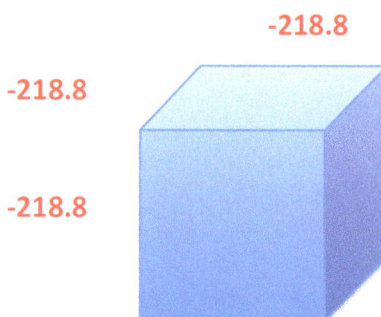

10,474,708.672 / 360 (1 year) Time in Seconds cube = 29,096.4129777
10,474,708.672 / 24 = 436,446.194666 /60 = 7,274.1-32444 /60= 121.235 /60= 2.00205842345679
10,474,708.672 / 144,000 = 72.7410324   / 24 = 3.03087635185 / 60 = / 60 = / 60 =
10,474,708.672 / 266.666 = 39,280.15   / 24 = / 60 = / 60 = / 60 =
10,474,708.672 / 6.666 = 1,571,206.3008   / 24 = / 60 = / 60 = / 60 =
10,474,708.672 / 5.555 = 1,885,447.56096 / 24 = / 60 = / 60 = / 60 =
10,474,708.672 / 4.444 = 2,356,8094512   / 24 = / 60 = / 60 = / 60 =
10,474,708.672 / 3.333 = 3,142,412.6016   / 24 = / 60 = / 60 = / 60 =
10,474,708.672 / 2.222 = 4,713,618.9024   / 24 = / 60 = / 60 = / 60 =
10,474,708.672 / 1.111 = 9,427,237.8048   / 24 = / 60 = / 60 = / 60 =

4.4.) HYDROCARBON MAIN FUEL TYPES DRY GAS, WET GAS, PETROL AND DIESEL.

4.5.) SUMMARY

5.) *AMMONIA HYDROXIDE NH4 (OH) – NITROGEN, HYDROGEN & OXYGEN.*

5.1.) NITROGEN CALCULATIONS.

NITROGENS MELTING POINT = - 210 so - 210 / 266.666 = 0.7875   1/x = 1.26984126984

- 210 / 11.111 = 18.9   1/x = 0.05291005291005291

- 210 / 1.333 = 157.5   1/x = 0.00634920634920634920

-210 added sum   = 2 +1 = [3] + 2 + 1 = 6

-210 multiplied product = 2 x 1 = (2) x 2 x 1 = 4

NITROGENS RIGHT ANGLE TRIANGLE MELTING POINT – 210

-210 x 280 = 58,800
58,800 / 2 = 29,400
29,400 / 24 Hours = 1,225 / 60 Minutes = 20.41666
20.41666 / 60 Seconds = 0.3402777
0.3402777 / 60 Arc Seconds = 0.005671296296296   1/x = 176.326530612244

-210 / 3 = 70

350 = 5 x 70

280 = 4 x 70

NITROGENS SQUARE MELTING POINT -210 x 4 = 840 / 3,456 = 0.2430555
1/x = 4.114228571

-210

NITROGENS BOILING POINT 195.8 C SO 195.8 / 266.666 = 0.73425   1/x = 1.36193394620

195.8 added sum

195.8 multiplied product

5..2.)   HYDROGEN CALCULATIONS.

4.)   HYDROCARBONS H(x)C(y) – HYDROGEN AND CARBON.

4.1.)   HYDROGEN CALCULATIONS.
HYDROGEN

HYDROGEN MELTING POINT = − 259.2 so - 259.2 / 266.666 = 0.972
- 259.2 / 11.111 = 23.328
- 259.2 / 1.333 = 194.4 // 97.2//48.6//24.3

- 259.2 added sum   = 2 + 5 + 9 + 2 [18] + 2 + 5 + 9 + 2 = 36

- 259.2 multiplied product = 2 x 5 x 9 x 2 (180) x 2 x 9 x 5 x 2 = 32,400

32,400 / 720 = 45   32,400 / 16 = 2,025   32,400 / 22.5 = 1,440   32,400 / 0.125 = 259,200

32,400 x 16 = 518,400   32,400 x 22.5 = 729,000   32,400 x .125 = 4,050

32,400 x 400 = 12,960,000   32,400 x 720 = 23,328,000   32,400 x 900 = 29,160,000

518,400 / 16 = 32,400   518,400 / 22.5 = 23,040   518,400 / 720 = 720

32,400 / 144 = 225   32,400 / 36 = 900   36 / 32,400 = 0.00111

144,000 / 32,400 = 4.4444   32,400 / 7.5 = 4,320 / 7.5 = 576 / 2 = 288 / 2 = 144

144 x 32,400 = 4,665,600 / 432 = 10,800 / 2,160 = 5

32,400 / 259.2 = 125
32,400 / 6.666 = 4,860   32,400 / 5.555 = 5,850   32,400 / 4.444 = 7,290
32,400 / 3.333 = 9,720   32,400 / 2.222 = 14,580   32,400 / 11.111 = 2,916

HYDROGEN RATIOS TO ALL OTHER NOBLE GASES

32,400 (Hydrogen -259.2) / 3,136 (Helium -272.2) = 10.33163265306122
3,136 (Helium -272.2) / 32,400 (Hydrogen -259.2) = 0.09679012345679

32,400 (Hydrogen -259.2) / 147,456 (Neon -248.6) = 0.2197265625
147,456 (Neon -248.6) / 32,400 (Hydrogen -259.2) = 4.55111

32,400 (Hydrogen -259.2) / 82,944 (Argon -189.4) = 0.390625
82,944 (Argon -189.4) / 32,400 (Hydrogen -259.2) = 2.56

32,400 (Hydrogen -259.2) / 19,600 (Krypton 157.4) = 1.653061224489796
19,600 (Krypton 157.4) / 32,400 (Hydrogen -259.2) = 0.604938271604938

32,400 (Hydrogen -259.2) / 64 (Xeon -111.8) = 506.25
64 (Xeon -111.8) / 32,400 (Hydrogen -259.2) = 0.00195308641975

32,400 (Hydrogen -259.2) / 1,225 (Radon -71.15) = 26.44897959183673
1,225 (Radon -71.15)/ 32,400 (Hydrogen -259.2) = 0.037808641975309 Argon/Helium

ALL SIX 864,000 COMBINATIONS

864,000	846,000	684,000	648,000	486,000	468,000
56,623,104*	55,443,456*	44,826,624*	42,467,328*	31,850,496*	30,670,848*
28,311,552*	27,721,728*	22,413,312*	21,233,664*	15,925,248*	15,335,424*
14,155,776*	13,860,864*	11,206,656*	10,616,832*	7,962,624*	7,667,712*
7,077,888*	6,930,432*	5,603,328*	5,308,416*	3,981,312*	3,833,856*
3,538,944*	3,465,216*	2,801,664*	2,654,208*	1,990,656*	1,916,928*
1,769,472*	1,732,608*	1,400,832*	1,327,104*	995,328*	958,464*
884,736*	866,304*	700,416*	663,552*	497,664*	479,232*
442,368*	433,152*	350,208*	331,776*	248,832*	239,616*
221,184*	216,576*	175,104*	165,888*	124,416	119,808*

110,592*	108,288*	87,552*	82,944*	62,208*	59,904*
55,296*	54,144*	43,776*	41,472*	31,104*	29,952*
27,684*	27,072*	21,888*	20,736*	15,552*	14,976*
13,824*	13,536*	10,944*	10,368*	7,776*	7,488*
6,912,000	6,768,000	5,472,000	5,184,000	3,888,000	3,744,000
3,456,000	3,384,000	2,736,000	2,592,000	1,944,000	1,872,000
1,728,000	1,692,000	1,368,000	1,296,000	972,000	936,000

864,000	846,000	684,000	648,000	486,000	468,000

432,000	423,000	342,000	H-324,000	243,000	234,000
216,000	211,500	171,000	162,000	121,500	117,000
108,000	105,000	85,500	81,000	60,750	58,500
54,000	52,875	42,750	40,500	30,375	29,250
27,000	26,437.5	21,375	20,250	15,187.5	14,625
13,500	13,218.75	10,687.5	10,125	7,593.75	7,312.5
6,750	*6,609.375*	*5,343.75*	*5,062.5*	*3,796.875*	
3,656.25					
3,375	3,304.6875	2,671.875	2,531.25	1,898.43	1,828.1
1,687.5	1,652.34375	1,335.9375	1,265.62	949.2187	914.06
843.75	826.171875	667.96875	632.812	474.6093	457.03
421.875	413.0859375	333.984375	316.406	237.3046	228.51
210.9375	206.54296875	166.99218	158.2031	118.6523	114.25
105.46875	103.271484375	83.496093	79.10156	59.32617	57.128
52.734375	51.6357421875	41.7480468	39.55078	29,66308	28.564
26.3671875	25.8187109375	20.8740234	19.77539	14.83154	14.282

864,000	846,000	684,000	648,000	486,000	468,000

HYDROGEN RATIOS TO ALL OTHER NOBLE GASES

32,400 (Hydrogen -259.2) / 3,136 (Helium -272.2) = 10.33163265306122
3,136 (Helium -272.2) / 32,400 (Hydrogen -259.2) = 0.09679012345679

32,400 (Hydrogen -259.2) / 147,456 (Neon -248.6) = 0.2197265625
147,456 (Neon -248.6) / 32,400 (Hydrogen -259.2) = 4.55111

32,400 (Hydrogen -259.2) / 82,944 (Argon -189.4) = 0.390625
82,944 (Argon -189.4) / 32,400 (Hydrogen -259.2) = 2.56

32,400 (Hydrogen -259.2) / 19,600 (Krypton 157.4) = 1.653061224489796
19,600 (Krypton 157.4) / 32,400 (Hydrogen -259.2) = 0.604938271604938

32,400 (Hydrogen -259.2) / 64 (Xeon -111.8) = 506.25
64 (Xeon -111.8) / 32,400 (Hydrogen -259.2) = 0.00195308641975

32,400 (Hydrogen -259.2) / 1,225 (Radon -71.15) = 26.44897959183673
1,225 (Radon -71.15)/ 32,400 (Hydrogen -259.2) = 0.037808641975309 Argon/Helium

HYDROGEN RIGHT TRIANGLE MELTING POINT = - 259.2 C (Earths Equinox 25,920)

864,000 / 259.2 = 3,333.3   864,000 / 345.6 = 2,500   864,000/ 432 = 2,000
2,160 / 259.2 = 8.333   259.2 / 2,160 = 0.972

Triangles Area / Time / Space calculation

259.2 x -345.6 x 432 = 38,698,352.6
259.2 x -345.6 = - 89,579.5 (Area of the square) / 2 = 44,789.76 (Area of the Right Triangle)

44,789.76 / 24 Hrs = 1,866.24   1,866.24 / 60 Min = 31.104   31.104 / 60 Sec = 0.5184
0.5184 / 60 Arc = 0.00864   0.00864 / 60 = 0.000144   0.000144 / 60 = 0.0000024
0.0000024 / 60 = 0.00000004   0.00000004 / 24 = 0.000000001666
0.00000004 / 60 = 0.0000000000666   0.0000000000666 / 60 = 0.00000000001111

0.5184 / 7.5 = 0.06912 / 0.266666 = 0.2592 x 400 = 103.68 / 16 = 6.48 / 22.5 = 0.288
0.288/ 0.125 = 2.304 x 400 = 921.9 / 1.333 = 691.2 / 1.333 = 518.4 /1.333 = 388.8 / 1.333 =
291.6

-259.2 / 3 = - 86.4

86.4 x 5 = 432

- 345.6 = - 86.4 x 4

HYDROGEN SQUARE MELTING POINT -259.2 (Earths Equinox) x 4 = 1,036.8

1,036.8/3,456 = 0.3
259.2 x 259.2 = 67,184.64 / 11.111 = 6,046.6   67,184.64 / 266.666 = 251.9424
67,184.64 / 144 = 466.56
67.184.64 / 7.5 = 8,957.952
67,184.64 / 2 = 33,592.32
33,592.32 / 3 = 11,197.44   11,197.44 x 4 = 44,789.76   11,197.44 x 5 = 55,987.2

HYDROGEN CUBE MELTING POINT CUBED = 259.2 x 259.2 x 259.2 = 17,414,258.688
1 year, 360 day Time in seconds cube 17,414,258.688 / 360 = 48,372.9408
48,372.9408/24 = 2,016/60 = 33.6/60 = 0.56/ 720 = 0.000777/ 16 = 0.0000486111/ 22.5 =
0000002160493827

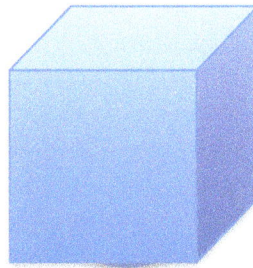

17,414,258.688 / 144,000 = 120.932352 / 7.5 = 16.1243136 / 16 = 1.0077696
 = 1.0077696 / 22.5 = 0.04478976
17,414,258.688 / 1.333 = 13,060,694.016
17,414,258.688 / 266.666 = 65,303.47008000002
17,414,258.688 / 6.666 = 2,612,138.8032
17,414,258.688 / 5.555 = 3,134,566.56384
17,414,258.688 / 4.444 = 3,918,208.2048
17,414,258.688 / 3.333 = 5,224,277.606400001
17,414,258.688 / 2.222 = 7,836,416.409600001
17,414,258.688 / 1.111 = 15,672,832.8192 / 9 = 1,741,425.8688

HYDROGEN RIGHT ANGLE TRIANGLE BOILING POINT = - 252.8 C

HYDROGEN Boiling point = - 252.87 so -252.87 / 266.666 = .9482625
                              - 252.87 so -252.87 / 11.111 = 22.7583
                              - 252.87 so -252.87 / 1.333 = 189.699925

252.87 = 2 + 5 + 2 + 8 + 7 (24) +7 + 8 + 2 + 5 + 2 = 48

252.87 = 2 x 5 x 2 x 8 x 7 (1,120) x 7 x 8 x 2 x 5 x 2 = 1,254,400 / 5.555 = 225,792

252.8 x 337.0666 = 85,210.4365 therefore triangle area 85,210.4365 / 2 = 42,605.2182
42,605.2182 / 24 Hours = 1,775.21743   1,775.21743 / 60 Minutes = 29.5869571
29.5869571 / 60 Seconds = 0.493115951
0.493115951 / 60 Arc seconds = 0.0821859918333

- 252.8 / 3 = 84.2666

421.333 = 84.2666 x 5

337.0666 = 84.2666 x 4

5.3.) *OXYGEN CALCULATIONS.*

OXYGEN

OXYGEN Melting point = - 218.8 / 266.666 = 0.8205   [1/x = 1.218769043266301]
              and = -218.8 / 11.111 = 19.692   [1/x = 0.050782043469429]
              and = -218.8 / 1.333 = 164.1   [1/x = 0.006093845216332]

and = -218.8 / 144 = 1.519444   [1/x = 0.658135283363803]

2 + 1 + 8 + 8 = (19) + 8 + 8 + 1 + 2 = 38

2 x 1 x 8 x 8 = (128) x 8 x 8 x 1 x 2 = 16,384

16,384 (Oxygen 218.8) / 22 = 2.909090
16,384 (Oxygen 218.8) x 22 = 1,408

16,384 (Oxygen 218.8) / 266.666 = 61.44   [1/x = 0.16276041666]
16,384 (Oxygen 218.8) / 11.111 = 1,474.56   [1/x = 0.000678168402778]
16,384 (Oxygen 218.8) / 1.333 = 122,288   [1/x = 0.000081380208333]
16,384 (Oxygen 218.8) / 144 = 113.777   [1/x = 0.0087890625]

OXYGEN RATIOS TO ALL OTHER NOBLE GASES INCLUDING HYDROGEN

OXYGEN RIGHT ANGLE TRIANGLE MELTING POINT – 218.8

-218.8 x 291.7333 = 63,831.25333
63,831.25333 / 2 = 31,915.62666
31,915.62666 / 24 Hours = 1,329.81 / 60 Minutes = 22.1636296296296
22.1636296296296 / 60 Seconds = 0.369393827160494
0.369393827160494 / 60 Arc Seconds = 0.006156563786008

-218.8 / 3 = 72.9333

364.666 = 5 x 72.9333

291.7333 = 4 x 72.9333

OXYGEN SQUARE MELTING POINT -218.8 x 4 = 875.2 / 3,456 = 0.25324074074074

-218.8

5.3.) *OXYGEN CALCULATIONS – RESONANCE FOR THE OXIDATION*

OXYGEN

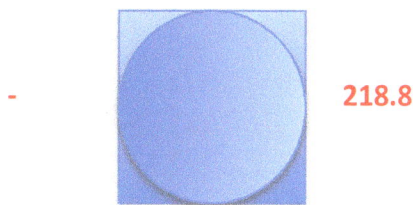

- 218.8

OXYGEN

OXYGEN CUBE MELTING POINT -218.8 CUBED = -10,474,708.672

1 year, 360 day Time in seconds cube 10,474,708.672 / 360 = 29,096.4129777
29,096.4129777 / 24 Hours = 1212.35054074074 / 60 Minutes = 20.20584234567901
20.20584234567901 / 60 Seconds = 0.33676403909465
0.33676403909465 / 60 Arc Seconds = 0.005612733984911
0.005612733984911 / 16 = 0.0003507958740570 / 22.5 = 0.000015509027736

-218.8

-218.8

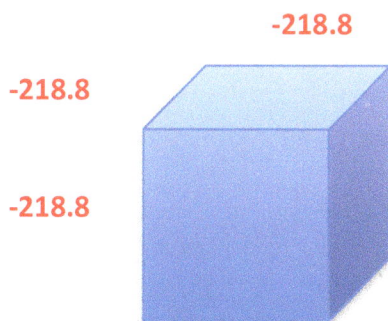

-218.8

10,474,708.672 / 360 (1 year) Time in Seconds cube = 29,096.4129777
10,474,708.672 / 24 = 436,446.194666 /60 = 7,274.1-32444 /60= 121.235 /60= 2.00205842345679
10,474,708.672 / 144,000 = 72.7410324   / 24 = 3.03087635185 / 60 = / 60 = / 60 =
10,474,708.672 / 266.666 = 39,280.15   / 24 = / 60 = / 60 = / 60 =
10,474,708.672 / 6.666 = 1,571,206.3008   / 24 = / 60 = / 60 = / 60 =
10,474,708.672 / 5.555 = 1,885,447.56096 / 24 = / 60 = / 60 = / 60 =
10,474,708.672 / 4.444 = 2,356,8094512   / 24 = / 60 = / 60 = / 60 =
10,474,708.672 / 3.333 = 3,142,412.6016   / 24 = / 60 = / 60 = / 60 =
10,474,708.672 / 2.222 = 4,713,618.9024   / 24 = / 60 = / 60 = / 60 =
10,474,708.672 / 1.111 = 9,427,237.8048   / 24 = / 60 = / 60 = / 60 =

5.4.) SUMMARY

6.) *316 STAINLESS STEEL -   Fe 63% Cr 18% Ni 14% Mo 3% & Mn 2%.*

6.1.)   IRON (Fe) CALCULATIONS.

IRON – Melting point – 1,538

Melting point =   / 266.666 =   [1/x =]
                      and =   / 11.111 =   [1/x =]
                      and =   / 1.333 =   [1/x =]
                      and =   / 144 =   [1/x =]

+ + + = () + + + + =

x x x = () x x x x =

 (O) / 22 =
 (O) x 22 =

(O) / 266.666 =   [1/x =]
(O) / 11.111 =   [1/x =]
(O) / 1.333 =   [1/x =]
(O) / 144 =   [1/x =]

6.2.)   CHROMIUM (Cr) CALCULATIONS.

CHROMIUM (Cr) – Melting point – 1,538

Melting point =   / 266.666 =   [1/x =]
                      and =   / 11.111 =   [1/x =]
                      and =   / 1.333 =   [1/x =]
                      and =   / 144 =   [1/x =]

+ + + = () + + + + =

x x x = () x x x x =

 (O) / 22 =
 (O) x 22 =

(O) / 266.666 =   [1/x =]
(O) / 11.111 =   [1/x =]
(O) / 1.333 =   [1/x =]
(O) / 144 =   [1/x =]

5.3.)   NICKLE (Ni) CALCULATIONS.

NICKLE (Ni) – Melting point –

Melting point = / 266.666 = [1/x =]
and = / 11.111 = [1/x =]
and = / 1.333 = [1/x =]
and = / 144 = [1/x =]

+ + + = () + + + + =

x x x = () x x x x =

(O) / 22 =
(O) x 22 =

(O) / 266.666 = [1/x =]
(O) / 11.111 = [1/x =]
(O) / 1.333 = [1/x =]
(O) / 144 = [1/x =]

5.4.) *MOLYBDENUM (Mo) CALCULATIONS*.

MOLYBDENUM (Mo) – Melting point –

Melting point = / 266.666 = [1/x =]
and = / 11.111 = [1/x =]
and = / 1.333 = [1/x =]
and = / 144 = [1/x =]

+ + + = () + + + + =

x x x = () x x x x =

(O) / 22 =
(O) x 22 =

(O) / 266.666 = [1/x =]
(O) / 11.111 = [1/x =]
(O) / 1.333 = [1/x =]
(O) / 144 = [1/x =]

5.5.) *MANGANESE (Mn) CALCULATIONS.*

MOLYBDENUM (Mo) – Melting point – 1,538

Melting point =   / 266.666 =   [1/x =]
                           and = / 11.111 =   [1/x =]
                           and = / 1.333 =   [1/x =]
                           and = / 144 =   [1/x =]

+ + + = () + + + + =

x x x = () x x x x =

 (O) / 22 =
 (O) x 22 =

(O) / 266.666 =   [1/x =]
(O) / 11.111 =   [1/x =]
(O) / 1.333 =   [1/x =]
(O) / 144 =   [1/x =]

5.6.) SUMMARY

PART 7.) ARGON

ARGON RIGHT ANGLE TRIANGLE MELTING POINT − 189.4

- 189.4 (3 side) x 252.5333 (4 side) = 47,829.81333 (Area of Square)
47,829.81333 / 2 = 23,914.90666 (Area of the Right Triangle)
23,914.90666 / 24 Hours = 996.45444
996.45444 / 60 Minutes = 16.60757407407
16.60757407407 / 60 Seconds = 0.276792901234568
0.276792901234568 / 60 Arc Seconds = 0.004613215020576

189.4 / 3 = 63.1333

315.666 = 5 x 63.1333

252.5333 = 4 x 63.1333

1 + 8 + 9 + 4 (22) + 4 + 9 + 8 + 1 = 44   1

1 x 8 x 9 x 4 (288) x 4 x 9 x 8 x 1 = 82,944

82,944 (Argon - 189.4) / 44 = 1,885.0909
82,944 (Argon - 189.4) x 44 = 3,649,536
82,944 (Argon - 189.4) / 266.666 = 311.0400000000001
82,944 (Argon - 189.4) / 11.111 = 7,464.960000000001
82,944 (Argon - 189.4) / 1.333 = 62,208.00000000002
82,944 (Argon - 189.4) / 144 = 576 //288//144//72//36//16//8//4//2//1

ARGON RATIOS TO ALL OTHER NOBLE GASES INCLUDING HYDROGEN

82,944 (Argon- 189.4) / 32,400 (Hydrogen – 259.2) = 2.56
2.56// 1.28//0.64// 0.32//0.16//0.08
32,400 (Hydrogen – 259.2) / 82,944 (Argon - 189.4)= 0.390625

82,944 (Argon - 189.4) / 3,136 (Helium - 272.2) = 26.44897959183673
3,136 (Helium - 272.2) / 82,944 (Argon - 189.4) = 0.037808641975309
        0.037808641975309 //+8 = 9.679012345679012

82,944 (Argon - 189.4) / 147,456 (-248.6 Neon) = 0.000140625//+5 = 0.144
147,456 (-248.6 Neon) / 82,944 (Argon - 189.4) = 1.777
1.777//0.888/0.444/0.222/0.111/0.0555

82,944 (Argon - 189.4) / 19,600 (Krypton 157.4) = 4.231836734693878
19,600 (Krypton 157.4) / 82,944 (Argon - 189.4) = 0.236304012345679 //+5 =3.7808

82,944 (Argon - 189.4) / 64 (Xeon 111.8) = 1,296
64 (Xeon 111.8) / 82,944 (Argon - 189.4) = 0.000771604938272

82,944 (Argon - 189.4) / 1,225 (Radon -71.15) = 67.70938775510204
1,225 (Radon -71.15) / 82,944 (Argon - 189.4) = 0.0147690007716

7.1.)

7.2.)

7.3.)

7.4.)

7.5.)

PART 8.) DESIGNER GAS - H, He, Ne, Ar, Kr, Xe, & Rn.

8.1.) PROTIUM [HYDROGEN (H)] – 259.2

PROTIUM [MEMBER OF THE HYDROGEN GROUP] MELTING POINT = – 259.2

– 259.2 / 266.666 = 0.972 (EARTH)
– 259.2 / 11.111 = 23.328 (SUN)
– 259.2 / 1.333 = 194.4 // 97.2//48.6//24.3// (POS)

– 259.2 / 2.4 = 108   – 259.2 / 51.84 = 5   – 259.2 / 144 = 1.8   – 259.2 / .432 = 600
– 259.2 / 79.2 = 3.272727   – 259.2 / 16 = 16.2   – 259.2 / 108 = 2.4   – 259.2 / 12 = 21.6
– 259.2 / 16 = 16.2   – 259.2 / 7.5 = 34.56

864,000 / – 259.2 = 3,333.333   864,000 / 3,136 = 275.5   864,000/ 26 = 33,230.769230
3,456 / – 259.2 = 13.333   3,456 / 108 = 32   3,456 / 144 = 24   3,456 / 12 = 288
3,456 / 16 = 216   3,456 / 36 = 96   3,456 / 72 = 48   3,456 / 900 = 3.84   3,456 / 22.5 = 153.6
2,160 / – 259.2 = 0.8333   – 259.2 / 2,160 = 0.12
 – 259.2 / 7.5 = 34.56 (SUN SQUARE)   34.56 / 7.5 = 4.608 (SUN)
– 259.2 x 7.5 = 1,944 (POS)   1,944 (POS) x 7.5 = 14,580 (518,400 TIME)

864,000 / – 259.2 = 3.333   864,000 / 345.6 = 2,500   864,000/ 432 = 2,000 (MS BATH)
2,160 / – 259.2 = 8.333 (1.333 = 16 / 12)   – 259.2 / 2,160 = 0.12

– 259.2 / 9 = 288   – 259.2 / 8.888 = 29.16 (RFEU)   – 259.2 / 7.777 = 33.3257
– 259.2 / 6.666 = 38.88   – 259.2 / 5.555 = 46.656   – 259.2 / 4.444 = 58.32
– 259.2 / 3.333 = 77.76   – 259.2 / 2.222 = 116.64   – 259.2 / 11.111 = 23.328

- 259.2 added sum   = 2 + 5 + 9 + 2 [18] + 2 + 5 + 9 + 2 = 36

- 259.2 multiplied product = 2 x 5 x 9 x 2 (180) x 2 x 9 x 5 x 2 = 32,400

32,400 / 720 = 45   32,400 / 16 = 2,025   32,400 / 22.5 = 1,440   32,400 / 0.125 = 259,200

32,400 x 16 = 518,400   32,400 x 22.5 = 729,000   32,400 x .125 = 4,050

32,400 x 400 = 12,960,000   32,400 x 720 = 23,328,000   32,400 x 900 = 29,160,000

518,400 / 16 = 32,400   518,400 / 22.5 = 23,040   518,400 / 720 = 720

32,400 / 144 = 225   32,400 / 36 = 900   36 / 32,400 = 0.00111

144,000 / 32,400 = 4.4444   32,400 / 7.5 = 4,320 / 7.5 = 576 / 2 = 288 / 2 = 144

144 x 32,400 = 4,665,600 / 432 = 10,800 / 2,160 = 5

32,400 / 259.2 = 125

32,400 / 6.666 = 4,860   32,400 / 5.555 = 5,850   32,400 / 4.444 = 7,290

32,400 / **3.333** = 9,720   32,400 / **2.222** = 14,580   32,400 / **11.111** = 2,916

ALL SIX 864,000 COMBINATIONS

864,000	846,000	684,000	648,000	486,000	468,000
56,623,104*	55,443,456*	44,826,624*	42,467,328*	31,850,496*	30,670,848*
28,311,552*	27,721,728*	22,413,312*	21,233,664*	15,925,248*	15,335,424*
14,155,776*	13,860,864*	11,206,656*	10,616,832*	7,962,624*	7,667,712*
7,077,888*	6,930,432*	5,603,328*	5,308,416*	3,981,312*	3,833,856*
3,538,944*	3,465,216*	2,801,664*	2,654,208*	1,990,656*	1,916,928*
1,769,472*	1,732,608*	1,400,832*	1,327,104*	995,328*	958,464*
884,736*	866,304*	700,416*	663,552*	497,664*	479,232*
442,368*	433,152*	350,208*	331,776*	248,832*	239,616*
221,184*	216,576*	175,104*	165,888*	124,416	119,808*
110,592*	108,288*	87,552*	82,944*	62,208*	59,904*
55,296*	54,144*	43,776*	41,472*	31,104*	29,952*
27,684*	27,072*	21,888*	20,736*	15,552*	14,976*
13,824*	13,536*	10,944*	10,368*	7,776*	7,488*
6,912,000	6,768,000	5,472,000	5,184,000	3,888,000	3,744,000
3,456,000	3,384,000	2,736,000	2,592,000	1,944,000	1,872,000
1,728,000	1,692,000	1,368,000	1,296,000	972,000	936,000

864,000	846,000	684,000	648,000	486,000	468,000
432,000	423,000	342,000	H-324,000	243,000	234,000
216,000	211,500	171,000	162,000	121,500	117,000
108,000	105,000	85,500	81,000	60,750	58,500
54,000	52,875	42,750	40,500	30,375	29,250
27,000	26,437.5	21,375	20,250	15,187.5	14,625
13,500	13,218.75	10,687.5	10,125	7,593.75	7,312.5
6,750	6,609.375	5,343.75	5,062.5	3,796.875	
3,656.25					
3,375	3,304.6875	2,671.875	2,531.25	1,898.43	1,828.1
1,687.5	1,652.34375	1,335.9375	1,265.62	949.2187	914.06
843.75	826.171875	667.96875	632.812	474.6093	457.03
421.875	413.0859375	333.984375	316.406	237.3046	228.51
210.9375	206.54296875	166.99218	158.2031	118.6523	114.25
105.46875	103.271484375	83.496093	79.10156	59.32617	57.128
52.734375	51.6357421875	41.7480468	39.55078	29,66308	28.564
26.3671875	25.8187109375	20.8740234	19.77539	14.83154	14.282

PART 9 OF 20 – MSAART PATENT APPLICATION DRAFT 518,400 KMV– 22:22:22 THUR 22ND SEPT 2022
© STRIKE FOUNDATION GUARANTEE LIMITED | MALCOLM BENDALL 2022 | GRAPHICS - STEVE EARL

864,000   846,000   684,000   648,000 486,000 468,000

8.1.) *PROTIUM [HYDROGEN (H)]*

HYDROGEN RATIOS TO ALL OTHER NOBLE GASES

32,400 (Hydrogen -259.2) / 3,136 (Helium -272.2) = **10.33163265306122**
3,136 (Helium -272.2) / 32,400 (Hydrogen -259.2) = **0.09679012345679**

32,400 (Hydrogen -259.2) / 147,456 (Neon -248.6) = **0.2197265625**
147,456 (Neon -248.6) / 32,400 (Hydrogen -259.2) = **4.55111**

32,400 (Hydrogen -259.2) / 82,944 (Argon -189.4) = **0.390625**
82,944 (Argon -189.4) / 32,400 (Hydrogen -259.2) = **2.56**

32,400 (Hydrogen -259.2) / 19,600 (Krypton 157.4) = **1.653061224489796**
19,600 (Krypton 157.4) / 32,400 (Hydrogen -259.2) = **0.604938271604938**

32,400 (Hydrogen -259.2) / 64 (Xeon -111.8) = **506.25**
64 (Xeon -111.8) / 32,400 (Hydrogen -259.2) = **0.00195308641975**

32,400 (Hydrogen -259.2) / 1,225 (Radon -71.15) = **26.44897959183673**
1,225 (Radon -71.15)/ 32,400 (Hydrogen -259.2) = **0.037808641975309 Argon/Helium**

HYDROGEN RIGHT TRIANGLE MELTING POINT = - 259.2 C (Earths Equinox **25,920**)

864,000 / **259.2** = 3,333.3   **864,000** / **345.6** = 2,500   **864,000/ 432** = 2,000
2,160 / **259.2** = 8.333   **259.2** / **2,160** = 0.972

Triangles Area / Time / Space calculation

259.2 x *-345.6* x *432* = 38,698,352.6
259.2 x *-345.6* = - 89,579.5 (Area of the square) / 2 = *44,789.76* (Area of the Right Triangle)

44,789.76 / *24* Hrs = *1,866.24*   *1,866.24* / *60* Min = *31.104*   *31.104* / *60* Sec = *0.5184*
0.5184 / *60* Arc = *0.00864*   *0.00864* / *60* = *0.000144*   *0.000144* / *60* = *0.0000024*
0.0000024 / *60* = *0.00000004*   *0.00000004* / *24* = *0.000000001666*
0.00000004 / *60* = *0.0000000000666*   *0.0000000000666* / *60* = *0.00000000001111*

0.5184 / **7.5** = 0.06912 / **0.266666** = **0.2592** x 400 = 103.68 / 16 = **6.48** / 22.5 = **0.288**
0.288/ 0.125 = **2.304** x 400 = 921.9 / 1.333 = **691.2** / 1.333 = **518.4** /1.333 = 388.8 / 1.333 =
291.6

-259.2 / 3 = - 86.4

86.4 x 5 = 432

- 345.6 = - 86.4 x 4

8.1.) PROTIUM [HYDROGEN (H)]

HYDROGEN SQUARE MELTING POINT -259.2 (Earths Equinox) x 4 = 1,036.8

1,036.8/3,456 = 0.3
259.2 x 259.2 = 67,184.64 / 11.111 = 6,046.6   67,184.64 / 266.666 = 251.9424
67,184.64 / 144 = 466.56
67.184.64 / 7.5 = 8,957.952
67,184.64 / 2 = 33,592.32
33,592.32 / 3 = 11,197.44   11,197.44 x 4 = 44,789.76   11,197.44 x 5 = 55,987.2

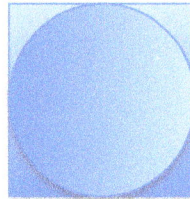

HYDROGEN CUBE MELTING POINT CUBED = 259.2 x 259.2 x 259.2 = 17,414,258.688
1 year, 360 day Time in seconds cube 17,414,258.688 / 360 = 48,372.9408
48,372.9408/24 = 2,016/60 = 33.6/60 = 0.56/ 720 = 0.000777/ 16 = 0.0000486111/ 22.5 = 0000002160493827

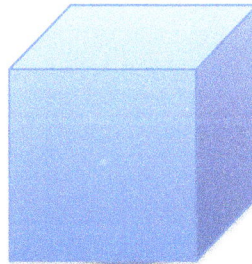

17,414,258.688 / 144,000 = 120.932352 / 7.5 = 16.1243136 / 16 = 1.0077696
 = 1.0077696 / 22.5 = 0.04478976
17,414,258.688 / 1.333 = 13,060,694.016
17,414,258.688 / 266.666 = 65,303.47008000002
17,414,258.688 / 6.666 = 2,612,138.8032
17,414,258.688 / 5.555 = 3,134,566.56384
17,414,258.688 / 4.444 = 3,918,208.2048
17,414,258.688 / 3.333 = 5,224,277.606400001
17,414,258.688 / 2.222 = 7,836,416.409600001
17,414,258.688 / 1.111 = 15,672,832.8192 / 9 = 1,741,425.8688

8.1.) *PROTIUM [HYDROGEN (H)]*

HYDROGEN RIGHT ANGLE TRIANGLE BOILING POINT = - 252.8 C

HYDROGEN Boiling point = - 252.87 so -252.87 / 266.666 = .9482625
- 252.87 so -252.87 / 11.111 = 22.7583
- 252.87 so -252.87 / 1.333 = 189.699925

252.87 = 2 + 5 + 2 + 8 + 7 (24) +7 + 8 + 2 + 5 + 2 = 48

252.87 = 2 x 5 x 2 x 8 x 7 (1,120) x 7 x 8 x 2 x 5 x 2 = 1,254,400 / 5.555 = 225,792

1,254,400 / 48 = 2,612.5

252.8 x 337.0666 = 85,210.4365 therefore triangle area 85,210.4365 / 2 = 42,605.2182
42,605.2182 / 24 Hours = 1,775.21743   1,775.21743 / 60 Minutes = 29.5869571
29.5869571 / 60 Seconds = 0.493115951
0.493115951 / 60 Arc seconds = 0.0821859918333

- 252.8 / 3 = 84.2666

421.333 = 84.2666 x 5

337.0666 = 84.2666 x 4

8.2.) HELIUM (He) -272.2 C

HELIUM MELTING POINT = -272.2 C

-272.2 / 266.666 = 1.02075   -272.2 / 11.111 = 24.498   -272.2 / 24 = 11.341666
-272.2 / 1.333 = 204,15   -272.2 / 51.84 = 5.25   -272.2 / 144 = 1.8902
-272.2 / 25.92 = 10.50154320987654   -272.2 / 2,160 = 0.1260185185185

864,000 / -272.2 = 3,174.13666   864,000 / 3,136 = 275.510   864,000 / 26 = 33,230
3,456 / -272.2 = 12.69654 3,456 / 26 =   3,456 / 3,136 =   3,456 / =   3,456 / =
2,160 / -272.2 =   -272.2 / 2,160 =

864,000 / -272.2 = 3,333.3   864,000 / 345.6 = 2,500   864,000 / 432 = 2,000
2,160 / -272.2 = 7.93534166   -272.2 / 2,160 = 0.1260185185

-272.2 / 9 = 30.24   -272.2 / 8.888 = 30.6225   -272.2 / 7.777 = 34.997142857
-272.2 / 6.666 = 40.83   -272.2 / 5.555 = 48.996   -272.2 / 4.444 = 61.245 (EARTH #)
-272.2 / 3.333 = 81.66   -272.2 / 2.222 = 122.49   -272.2 / 1.111 = 244.98

2 + 7 + 2 + 2 (13) + 2 + 2 + 7 + 2 = 26

2 x 7 x 2 x 2 (56) x 2 x 2 x 7 x 2 = 3,136 / 5.555 = 5,648

3,136 / 26 = 120.615385

3,136 / 5.555 = 564.48   3,136 / 266.666 = 11.76   3,136 / 144 = 21.777

3,136 / -272.2 = 11.5209405 (1/x = 0.086798469387755)
3,136 / 5.555 = 564.48 / 11.111 = 50.80320000000001
32,400 (Hydrogen -259.2) / 3,136 (Helium -272.2) = 10.33163265306122

3,136 / -272.2 = 11.52094048493775
3,136 / 6.666 =   3,136 / 5.555 =   3,136 / 4.444 =
3,136 / 3.333 =   3,136 / 2.222 =   3,136 / 11.111 =

HELIUM RIGHT ANGLE TRIANGLE MELTING POINT – 272.2

272.2 x 362.9333 = 98,790.45333
98,790.45333 / 2 = 49,395.22666

49,395.22666 / 24 Hours = 2,058.13444   2,058.13444 / 60 Minutes = 34.30224074
34.30224074 / 60 Seconds = 0.571704012345679
0.571704012345679 / 60 Arc Seconds = 0.009528400205761

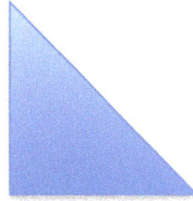

- 272.2 / 3 = 90.7333

453.666 = 5 x 90.7333

362.9333 = 4 x 90.7333

8.2.)   HELIUM (He) -272.2 C

HELIUM RATIOS TO ALL OTHER NOBLE GASES INCLUDING HYDROGEN

3,136 (Helium - 272.2) / 32,400 (Hydrogen -259.2) = 0.09679012345679
  0.09679012345679 //- 9 = 0.000378086419753 = Helium to Argon 8th Octave
32,400 (Hydrogen -259.2) / 3,136 (Helium -272.2) = 10.33163265306122

3,136 (Helium - 272.2) / 147,456 (-248.6 Neon) = 0.02126736111
147,456 (-248.6 Neon) / 3,136 (Helium - 272.2) = 47.02040816326531

3,136 (Helium - 272.2) / 82,944 (Argon) = 0.037808641975309
          0.037808641975309 //+8 = 9.679012345679012
82,944 (Argon) / 3,136 (Helium - 272.2) = 26.44897959183673

3,136 (Helium - 272.2) / 19,600 (Krypton 157.4) = 0.16
19,600 (Krypton 157.4) / 3,136 (Helium - 272.2) = 6.25

3,136 (Helium - 272.2) / 64 (Xeon 111.8) = 49
64 (Xeon 111.8) / 3,136 (Helium - 272.2) = 0.020408163265306

 3,136 (Helium - 272.2) / 1,225 (Radon -71.15) = 2.56
1,225 (Radon -71.15) / 3,136 (Helium - 272.2) = 0.390625 //+ 12 = 1,600 – 25,600

8.3.) *NEON (Ne)* -248.6 C
NEON RIGHT ANGLE TRIANGLE MELTING POINT -248.6

-248.6 / 266.666 =   -248.6 / 11.111 =   -248.6 / 24 =   -248.6 / 1.333 =
 -248.6 / 51.84 =   -248.6 / 144 =   -248.6 / =   -248.6 / 2,160 =

864,000 / -248.6 =   864,000 / 147,456 =   864,000/ 26 =
3,456 / -248.6 = 3,456 / = 3,456 / = 3,456 / = 3,456 / =
2,160 / -248.6 =   -248.6 / 2,160 =

864,000 / -248.6 =   864,000 / 345.6 = 2,500   864,000/ 432 = 2,000
2,160 / -248.6 =   -248.6 / 2,160 =

-248.6 / 26 =
-248.6 / 9 =   -248.6 / 8.888 =   -248.6 / 7.777 =
-248.6 / 6.666 =   -248.6 / 5.555 =   -248.6 / 4.444 =
-248.6 / 3.333 =   -248.6 / 2.222 =   -248.6 / 11.111 =

864,000 / -248.6 =   864,000 / 345.6 =   864,000/ 432 =
2,160 / -248.6 =   -248.6 / 2,160 =

 / =
 / 6.666 =   / 5.555 =   / 4.444 =
 / 3.333 =   / 2.222 =   / 11.111 =

864,000 / 259.2 = 3,333.3   864,000 / 345.6 = 2,500   864,000/ 432 = 2,000
2,160 / 259.2 = 8.333   259.2 / 2,160 = 0.972

-248.6 (3 side) x 331.4666 (4 side) = 82,402.61333 (Area of Square)
82,402.61333 / 2 = 41,201.30666 (Area of the Right Triangle)
41,201.30666 / 24 Hours = 1,716.72111
1,716.72111 / 60 Minutes = 28.6120185185185
 28.6120185185185 / 60 Seconds = 0.476866975308642

-248.6/ 3 = 82.8666

414.333 = 5 x 82.8666

331.4666 = 4 x 82.8666

8.3.) NEON (Ne) -248.6 C

2 + 4 + 8 + 6 (20) + 6 + 8 + 4 + 2 = 40
2 x 4 x 8 x 6 (384) x 6 x 8 x 4 x 2 = 147,456 /

147,456 / 40 = 3,684

147,456 / 1.111 = 132,710.4
147,456 (Neon 248.6) / 40 = 3,686.4
147,456 (Neon 248.6) x 40 = 5,898,240
147,456 (Neon 248.6) / 266.666 = 552.9600000000001
147,456 (Neon 248.6) / 11.111 = 13,271.04
147,456 (Neon 248.6) / 1.333 = 110,592
147,456 (Neon 248.6) / 144 = 1,024

147,456 / 259.2 = 568.888 // Octave of 1.111//
147,456 / 9 = 16,384   147,456 / 8.888 = 16,588.8   147,456 / 7.777 = 18,958.628
147,456 / 6.666 = 22,118.4   147,456 / 5.555 = 26,542.08   147,456 / 4.444 = 33,177.6
147,456 / 3.333 = 44,236.8 (POS) 147,456 / 2.222 = 66,355.2 147,456 / 11.111 = 13,271.04

NEON RATIOS TO ALL OTHER NOBLE GASES INCLUDING HYDROGEN

147,456 (Neon -248.6) / 32,400 (Hydrogen -259.2) = 4.5111
32,400 (Hydrogen -259.2) / 147,456 (Neon 248.6)= 0.2197265625

147,456 (Neon -248.6) / 3,136 (Helium - 272.2) = 47.02040816326531
3,136 (Helium - 272.2) / 147,456 (Neon 248.6) = 0.02126736111

147,456 (Neon -248.6) / 147,456 (-248.6 Neon) = 1
147,456 (-248.6 Neon) / 147,456 (Neon 248.6) = 1

147,456 (Neon - 248.6) / 82,944 (Argon -189.4) = 1.777
82,944 (Argon 189.4) / 147,456 (Neon - 248.6) = 0.5625

147,456 (Neon - 248.6) / 19,600 (Krypton 157.4) = 7.523265306122449
19,600 (Krypton 157.4) / 147,456 (Neon - 248.6) = 0.1329210069444

147,456 (Neon 248.6) / 64 (Xeon 111.8) = 2,304
64 (Xeon 111.8) / 147,456 (Neon 248.6) = 0.00043402777

147,456 (Neon -248.6) / 1,225 (Radon -71.15) = 120.3722448979592
1,225 (Radon -71.15) / 147,456 (Neon 248.6) = 0.008307562934028

8.4.) ARGON (Ar) – 189.4

ARGON RIGHT ANGLE TRIANGLE MELTING POINT – 189.4

– 189.4 / 266.666 = – 189.4 / 11.111 = – 189.4 / 24 = – 189.4 / 1.333 =
– 189.4 / 51.84 = – 189.4 / 144 = – 189.4 / = – 189.4 / 2,160 =

864,000 / – 189.4 = 864,000 / 82,944 = 864,000 / 44 =
3,456 / – 189.4 = 3,456 / = 3,456 / = 3,456 / = 3,456 / =
2,160 / – 189.4 = – 189.4 / 2,160 =
864,000 / – 189.4 = 864,000 / 345.6 = 2,500 864,000 / 432 = 2,000
2,160 / – 189.4 = – 189.4 / 2,160 = 0.972
– 189.4 / 44 = 4.30454545
– 189.4 / 9 = 21.0444 – 189.4 / 8.888 = 21.3075 – 189.4 / 7.777 = 24.35142857142857
– 189.4 / 6.666 = 28.41 – 189.4 / 5.555 = 34.092 – 189.4 / 4.444 = 42.615
– 189.4 / 3.333 = 56.82 – 189.4 / 2.222 = 85.23 – 189.4 / 1.111 = 170.46

864,000 / – 189.4 = 864,000 / 345.6 = 2,500 864,000 / 432 = 2,000
2,160 / – 189.4 = – 189.4 / 2,160 = 0.972

- 189.4 (3 side) x 252.5333 (4 side) = 47,829.81333 (Area of Square)
47,829.81333 / 2 = 23,914.90666 (Area of the Right Triangle)
23,914.90666 / 24 Hours = 996.45444
996.45444 / 60 Minutes = 16.60757407407
16.60757407407 / 60 Seconds = 0.276792901234568
0.276792901234568 / 60 Arc Seconds = 0.004613215020576

189.4 / 3 = 63.1333

315.666 = 5 x 63.1333

252.5333 = 4 x 63.1333

1 + 8 + 9 + 4 (22) + 4 + 9 + 8 + 1 = 44

1 x 8 x 9 x 4 (288) x 4 x 9 x 8 x 1 = 82,944

82,944 / 44 = 1,885.090909

82,944 (Argon - 189.4) / 44 = 1,885.0909

82,944 (Argon - 189.4) x 44 = 3,649,536
82,944 (Argon - 189.4) / 266.666 = 311.0400000000001
82,944 (Argon - 189.4) / 11.111 = 7,464.960000000001
82,944 (Argon - 189.4) / 1.333 = 62,208.00000000002
82,944 (Argon - 189.4) / 144 = 576 //288//144//72//36//16//8//4//2//1
82,944 (Argon - 189.4) / 24 = 3,456
82,944 (Argon - 189.4) / 16 = 5,184
82,944 (Argon - 189.4) / 12 = 6,912
82,944 (Argon - 189.4) / 108 = 768 (24 OCTAVE SERIES)

8.4.)   ARGON (Ar) – 189.4

ARGON RATIOS TO ALL OTHER NOBLE GASES INCLUDING HYDROGEN

82,944 (Argon- 189.4) / 32,400 (Hydrogen – 259.2) = 2.56
2.56// 1.28//0.64// 0.32//0.16//0.08
32,400 (Hydrogen – 259.2) / 82,944 (Argon - 189.4)= 0.390625

82,944 (Argon - 189.4) / 3,136 (Helium - 272.2) = 26.44897959183673
3,136 (Helium - 272.2) / 82,944 (Argon - 189.4) = 0.037808641975309
0.037808641975309 //+8 = 9.679012345679012

82,944 (Argon - 189.4) / 147,456 (-248.6 Neon) = 0.000140625//+5 = 0.144
147,456 (-248.6 Neon) / 82,944 (Argon - 189.4) = 1.777
1.777//0.888/0.444/0.222/0.111/0.0555

82,944 (Argon - 189.4) / 19,600 (Krypton 157.4) = 4.231836734693878
19,600 (Krypton 157.4) / 82,944 (Argon - 189.4) = 0.236304012345679 //+5 =3.7808

82,944 (Argon - 189.4) / 64 (Xeon 111.8) = 1,296
64 (Xeon 111.8) / 82,944 (Argon - 189.4) = 0.000771604938272

82,944 (Argon - 189.4) / 1,225 (Radon -71.15) = 67.70938775510204
1,225 (Radon -71.15) / 82,944 (Argon - 189.4) = 0.0147690007716

8.5.) *KRYPTON (Kr))* − 157.4 C

KRYPTON RIGHT ANGLE TRIANGLE MELTING POINT − 157.4 C

− 157.4 C / 266.666 =   − 157.4 C / 11.111 =   − 157.4 C / 24 =   − 157.4 C / 1.333 =
− 157.4 C / 51.84 =   − 157.4 C / 144 =   − 157.4 C / =   − 157.4 C / 2,160 =

864,000 / − 157.4 C =   864,000 / 3,136 =   864,000/ 26 =
3,456 / − 157.4 C = 3,456 / =   3,456 / =   3,456 / =   3,456 / =
2,160 / − 157.4 C =   − 157.4 C / 2,160 =

864,000 / − 157.4 C =   864,000 / 345.6 = 2,500   864,000/ 432 = 2,000
2,160 / − 157.4 C = 8.333   − 157.4 C / 2,160 = 0.972

− 157.4 C / 26 =
− 157.4 C / 9 =   − 189.4 / 8.888 =   − 189.4 / 7.777 =
− 157.4 C / 6.666 =   − 157.4 C / 5.555 =   − 157.4 C / 4.444 =
− 157.4 C / 3.333 =   − 157.4 C / 2.222 =   − 157.4 C / 11.111 =

864,000 / − 157.4 C =   864,000 / 345.6 =   864,000/ 432 =
2,160 / − 157.4 C =   − 157.4 C / 2,160 =

864,000 / − 157.4 C =   864,000 / 345.6 = 2,500   864,000/ 432 = 2,000
2,160 / − 157.4 C =   − 157.4 C / 2,160 =

− 157.4 C (3 side) x 209.8666 (4 side) = 33,033.01333 (Area of Square)
33,033.01333 / 2 = 16,516.50666 (Area / Time / Space of the Right Triangle)

16,516.50666 / 24 Hrs = 688.18777
688.18777/ 60 Min = 11.469796296296296
11.469796296296296 / 60 Sec = 0.191163271604938
0.191163271604938 / 60 Arc Sec = 0.003186054526749

-157.4 / 3 = 52.4666

262.333 = 5 x 52.4666

209.8666 = 4 x 52.4666

8.5.)   KRYPTON (Kr) – 157.4 C

1 + 5 + 7 + 4 (17) + 4 + 7 + 5 + 1 = 34

1 x 5 x 7 x 4 (140) x 4 x 7 x 5 x 1 = 19,600

19,600 / 34 = 576.4705882352941

19,600 (Krypton -157.4) / 34 = 576.4705882352941
19,600 (Krypton -157.4) x 34 = 666,400
19,600 (Krypton -157.4) / 266.666 = 73.50000000000002
19,600 (Krypton -157.4) / 11.111 = 1,764
19,600 (Krypton -157.4) / 1.333 = 14,700
19,600 (Krypton -157.4) / 144 = 136.111

19,600 / 34 =
19,600 / 9 =   19,600 / 8.888 =   19,600 / 7.777 =
19,600 / 6.666 =   19,600 / 5.555 =   19,600 / 4.444 =
19,600 / 3.333 =   19,600 / 2.222 =   19,600 / 11.111 =

864,000 / 19,600 =   864,000 / 345.6 = 2,500   864,000/ 432 = 2,000
2,160 / 19,600 =   19,600 / 2,160 =

KRYPTON RATIOS TO ALL OTHER NOBLE GASES INCLUDING HYDROGEN

19,600 (Krypton -157.4) / 32,400 (Hydrogen -259.6) = 0.604938271604938
32,400 (Hydrogen – 259.6) / 19,600 (Krypton -157.4) = 1.653061224489796

19,600 (Krypton -157.4) / 3,136 (Helium – 272.2) = 6.25
3,136 (Helium – 272.2) / 19,600 (Krypton -157.4) = 0.16

19,600 (Krypton -157.4) / 147,456 (-248.6 Neon) = 0.1329210069444
147,456 (-248.6 Neon) / 19,600 (Krypton -157.4) = 7.523265306122449

19,600 (Krypton -157.4) / 19,600 (Krypton - 157.4) = 1
19,600 (Krypton - 157.4) / 19,600 (Krypton -157.4) = 1

19,600 (Krypton -157.4) / 64 (Xeon - 111.8) = 306.25
64 (Xeon 111.8) / 19,600 (Krypton -157.4) = 0.003265306122449

19,600 (Krypton -157.4) / 1,225 (Radon -71.15) = 16
1,225 (Radon -71.15) / 19,600 (Krypton -157.4) = 0.0625

8.6.) XEON (Xe) -111.8 C

XEON RIGHT ANGLE TRIANGLE MELTING POINT -111.8 C

SO -111.8 C / 266.666 =   -111.8 C / 11.111 =   -111.8 C / 24 =   -111.8 C / 1.333 =
SO -111.8 C / 51.84 =   -111.8 C / 144 =   -111.8 C / =   -111.8 C / 2,160 =

864,000 / -111.8 C =   864,000 / 3,136 =   864,000 / 26 =
3,456 / -111.8 C = 3,456 / = 3,456 / = 3,456 / = 3,456 / =
2,160 / -111.8 C =   -111.8 C / 2,160 =

864,000 / -111.8 C =   864,000 / 345.6 = 2,500   864,000 / 432 = 2,000
2,160 / -111.8 C =   -111.8 C / 2,160 =

-111.8 C / 26 =
-111.8 C / 9 =   -111.8 C / 8.888 =   -111.8 C / 7.777 =
-111.8 C / 6.666 =   -111.8 C / 5.555 =   -111.8 C / 4.444 =
-111.8 C / 3.333 =   -111.8 C / 2.222 =   -111.8 C / 11.111 =

864,000 / -111.8 C =   864,000 / 345.6 =   864,000 / 432 =
2,160 / -111.8 C =   -111.8 C / 2,160 =

-111.8 (3 side) x 149.0666(4 side) = 16,665.65333 (Area of Square)
 16,665.65333 / 2 = 8,332.82666 (Area of the Right Triangle)

8,332.82666 / 24 Hours = 347.20111
347.20111 / 60 Minutes = 5.786685185185183
5.786685185185183 / 60 Seconds = 0.09644475308642
0.09644475308642 / 60 Arc Seconds = 0.00160741255144

-111.8/ 3 = 37.2666

186.333 = 5 x 37.2666

149.0666 = 4 x 37.2666

8.6.)   *XEON (Xe)* -111.8 C

1 + 1 + 1 + 8 (11) + 8 + 1 + 1 + 1 = 22

1 x 1 x 1 x 8 (8) x 8 x 1 x 1 x 1 = 64

64 / 22 = 2.9090909

64 (Xeon -111.8) / 22 = 2.909090
64 (Xeon -111.8) x 22 = 1,408

64 (Xeon -111.8) / 266.666 = 0.24   (1/x = 4.1666)
64 (Xeon -111.8) / 11.111 = 5.76   (1/x = 0.1736111)
64 (Xeon -111.8) / 1.333 = 48   (1/x = 0.0208333)
64 (Xeon -111.8) / 144 = 0.444   (1/x = 2.25)

64 / =
64 / 9 =   64 / 8.888 =   64 / 7.777 =
64 / 6.666 =   64 / 5.555 =   64 / 4.444 =
64 / 3.333 =   64 / 2.222 =   64 / 11.111 =

864,000 / 64 =   864,000 / 345.6 = 2,500   864,000/ 432 = 2,000
2,160 / 64 =   64 / 2,160 = 0.972

XENON RATIOS TO ALL OTHER NOBLE GASES INCLUDING HYDROGEN

64 (Xeon -111.8) / 32,400 (Hydrogen) = 0.001975308641975
32,400 (Hydrogen) / 64 (Xeon -111.8) = 506.25

64 (Xeon -111.8) / 3,136 (Helium – 272.2) = 0.02408163265306
3,136 (Helium – 272.2) / 64 (Xeon -111.8) = 49

64 (Xeon -111.8) / 147,456 (-248.6 Neon) = 0.00043402777
147,456 (-248.6 Neon) / 64 (Xeon -111.8) = 2,304

64 (Xeon -111.8) / 19,600 (Krypton 157.4) = 0.003265306122449
19,600 (Krypton 157.4) / 64 (Xeon -111.8) = 306.25

 64 (Xeon -111.8) / 64 (Xeon 111.8) = 1
64 (Xeon 111.8) /64 (Xeon -111.8) = 1

64 (Xeon -111.8) / 1,225 (Radon -71.15) = 0.052244897959184
1,225 (Radon -71.15) / 64 (Xeon -111.8) = 19.140625

8.7.) RADON (Rn) -71.15

RADON RIGHT ANGLE TRIANGLE MELTING POINT -71.15

-71.15 / 266.666 =   -71.15 / 11.111 =   -71.15 / 24 =   -71.15 / 1.333 =
-71.15 / 51.84 =   -71.15 / 144 =   -71.15 / =   -71.15 / 2,160 =

864,000 / -71.15 =   864,000 / 3,136 =   864,000/ 26 =
3,456 / -71.15 =   3,456 / =   3,456 / =   3,456 / =   3,456 / =
2,160 / -71.15 =   -71.15 / 2,160 =

864,000 / -71.15 =   864,000 / 345.6 = 2,500   864,000/ 432 = 2,000
2,160 / -71.15 =   -71.15 / 2,160 = 0.972

-71.15 / 26 =
-71.15 / 9 =   -71.15 / 8.888 =   -71.15 / 7.777 =
-71.15 / 6.666 =   -71.15 / 5.555 =   -71.15 / 4.444 =
-71.15 / 3.333 =   -71.15 / 2.222 =   -71.15 / 11.111 =

864,000 / -71.15 =   864,000 / 345.6 =   864,000/ 432 =
2,160 / -71.15 =   -71.15 / 2,160 =

-71.15 (3 side) x (4 side) = 6,749.76333 (Area of Square)
6,749.76333 / 2 = 3,374.881666 (Area of the – 71.15 Right Triangle)
3,374.881666 / 24 Hours = 140.620069444
140.620069444 / 60 Minutes = 2.343667824074074
2.343667824074074 / 60 Seconds = 0.039061130401235
0.039061130401235 / 60 Arc Seconds = 0.000651018840021

-71.15 / **3** = **23.71666**

118.58333 = **5** x **23.1666**

94.8666= **4** x **23.71666**

7 + 1 + 1 + 5 (14) + 5 + 1 + 1 + 7 = 28

7 x 1 x 1 x 5 (35) x 5 x 1 x 1 x 7 = 1,225

1,225 / 28 = 43.75

1,225 (Radon -71.15) / **28** = **43.75**
1,225 (Radon -71.15) x **28** = **34,300**
1,225 (Radon -71.15) / **266.666** = **4.593750000000001**
1,225 (Radon -71.15) / **11.111** = **110.25**
1,225 (Radon -71.15) / **1.333** = **918.75**
1,225 (Radon -71.15) / **144** = **8.5069444**
1,225 (Radon -71.15)
1,225 (Radon -71.15)

8.7.) RADON (Rn) -71.15

1,225 / =
1,225 / **6.666** =   1,225 / **5.555** =   1,225 / **4.444** =
1,225 / **3.333** =   1,225 / **2.222** =   1,225 / **11.111** =

RADON RATIOS TO ALL OTHER NOBLE GASES INCLUDING HYDROGEN

1,225 (Radon -71.15) / **32,400** (Hydrogen) = 0.037808641975309
32,400 (Hydrogen) / 1,225 (Radon -71.15)= 26.44897959183673

1,225 (Radon -71.15) / **3,136** (Helium – 272.2) = 0.390625
3,136 (Helium – 272.2) / 1,225 (Radon -71.15) = 2.56

1,225 (Radon -71.15) / **147,456** (-248.6 Neon) = 0.008307562934028
147,456 (-248.6 Neon) / 1,225 (Radon -71.15) = 120.3722448979592

1,225 (Radon -71.15) / **19,600** (Krypton 157.4) = 0.0625
19,600 (Krypton 157.4) / 1,225 (Radon -71.15) = 16

1,225 (Radon -71.15) / **64** (Xeon 111.8) = 19.140625
64 (Xeon 111.8) / 1,225 (Radon -71.15) = 0.052244897959184

1,225 (Radon -71.15) / **1,225** (Radon -71.15) = 1
1,225 (Radon -71.15) / 1,225 (Radon -71.15) = 1

8.8.) *NOBLE GASES FULL COMBINED TABLES*

NOBLE GASES FULL COMBINED TABLES

3,136 (Helium – 272.2) / 64 (Xeon 111.8) = 49

3,136 (Helium – 272.2) / 19,600 (Krypton 157.4) = 0.16 [1/x = 6.25]

32,400 (Hydrogen) / 64 (Xeon 111.8) = 506.25

3,136 (Helium – 272.2) / 64 (Xeon 111.8) = 49
64 (Xeon 111.8) / 3,136 (Helium – 272.2) = 0.020408163265306

3,136 (Helium – 272.2) / 82,944 (Argon) = 0.037808641975309
        0.037808641975309 //+8 = 9.679012345679012
82,944 (Argon) / 3,136 (Helium – 272.2) = 26.44897959183673

3,136 (Helium – 272.2) / 19,600 (Krypton 157.4) = 0.16
19,600 (Krypton 157.4) / 3,136 (Helium – 272.2) = 6.25

3,136 (Helium – 272.2) / 147,456 (-248.6 Neon) = 0.02126736111
                [1/x = 47.020816326531]

147,456 (-248.6 Neon) / 3,136 (Helium – 272.2) = 47.02040816326531

3,136 (Helium – 272.2) / 82,944 (Argon) = 0.037808641975309

0.037808641975309 //+8 = 9.679012345679012
82,944 (Argon) / 3,136 (Helium – 272.2) = 26.44897959183673

3,136 (Helium – 272.2) / 32,400 (Hydrogen) = 0.09679012345679
0.09679012345679 //- 9 = 0.000378086419753 = Helium to Argon 8 th Octave

32,400 (Hydrogen) / 3,136 (Helium -272.2) = 10.33163265306122
3,136 (Helium -272.2) / 32,400 (Hydrogen) = 0.09679012345679

32,400 (Hydrogen) / 147,456 (Neon -248.6) = 0.2197265625
147,456 (Neon -248.6) / 32,400 (Hydrogen) = 4.55111

32,400 (Hydrogen) / 82,944 (Argon -189.4) = 0.390625
82,944 (Argon -189.4) / 32,400 (Hydrogen) = 2.56

32,400 (Hydrogen) / 19,600 (Krypton 157.4) = 1.653061224489796
19,600 (Krypton 157.4) / 32,400 (Hydrogen) = 0.604938271604938

32,400 (Hydrogen) / 64 (Xeon -111.8) = 506.25
64 (Xeon -111.8) / 32,400 (Hydrogen) = 0.00195308641975

32,400 (Hydrogen) / 1,225 (Radon -71.15) = 26.44897959183673
1,225 (Radon -71.15) / 32,400 (Hydrogen) = 0.037808641975309 Argon/Helium ratio

8.9.) DESIGNER GAS SECRETS

HYDROGEN

HYDROGEN RATIOS TO ALL OTHER NOBLE GASES

32,400 (Hydrogen -259.2) / 3,136 (Helium -272.2) = 10.33163265306122
3,136 (Helium -272.2) / 32,400 (Hydrogen -259.2) = 0.09679012345679

32,400 (Hydrogen -259.2) / 147,456 (Neon -248.6) = 0.2197265625
147,456 (Neon -248.6) / 32,400 (Hydrogen -259.2) = 4.55111

32,400 (Hydrogen -259.2) / 82,944 (Argon -189.4) = 0.390625
82,944 (Argon -189.4) / 32,400 (Hydrogen -259.2) = 2.56

32,400 (Hydrogen -259.2) / 19,600 (Krypton 157.4) = 1.653061224489796
19,600 (Krypton 157.4) / 32,400 (Hydrogen -259.2) = 0.604938271604938

32,400 (Hydrogen -259.2) / 64 (Xeon -111.8) = 506.25
64 (Xeon -111.8) / 32,400 (Hydrogen -259.2) = 0.00195308641975

32,400 (Hydrogen -259.2) / 1,225 (Radon -71.15) = 26.44897959183673
1,225 (Radon -71.15) / 32,400 (Hydrogen -259.2) = 0.037808641975309
Argon/Helium ratio

ARGON

ARGON RATIOS TO ALL OTHER NOBLE GASES INCLUDING HYDROGEN

82,944 (Argon – 189.4) / 32,400 (Hydrogen -259.2) = 2.56
2.56 // 1.28//0.64// 0.32//0.16//0.08 //0.04//0.02//
32,400 (Hydrogen -259.2) / 82,944 (Argon – 189.4)= 0.390625

82,944 (Argon – 189.4) / 3,136 (Helium -272.2) = 26.44897959183673
3,136 (Helium -272.2) / 82,944 (Argon – 189.4) = 0.037808641975309
        0.037808641975309 //+8 = 9.679012345679012

82,944 (Argon – 189.4) / 147,456 (-248.6 Neon) = 0.5625
147,456 (-248.6 Neon) / 82,944 (Argon – 189.4) = 1.777
1.777//0.888//0.444//0.222/0.111/0.0555

82,944 (Argon – 189.4) / 19,600 (Krypton 157.4) = 4.231836734693878
19,600 (Krypton 157.4) / 82,944 (Argon – 189.4) = 0.236304012345679 //+5 =3.7808

82,944 (Argon – 189.4) / 64 (Xeon 111.8) = 1,296
64 (Xeon 111.8) / 82,944 (Argon – 189.4) = 0.000771604938272

82,944 (Argon – 189.4) / 1,225 (Radon -71.15) = 67.70938775510204
1,225 (Radon -71.15) / 82,944 (Argon – 189.4) = 0.0147690007716

9.) STEPS TO CALCULATE THE RESONANT FREQUENCIES AND GEOMETRY TO ACHIEVE COLD FUSION LOW ENERGY ATOMIC TRANSMUTATIONS (CFLEAR)

AN INTRODUCTION TO PLASMOID VORTEX MATHS LAWS

LAW 1.) – *A number is primarily used to identify the dimension it represents; 0 & 1 are out.*

Example 1 A :- The Sun's diameter in miles (Aether) = 864,000
864 Octaves down // 432 // 216 // 108 //54 // 27 // 13.5
864 Octaves up // 1,728 // 3,456 // 6,912 // 13,824 //27,648 // 55,296 // 110,592

Example 1 B :- The Sun's square in miles (Matter) = 3,456,000
3,456 Octaves down // 1,728 // 864 // 432 // 216 // 108 // 54 // 27 // 13.5 // 6.75 // 3.375 // 1.6875
3,456 Octaves up // 6,912 // 13,824 // 27,648 // 55,296 // 110,592 // 221,184 // 442,368 // 884,736

Example 1 C :- The Moon's diameter in miles (Matter) = 2,160
2,160 Octaves down // *1,080 Moon's R* // 540 // 270 // 135 // 67.5 // 33.75 // 16.875 // 8.4375
2,160 Octaves up // 4,320 Sun's R // 8,640 Sun's D // 17,280 // *34,560 Matter* // 69,120 // 138,240 // 276,480

Example 1 D :- The Sun's PROTIUM (Hydrogen H) (Aether to Matter product) = Melting point minus 259.2
259.2 Octaves down // 129.6 // 64.8 (SUN D) // 32.4 (SUN R) // 16.2 (MOON D) // 8.1 // 4.05 // 2.025
259.2 Octaves up // 518.4 // 1,036.8 // 2,073.6 // 4,147.2 // 8,294.4 // 16,588.8 // 33,177.6

LAW 2.) – *Product of the numbers within a number are the Plasmoid Unification 1st Key.*

Example 2 A :- Sun Aether = 864 so = 8 x 6 x 4 = 192   // 96 // 48 // 24 // 12 // 6 // 3 // 1.5 // 7.5

Example 2 B :- Sun Matter = 3,456 so = 3 x 4 x 5 x 6 = 360 Compass // 180 // 90 // 45 // 22.5 / /11.25

*Example 2 C :- **Moon** Matter = 2,160 so = 2 x 1 x 6 x 0 = 12 Compass // 24 // 48 // 96 // 192 // 384*

Example 2 D :- Sun Protium (H) = 259.2 so = 2 x 5 x 9 x 2 = 180 Compass // 90 // 45 // 22.5 // 11.25

LAW 3.) - **Plasmoid Unification 1st Key numbers products are then multiplied by their Plasmoid Unification 2nd Key mirror numbers product, squaring law two's result.**

Example 3 A :- Sun Aether = 192 x 4 x 6 x 8 = 36,864 [36 Matter, 864 Aether]
36,864 Octaves down // 18,432 // 9,216 // 4,608 // 2,304 // 1,152 // 576 // 288 // 144 // 72 // 36 // 18 // 9
36,864 Octaves up //

Example 3B :- Sun Matter = 360 x 6 x 5 x 4 x 3 = 129,600 [Resonant Frequency Energy Unit]
129,600 Octaves down // 64,800 // 32,400 // 162 // 81 // 40.5 // 20.25 // 10.125 // 5.0625 // 2.53125
129,600 Octaves up // 259,200 // 518,400 // 1,036,800 // 2,073,600 // 4,147,200 // 8,294,400 // 16,588,800

Example 3C :- Moon Matter = 12 x 0 x 6 x 1 x 2 = 144 [light]
144 [light] Octaves down // 72 // 36 // 18 // 9
144 [light] Octaves up // 288 // 576 // 1,152 // 2,304 // 4,608 // 9,216 // 18,432 // 36,864 // 73,728

Example 3D :- Hydrogen (Aether to Matter product) = 180 x 2 x 9 x 5 x 2 = 32,400 [r]
32,400 [r] Octaves down // 18,432 // 9,216 // 4,608 // 2,304 // 1,152 //
32,400 [r] Octaves up // 64,800 // 129,600 // 259,200 // 518,400 // 1,036,800 // 2,073,600 // 4,147,200

- 259.2 multiplied product = 2 x 5 x 9 x 2 (180) x 2 x 9 x 5 x 2 = 32,400

9.) *OCTAVE TABLES FOR HYDROGEN (H), NEON (Ne) AND ARGON (Ar)*

	1,061,683,200	1,207,959,552	679,477,248	521,288
	530,841,600	603,979,776	339,738,624	262,144
	265,420,800	301,989,888	169,869,312	131,072
	132,710,400	150,944,944	84,934,656	65,536
	16,588,800	75,497,472	42,467,328	32,768
Ar	8,294,400	37,748,736	21,233,664	16,384
	4,147,200	18,874,368	10,616,832	8,192
	2,073,600	9,437,184	5,308,416	4,096
	1,036,800	4,718,592	2,654,208	2,048
	518,400	2,359,296	1,327,104	1,024
	259,200	1,179,648	663,552	512
	129,600	589,824	311,776	256
	64,800	294,912	165,888	128
	32,400	147,456	82,944	64

PROTIUM (H)	NEON (Ne)	ARGON (Ar)	XEON (Xe)
16,200	73,728	41,472	32
8,100 SUN MATTER	36,864 SUN AETHER	20,736	16
4,050	18,432	10,368	8
2,025	9,216	5,184	4
1,012.5	4,608	2,592	2
506.25	2,304	1,296	1
253.125	1,152	648	0.5
126.5625	576	324 PROTIUM	0.25
63.28125	288	162	0.125
31.640625	144	81	0.0625
15.8203125	72	40.5	0.03125
7.91015625	36	20.25	0.015625

9.) - OCTAVE TABLES - HELIUM (He), KRYPTON (Kr) AND RADON (Rn)

25,690,112	160,563,200	1,0035,200
12,845,056	80,281,600	5,017,600
6,422,528	40,140,800	2,508,800
3,211,642	20,070,400	1,254,400
1,605,632	10,035,200	627,200
802,816	5,017,600	Helium 313,600 He
401,408	2,508,800	156,800
200,704	1,254,400	78,400
100,352	627,200	39,200
50,176	HELIUM 313,600 He	KRYPTON 19,600 Kr
25,088	156,800	9,800
12,544	78,400	4,900
6,272	39,200	2,450

3,136 HELIUM (He)	19,600 KRYPTON (Kr)	1,225 RADON (Rn)
1,568	9,800	612.5
784	4,900	306.25
392	2,450	153.125
Kr 196 KRYPTON	Rn 1,225 RADON	76.5625
98	612.5	38.28125
49	153.125	19.140625
24.5	38.28125	9.5703125
Rn 12.25 RADON	19.140625	4.78515625
6.125	9.5703125	2.392578125
3.0625	4.78515625	1.1962890625
1.53125	2.39257812	0.59814453125
0.76562	1.19628906	0.299072265625

11.)

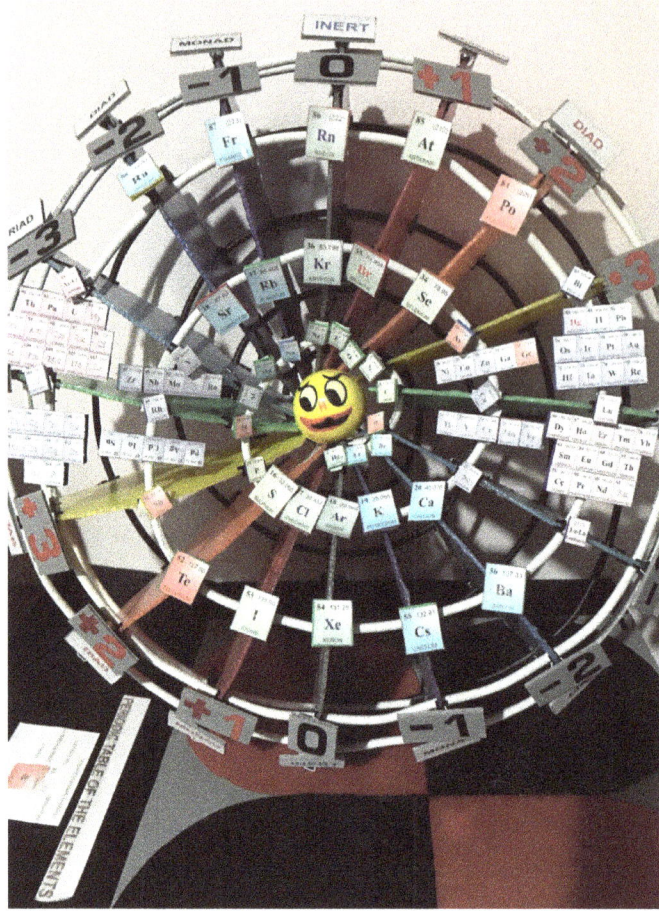

FIG 215 : - AETHER AND MATTER ELEMENT AND DIMENSIONAL NUMBERS

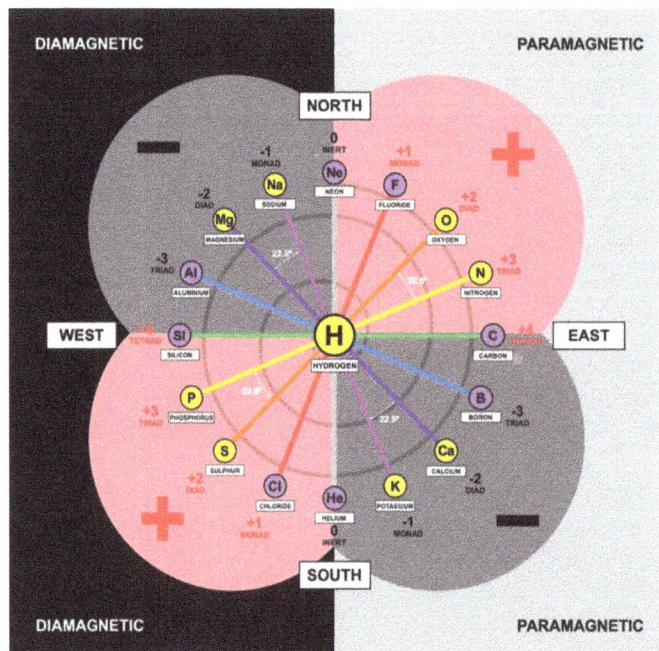

FIG 216 : - AETHER AND MATTER ELEMENT AND DIMENSIONAL NUMBERS

12.) – DESIGNER NOBLE GASES FULL OCTAVE PLANE

12.1) RESONANT IMPLOSION EMPIRICAL MELTING POINT TOTALS

PROTIUM (H) –   - 259.2

HELIUM (He) -   - 272.2   GAP = 13

NEON (Ne) -   -248.6   GAP = 23.6

ARGON (Ar) -   -189.4   GAP = 59.2

KRYPTON (Kr) –   -157.4   GAP = 32

XEON (Xe) –   -111.8   GAP = 45.6

RADON (Rn) –   - 71.15   GAP = 40.65

TOTAL = 1,309.75   TOTAL = 214.05

Inert Gas Clusters
(Rare Gas, Noble Gas)

He	Helium
Ne	Neon
Ar	Argon
Kr	Krypton
Xe	Xeon
Rn	Radon

FIG 217 : - AETHER AND MATTER ELEMENT AND DIMENSIONAL NUMBERS

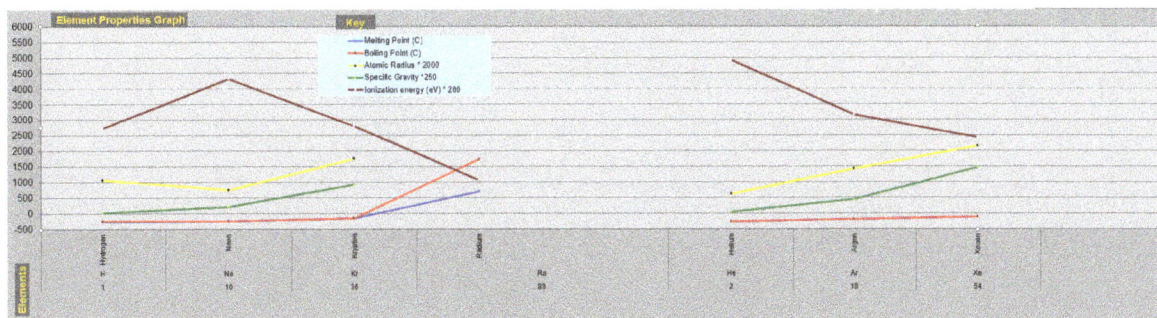

FIG 218 : - AETHER AND MATTER ELEMENT AND DIMENSIONAL NUMBERS

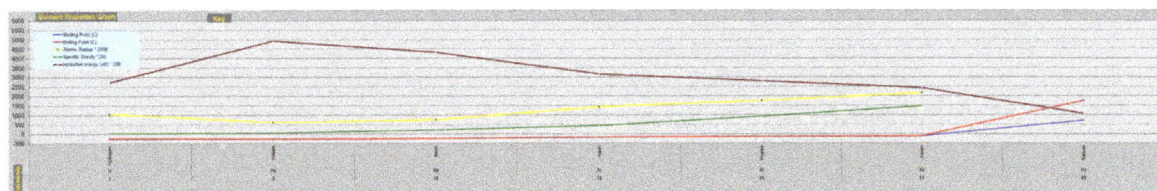

FIG 219 : - AETHER AND MATTER ELEMENT AND DIMENSIONAL NUMBERS

12.2) *RESONANT IMPLOSION MELTING POINT NUMBER SUM TOTALS*

PROTIUM (H) –   36

HELIUM (He) -   26   GAP = 10

NEON (Ne) -   40   GAP = 14

ARGON (Ar) -   44   GAP = 4

KRYPTON (Kr) –   34   GAP = 10

XEON (Xe) –   22   GAP = 12

RADON (Rn) –   28   GAP = 6

TOTAL = 230   TOTAL = 56

12.3) RESONANT IMPLOSION MELTING POINT NUMBER PRODUCTS TOTALS

1.) PROTIUM (H) –   **32,400**   GAP 36,000 / 0 - Zero Point to 32,400 = 3,600

2.) HELIUM (He) -   **3,136**   GAP H to He = 29,264   GAP He to Kr

3.) NEON (Ne) -   **147,456**   GAP He to Ne = 144,320

4.) ARGON (Ar) -   **82,944**   GAP Ne to Ar = 64,512

5.) KRYPTON (Kr) –   **19,600**   GAP Ar to Kr = 63,344   GAP He to Kr = 18,375

6.) XEON (Xe) –   **64**   GAP Kr to Xe = 19,536

7.) RADON (Rn) –   **1,225**   GAP Xe to Rn = 1,161   GAP Kr to Rn = 294,000

     TOTAL = 286,825   **TOTAL = 322,137**   **TOTAL = 312,375**

PROTIUM (H) 32,400 (82,944 sq rt = 180) / NEON (Ne) 3,136 = 10.3316 (3,136 sq rt = 56)

PROTIUM (H) 32,400 / 82,944 (82,944 square root = 288)

NEON (Ne) 147,456 (82,944 sq rt = 288) / ARGON (Ar) (82,944 sq rt = 288) 82,944 = 1.777 $1/x$ =0.5625

82,944 ARGON (Ar) / 64 XEON (Xe) = 2,304 $1/x$ = 0.000852878464

RADON (Rn) – KRYPTON (Kr) – HELIUM (He) - OCTAVE SEQUENCE

Rn - 1,225 // 2,450 // 4,900 // 9,800 // Kr - 19,600 // 39,200 // 78,400 // 156,800 // He - 313,600 //

1................GAP - 18,375...............2................. GAP – 294,000............... 3

....// 18,375 // 36,750 // 73,500 // 147,000 // 294,000 //...........................

12.3) RESONANT IMPLOSION MELTING POINT NUMBER PRODUCTS TOTALS

PROTIUM (H) to ARGON (Ar) - OCTAVE SEQUENCE

H - 324 // 684 // 1,296 // 2,592 // 5,184 // 10,364 // 20,734 // 41,472 // Ar – 82,944 //

32,400.....................................GAP – 50,544   82,944

394.87 // 789.75 // 1,579.5 // 3,159 // 6,318 // 12,636 // 25,272 // 50,544 // 101.088 // 202,176 // 404,352

// 808,704 // 1,617,408 // 3,234,816 //

PART 10

PLASMOID ZERO POINT ENABLED ELEMENTAL DISSASSOCIATION, ASSIMILATION, TRANSFORMATION AND TRANSMUTATION

MSAART PATENT APPLICATION NOTES

THE EXPLANATION AND DOCUMENTATION OF

PLASMOID ZERO POINT ENABLED ELEMENTAL

DISSASSOCIATION, ASSIMILATION, TRANSFORMATION

AND TRANSMUTATION BY PROCESSES OF

LOW ENERGY ATOMIC TRANSMUTATIONS (LEAT)

AND

COLD FUSION ELEMENTAL TRANSFORMATION

SUMMARY CONCLUSION
PART TEN OF TWENTY

DRAFT 518,400 B KMV – THURSDAY 22ND SEPT 2022

APPLICATIONS FOR A PLASMOIDS FORM AND FUNCTIONS

ABSTRACT

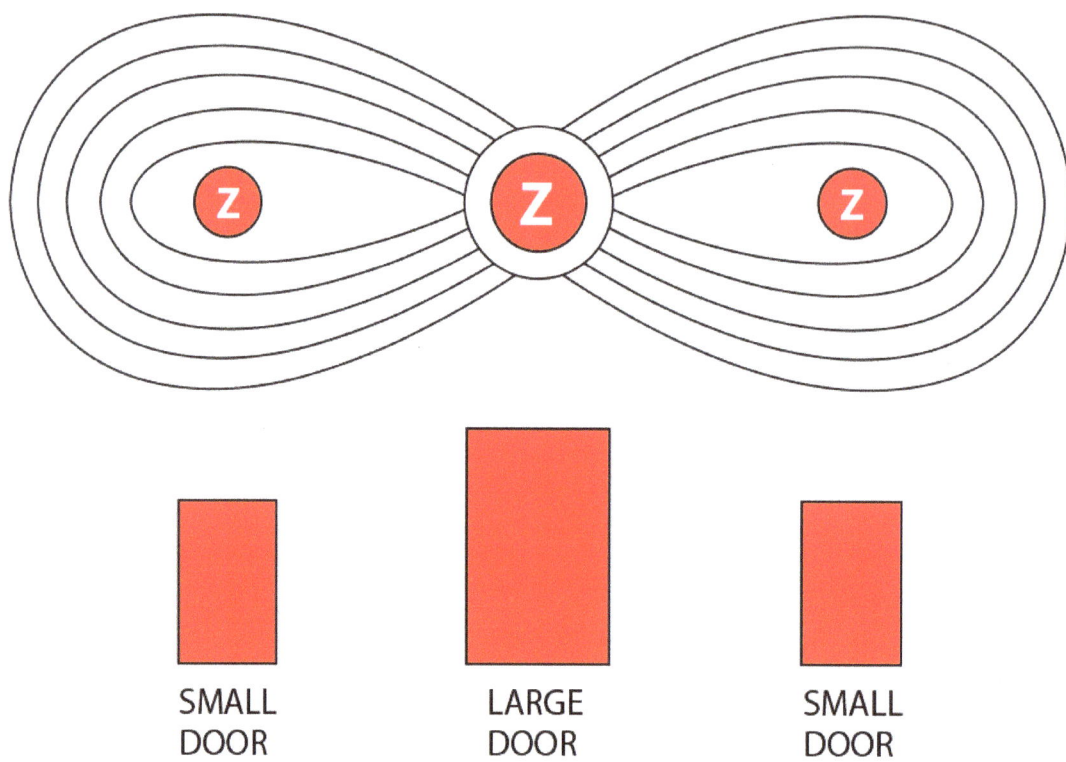

SMALL DOOR

LARGE DOOR

SMALL DOOR

FIG 219: - AETHER AND MATTER ELEMENT AND DIMENSIONAL

APPLICATIONS FOR A PLASMOIDS FORM AND FUNCTIONS

PART 10 OF 20 – MSAART PATENT APPLICATION NOTES - DRAFT 518,400 – 22:22:22 THUR 22ND SEPT 2022
© STRIKE FOUNDATION GUARANTEE LIMITED | MALCOLM BENDALL 2022 | GRAPHICS - STEVE EARL

ABSTRACT

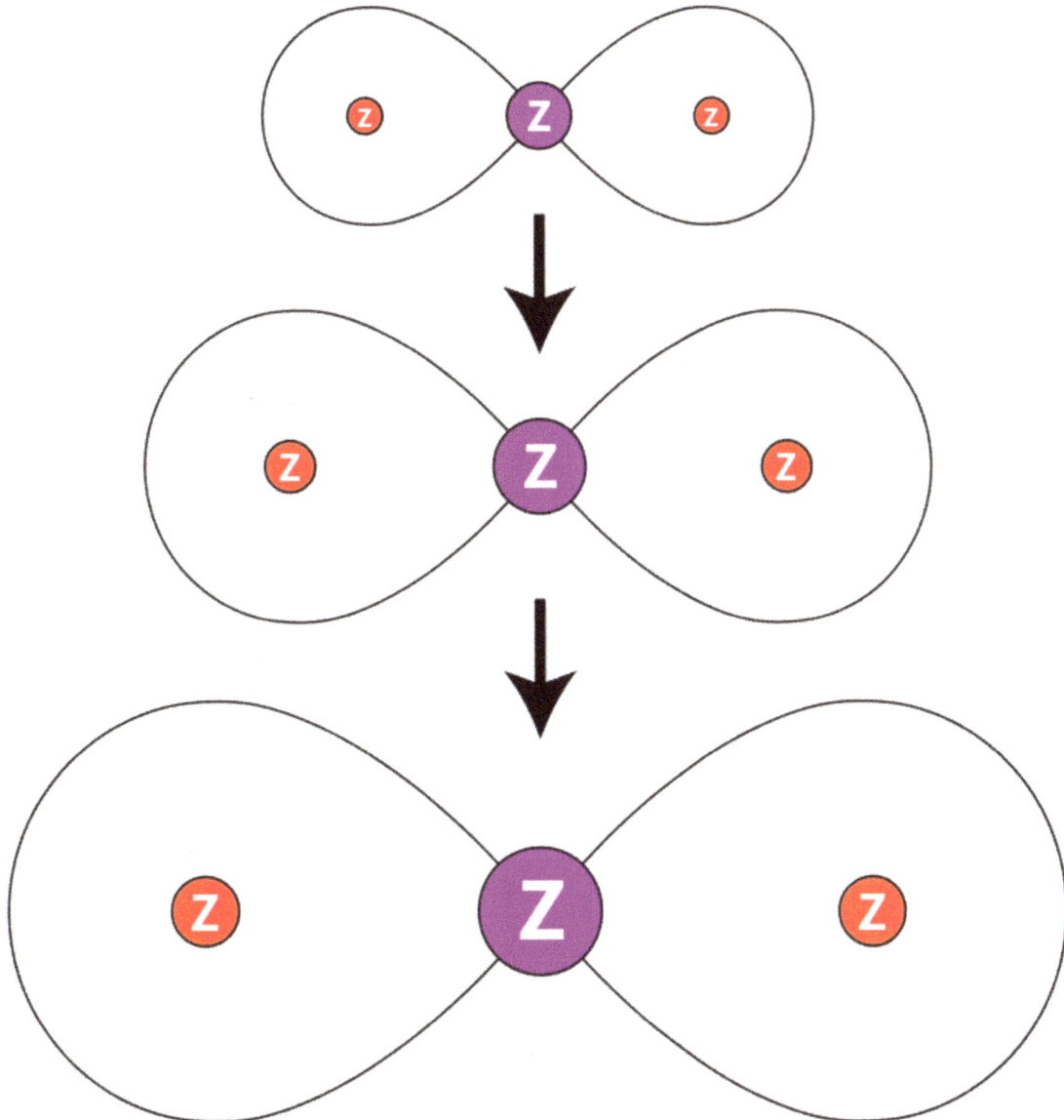

FIG 220: - AETHER AND MATTER ELEMENT AND DIMENSIONAL

APPLICATIONS FOR A PLASMOIDS FORM AND FUNCTIONS

ABSTRACT

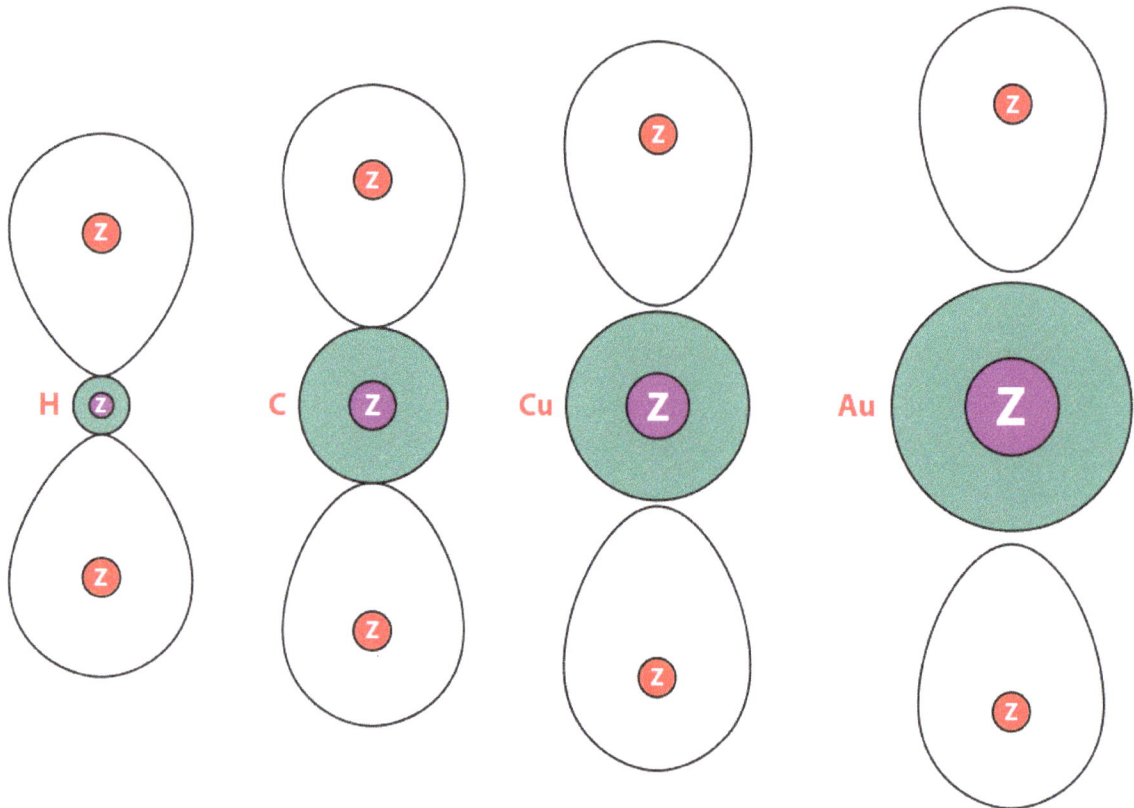

FIG 221: - AETHER AND MATTER ELEMENT AND DIMENSIONAL

APPLICATIONS FOR A PLASMOIDS FORM AND FUNCTIONS

ABSTRACT

FIG 222 AND 223: - AETHER AND MATTER ELEMENT AND DIMENSIONAL

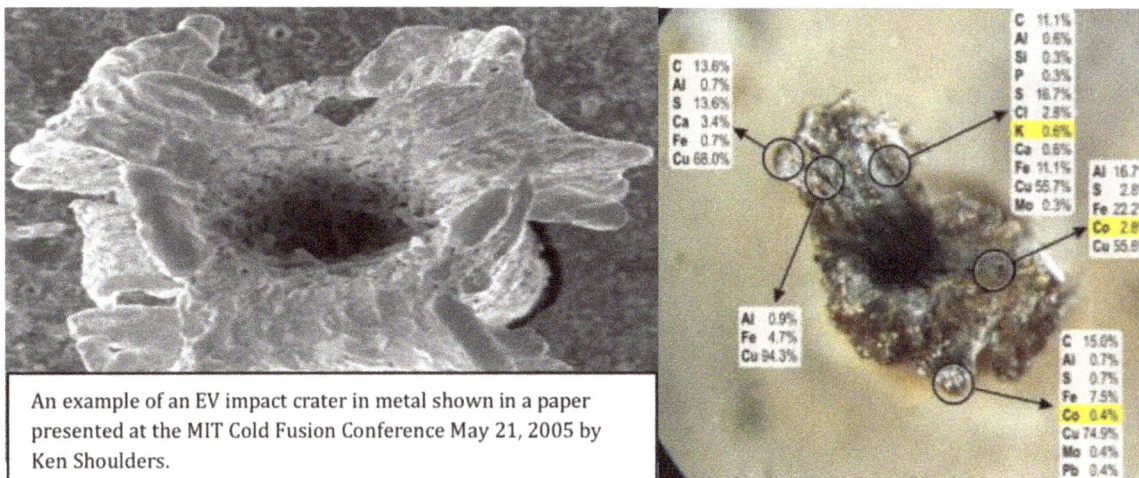

An example of an EV impact crater in metal shown in a paper presented at the MIT Cold Fusion Conference May 21, 2005 by Ken Shoulders.

FIG 224 AND 225: - AETHER AND MATTER ELEMENT AND DIMENSIONAL

APPLICATIONS FOR A PLASMOIDS FORM AND FUNCTIONS

ABSTRACT

FIG 226 AND 227 : - AETHER AND MATTER ELEMENT AND DIMENSIONAL

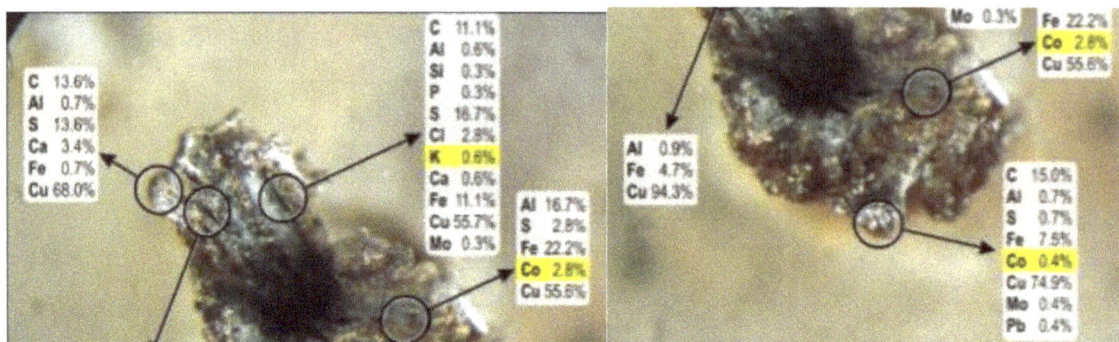

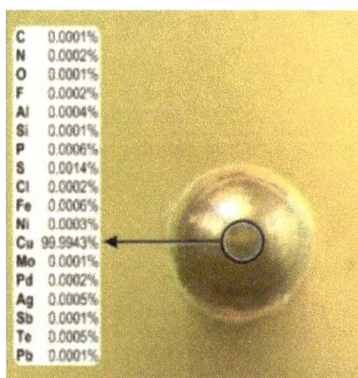

FIG 228: - AETHER AND MATTER ELEMENT AND DIMENSIONAL

APPLICATIONS FOR A PLASMOIDS FORM AND FUNCTIONS

INTRODUCTION

FIG 229 AND 230 : - AETHER AND MATTER ELEMENT AND DIMENSIONAL

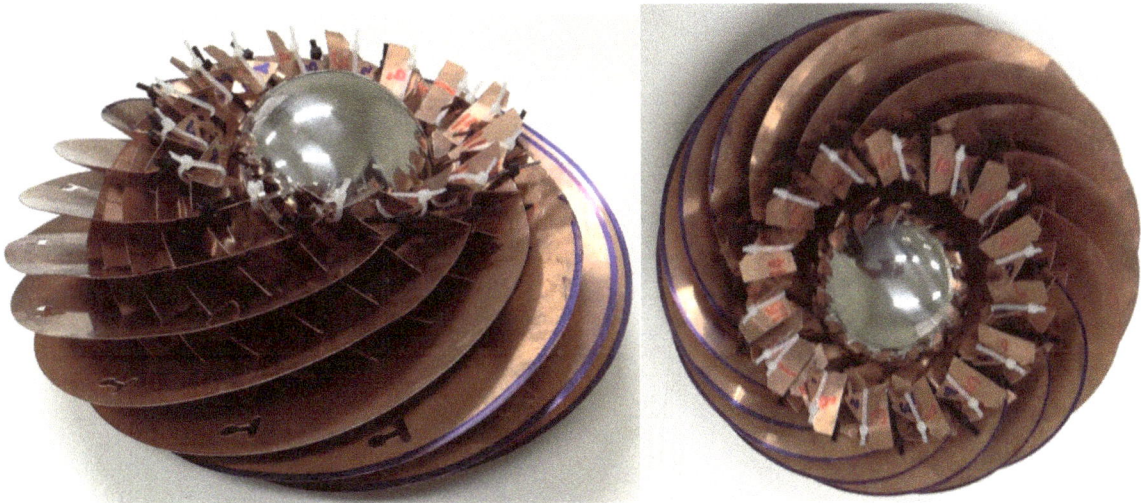

FIG 231 AND 232 : - AETHER AND MATTER ELEMENT AND DIMENSIONAL

APPLICATIONS FOR A PLASMOIDS FORM AND FUNCTIONS

INTRODUCTION

a = 2,160
b = 2,880
c = 3,600
P = 8,640
S = 4,320
K = 3,110,400
ha = 2,880
hb = 2,160
hc = 1,728

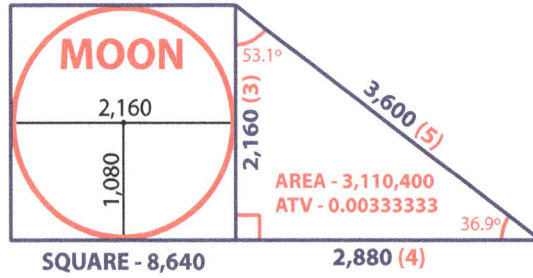

MOON

2,160

1,080

2,160 (3)

53.1°

3,600 (5)

AREA - 3,110,400
ATV - 0.00333333

36.9°

SQUARE - 8,640

2,880 (4)

a = 7,920
b = 10,560
c = 13,200
P = 31,680
S = 15,840
K = 41,817,600
ha = 10,560
hb = 7,920
hc = 6,336

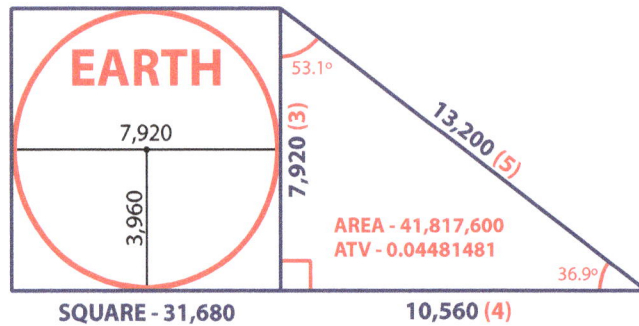

EARTH

7,920

3,960

7,920 (3)

53.1°

13,200 (5)

AREA - 41,817,600
ATV - 0.04481481

36.9°

SQUARE - 31,680

10,560 (4)

a = 864,000
b = 1,152,000
c = 1,440,000
P = 3,456,000
S = 1,728,000
K = 497,664,000,000
ha = 1,152,000
hb = 864,000
hc = 691,200

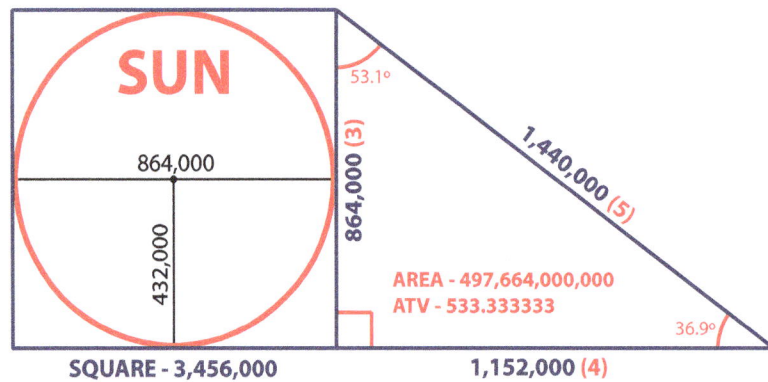

SUN

864,000

432,000

864,000 (3)

53.1°

1,440,000 (5)

AREA - 497,664,000,000
ATV - 533.333333

36.9°

SQUARE - 3,456,000

1,152,000 (4)

DIAGRAM 233 – SUN, MOON AND EARTH ATV CALCULATIONS WITH RIGHT TRIANGLE CALCULATIONS

APPLICATIONS FOR A PLASMOIDS FORM AND FUNCTIONS

INTRODUCTION

EXPLANAATION OF THE CHANGE IN VARIABLE LIGHT SPEED AND TIME

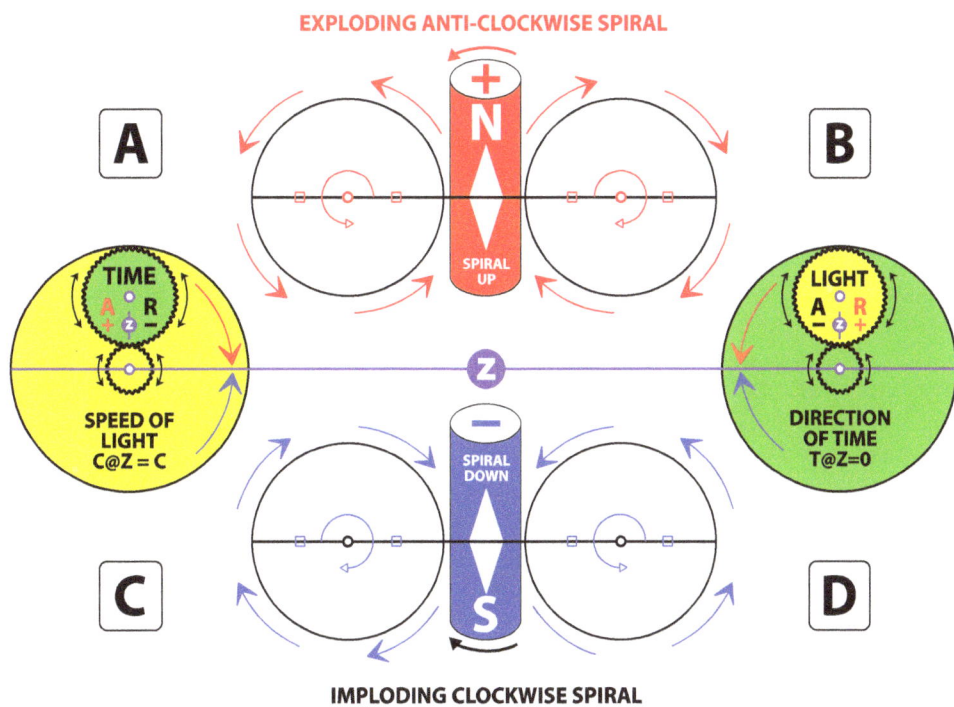

EXPLODING ANTI-CLOCKWISE SPIRAL

IMPLODING CLOCKWISE SPIRAL

A

RED A = Time acceleration caused by an anti-clockwise spiral explosive force repelled by the high charge density zero point and attracted to and seeking a lower charge density.

BLUE R = Time reversal caused by a clockwise imploding spiral repelled by a low charge density and attracted to and seeking the high charge density zero point.

B

- As the Negative charge increases, time reverses.
- As the Positive increases, time accelerates
- As the Negative charge increases, the speed of Light increases.
- As the Positive charge increases, the speed of Light decreases.

C

C is only constant at the Zero Point. Where the [AC] left hand clockwise spin [L] is zeroed out by the [AC] right hand anti-clockwise spin [R], thereby creating a zero frequency and consequently a DC point.

Therefore, E only equals M (mass) times the speed of light squared when L/R equals one at zero point.

Therefore E = MC²
E = L/R (C x L/R)²

D

Arrow Tail Energy going from you

Arrowhead Energy coming into you

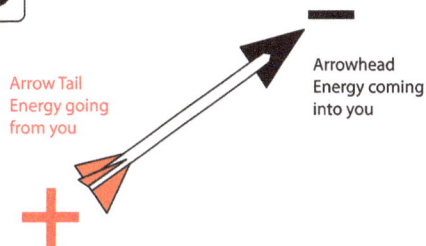

DIAGRAM 234 – VARIABLE LIGHT SPEED AND TIME CALCULATOR

APPLICATIONS FOR A PLASMOIDS FORM AND FUNCTIONS

FIG 235: - AN INTRODUCTION TO PLASMOID VORTEX MATHS LAWS

LAW 1.) – *A number is primarily used to identify the dimension it represents; 0 & 1 are out.*

Example 1 A :- The Sun's diameter in miles (Aether) = 864,000
864 Octaves down // 432 // 216 // 108 //54 // 27 // 13.5
864 Octaves up // 1,728 // 3,456 // 6,912 // 13,824 //27,648 // 55,296 // 110,592

Example 1 B :- The Sun's square in miles (Matter) = 3,456,000
3,456 Octaves down // 1,728 // 864 // 432 // 216 // 108 // 54 // 27 // 13.5 // 6.75 // 3.375 // 1.6875
3,456 Octaves up // 6,912 // 13,824 // 27,648 // 55,296 // 110,592 // 221,184 // 442,368 // 884,736

Example 1 C :- The Moon's diameter in miles (Matter) = 2,160
2,160 Octaves down // **1,080 Moon's R** // 540 // 270 // 135 // 67.5 // 33.75 // 16.875 // 8.4375
2,160 Octaves up // 4,320 Sun's R // 8,640 Sun's D // 17,280 // **34,560 Matter** // 69,120 // 138,240 // 276,480

Example 1 D :- The Sun's PROTIUM (Hydrogen H) (Aether to Matter product) = Melting point minus 259.2
259.2 Octaves down // **129.6** // **64.8 (SUN D)** // **32.4 (SUN R)** // **16.2 (MOON D)** // 8.1 // 4.05 // 2.025
259.2 Octaves up // 518.4 // 1,036.8 // 2,073.6 // 4,147.2 // 8,294.4 // 16,588.8 // 33,177.6

LAW 2.) – *Product of the numbers within a number are the Plasmoid Unification 1st Key.*

*Example 2 A :- Sun Aether = 864 so = 8 x 6 x 4 = **192** // 96 // 48 // 24 // 12 // 6 // 3 // 1.5 // 7.5*

*Example 2 B :- Sun Matter = 3,456 so = 3 x 4 x 5 x 6 = **360 Compass** // 180 // 90 // 45 // 22.5 / /11.25*

*Example 2 C :- Moon Matter = 2,160 so = 2 x 1 x 6 x 0 = **12 Compass** // 24 // 48 // 96 // 192 // 384*

*Example 2 D :- Sun Protium (H) = 259.2 so = 2 x 5 x 9 x 2 = **180 Compass** // 90 // 45 // 22.5 // 11.25*

LAW 3.) - Plasmoid Unification 1st Key numbers products are then multiplied by their Plasmoid Unification 2nd Key mirror numbers product, squaring law two's result.

*Example 3 A :- Sun Aether = **192** x 4 x 6 x 8 = **36,864** [36 Matter,864 Aether]*
36,864 Octaves down // 18,432 // 9,216 // 4,608 // 2,304 // 1,152 // 576 // 288 // 144 // 72 // 36 // 18 // 9
36,864 Octaves up // 73,728 // **147,456 (ARGON)** // 294,912 // 589,824 // 1,179,648 // 2,359,296 // 4,718,592

*Example 3B :- Sun Matter = **360** x 6 x 5 x 4 x 3 =**129,600** [Resonant Frequency Energy Unit]*
129,600 Octaves down // **64,800** // **32,400** // **162** // 81 // 40.5 // 20.25 // 10.125 // 5.0625 // 2.53125
129,600 Octaves up // **259,200** // **518,400** // 1,036,800 // 2,073,600 // 4,147,200 // 8,294,400 // 16,588,800

*Example 3C :- Moon Matter = **12** x 0 x 6 x 1 x 2 = **144** [light]*
144 [light] Octaves down // 72 // 36 // 18 // 9
144 [light] Octaves up // 288 // 576 // 1,152 // **2,304** // **4,608** // 9,216 // 18,432 // **36,864 SUN Z** // 73,728

*Example 3D :- Hydrogen (Aether to Matter product) = **180** X 2 x 9 x 5 x 2 = **32,400** [r]*
32,400 [r]Octaves down // 18,432 // 9,216 // 4,608 // 2,304 // 1,152 //
32,400 [r] Octaves up // 64,800 // 129,600 // 259,200 // 518,400 // 1,036,800 // 2,073,600 // 4,147,200

- **259.2 multiplied product = 2 x 5 x 9 x 2 (180) x 2 x 9 x 5 x 2 = 32,400**

APPLICATIONS FOR A PLASMOIDS FORM AND FUNCTIONS

FIG 235: - AN INTRODUCTION TO PLASMOID VORTEX MATHS LAWS

- 259.2 **added sum** = 2 + 5 + 9 + 2 [18] + 2 + 5 + 9 + 2 = 36

- 259.2 **multiplied product** = 2 x 5 x 9 x 2 (180) x 2 x 9 x 5 x 2 = 32,400

32,400 / 720 = 45   32,400 / 16 = 2,025   32,400 / 22.5 = 1,440   32,400 / 0.125 = 259,200

32,400 x 16 = 518,400   32,400 x 22.5 = 729,000   32,400 x .125 = 4,050

32,400 x 400 = 12,960,000   32,400 x 720 = 23,328,000   32,400 x 900 = 29,160,000

518,400 / 16 = 32,400   518,400 / 22.5 = 23,040   518,400 / 720 = 720

32,400 / 144 = 225   32,400 / 36 = 900   36 / 32,400 = 0.00111

144,000 / 32,400 = 4.4444   32,400 / 7.5 = 4,320 / 7.5 = 576 / 2 = 288 / 2 = 144

144 x 32,400 = 4,665,600 / 432 = 10,800 / 2,160 = 5

32,400 / 259.2 = 125

32,400 / 6.666 = 4,860   32,400 / 5.555 = 5,850   32,400 / 4.444 = 7,290

32,400 / 3.333 = 9,720   32,400 / 2.222 = 14,580   32,400 / 11.111 = 2,916

LAW 4.) - *Plasmoid Unification 2nd Key mirror numbers Octaves identify Model of the Element (MOE) and Plasmoid Unification Model (PUM) Octave planes. Those planes*

Example : Sun Aether = 36,864
36,864 // 18,432// 9,216// 4,608// 2,304 // 1,152 // 576 // 288 // 144 // 72 // 36 //18 // 9

Example : Sun Matter = 129,600
129,600 // 64,800 // 32,400 // 16,200 // 8,100 // 4,050 // 2,025 // 1,012.5 // 506.25 // 253.125

STEP 5.) -

Example the Sun's diameter in miles Aether = 864,000
Example the Sun's square in miles Matter = 3,456,000

STEP 6.) -

Example the Sun's diameter in miles Aether = 864,000
Example the Sun's square in miles Matter = 3,456,000

STEP 7.) -

APPLICATIONS FOR A PLASMOIDS FORM AND FUNCTIONS

ALL SIX 864,000 COMBINATIONS

864,000	846,000	684,000	648,000	486,000	468,000
56,623,104*	55,443,456*	44,826,624*	42,467,328*	31,850,496*	30,670,848*
28,311,552*	27,721,728*	22,413,312*	21,233,664*	15,925,248*	15,335,424*
14,155,776*	13,860,864*	11,206,656*	10,616,832*	7,962,624*	7,667,712*
7,077,888*	6,930,432*	5,603,328*	5,308,416*	3,981,312*	3,833,856*
3,538,944*	3,465,216*	2,801,664*	2,654,208*	1,990,656*	1,916,928*
1,769,472*	1,732,608*	1,400,832*	1,327,104*	995,328*	958,464*
884,736*	866,304*	700,416*	663,552*	497,664*	479,232*
442,368*	433,152*	350,208*	331,776*	248,832*	239,616*
221,184*	216,576*	175,104*	165,888*	124,416	119,808*
110,592*	108,288*	87,552*	82,944*	62,208*	59,904*
55,296*	54,144*	43,776*	41,472*	31,104*	29,952*
27,684*	27,072*	21,888*	20,736*	15,552*	14,976*
13,824*	13,536*	10,944*	10,368*	7,776*	7,488*
6,912,000	6,768,000	5,472,000	5,184,000	3,888,000	3,744,000
3,456,000	3,384,000	2,736,000	2,592,000	1,944,000	1,872,000
1,728,000	1,692,000	1,368,000	1,296,000	972,000	936,000

864,000	846,000	684,000	648,000	486,000	468,000
432,000	423,000	342,000	H-324,000	243,000	234,000
216,000	211,500	171,000	162,000	121,500	117,000
108,000	105,000	85,500	81,000	60,750	58,500
54,000	52,875	42,750	40,500	30,375	29,250
27,000	26,437.5	21,375	20,250	15,187.5	14,625
13,500	13,218.75	10,687.5	10,125	7,593.75	7,312.5
6,750	6,609.375	5,343.75	5,062.5	3,796.875	3,656.25
3,375	3,304.6875	2,671.875	2,531.25	1,898.43	1,828.1
1,687.5	1,652.34375	1,335.9375	1,265.62	949.2187	914.06
843.75	826.171875	667.96875	632.812	474.6093	457.03
421.875	413.0859375	333.984375	316.406	237.3046	228.51
210.9375	206.54296875	166.99218	158.2031	118.6523	114.25
105.46875	103.271484375	83.496093	79.10156	59.32617	57.128
52.734375	51.6357421875	41.7480468	39.55078	29,66308	28.564
26.3671875	25.8187109375	20.8740234	19.77539	14.83154	14.282

864,000	846,000	684,000	648,000	486,000	468,000

APPLICATIONS FOR A PLASMOIDS FORM AND FUNCTIONS

ABSTRACT

PART 10 OF 20 – MSAART PATENT APPLICATION NOTES - DRAFT 518,400 – 22:22:22 THUR 22ND SEPT 2022
© STRIKE FOUNDATION GUARANTEE LIMITED | MALCOLM BENDALL 2022 | GRAPHICS - STEVE EARL

FIG 236: - *STEP BY STEP DESCRIPTION OF THE METHODOLOGY TO CALCULATE THE RESONANCE OF ELEMENTS AND MOLECULES SO AS TO INDUCE IMPLOSIVE STATES THAT WILL ASSIST PLASMOIDS TO INDUCE FUSION TRANSMUTATIONS*

STEP 1.) – *Determine the melting point of the element in degrees Celcius.*

Protium H, Ne and Ar resonant group example = H = - 259.2 C.
Helium, Kr and Rn resonant group example = He - 272.2, Kr – 157.4 and Rn 71.15
Tungsten example =
Palladium example =
Potassium K,

STEP 2.) – *Multiply the numbers*

Protium Example =

STEP 3.) -
Protium Example =

STEP 4.)
Protium Example =

STEP 5.)
Protium Example =

STEP 6.)
Protium Example =

STEP 7.)
Protium Example =

STEP 8.)
Protium Example =

STEP 9.)
Protium Example =

STEP 10.)
Protium Example =

APPLICATIONS FOR A PLASMOIDS FORM AND FUNCTIONS

INTRODUCTION

FIG 237: - AETHER AND MATTER ELEMENT AND DIMENSIONAL

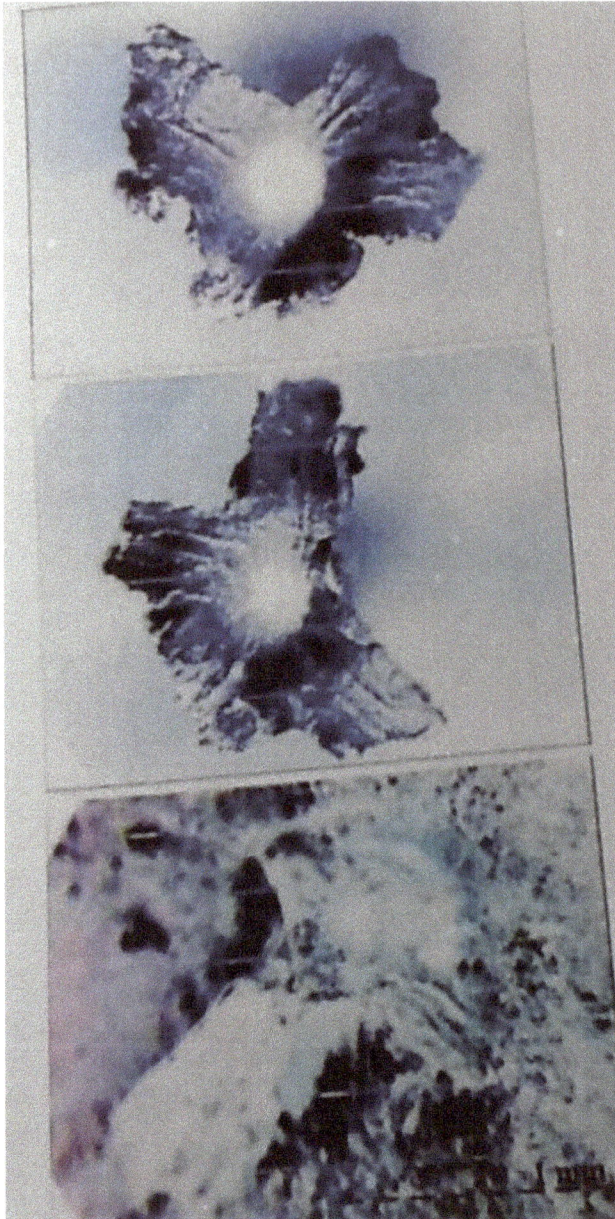

APPLICATIONS FOR A PLASMOIDS FORM AND FUNCTIONS

INTRODUCTION

FIG 238: - AETHER AND MATTER ELEMENT AND DIMENSIONAL

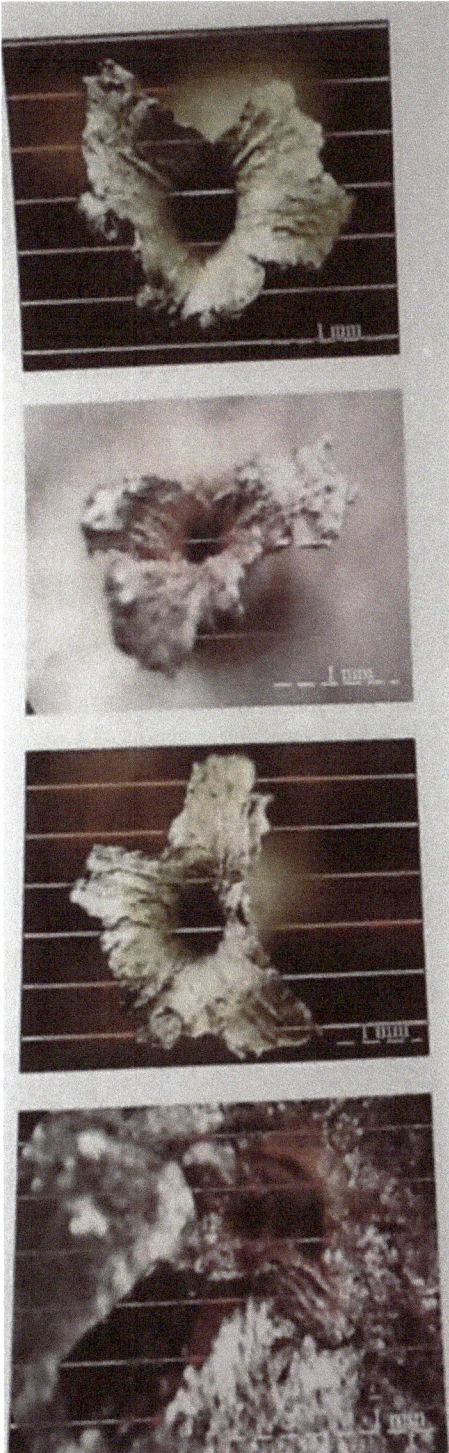

APPLICATIONS FOR A PLASMOIDS FORM AND FUNCTIONS

INTRODUCTION

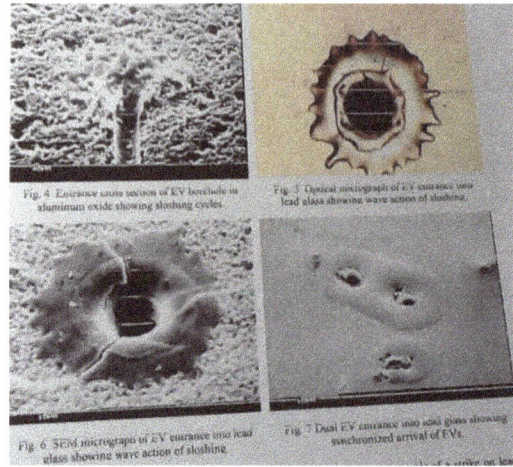

FIG 239: - AETHER AND MATTER ELEMENT AND DIMENSIONAL

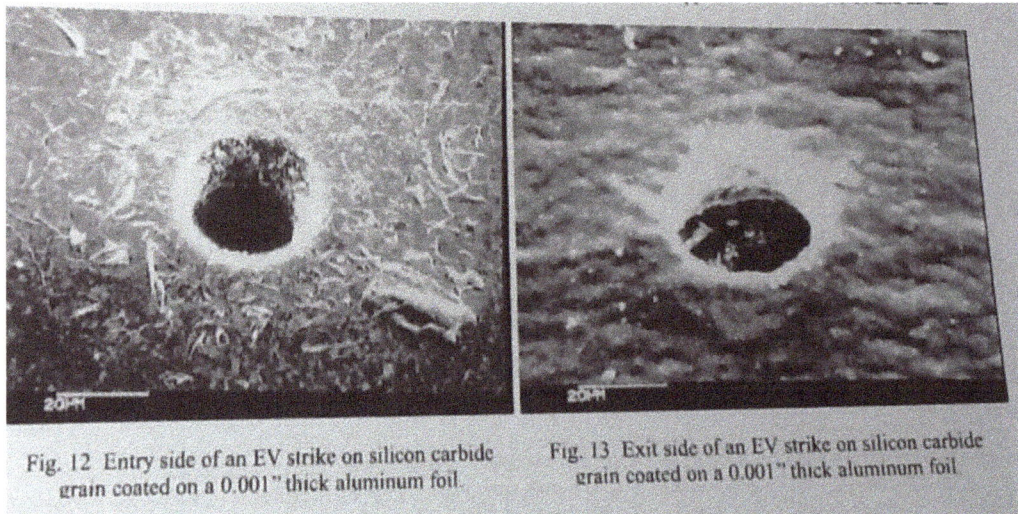

FIG 240: - AETHER AND MATTER ELEMENT AND DIMENSIONAL

APPLICATIONS FOR A PLASMOIDS FORM AND FUNCTIONS

INTRODUCTION

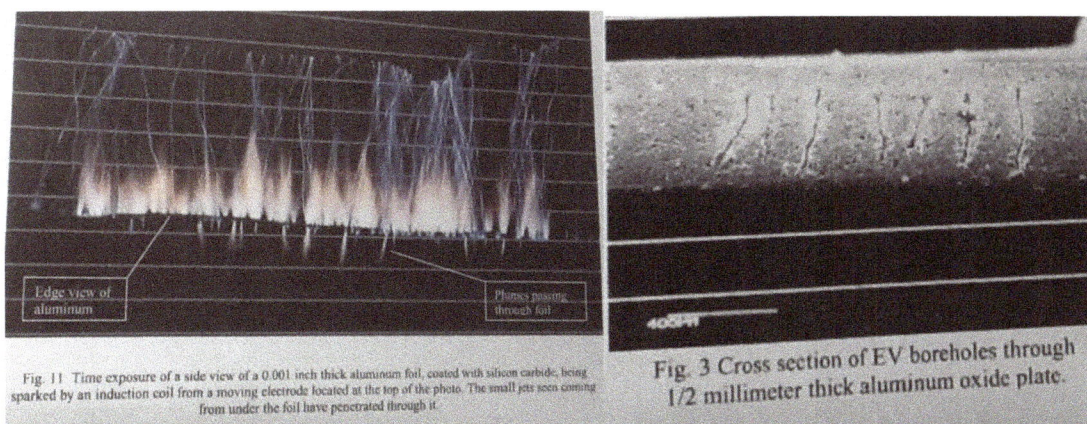

Fig. 11 Time exposure of a side view of a 0.001 inch thick aluminum foil, coated with silicon carbide, being sparked by an induction coil from a moving electrode located at the top of the photo. The small jets seen coming from under the foil have penetrated through it.

Fig. 3 Cross section of EV boreholes through 1/2 millimeter thick aluminum oxide plate.

FIG 241 AN 242: - AETHER AND MATTER ELEMENT AND DIMENSIONAL

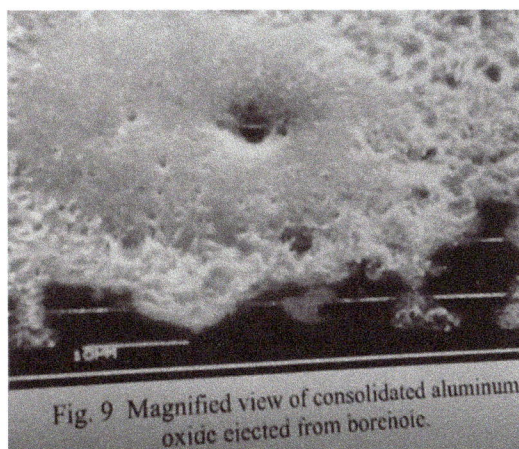

Fig. 9 Magnified view of consolidated aluminum oxide ejected from borehole.

FIG 243: - AETHER AND MATTER ELEMENT AND DIMENSIONAL

APPLICATIONS FOR A PLASMOIDS FORM AND FUNCTIONS

INTRODUCTION

FIG 244 AND 245 : - AETHER AND MATTER ELEMENT AND DIMENSIONAL

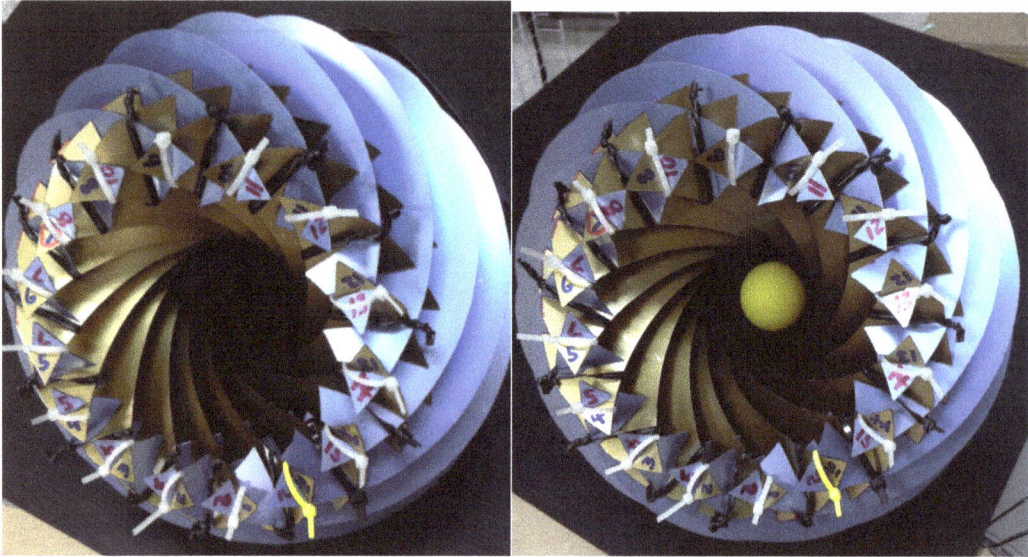

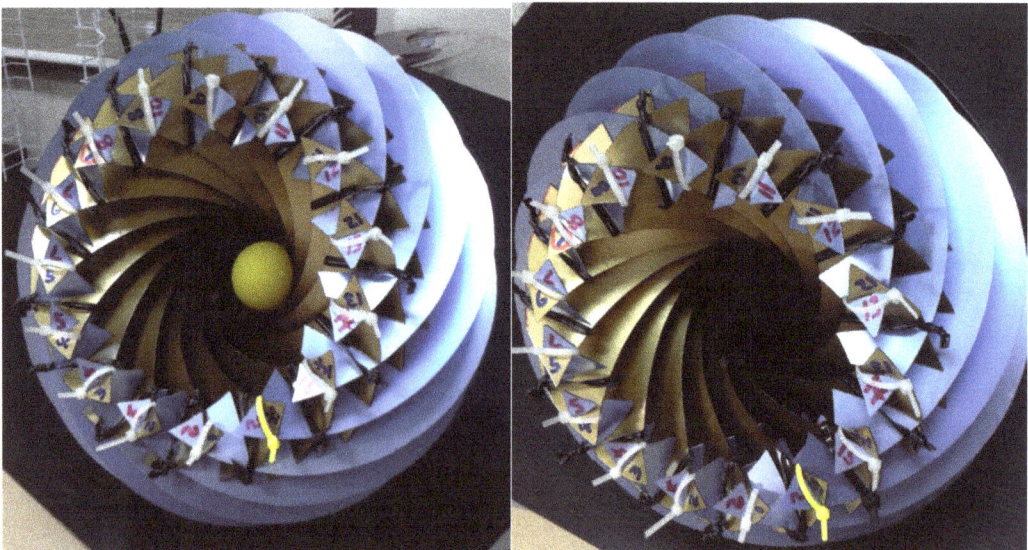

FIG 246 AND 247 : - AETHER AND MATTER ELEMENT AND DIMENSIONAL

APPLICATIONS FOR A PLASMOIDS FORM AND FUNCTIONS

INTRODUCTION

FIG 248 AND 249 : - AETHER AND MATTER ELEMENT AND DIMENSIONAL

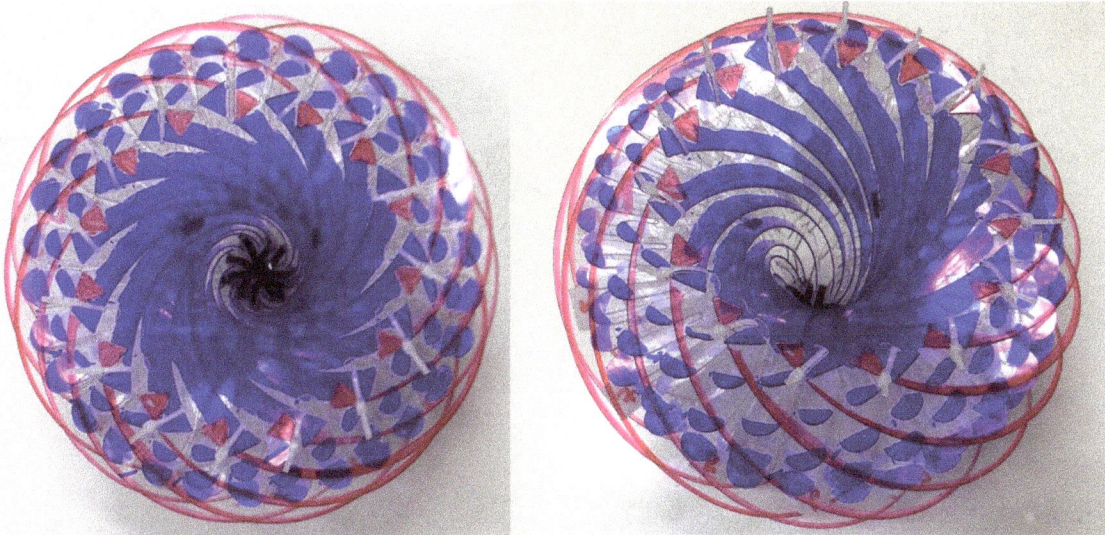

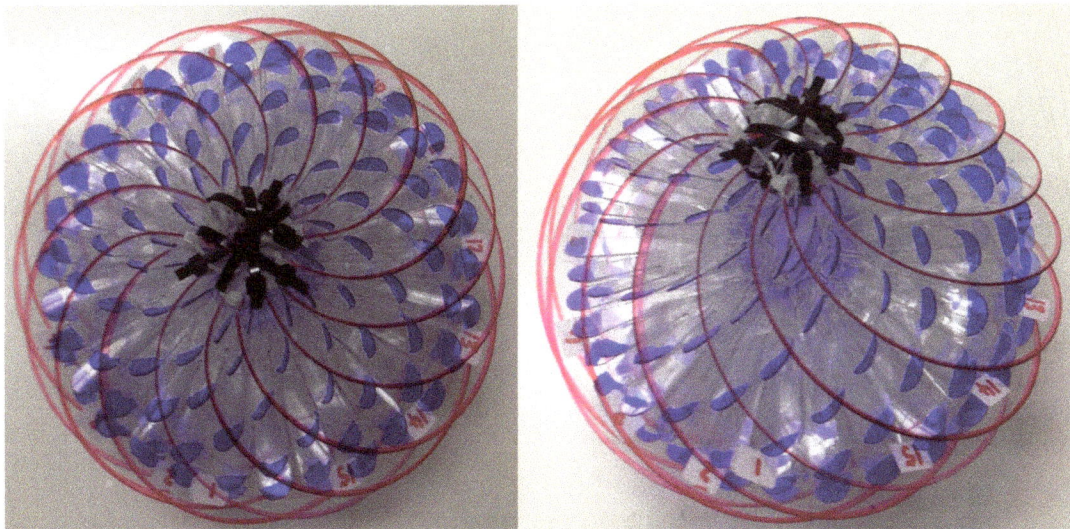

FIG 250 AND 251: - AETHER AND MATTER ELEMENT AND DIMENSIONAL

APPLICATIONS FOR A PLASMOIDS FORM AND FUNCTIONS

INTRODUCTION

FIG 252 AND 253 : - AETHER AND MATTER ELEMENT AND DIMENSIONAL

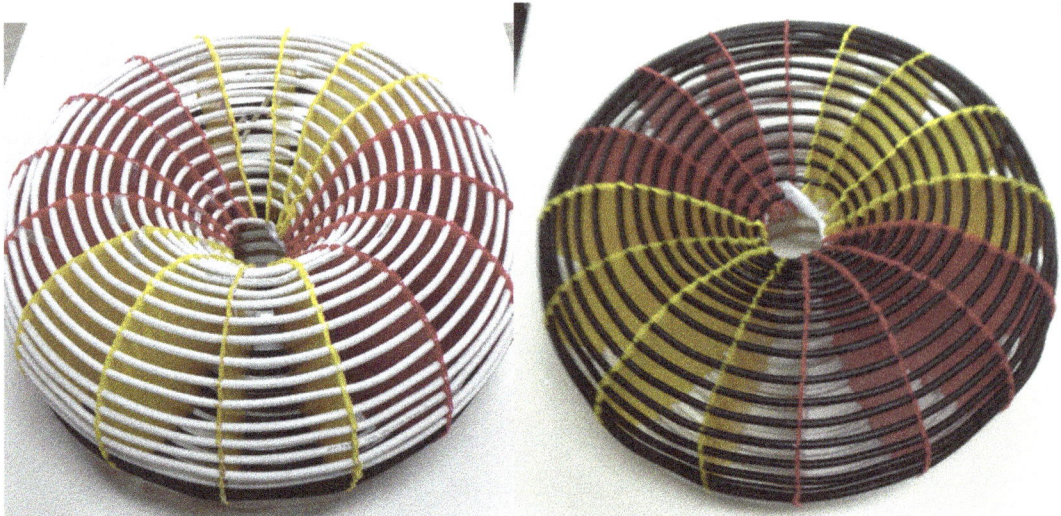

FIG 254 AND 255 : - AETHER AND MATTER ELEMENT AND DIMENSIONAL

APPLICATIONS FOR A PLASMOIDS FORM AND FUNCTIONS

INTRODUCTION

FIG 256 : - AETHER AND MATTER ELEMENT AND DIMENSIONAL

APPLICATIONS FOR A PLASMOIDS FORM AND FUNCTIONS

GROUP ONE

**PLASMOID ZP INDUCED LANDFILL LEACHATE LIQUID
LOW ENERGY ATOMIC TRANSMUTATIONS (LEAT) &
COLD FUSION TRANSMUTATIONS**

Elemental pairs Na – Mg and K – Ca melting points and AVT's.

Na (11) – Sodium's melting point = 97.79 C

9 x 7 x 7 x 9 = 3,969 x 9 x 7 x 7 x 9 = **15,752,961**

9 + 7 + 7 + 9 = 32 + 9 + 7 + 7 + 9 = **64**

AVT = 15,752,961 / 24 Hours / 60 Min / 60 sec / 60 Arc sec = 3.038765625

3.038765625 / 16 / 22.5 x 400 =

Mg (13) – Magnesium's melting point = 650 C

6 x 5 = 30 x 6 x 5 = **900**

// 921,6 // 460,8 // 230,4 //115,2 // 57,6 // 28,8 // 14,4 // 7,2 // 3,6 // 1,8 // **900**

6 + 5 = **11**

AVT = **900** / 24 = 37.5/60 = 0.625/60 = 0.01041667/60 = 0.0001736111   //////0.0111

K (19) – Potassium's melting point = 63.5 C

[Cl, Ca & Ni]

6 x 3 x 5 = 90 x 6 x 3 x 5 = *8,100*

Mirror Multiplied = 518,400//259,200//129,600//64,800//32,400//16,200// *8,100* //4,050//2,025//1,012.5

Mirror Added = *28*

AVT =

Ca (20) – Calcium's melting point = 842 C

842 Calcium mirror multiplied = *4,096* // 2,048// 1,024// 512// 256// 128 // 64// 32// 16

842 Calcium mirror added = *28*

AVT =

APPLICATIONS FOR A PLASMOIDS FORM AND FUNCTIONS

GROUP ONE

PLASMOID ZP INDUCED
LOW ENERGY ATOMIC TRANSMUTATIONS (LEAT) &
COLD FUSION TRANSMUTATIONS

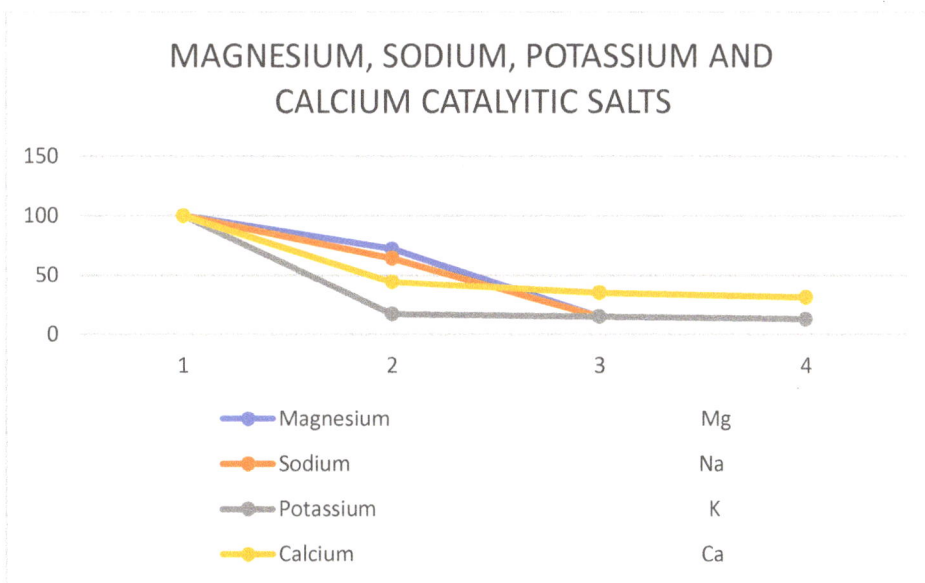

FIG 257 : - AETHER AND MATTER ELEMENT AND DIMENSIONAL

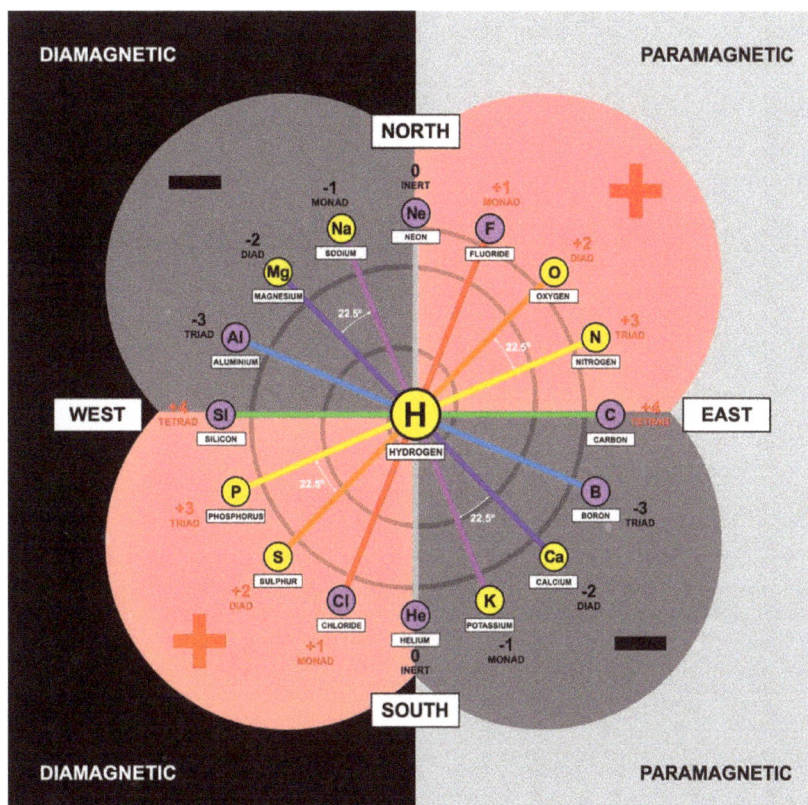

FIG 258 : - AETHER AND MATTER ELEMENT AND DIMENSIONAL

APPLICATIONS FOR A PLASMOIDS FORM AND FUNCTIONS

GROUP ONE

PLASMOID ZP INDUCED
LOW ENERGY ATOMIC TRANSMUTATIONS (LEAT) &
COLD FUSION TRANSMUTATIONS

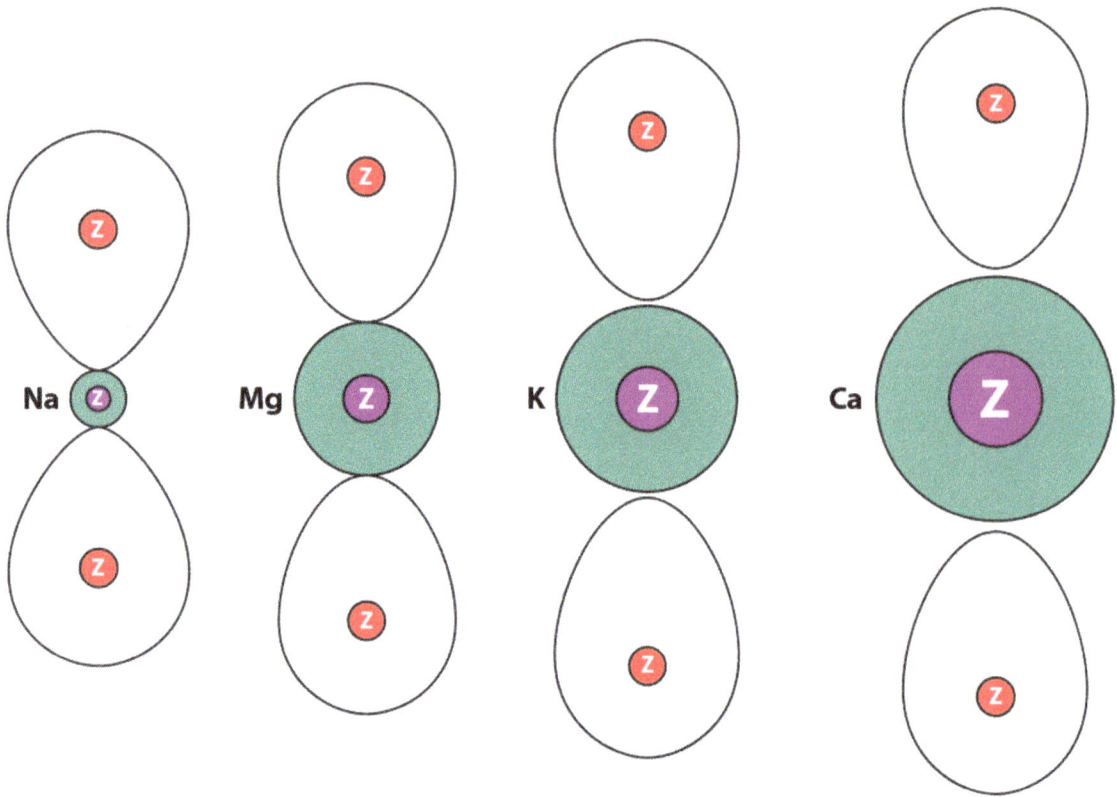

FIG 259 : - AETHER AND MATTER ELEMENT AND DIMENSIONAL

APPLICATIONS FOR A PLASMOIDS FORM AND FUNCTIONS
ELEMENTS AND MOLECULES SUBTRACTED FROM THE NORLANDS TIP   LEACHATE FLUID
AFTER 3 MINUTES WITHIN THE PLASMOID CREATOR

ELEMENTS SUBTRACTED FROM THE LEACHATE AFTER 3 MINUTES OF OPERATION

MAGNESIUM (86 – 24 mg/L) ....DOWN 72%

SODIUM (1,279 – 462 mg/L) ....DOWN 64%
POTASSIUM (638 – 532 mg/L) ....…..DOWN 17%
CALCIUM (72 – 32 mg/L)... DOWN 44%

*CHLORIDE (*1796 -504)...DOWN 70%
ARSENIC (100 – 52ug/L)...DOWN 48%
CHROMIUM (53 – 26 ug/L)...DOWN 49%
NICKLE (140 – 64 ug/L)... DOWN 54%
IRON (2.20 – 2.00 ug/L)...DOWN 10%
MANGANESE (150 – 120 ug/L)...DOWN 20%

MOLECULES *SUBTRACTED* FROM THE LEACHATE AFTER *3 MINUTES* OF OPERATION

DISSOLVED *METHANE* (0.59 – 0.53 mg/L)...............................DOWN 10%
AMMONIA [AS NH4] (1,787 – 576)............................... DOWN 68%
AMMONIACAL NITROGEN [AS N] (1,386 – 446 mg/L) DOWN 68%
TOTAL *OXIDISED NITROGEN* (3.6 – 2.6 mg/L)............... DOWN 28%
NITRITE (0.01 – 0.59 mg/L)...UP 98%
NITRATE (21.3 – 26.1mg/L)...UP 18%
CHEMICAL OXYGEN DEMAND (2,200 – 770 mg/L)..............DOWN 65%
BIOCHEMICAL OXYGEN DEMAND (8.4 – 8.0 mg/L)...............DOWN 5%
SULPHATE (136 – 45 mg/L)...DOWN 67%
SULPHIDE (0.10 – 0.03)...DOWN 70%
PHOSPHATE (16.49 – 5.95)...DOWN 65%
TOTAL ORGANIC CARBON (990 – 355 mg/L)...............................DOWN 65%
ALKALINITY – CARBONATE as CaCO3 (7,906 – 2,825 mg/L).....DOWN 65%
TOTAL CYANIDE (1.43 – 0.53 mg/L)...DOWN 63%
ELECTRICAL CONDUCTIVITY (12,445 -5,375 uS/cm)...............DOWN 43%

ELEMENTS *ADDED* TO THE LEACHATE AFTER *3 MINUTES* OF OPERATION

COPPER STANDARD (40 – 1,800 ug/L)...............................*UP 4,500%*
ZINC (220 – 270 ug/L)...UP 18%
LEAD (8 – 9 ug/L)...UP 11%

MOLECULES *ADDED* TO THE LEACHATE AFTER *3 MINUTES* OF OPERATION

NITRITE (0.01 – 0.59 mg/L)...UP 98%
NITRATE (21.3 – 26.1mg/L)...UP 18%
SUSPENDED SOLIDS (40 – 108 mg/L)...............................UP 63%
PH (PH 8.7 - PH 8.8)...UP 1%

APPLICATIONS FOR A PLASMOIDS FORM AND FUNCTIONS

GROUP TWO

PLASMOID ZP INDUCED POTASSIUM LOW ENERGY ATOMIC TRANSMUTATIONS (LEAT) & COLD FUSION TRANSMUTATIONS

*****Ni (28)** – Nickel's melting point = 1,455 C

*****Ca (20)** – Calcium's melting point = 842 C

*******K (19)** – Potassium Nucleus Potassium's melting point = 63.5 C

*****Cl (17)** - Chlorine's melting point = 101.5 C

Nickel (Ni), Calcium (Ca) and **Chlorine Cl** are transmuted by the Plasmoid's capture of electrons, a Proton and Hydroxide (OH) using a Potassium (K) by mechanism of sharing the Plasmoid's Primary large central Zero Point (the large open door) with Potassium's Zero Point.

The Ring Structures Plasmoids often form are made from the capture of other Plasmoids or Elements into the outer Secondary peripheral Circular Zero point located on the Zero point Event Horizon Plane (The two small outer closed doors.) at the centre of the outer toroidal ring.

*******K (19)** – Potassium Nucleus - Potassium's melting point = 63.5 C

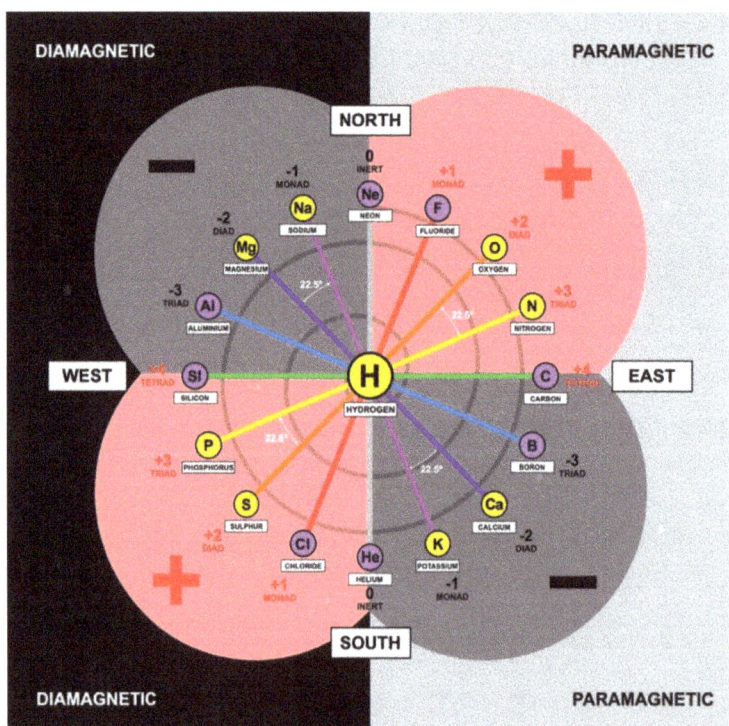

FIG 260 : - AETHER AND MATTER ELEMENT AND DIMENSIONAL
APPLICATIONS FOR A PLASMOIDS FORM AND FUNCTIONS

ELEMENTS AND MOLECULES SUBTRACTED FROM THE NORLANDS TIP   LEACHATE FLUID AFTER 3 MINUTES WITHIN THE PLASMOID CREATOR

ELEMENTS SUBTRACTED FROM THE LEACHATE AFTER 3 MINUTES OF OPERATION

MAGNESIUM (86 – 24 mg/L) ....DOWN 72%
SODIUM (1,279 – 462 mg/L) ....DOWN 64%
POTASSIUM (638 – 532 mg/L) ....DOWN 17%

CALCIUM (72 – 32 mg/L).. DOWN 44%

CHLORIDE (1796 -504)...DOWN 70%
ARSENIC (100 – 52ug/L)..DOWN 48%
CHROMIUM (53 – 26 ug/L)..DOWN 49%
NICKLE (140 – 64 ug/L)...................................... DOWN 54%
IRON (2.20 – 2.00 ug/L)...DOWN 10%
MANGANESE (150 – 120 ug/L)..DOWN 20%

MOLECULES SUBTRACTED FROM THE LEACHATE AFTER 3 MINUTES OF OPERATION

DISSOLVED METHANE (0.59 – 0.53 mg/L)................................DOWN 10%
AMMONIA [AS NH4] (1,787 – 576)............................... DOWN 68%
AMMONIACAL NITROGEN [AS N] (1,386 – 446 mg/L) DOWN 68%
TOTAL OXIDISED NITROGEN (3.6 – 2.6 mg/L)................. DOWN 28%
NITRITE (0.01 – 0.59 mg/L)...UP 98%
NITRATE (21.3 – 26.1mg/L)..UP 18%
CHEMICAL OXYGEN DEMAND (2,200 – 770 mg/L)..................DOWN 65%
BIOCHEMICAL OXYGEN DEMAND (8.4 – 8.0 mg/L)..................DOWN 5%
SULPHATE (136 – 45 mg/L)..DOWN 67%
SULPHIDE (0.10 – 0.03)..DOWN 70%
PHOSPHATE (16.49 – 5.95)...DOWN 65%
TOTAL ORGANIC CARBON (990 – 355 mg/L)...........................DOWN 65%
ALKALINITY – CARBONATE as CaCO3 (7,906 – 2,825 mg/L).....DOWN 65%
TOTAL CYANIDE (1.43 – 0.53 mg/L)......................................DOWN 63%
ELECTRICAL CONDUCTIVITY (12,445 -5,375 uS/cm)..................DOWN 43%

ELEMENTS ADDED TO THE LEACHATE AFTER 3 MINUTES OF OPERATION

COPPER STANDARD (40 – 1,800 ug/L)..................................UP 4,500%
ZINC (220 – 270 ug/L)..UP   18%
LEAD (8 – 9 ug/L)..UP   11%

MOLECULES ADDED TO THE LEACHATE AFTER 3 MINUTES OF OPERATION

NITRITE (0.01 – 0.59 mg/L)..UP 98%
NITRATE (21.3 – 26.1mg/L)..UP 18%
SUSPENDED SOLIDS (40 – 108 mg/L)...............................UP 63%
PH (PH 8.7 - PH 8.8)..UP 1%

GROUP TWO

PLASMOID ZP INDUCED POTASSIUM
LOW ENERGY ATOMIC TRANSMUTATIONS (LEAT)
&
COLD FUSION TRANSMUTATIONS

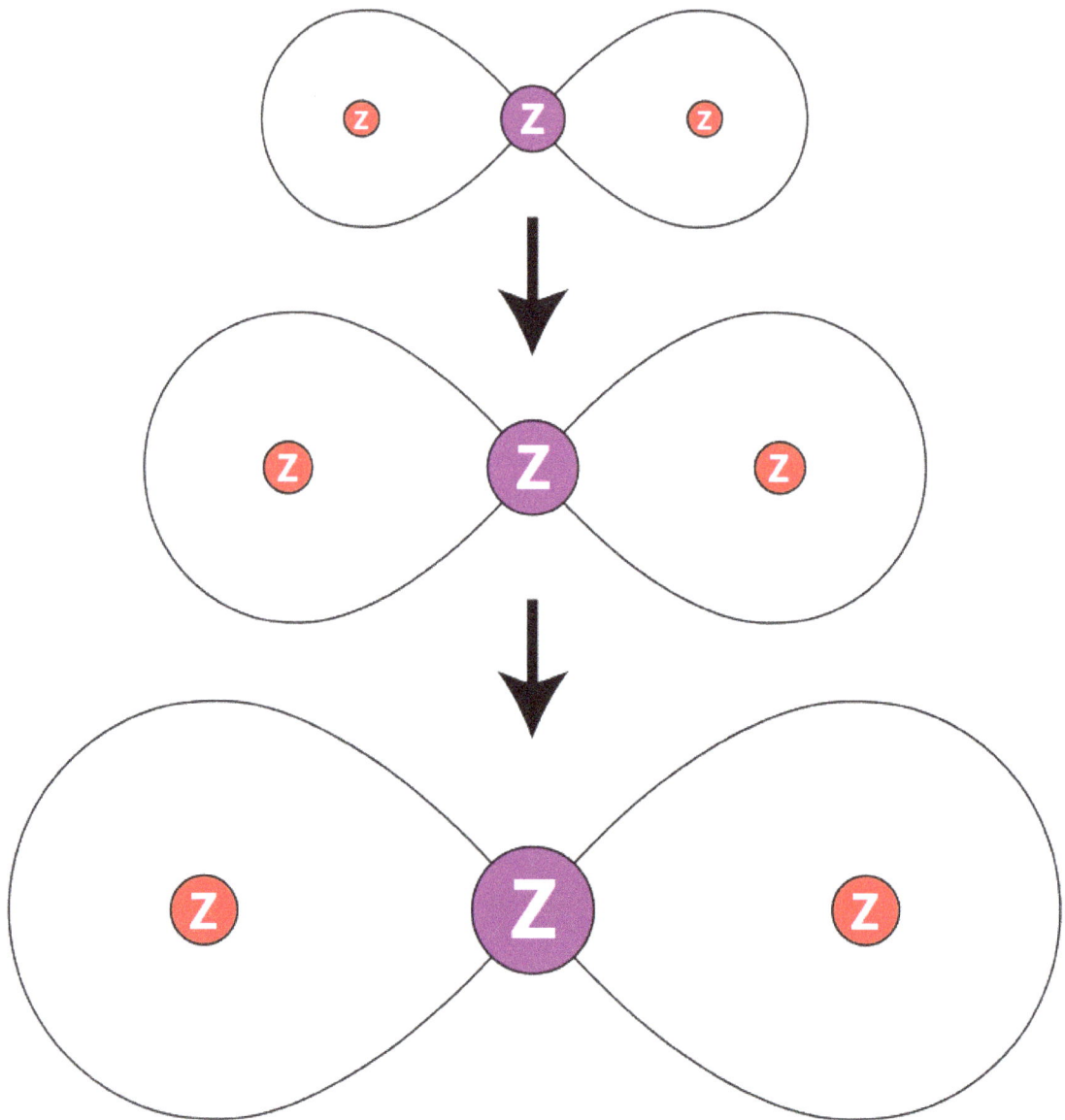

FIG 261 : - AETHER AND MATTER ELEMENT AND DIMENSIONAL

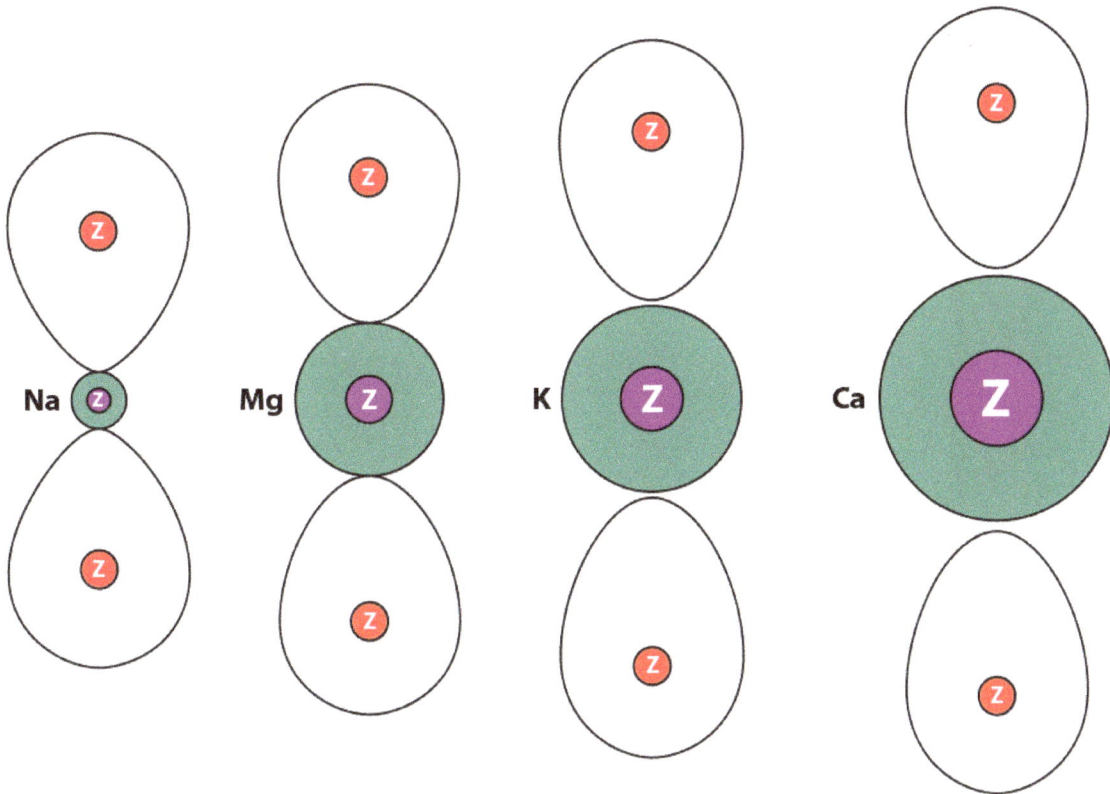

FIG 262 : - AETHER AND MATTER ELEMENT AND DIMENSIONAL

GROUP THREE

PLASMOID ZP INDUCED PALLADIUM COLD FUSION
&
LOW ENERGY ATOMIC REACTIONS (LEAR) TRANSMUTATIONS

***Cd (48)** - Cadmium's melting point = 321.1 C

*****Pd (46)** - Palladium Nucleus. Palladium's melting point = 1,555 C

***I (53)** – Iodine's melting point = 113.7 C

Cadmium (Cd) and **Iodine (I)** are transmuted by the Plasmoid's capture of a Proton and Oxygen (O) using a Palladium (Pd) by mechanism of sharing the Plasmoid's Primary large central Zero Point (the large open door) with Palladiums Zero Point.

The Ring Structures Plasmoids often form are made from the capture of other Plasmoids or elements into the outer Secondary ring peripheral Circular Zero point located on the Zero point Event Horizon Plane (The two small outer closed doors.) at the centre of the outer toroidal ring.

*****Pd (46)** - Palladium Nucleus. Palladium's melting point = 1,555 C

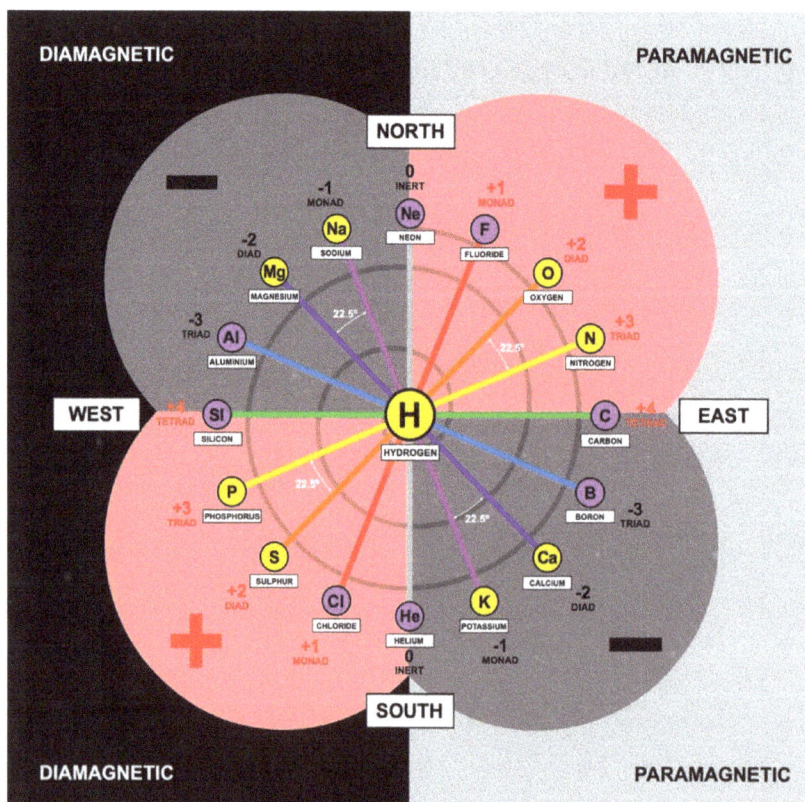

FIG 263: - AETHER AND MATTER ELEMENT AND DIMENSIONAL
GROUP THREE

PLASMOID ZP INDUCED PALLADIUM CENTRED COLD FUSION

&

LOW ENERGY ATOMIC REACTIONS (LEAR) TRANSMUTATIONS

***Ni (28) – Nickel's melting point = 1,455 C

***Ca (20) – Calcium's melting point = 842 C

***Cl (17) - Chlorine's melting point = 101.5 C

***K. (19) – Potassium Nucleus Potassium's melting point = 63.5 C

***Cd (48) - Cadmium's melting point = 321.1 C

***I (53) [126.9] – Iodine's melting point = 113.7 C

*****Pd (46) - Palladium Nucleus Palladium's melting point = 1,555 C

GROUP FOUR

THE PLASMOID ZP INDUCED 316 STAINLESS STEEL
LOW ENERGY ATOMIC TRANSMUTATIONS (LEAT)
&

COLD FUSION ELEMENTAL LIST

26 Fe 64% - Melting point = 1,538 C

28 **Ni 18% – Nickel's melting point = 1,455 C

24 Cr 14% – Chromium's Melting point = 1,907 C

42 Mo 3% – Molybdenum's Melting point = 2,623 C

25 Mn 2% – Manganese Melting point = 1,246 C

316 STAINLESS STEEL - Fe 63%, Cr 18%, Ni 14%, Mo 3% & Mn 2%.

4.1.) 26 IRON (Fe) CALCULATIONS.

IRON – Melting point – 1,538

IRON Melting point = 1,538 / 266.666 = 5.7675
and = 1,538 / 11.111 = 138.42
and = 1,538 / 1.333 = 1,153.5
and = 1,538 / 144 = 10.680555
and = 1,538 / 51.84 = 10.680555

1 + 5 + 3 + 8 = (17) + 8 + 3 + 5 + 1 = 34

1 x 5 x 3 x 8 = (120) x 8 x 3 x 5 x 1 = 14,400

(14,400) / 34 = 423.529412

(14,400) x 34 = 489,600

(14,400) / 266.666 = 54 // 108 // 216 // 432 // 864 // 1,728 // 3,456 // 6,912

(14,400) / 11.111 = 1,296 - // 648 // 324 // 162 // 81 // + 1,296 // 2,592 // 5,184 // 10,368

(14,400) / 1.333 = 10,800

(14,400) / 144 = 100

(14,400) / 51.84 = 277.777 // 555.555 // 1,111.111 // 2,222.222 // 4,444.444 // 8,888.888

4.2.) 24 CHROMIUM (Cr) CALCULATIONS.

CHROMIUM (Cr) – Melting point – 1,907 C

Melting point = 1,907 C / 266.666 = 7.15125   [1/x =]
and = 1,907 C / 11.111 = 171.63   [1/x =]

and = **1,907 C** / 1.333 = **1,430.25**   [1/x =]
and = **1,907 C** / 144 = **13.2430556**   [1/x =]
and = **1,907 C** / 51.84 =

1 + 9 + 0 + 7 = (17) + 7 + 0 + 9 + 1 = 34

1 x 9 x 0 x 7 = (63) x 7 x 0 x 9 x 1 = 3,969

(3,969) / 34 = **116.735294**
(3,969) x 34 = **134,946**

(3,969) / 266.666 = **14.88375**
(3,969) / 11.111 = **357.21**
(3,969) / 1.333 = **2,976.75**
(3,969) / 144 = **27.5625**
(3,969) / 51.84 = **76.5625**
(3,969) / 25.92 = **153.125**

4.3.) 28 NICKLE (Ni) CALCULATIONS.

***NICKLE (Ni) – Melting point – 1,455 C

Melting point = **1,455 C** / 266.666 = **5.45625**
and = **1,455 C** / 11.111 = **130.95**
and = **1,455 C** / 1.333 = **1.091.25**
and = **1,455 C** / 144 = **10.1041666**
and = **1,455 C** / 51.84 = **28.0671296**

1 + 4 + 5 + 5 = (15) + 5 + 5 + 4 + 1 = 30

1 x 4 x 5 x 5 = (100) x 5 x 5 x 4 x 1 = 10,000

(10,000) / 30 = **333.333**
(10,000) x 30 = **300,000**

(10,000) / 266.666 = **37.5**
(10,000) / 11.111 = **900**
(10,000) / 1.333 = **7,500**
(10,000) / 144 = **69.444**
(3,969) / 51.84 = **76.5625**
(3,969) / 25.92 = **153.125**
(5,184) / 12.96 = **400**

4.4.) 42 MOLYBDENUM (Mo) CALCULATIONS.

MOLYBDENUM (Mo) – Melting point – 2,623 C

Melting point = **2,623** / 266.666 = **9.83625**   [1 / x = 0.10166470452408]
and = **2,623** / 11.111 = **236.07**   [1 / x 0.004236031685517]

and = **2,623** / 1.333 = **1,967.25**   [1 / x = 0.000508323802262]
and = **2,623** / 144 = **18.2152777**   [1 / x = 0.0548989706443]
and = **2,623** / 51.84 = **50.59799382716049**

2 + 6 + 2 + 3 = (13) + 3 + 2 + 6 + 2 = 26

2 x 6 x 2 x 3 = (72) x 3 x 2 x 6 x 2 = 5,184

(5,184) / 26 = 199.3846
(5,184) x 26 = 134,784

(5,184) / 266.666 = 19.44 // 9.72 // 4.86 // 2.43 //
(5,184) / 11.111 = 466.56
(5,184) / 1.333 = 3,888
(5,184) / 144 = 36
(5,184) / 51.84 = 100
(5,184) / 12.96 = 400

4.5.) 25 MANGANESE (Mn) CALCULATIONS.

MANGANESE (Mn) – Melting point – 1,246 C

Melting point = **1,246 C** / 266.666 = **4.6725**
                and = **1,246 C** / 11.111 = **112.14**
                and = **1,246 C** / 1.333 = **934.5**
                and = **1,246 C** / 144 = **8.652777**
                and = **1,246 C** / 51.84 = **24.03549382716049**

1 + 2 + 4 + 6 = (13) + 6 + 4 + 2 + 1 = 26

1 x 2 x 4 x 6 = (48) x 6 x 2 x 4 x 6 = 2,304

(2,304) / 26 = 88.615384
(2,304) x 26 = 59,904

(2,304) / 266.666 = 8.64
(2,304) / 11.111 = 207.36 // 103.68 // 51.84 // 25.92 //12.96//6.48//3.24//1.62//0.81
(2,304) / 1.333 = 1,728 // 864 // 432 // 216 // 108 // 54 // 27 // 13.5 //
(2,304) / 144 = 16
(2,304) / 51.84 = 4.444

4.6.) SUMMARY

ELEMENTS SUBTRACTED AND ADDED FROM LEACHATE AFTER 3 MINUTES OF OPERATION

CHROMIUM (53 – 26 ug/L)..DOWN 49%
NICKLE (140 – 64 ug/L).. DOWN 46%

MANGANESE (150 – 120 ug/L)...DOWN 20%
IRON (2.20 – 2.00 ug/L)..DOWN 10%
ARSENIC (100 – 52ug/L)..DOWN 48%

ELEMENTS *SUBTRACTED* AND ADDED FROM LEACHATE AFTER *5 MINUTES* OF OPERATION

CHROMIUM (53 – 75 ug/L)..UP 30%
NICKLE (140 – 130 ug/L)..DOWN 7%
MANGANESE (150 – 74 ug/L).. DOWN 50%
IRON (2.20 – 1.20 ug/L)..DOWN 54%
ARSENIC (100 – 85 ug/L)...DOWN 15%

ELEMENTS *SUBTRACTED* AND *ADDED* FROM LEACHATE AFTER *8 MINUTES* OF OPERATION

CHROMIUM DECREASE (53 - 42 ug/L).......................................DOWN 21%
NICKEL DECREASE (140 – 110 ug/L)..DOWN 21%
MANGANESE DECREASE (150 – 120 ug/L)...............................DOWN 21%
IRON DECREASE (2.20 – 2.10 ug/L)..DOWN 5%
ARSENIC DECREASE (100 - 98 ug/L)..DOWN 2%

FIG 264 : - CHROMIUM, NICKEL, MANGANESE AND IRON.

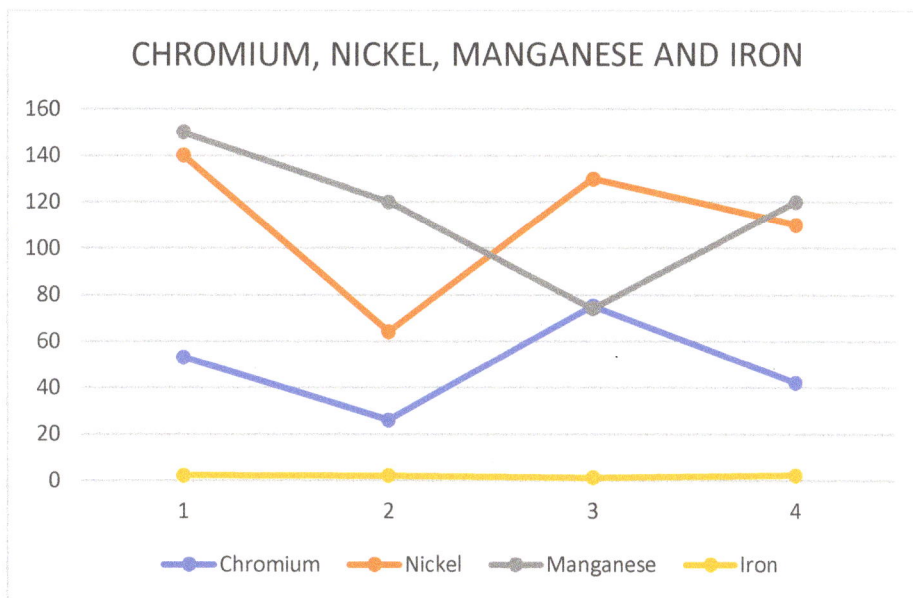

Chromium	Cr	53	26	75	42
Nickel	Ni	140	64	130	110
Manganese	Mn	150	120	74	120
Iron	Fe	2.2	2	1.2	2.1

GROUP FIVE

THE PLASMOID INDUCED COLD FUSION ELEMENTAL LIST

26   **Fe** 64% - Melting point = 1,538 C

28 **Ni** 18% – Nickel's melting point = 1,455 C

24   **Cr** 14% – Chromium's Melting point = 1,907 C

42   **Mo** 3% – Molybdenum's Melting point = 2,623 C

25   **Mn** 2% – Manganese Melting point = 1,246 C

Ti (22) – Titanium's melting point = 1,668 C

Na (11) - Mirror melting point = 97.79 C

1 x 5 x 3 x 8 = (120) x 8 x 3 x 5 x 1 = 14,400
1 + 5 + 3 + 8 = (17) 8 + 3 + 5 + 1 = 34

Al (13) - Aluminum's melting point = 660.3 C

A Plasmoid turns 0.5 of a turn in one second x 60 Min = 30 x 60 min = 1,800 turns in one hour x 24 hours = 43,200 turns a day x 6 days equals creation time equivalent of 259,200.
Therefore the resonance of the Plasmoid determines the procession of the Earths Equinox and is the active action Radius of 5,184,000 Time Diameter it's-self.
Therefore also the resonance of the Plasmoid sets the seconds in a day.

259,200 Plasmoid / **14,400** Iron = 18   18 // 36 // 72 // 144 // 288 //

GROUP FIVE

GROUP FIVE

GROUP FIVE

GROUP FIVE

GROUP FIVE

GROUP FIVE

GROUP FIVE

GROUP FIVE

GROUP FIVE

GROUP FIVE

GROUP FIVE

PART 11

CHARGE SEPARATION AND
AMPLIFICATION

CHARGE SEPARATION AND AMPLIFICATION BY PROTON AND ELECTRON GRAVITY SEPARATION THRU SPIRALS, CONES AND IMPLODED SPHERE TOROIDAL MSAARTS.

CHARGE SEPARATION AND AMPLIFICATION

NEW PART ELEVEN OF TWENTY

DRAFT 518,400 B KMV – THURSDAY 22ND SEPTEMBER 2022

APPLICATIONS FOR A PLASMOIDS FORM AND FUNCTIONS

CHARGE SEPARATION WITHIN WATER BY CENTRIFUGE

Positive and Negative Charges can be separated within water using Gravity by creating an implosive spiral which creates a strong centrifugal force pushing the Protons to the outside and the Electrons and dissolved gases to the inside. This is made possible because Protons are 8,640 times heavier than Electrons. In mineral processing a specific gravity (SG) difference, (water as one compared to the Matter) of only 0.5 SG is sufficient to be able to separate Matter and individual elements from each other by using standard industry mineral processing spirals.

In cases of differences in magnetic susceptibility using magnetism the same principles apply in mineral processing using most commonly a rotating drum magnetic separator.

CHARGE SEPARATION AND PLASMOID GENERATION WITHIN WATER BY IMPLOSION

With a clockwise implosive flow the heavy Protons move to the outside of the pipe displacing the Electrons to the inside of the pipes tightening curve. As the spiral tightens further and greater centrifugal forces push more Protons to the outside of the pipe and therefore causing their further displacement, concentration and holding of the Electrons by the Calandra Effect to the inside wall of the pipe. A critical effect is then initiated at these higher water velocities, as the net Negative charged water held by the wall is subject to gassing and degassing, of dissolved Elements in their Gas Phase. Bubbles appear and are rolled along the wall experiencing both compression and then expansion on that event horizon. This is caused by the increasing Calandra Effect, caused by the waters increasing speed, fighting against the centrifugal forces trying to pull the water away from the wall. These chaotic events implode the formed bubbles into Plasmoids which find themselves in a Net Negative charged environment with concentrated free electrons to absorb.

CHARGE SEPARATION AND NO PLASMOID GENERATION WITHIN WATER BY EXPLOSION

The opposite is true for expanding, anti-clockwise spiral, explosive
When the spiral expands in its diameter at a point the centrifugal forces diminish and the Protons vector down yielding to the Earths gravity's greater pull. They vacate their horizontal positions at the pipes periphery to a position at 90 degrees to that at the bottom of the pipe.

On the explosive anti-clockwise outside the gravity pulls them puts them to the outside

APPLICATIONS FOR A PLASMOIDS FORM AND FUNCTIONS

FOR THE EXHAUST three metal cones

With Cat G3508E pipes separate them

Cut plastic, cement or Silica spiral

FOR WATER

Three 51.84 degrees metal cones with Spiral plastic spiral tubing separating them.

PARTS LIST FOR METAL DEVICE DIRECT ELECTRICITY FROM WATER.
Steve Earl to do diagrams

1.) Three metal cones 51.84 Degrees

2.) silastic gel

3.) clip on water attachments

CONSTRUCTION TECHNIQUES FOR METAL SPIRALS.

1AA) Mark Fibonacci spiral onto the metal cones.

2AA)

PARTS LIST FOR WITCHES HATS ONE METRE HIGH.

PARTS NEEDED

1.) uncoated copper foil

2.) three witches hats

3.) silicon gel

4.) water fittings

CONSTRUCTION TECHNIQUES

1A) Mark Fibonacci spiral Onto the Orange witches hats.

2B) Apply a full thick bead of silicon gel
to the witches hat upon the marked spiral lines.

3C) Apply the copper foil between the beads of silicon.

IN THE Case of elements in their gaseous states, including exhaust gas,
The separation of protons and electrons by spirals

APPLICATIONS FOR A PLASMOIDS FORM AND FUNCTIONS

CHARGE SEPARATION WITHIN WATER BY CENTRIFUGE

DIAGRAM 265   *DIAGRAM 266*

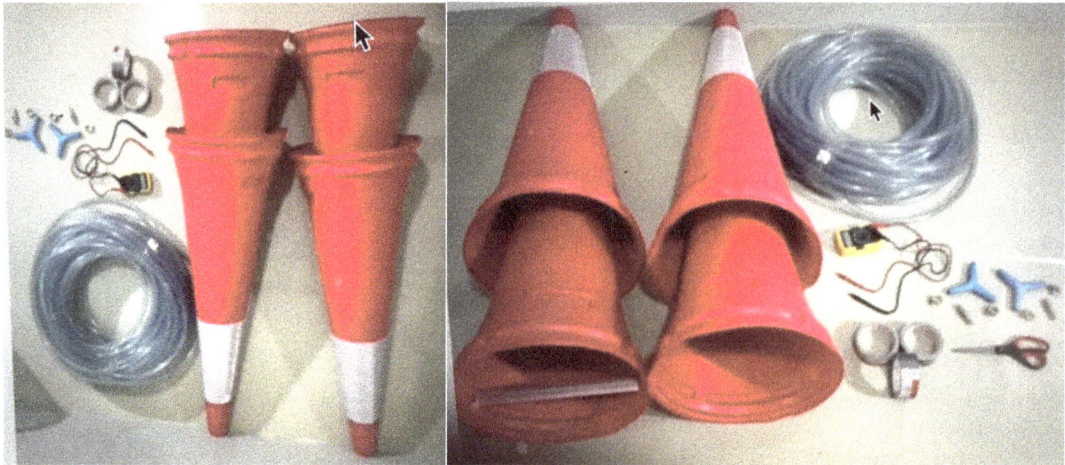

DIAGRAM 267   **DIAGRAM 268**

APPLICATIONS FOR A PLASMOIDS FORM AND FUNCTIONS

CHARGE SEPARATION BY CENTRIFUGE

DIAGRAM 269   **DIAGRAM 270**

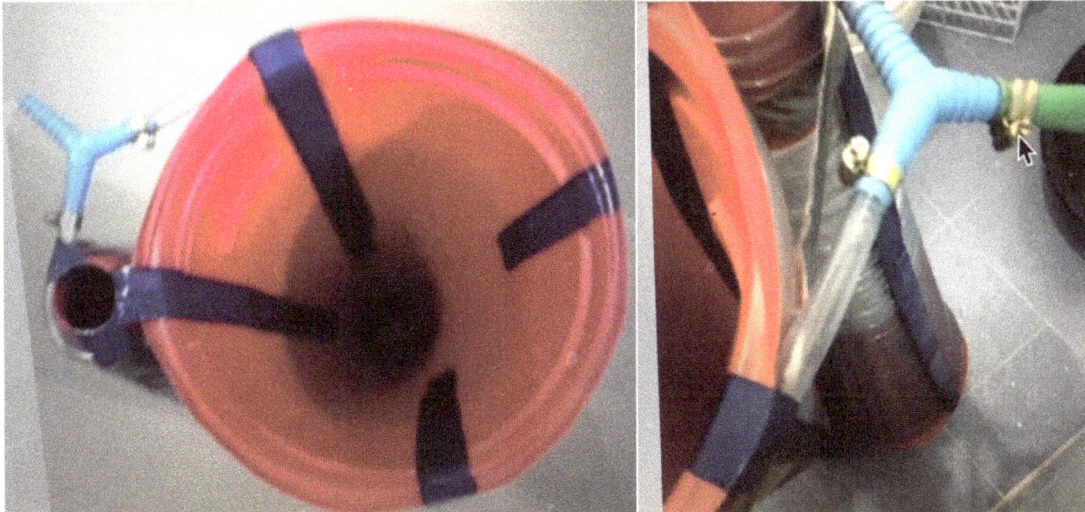

DIAGRAM 271

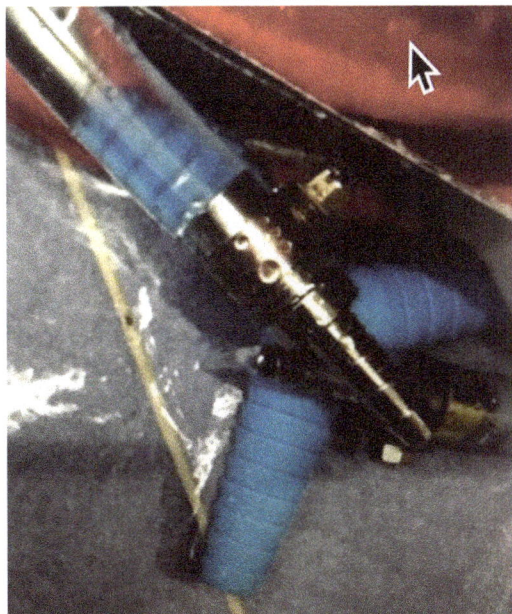

APPLICATIONS FOR A PLASMOIDS FORM AND FUNCTIONS

CHARGE SEPARATION BY CENTRIFUGE

APPLICATIONS FOR A PLASMOIDS FORM AND FUNCTIONS

CHARGE SEPARATION BY CENTRIFUGE

APPLICATIONS FOR A PLASMOIDS FORM AND FUNCTIONS

CHARGE SEPARATION BY CENTRIFUGE

APPLICATIONS FOR A PLASMOIDS FORM AND FUNCTIONS

CHARGE SEPARATION BY CENTRIFUGE

APPLICATIONS FOR A PLASMOIDS FORM AND FUNCTIONS

CHARGE SEPARATION BY CENTRIFUGE

APPLICATIONS FOR A PLASMOIDS FORM AND FUNCTIONS

CHARGE SEPARATION BY CENTRIFUGE

APPLICATIONS FOR A PLASMOIDS FORM AND FUNCTIONS

CHARGE SEPARATION BY CENTRIFUGE

APPLICATIONS FOR A PLASMOIDS FORM AND FUNCTIONS

CHARGE SEPARATION BY CENTRIFUGE

PART 12

RADIATOR PLASMOID
SYSTEM FOR 2005 FORD
FAIRLANE FUTURA

MSAART PATENT APPLICATION NOTES

RADIATOR PLASMOID SYSTEM FOR 2005 FORD FAIRLANE FUTURA

PART TWELVE OF TWENTY

DRAFT 518,400 B KMV BY

MALCOLM V of SCOTLAND | MALCOLM BENDALL
THURSDAY 22ND SEPTEMBER 2022

APPLICATIONS FOR A PLASMOIDS FORM AND FUNCTIONS

RADIATOR PLASMOID SYSTEM FOR 2005 FORD FAIRLANE FUTURA

PART ONE

APPLICATIONS FOR A PLASMOIDS FORM AND FUNCTIONS

RADIATOR PLASMOID SYSTEM FOR 2005 FORD FAIRLANE FUTURA PART 1

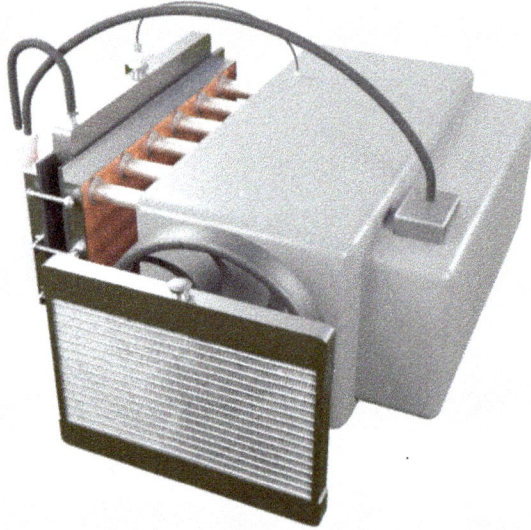

DIAGRAM 272

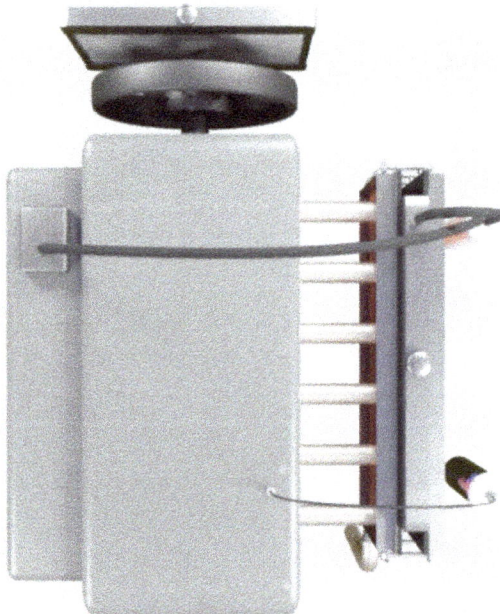

DIAGRAM 273

APPLICATIONS FOR A PLASMOIDS FORM AND FUNCTIONS

RADIATOR PLASMOID SYSTEM FOR 2005 FORD FAIRLANE FUTURA PART 1

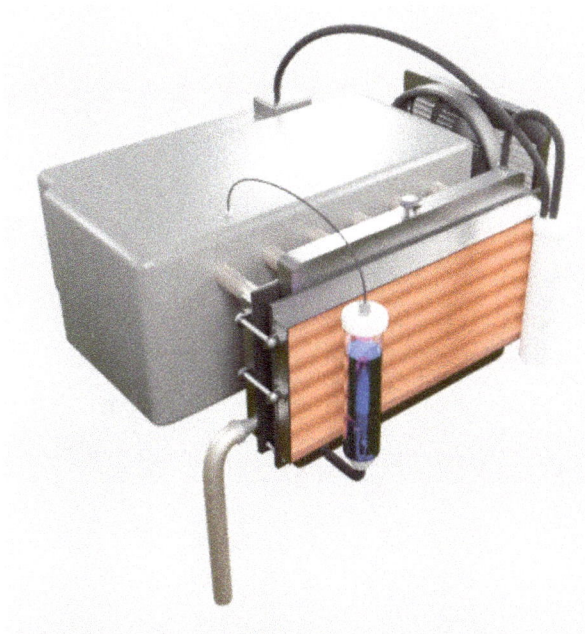

DIAGRAM 274

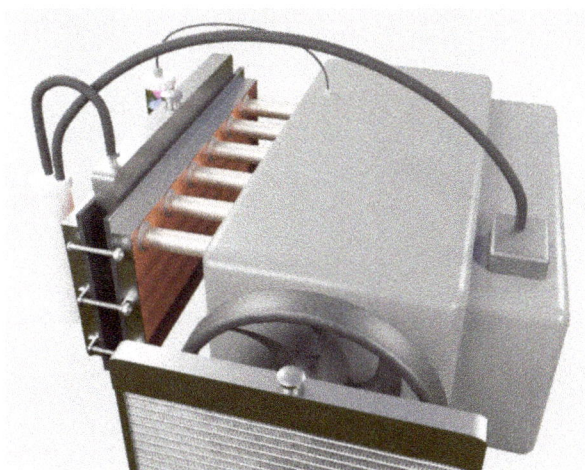

DIAGRAM 275

APPLICATIONS FOR A PLASMOIDS FORM AND FUNCTIONS

RADIATOR PLASMOID SYSTEM FOR 2005 FORD FAIRLANE FUTURA PART 1

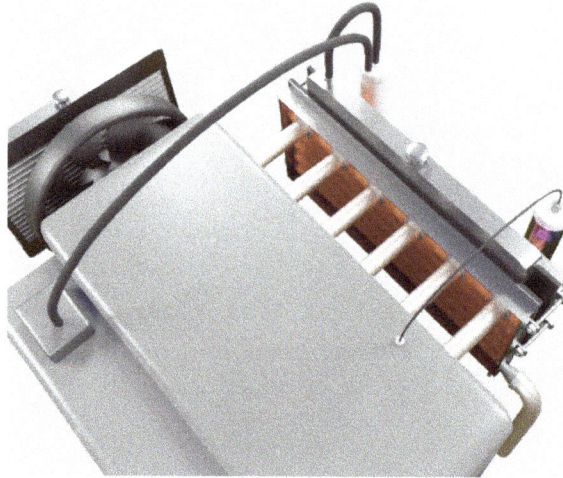

DIAGRAM 276

DIAGRAM 277

APPLICATIONS FOR A PLASMOIDS FORM AND FUNCTIONS

RADIATOR PLASMOID SYSTEM FOR 2005 FORD FAIRLANE FUTURA PART 1

DIAGRAM 278

DIAGRAM 279

APPLICATIONS FOR A PLASMOIDS FORM AND FUNCTIONS

RADIATOR PLASMOID SYSTEM FOR 2005 FORD FAIRLANE FUTURA PART 1

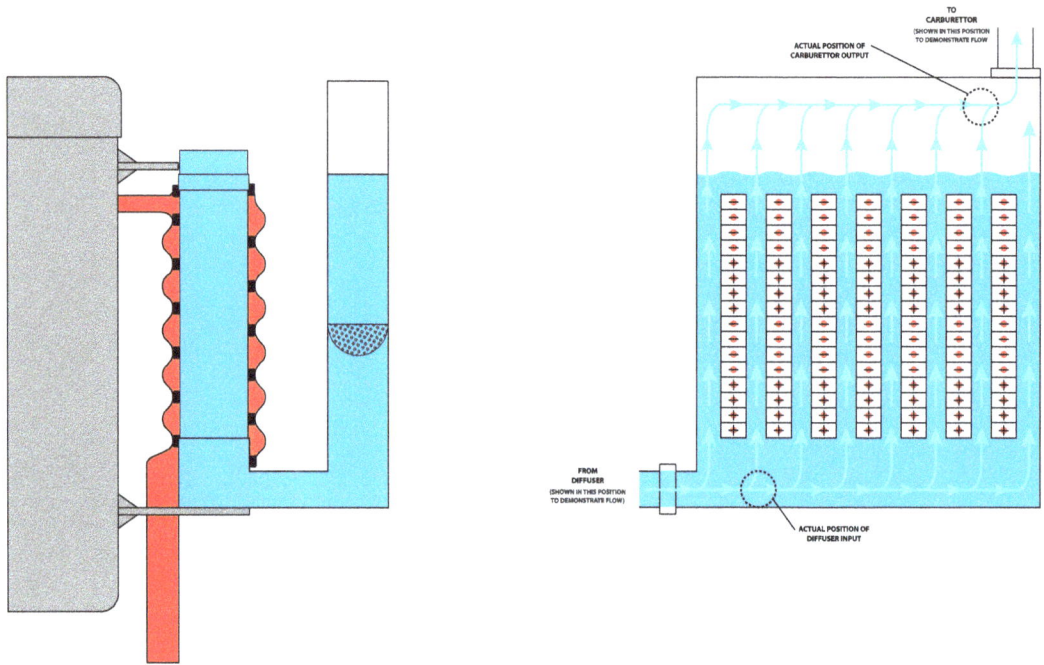

DIAGRAM 280, 281 AND 282

APPLICATIONS FOR A PLASMOIDS FORM AND FUNCTIONS

RADIATOR PLASMOID SYSTEM FOR 2005 FORD FAIRLANE FUTURA

PART TWO

APPLICATIONS FOR A PLASMOIDS FORM AND FUNCTIONS

RADIATOR PLASMOID SYSTEM FOR 2005 FORD FAIRLANE FUTURA PART 2

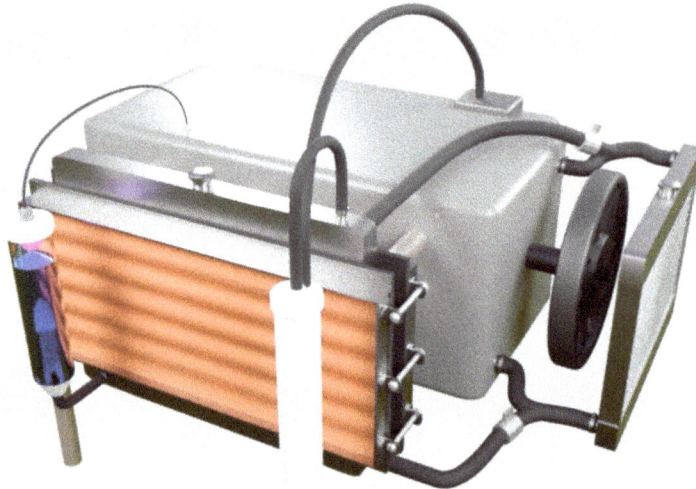

DIAGRAM 283

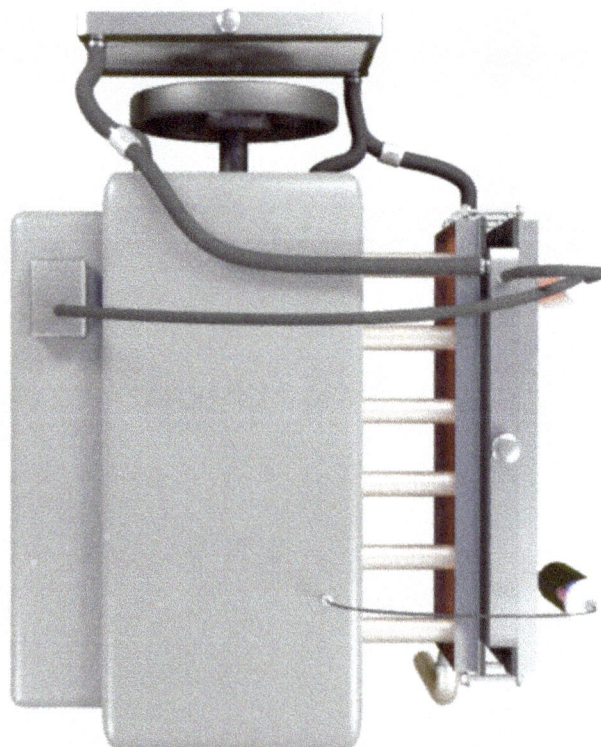

DIAGRAM 284

APPLICATIONS FOR A PLASMOIDS FORM AND FUNCTIONS

RADIATOR PLASMOID SYSTEM FOR 2005 FORD FAIRLANE FUTURA PART 2

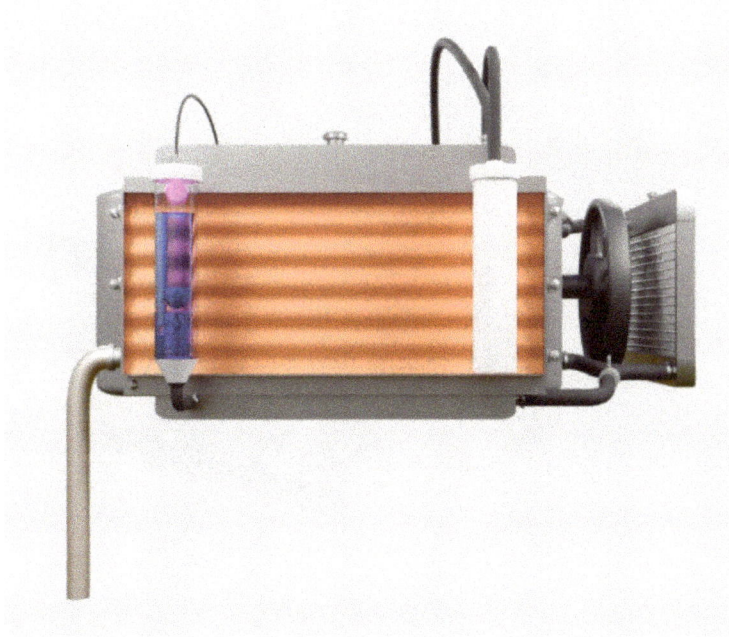

DIAGRAM 285

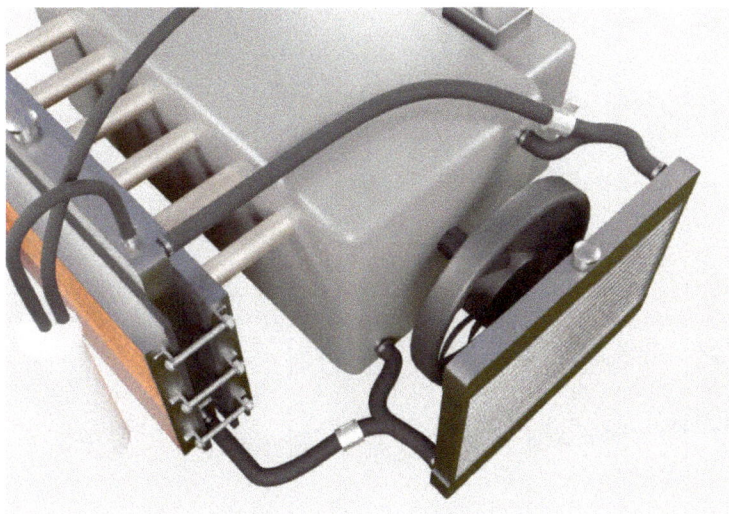

DIAGRAM 286

APPLICATIONS FOR A PLASMOIDS FORM AND FUNCTIONS

RADIATOR PLASMOID SYSTEM FOR 2005 FORD FAIRLANE FUTURA PART 2

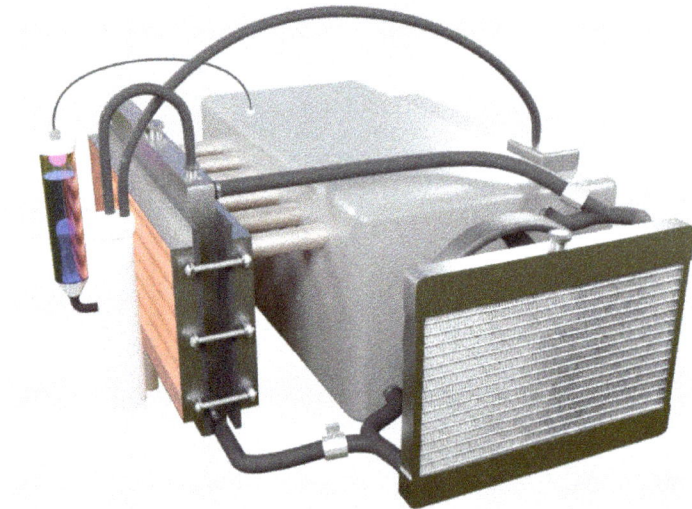

DIAGRAM 287

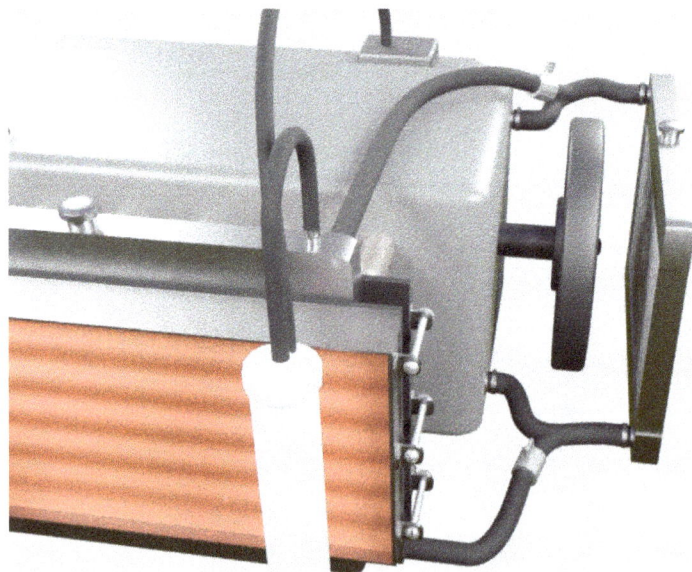

DIAGRAM 288

APPLICATIONS FOR A PLASMOIDS FORM AND FUNCTIONS

RADIATOR PLASMOID SYSTEM FOR 2005 FORD FAIRLANE FUTURA PART 2

DIAGRAM 289

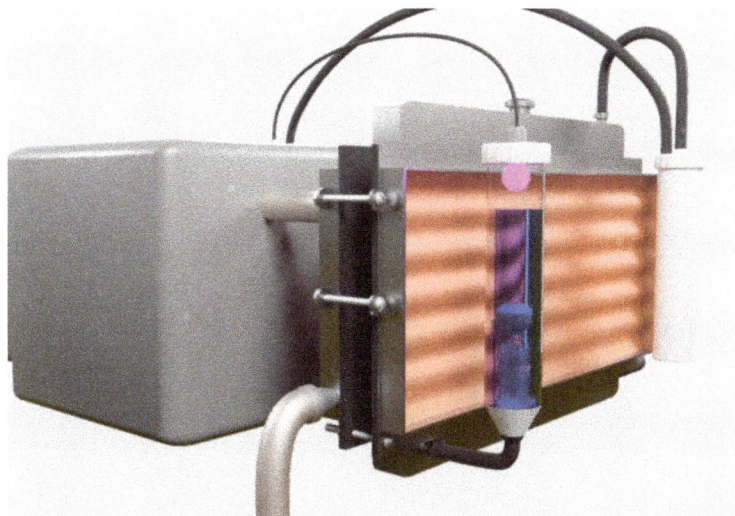

DIAGRAM 290

APPLICATIONS FOR A PLASMOIDS FORM AND FUNCTIONS

RADIATOR PLASMOID SYSTEM FOR 2005 FORD FAIRLANE FUTURA PART 2

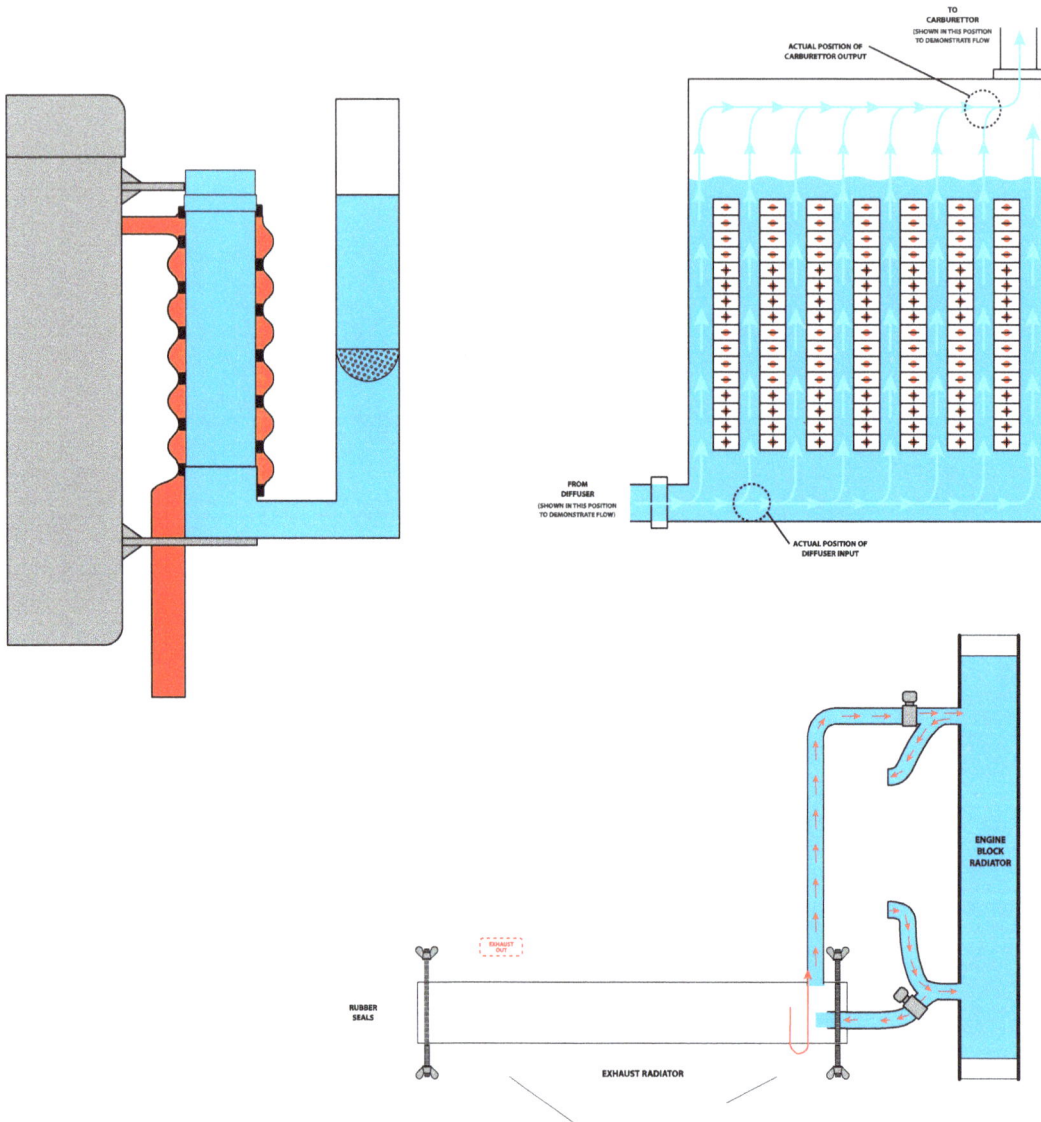

DIAGRAM 291, 292 AND 293

APPLICATIONS FOR A PLASMOIDS FORM AND FUNCTIONS

RADIATOR PLASMOID SYSTEM FOR 2005 FORD FAIRLANE FUTURA

PART THREE

APPLICATIONS FOR A PLASMOIDS FORM AND FUNCTIONS

APPLICATIONS FOR A PLASMOIDS FORM AND FUNCTIONS

RADIATOR PLASMOID SYSTEM FOR 2005 FORD FAIRLANE FUTURA PART 3

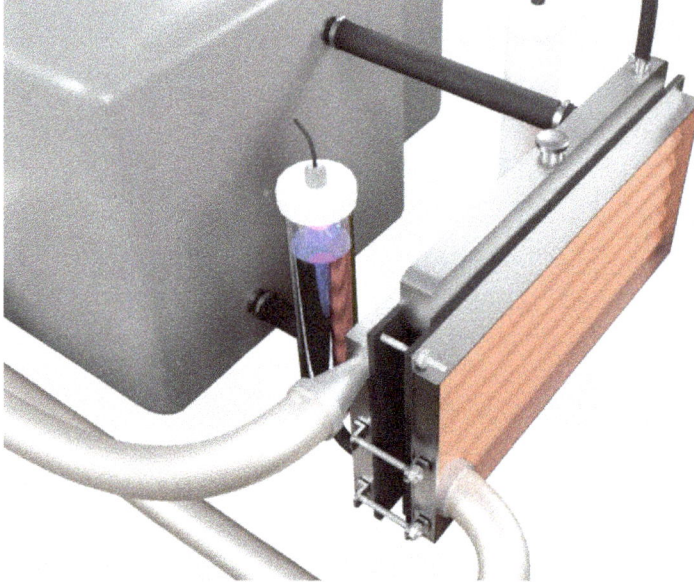

DIAGRAM 294

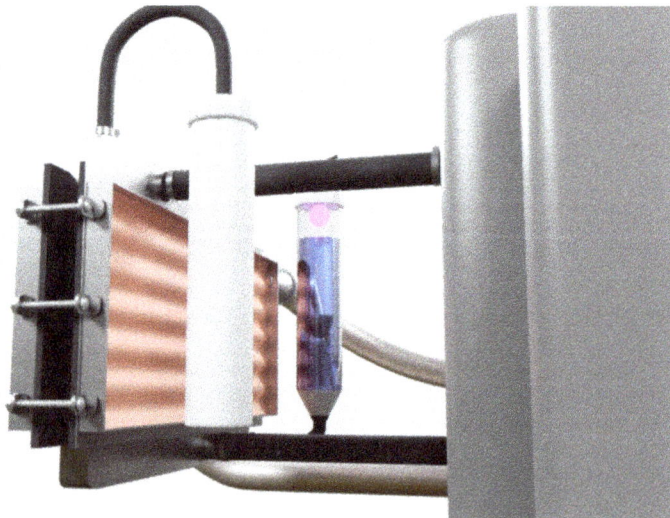

DIAGRAM 295

APPLICATIONS FOR A PLASMOIDS FORM AND FUNCTIONS

RADIATOR PLASMOID SYSTEM FOR 2005 FORD FAIRLANE FUTURA PART 3

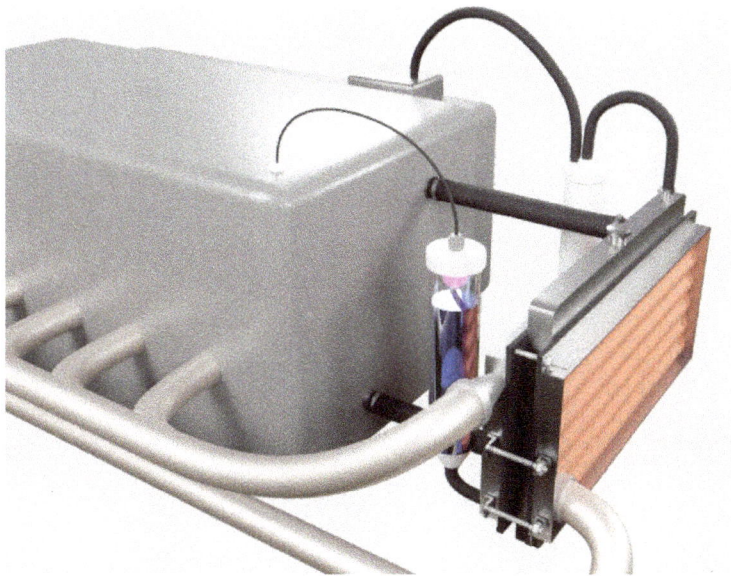

DIAGRAM 296

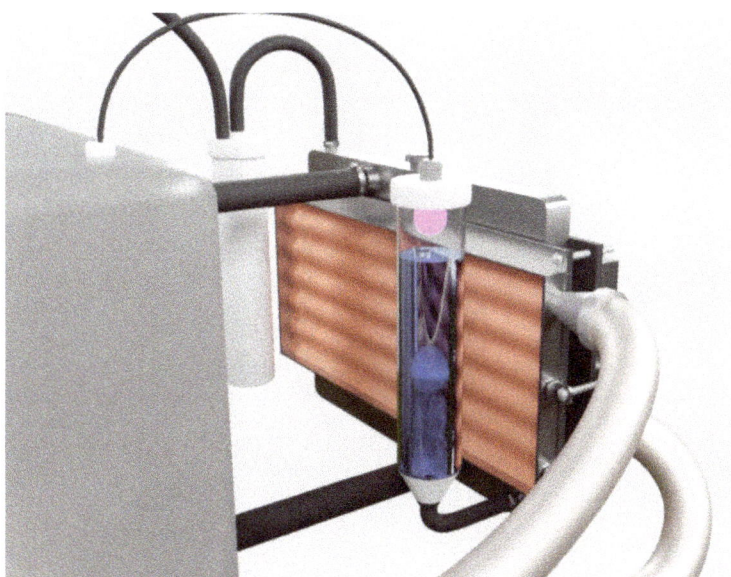

DIAGRAM 297

APPLICATIONS FOR A PLASMOIDS FORM AND FUNCTIONS

RADIATOR PLASMOID SYSTEM FOR 2005 FORD FAIRLANE FUTURA PART 3

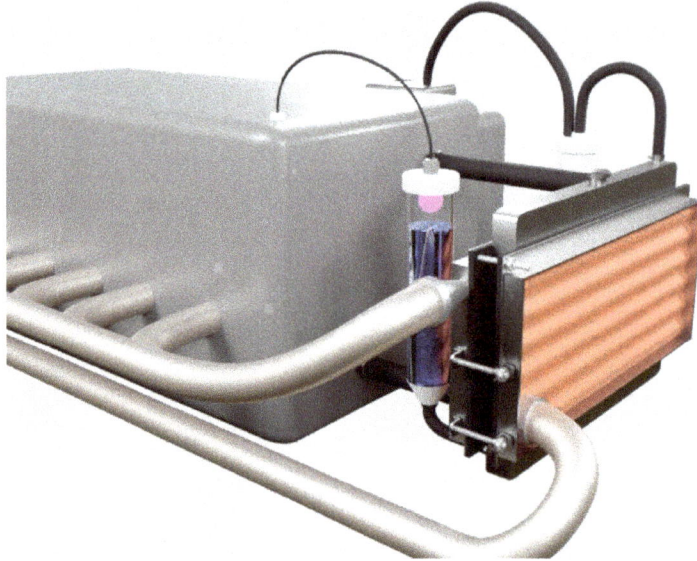

DIAGRAM 298

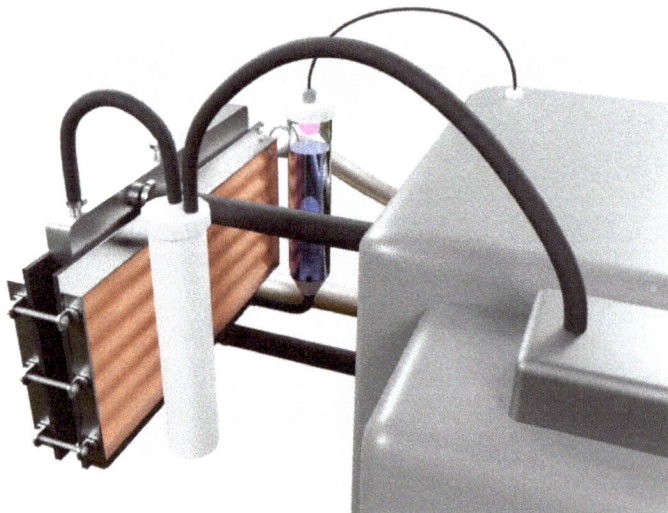

DIAGRAM 299

APPLICATIONS FOR A PLASMOIDS FORM AND FUNCTIONS

RADIATOR PLASMOID SYSTEM FOR 2005 FORD FAIRLANE FUTURA PART 3

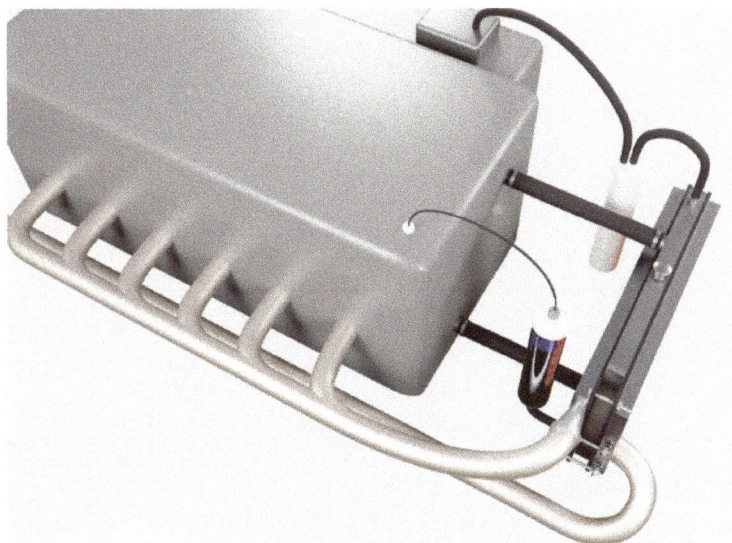

DIAGRAM 300

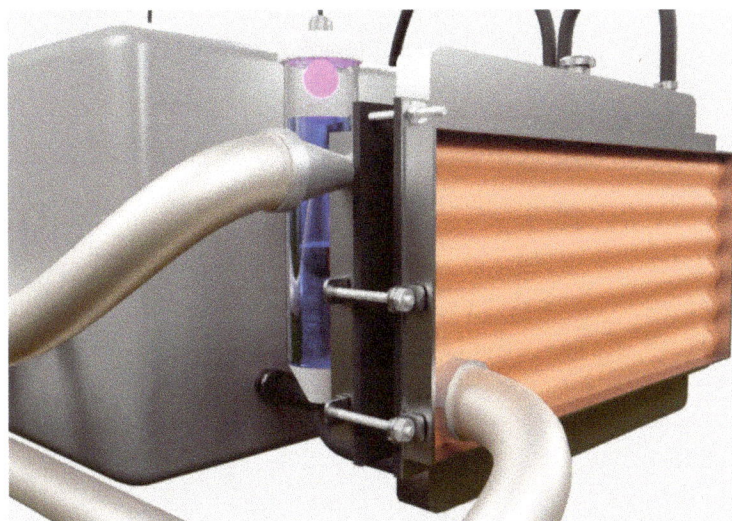

DIAGRAM 301

APPLICATIONS FOR A PLASMOIDS FORM AND FUNCTIONS

RADIATOR PLASMOID SYSTEM FOR 2005 FORD FAIRLANE FUTURA PART 3

DIAGRAM 302

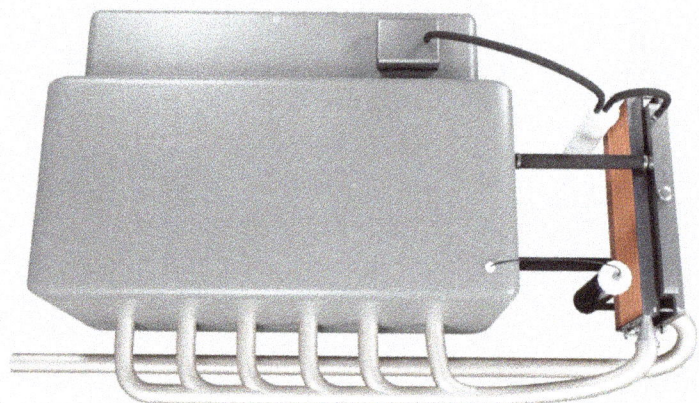

DIAGRAM 303

APPLICATIONS FOR A PLASMOIDS FORM AND FUNCTIONS

RADIATOR PLASMOID SYSTEM FOR 2005 FORD FAIRLANE FUTURA PART 3

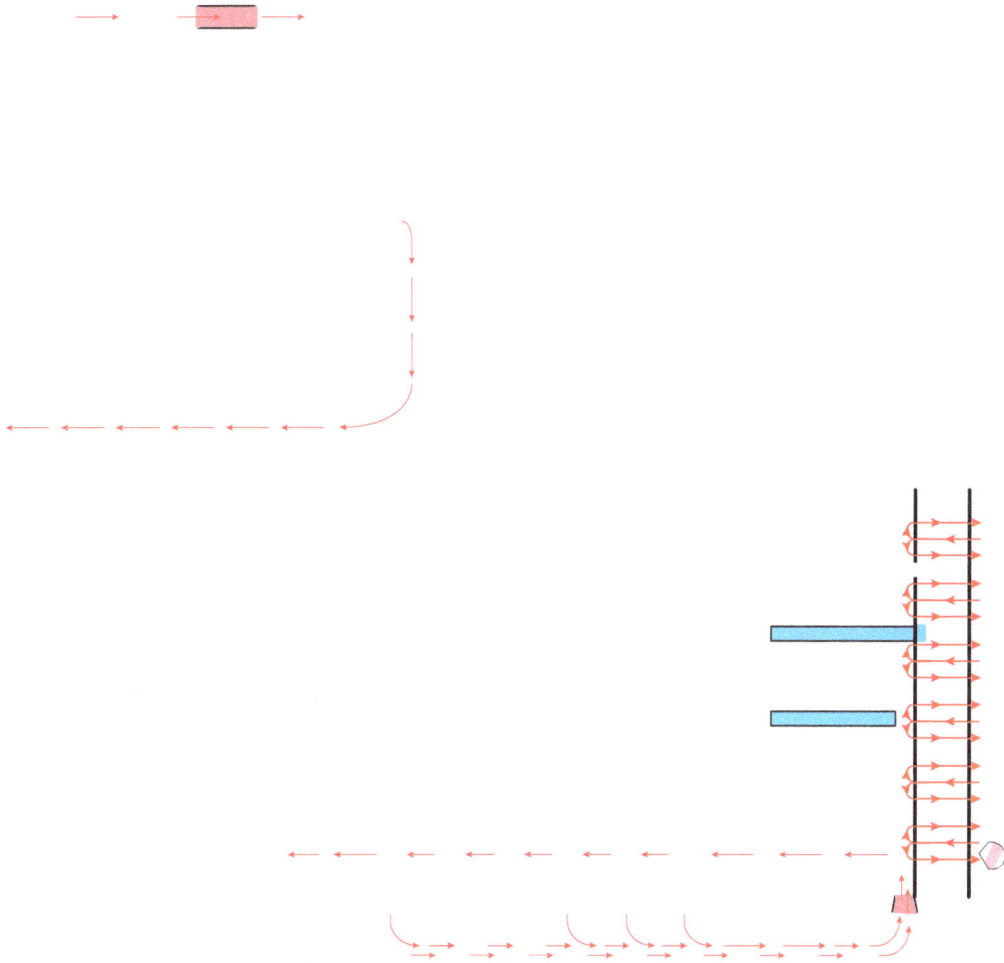

DIAGRAMS 304, 305 AND 306

APPLICATIONS FOR A PLASMOIDS FORM AND FUNCTIONS

RADIATOR PLASMOID SYSTEM FOR 2005 FORD FAIRLANE FUTURA

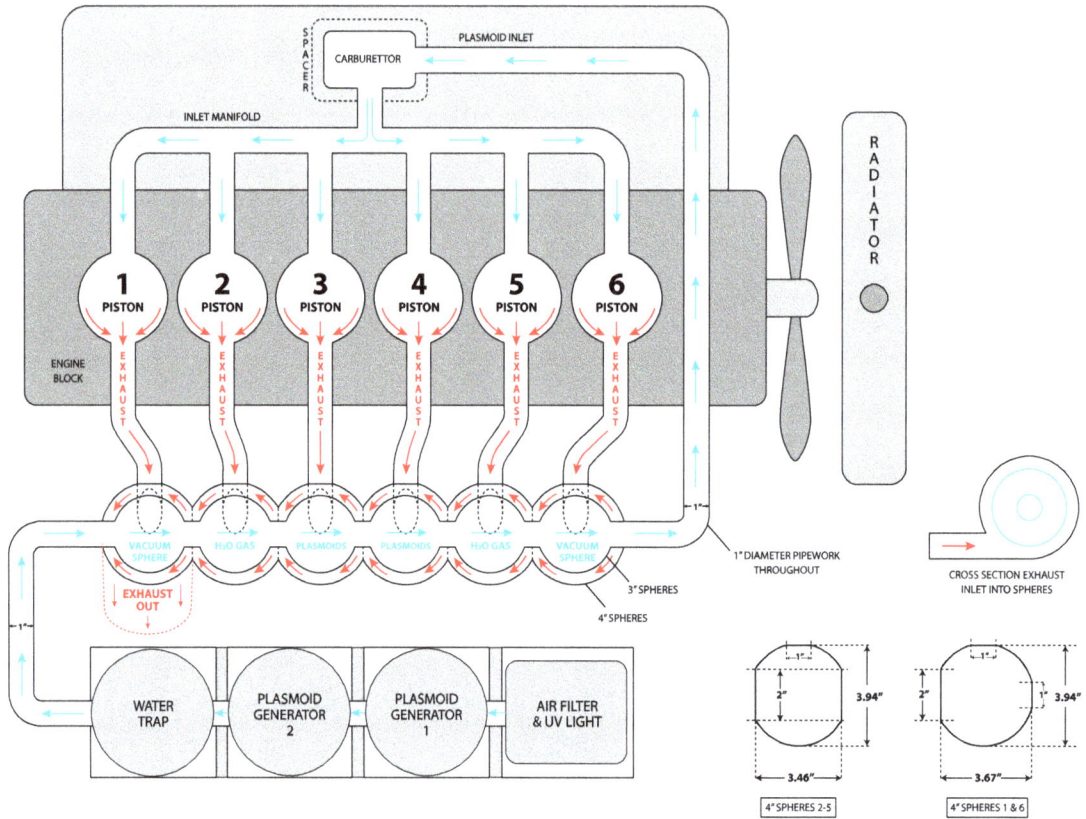

DIAGRAM 307

APPLICATIONS FOR A PLASMOIDS FORM AND FUNCTIONS

RADIATOR PLASMOID SYSTEM FOR 2005 FORD FAIRLANE FUTURA

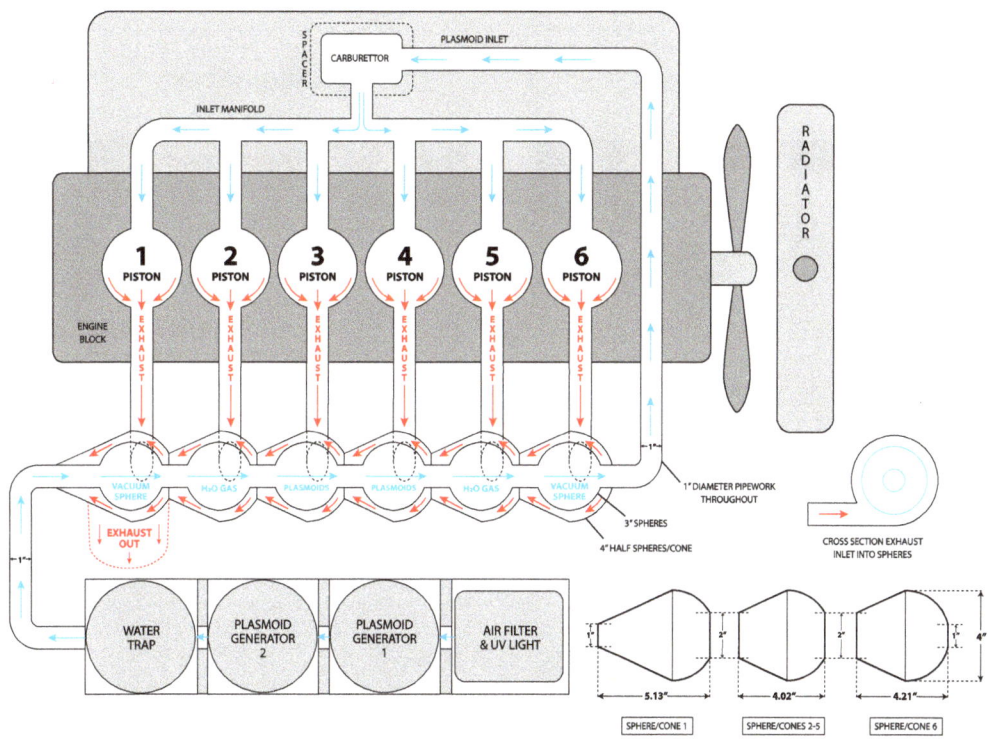

DIAGRAM 308

APPLICATIONS FOR A PLASMOIDS FORM AND FUNCTIONS

RADIATOR PLASMOID SYSTEM FOR 2005 FORD FAIRLANE FUTURA

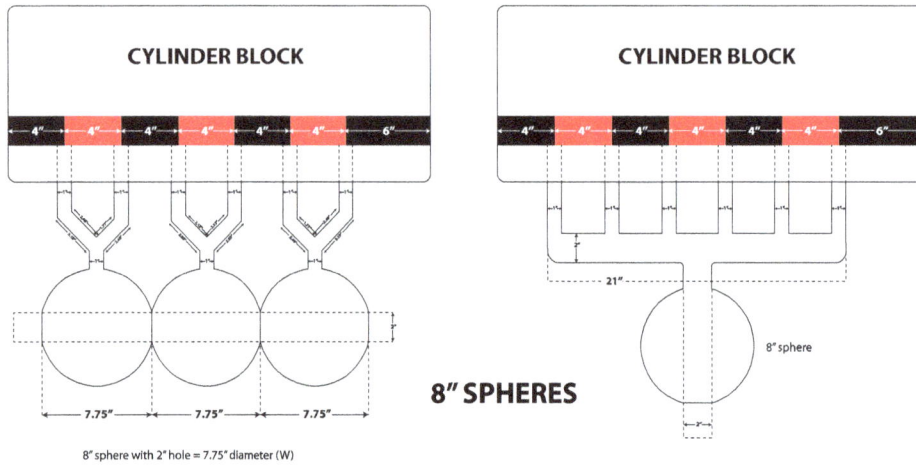

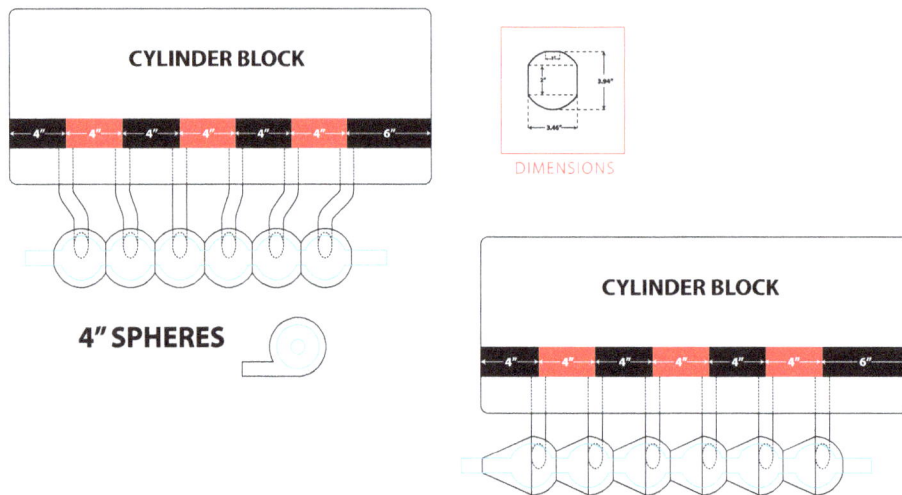

DIAGRAM 309, 310, 311 & 312

PART 13

SOLOMON'S MOLTEN SEA LENSE PLASMOID GENERATOR AND ARC CHARGER

MSAART PATENT APPLICATION NOTES

APPENDIX 11:- SOLOMON'S MOLTEN SEA LENSE

PLASMOID GENERATOR AND ARC CHARGER

PART THIRTEEN OF TWENTY

DRAFT 518,400 B KMV

BY

MALCOLM V of SCOTLAND | MALCOLM BENDALL
THURSDAY 22ND SEPTEMBER 2022

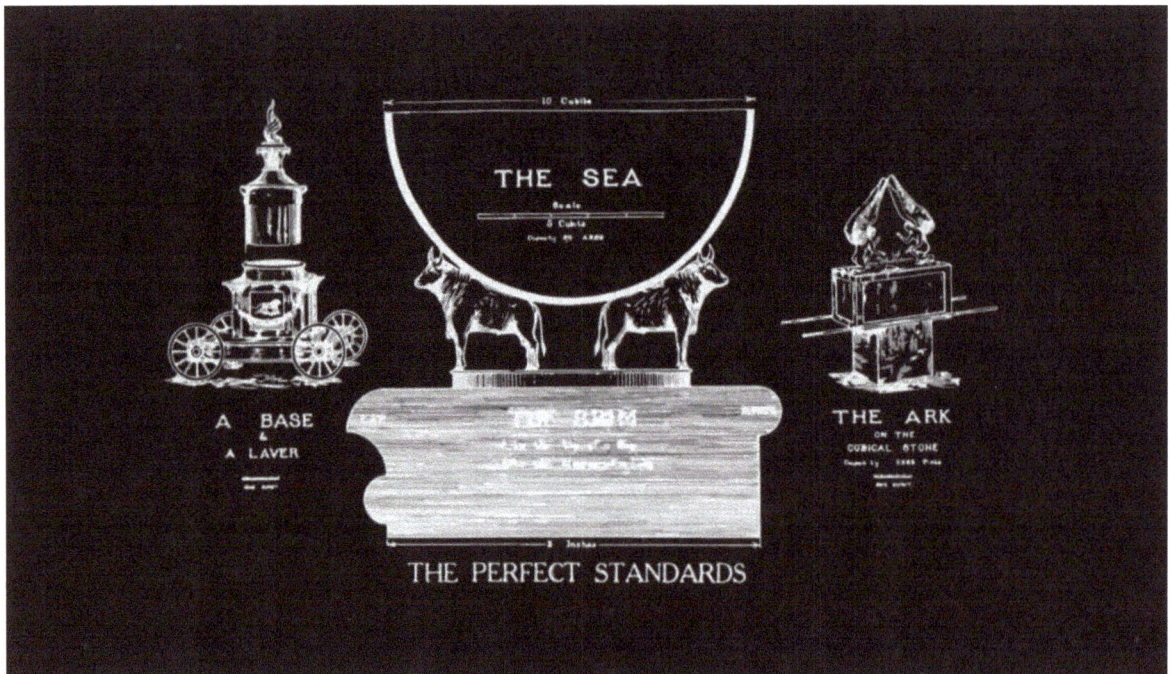

FIG 313 : -SOLOMONS MOLTEN SEA

APPLICATIONS FOR A PLASMOIDS FORM AND FUNCTIONS

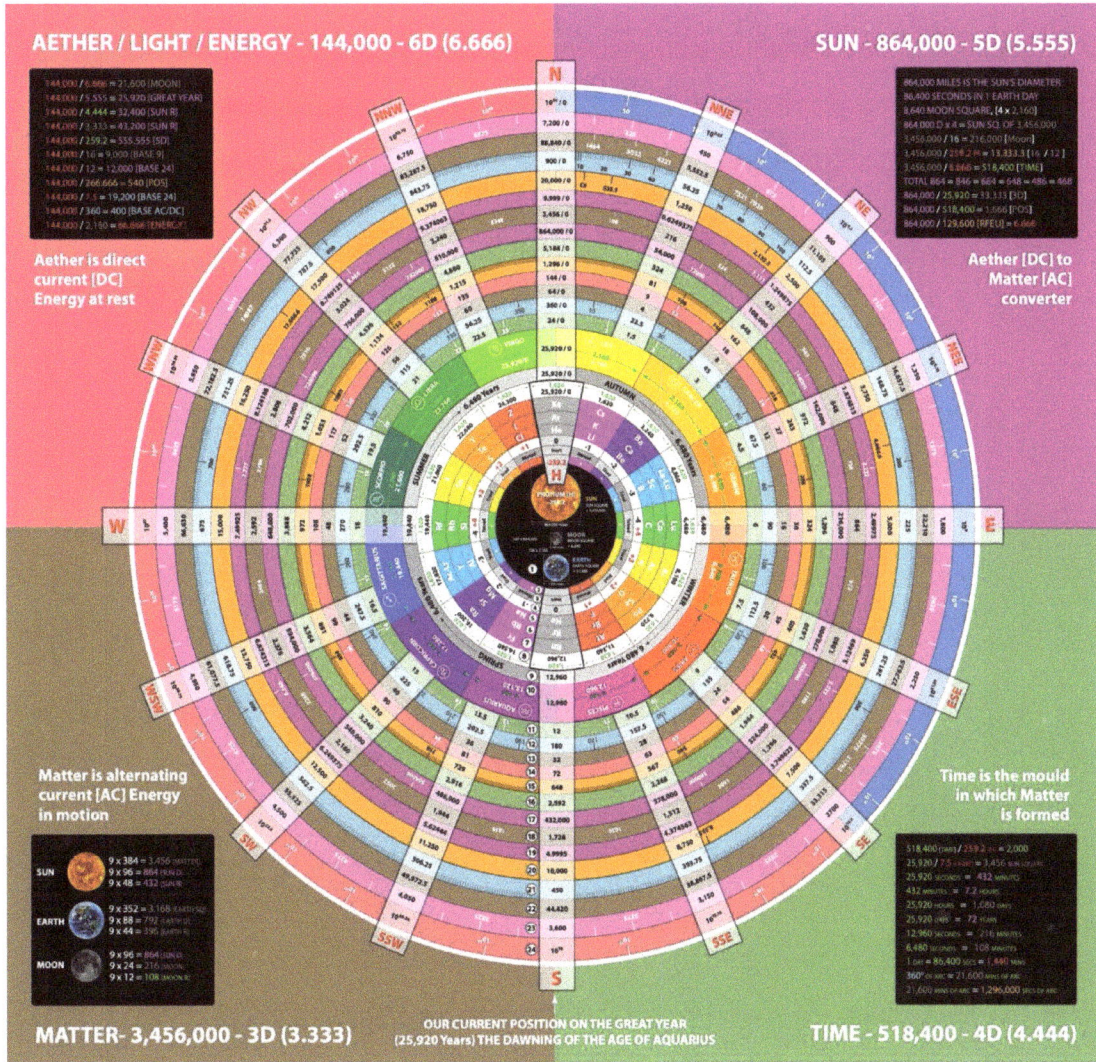

AETHER / LIGHT / ENERGY - 144,000 - 6D (6.666)

144,000 / 6.666 = 21,600 [MOON]
144,000 / 5.555 = 25,920 [GREAT YEAR]
144,000 / 4.444 = 32,400 [SUN R]
144,000 / 3.333 = 43,200 [SUN R]
144,000 / 259.2 = 555.555 [5D]
144,000 / 16 = 9,000 [BASE 9]
144,000 / 12 = 12,000 [BASE 24]
144,000 / 266.666 = 540 [POS]
144,000 / 7.5 = 19,200 [BASE 24]
144,000 / 360 = 400 [BASE AC/DC]
144,000 / 2,160 = 66.666 [ENERGY]

Aether is direct current [DC] Energy at rest

SUN - 864,000 - 5D (5.555)

864,000 MILES IS THE SUN'S DIAMETER
86,400 SECONDS IN 1 EARTH DAY
8,640 MOON SQUARE, [4 x 2,160]
864,000 D x 4 = SUN SQ. OF 3,456,000
3,456,000 / 16 = 216,000 [Moon]
3,456,000 / 259.2 = 13,333.3 [16 / 12]
3,456,000 / 6.666 = 518,400 [TIME]
TOTAL 864 + 846 = 664 + 648 + 486 = 468
864,000 / 25,920 = 33.333 [3D]
864,000 / 518,400 = 1.666 [POS]
864,000 / 129,600 [RFEU] = 6.666

Aether [DC] to Matter [AC] converter

Matter is alternating current [AC] Energy in motion

SUN   9 x 384 = 3,456 [MATTER]
      9 x 96 = 864 [SUN D]
      9 x 48 = 432 [SUN R]

EARTH 9 x 352 = 3,168 [EARTH D]
      9 x 88 = 792 [EARTH D]
      9 x 44 = 396 [EARTH R]

MOON 9 x 96 = 864 [MOON D]
      9 x 24 = 216 [MOON D]
      9 x 12 = 108 [MOON R]

Time is the mould in which Matter is formed

518,400 [TIME] / 259.2 = 2,000
25,920 / 7.5 = 3,456 sec of time
25,920 SECONDS = 432 MIN/PD
432 MINUTES = 7.2 HOURS
25,920 MINUTES = 1,080 DAYS
25,920 HOURS = 1,080 DAYS
25,920 DAYS = 72 YEARS
12,960 SECONDS = 216 MINUTES
6,480 SECONDS = 108 MINUTES
1 DAY = 86,400 SECS = 1,440 MIN
360° OR SEC = 21,600 MIN OF ARC
21,600 MINS OF ARC = 1,296,000 SECS OF ARC

MATTER- 3,456,000 - 3D (3.333)

OUR CURRENT POSITION ON THE GREAT YEAR
(25,920 Years) THE DAWNING OF THE AGE OF AQUARIUS

TIME - 518,400 - 4D (4.444)

PLASMOID UNIFICATION MODEL

Plasmoids are doughnut or toroidal shaped clusters of net Protons or net Electrons that once captured and placed into a Toroidal orbit are capable of absorbing, storing and releasing enormous amounts of energy present within their self-generated and structured electro-magnetic containment field. Plasmoids, in effect, function as an atomic battery that can be self-charging due to its ability to convert matter to available clean energy. Plasmoids by their unique geometry cause a consequential electro-magnetic containment field to generate a Zero point naturally and casually, without much effort, have the ability to convert the nuclear Mass of Protium (Atoms) into energy.

The Plasmoid Unification Model (PUM) posits that Plasmoids are epoch-making and that knowledge of them has been hidden in plain sight for centuries. This PUM 'slide rule' reveals the algorithmic relationships between life's elements critical to mankind's existence and development. It starts with Protium [H] which has a melting point of -259.2°C and is the most abundant element in our Solar System. Protium determines the 25,920 Great Year frequency of our Solar System. The resonant frequencies of all other elements can then be calculated when 25,920 years is reduced from years to days, hours and seconds.

The PUM is evidence that the Universe is an intelligent design. That design is in perfect octave harmonic resonance with itself. Therefore, all of creation from Galaxies to Planets to Elements all resonate in unison with a collective chord 'As Above So Below'. This is interconnected with an Energy 'web', the 24 components and laws of which are all based and governed on the same 16 sector Torus Plasmoid precepts shown. The concepts and ruling principles of the PUM can, and have, been applied to make Energy to Matter and Matter to Energy conversions. When applied to the modern hydrocarbon powered internal combustion engine, PUM technology removes exhaust toxic waste products and increases the engine power output by transforming waste energy back into fuel. Plasmoids employed in conjunction with the Plasmoid Toroidal Implosive Turbine provide a new novel Matter to Energy and Energy to Matter propulsion device for water, land, air and space travel.

LEGEND

■ Aether/Light ■ Sun ■ Time ■ Matter ■ Frequency ■ Degrees

1	**Sun** (864,000) [DC], Earth (7,920) [AC], Moon (2,160) [AC]
2	**Music** [AC] (Do = C = 24 x 11.111 = 266.666)
3	Elemental Crystal Forms + / - Monad, Diad, Triad, Tetrad
4	Elemental Valencies (0, - 1, - 2, - 3, - 4 and 0, +1, +2, +3, +4)
5	Elements 1-16 (He - Cl)
6	Elements 17-32 (Ar - I)
7	Elements 33-48 (Xe - 'Z')
8	Elemental Frequencies (1,620 x 16 = 25,920 light frequency of -259.2 C)
9	Seasons of Great Year (25,920 / 0 years)
10	Zodiac Great Year (25,920 / 0 years)
11	Clock 24 Hour / 0 Hour [AC] / 0 Hour [DC] Clock
12	Compass 360° Degrees [AC] / 0° Degrees [DC]
13	Matter 64 / 0 64 Points being 32 Resonant Planes / 0
14	Light 144 / 0
15	Resonate Frequency Energy Unit (RFEU) - 1,296 (129,600)
16	All Time 5,184 (518,400 secs)
17	Aether- Sun (864,000 miles diameter, 432,000 miles radius)
18	Matter 3,456 (3,456,000 miles - Sun Square)
19	Dimensions 3D Matter = 3.33, 4D Time = 4.44, 5D Aether = 5.55, 6D Light = 6.666
20	Sound and Music (0 - 20,000 Hz)
21	Language 1-9, 10-90, 100-900 (111,222,333,444,555,666,777,888,999 / 45,450,4,500)
22	Solar System Sun & Planetary Diameters and Radii in miles
23	Plasmoids 7,200 Degrees / 0, 32 Planes 64 Radial Points, One Zero Point
24	**All Plasmoid Energy** = All Alternating Current [AC] Frequencies
	[■■■ = Non Ionizing] [■■■ = Ionizing]

FIG 314 : -16 = AETHER SQUARES, MOLTEN SEA AETHER TO MATTER GEOMETRY

APPLICATIONS FOR A PLASMOIDS FORM AND FUNCTIONS

12 = AETHER SQUARES, MOLTEN SEA AETHER TO MATTER GEOMETRY

12 = AETHER [DC] - PART 1

Grand Total = 77,760

311,040 // 155,520 // 77,760 // 38,880 // 19,440 // **9,720** // **4,860** // **2,430** // **1,215** //

Earth Sun Sun Moon

12° = 43,200 Seconds of Arc

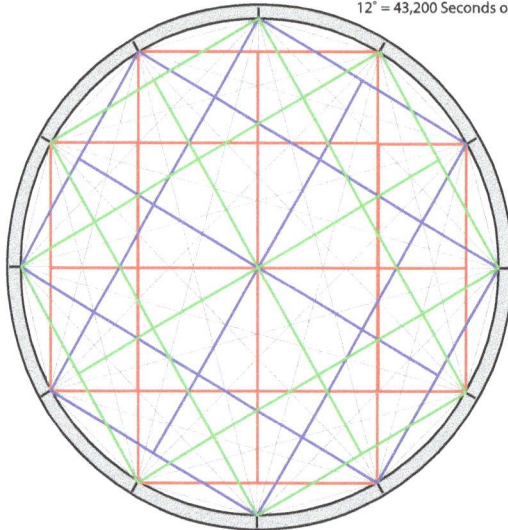

8 rectangles + 4 squares = 2,160° x 12 = 25,920°
8 rectangles + 4 squares = 2,160° x 12 = 25,920°
8 rectangles + 4 squares = 2,160° x 12 = 25,920°

24 x 2,160° = 51,840°   12 x 2,160° = 25,920°   TOTAL = 77,760° (1)

77,760 (1)
+ 84,240 (2)
GRAND TOTAL = 162,000

FIG 315 : - 12 = AETHER = AETHER SQUARES, MOLTEN SEA AETHER TO MATTER GEOMETRY

12 = AETHER [DC]

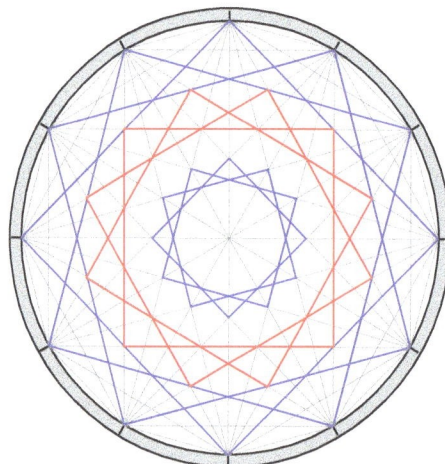

6 x 2,160 = 12,960 x 4 = 51,840
3 x 2,160 = 6,480 x 4 = 25,920
9 x 2,160 =

FIG 316 : - 12 = AETHER SQUARES, MOLTEN SEA AETHER TO MATTER GEOMETRY

APPLICATIONS FOR A PLASMOIDS FORM AND FUNCTIONS

12 = AETHER MOLTEN SEA AETHER TO MATTER GEOMETRY

12 = AETHER [DC] - PART 2

5,184,000 // 2,592,000 // 1,296,000 // 648,000 // 324,000 // **162,000** //
81,000 // 40,500 // 20,250 // 10,125 // 5,062.5 // 5,184,000

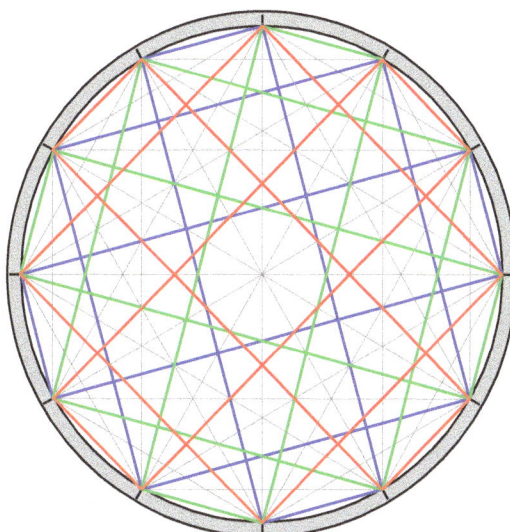

3 x 34,560 = 103,680 + 77,760 = 181,400   13Sq x 2,160° = 28,080°
[+4 Large Rectangles 12+4 = 16 x 360 = 5,760 , 16 x 2,160 = 34,560] 13Sq x 360° = 4,680°
4 Half Squares [2 Sq], 8 Small Squares, 2 Large Squares = 11Sq x 360° = 3,960°
4 x 360° = 1,440° , 8 x 360° = 2,880° , 8 x 2,160° = 17,280° = 11Sq x 2,160° = 23,760°
4 x 2,160° = 8,640°

11 x 3 = 33 x 360° = 11,880°, 11 x 3 = 33 x 2,160° = 71,280°
13 x 3 = 39 x 360° = 14,040°, 13 x 3 = 39 x 2,160° = 84,240°

77,760 (1)
+ 84,240 (2)
162,000

FIG 317 : - 12 = AETHER MOLTEN SEA AETHER TO MATTER GEOMETRY

12

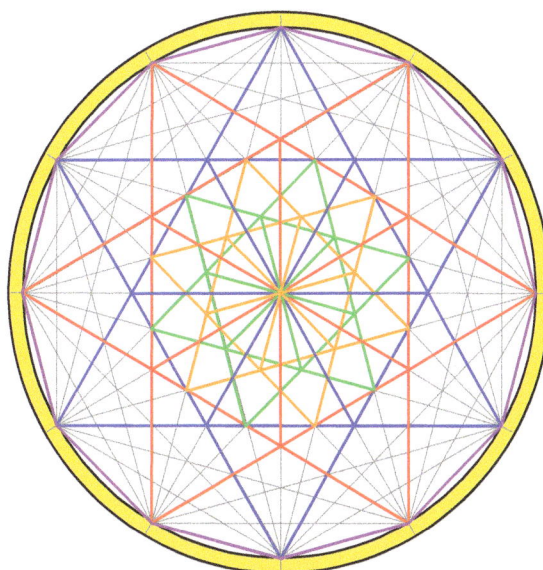

FIG 318 : - 12 = AETHER, MOLTEN SEA AETHER TO MATTER GEOMETRY

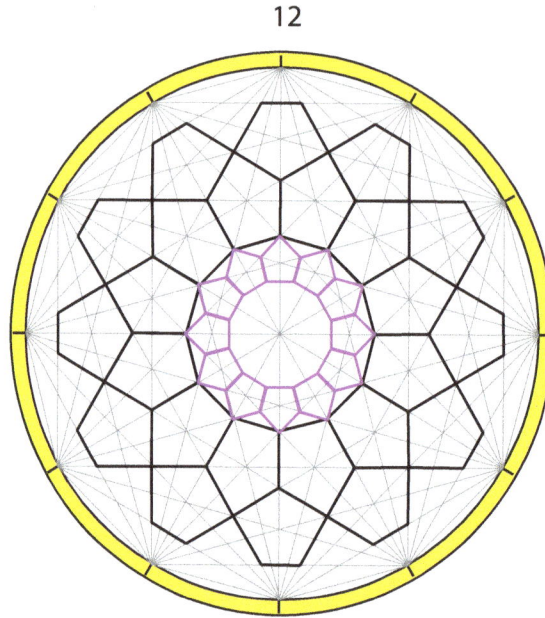

APPLICATIONS FOR A PLASMOIDS FORM AND FUNCTIONS

12 = AETHER MOLTEN SEA AETHER TO MATTER GEOMETRY

12

FIG 319 : - 16 = MATTER - MOLTEN SEA MATTER TO AETHER GEOMETRY

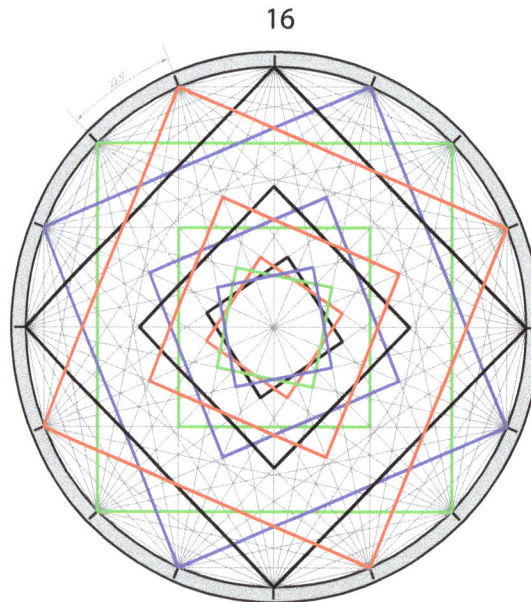

16

3 squares x 360° = 1,080	3 x 2,160 = 6,840
3 squares x 360° = 1,080	3 x 2,160 = 6,840
3 squares x 360° = 1,080	3 x 2,160 = 6,840
3 squares x 360° = 1,080	3 x 2,160 = 6,840
12 squares x 360° = 4,320	12 x 2,160 = 25,920
48 corners	

FIG 320 : - 16 = MATTER - MOLTEN SEA AETHER TO MATTER GEOMETRY

APPLICATIONS FOR A PLASMOIDS FORM AND FUNCTIONS

16 = MATTER - MOLTEN SEA AETHER TO MATTER GEOMETRY

24 Boxes x 360° = 8,640° x 4 = 34,560° [MATTER]
24 Boxes x 360° = 8,640°
24 Boxes x 360° = 8,640°   34,560° = 124,416,000 secs of Arc
24 Boxes x 360° = 8,640°   34,560° = 2,073,600 mins of Arc [792]
                            Earth diameter

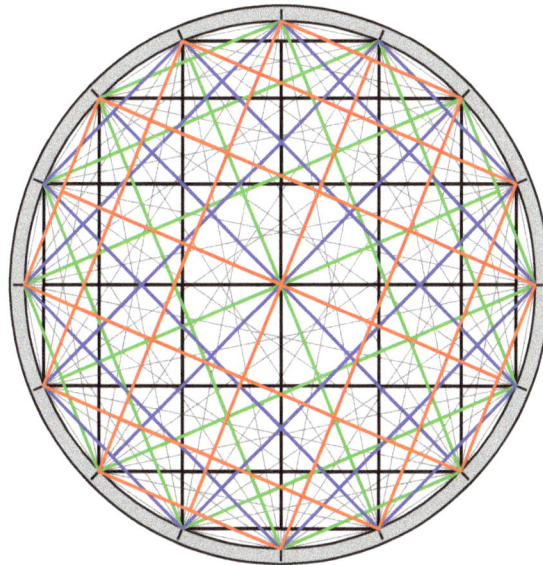

24 Boxes x 2,160 = 51,840
24 Boxes x 2,160 = 51,840
24 Boxes x 2,160 = 51,840
24 Boxes x 2,160 = 51,840
96   **207,360**

8 small rectangles = 8 x 2,160 = 17,280 [SUN].   8 x 360° = 2,880 [1,440 x 2 LIGHT]
12 large squares = 12 x 2,160 = 25,920 [EQUI]. 12 x 360 = 4,320
4 large squares = 4 x 2,160 = 8,640 [SUN D].   4 x 360° = 1,440

FIG 321 : - 16 = MATTER - MOLTEN SEA AETHER TO MATTER GEOMETRY

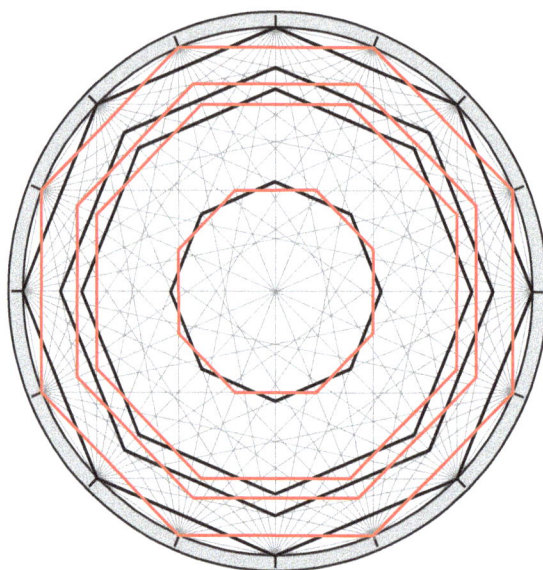

FIG 322 : - *16 = MATTER - MOLTEN SEA AETHER TO MATTER GEOMETRY*
APPLICATIONS FOR A PLASMOIDS FORM AND FUNCTIONS

16 = MATTER - MOLTEN SEA AETHER TO MATTER GEOMETRY

11.25 = 40,500 seconds of Arc
22.5 = 81,000 seconds of Arc

= 576,000 seconds of Arc
= 960 minutes of Arc
= 40 hours of Arc

16

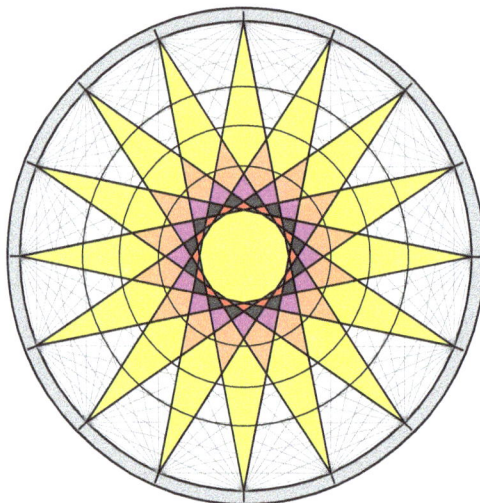

16 points making 5 circles. 4+1
5 arrows [5x16 = 80]

FIG 323 : -SOLOMONS MOLTEN SEA

15 = TIME

A 15 point circle misses
the centre, thereby creating
a Zero Point centre

7.5 = 27,000 Arc seconds
51.84 = 186,624 Arc seconds

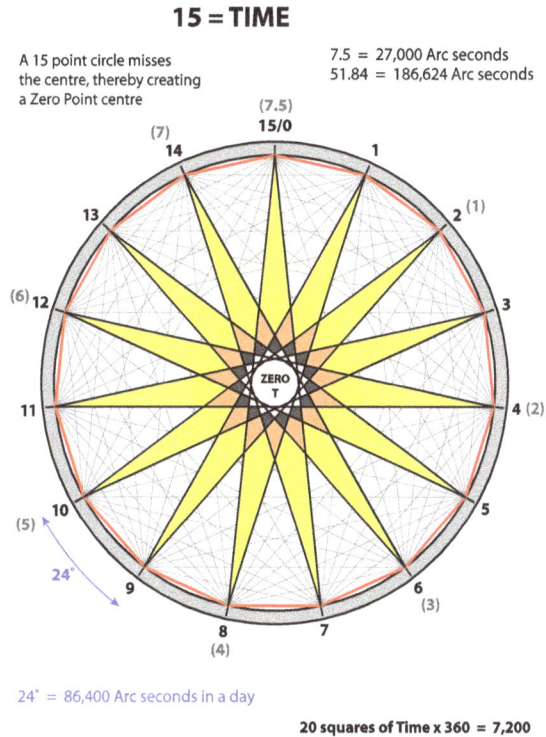

24° = 86,400 Arc seconds in a day

20 squares of Time x 360 = 7,200

FIG 324 : -SOLOMONS MOLTEN SEA

APPLICATIONS FOR A PLASMOIDS FORM AND FUNCTIONS

MOLTEN SEA GEOMETRY

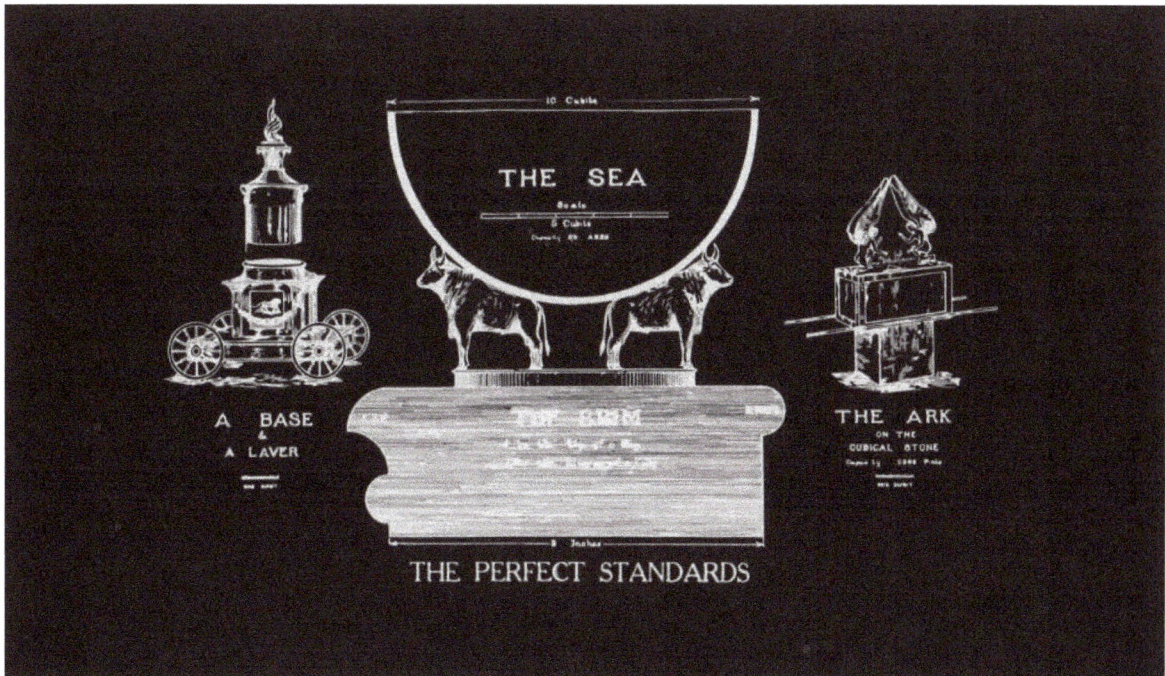

FIG 325 : -SOLOMONS MOLTEN SEA

10 CUBITS IN DIAMETER FULL OF 2,000 BATHS OF WATER WHICH IS HIN AND 144,000 LOGS

Aether 150 Suns x 864,000 = 129,600,000 150 Suns x 12 = 1,800   12 x 864,000 = 10,368,000

Matter 150 Suns x 3,456,000 = 518,400,000 TIME

Matter 150 Suns x 144,000 = 21,600,000

864,000 x 144,000 = 124,416,000,000

Aether 150 Moons x 2,160 = 324,000

Aether 2,160 x 144,000 LIGHT = TIME 311,040,000

Matter 150 Moons x 8,640 = 1,296,000   129,000 x 150 = 19,440,000 //972 //486//243

Aether 150 Moons x 16 = 900   2,160 x 16 = 34,560

Matter 150 Moons x 51,840 miles in 24 hours = 7,776,000

MOLTEN SEA NUMBERS

NUMBERS FOR THE OUTSIDE OF THE MOLTEN SEA TIMES 11.111

Earth 7,920 x 11.111 = 87,999.99999999912 [88,000]   1 / x = 0.00001136363

88,000 / 266.666 = 330   1 / x = 0.00303030

Earth 7,920 / 266.666 = 29.7   1/x = 0.12345679012345

 7,920 x 266.666 = 2,112,000

Sun 864,000 x 11.111 = 9,599,999.999999904 [9,600,000]   1 / x = 0.0000001041667

9,600,000 / 266.666 = 36,000

Moon 2,160 x 11.111 = 23,999.99999999976 [24,000]   1/x =0.000041666

2,160 / 266.666 = 8.10000002025   1 / x = 0.123456789814815

(432,000 x 11.111 = (300 x 11.111 = 3,333.333)   (150 x 11.111 = 1,666.666)   (50 x 11.111 = 555.555)

(6 x 11.111 = 66.666) (3,333.333 / 0.125 = 26,666.666) (1,666.666 / 0.125 = 13333.33)

(555,555 / 0.125 = 4,444,440) (4,444,440 / 1.3333 = 3,333,330.000000083333)

(4,444,440 / 0.125 = 35,555,520 / 0.125 = 284,444,160 / 0.125 = 2,275,553,280 / 0.125 =)

NUMBERS FOR THE INSIDE OF THE MOLTEN SEA IN HANDS TIMES 11.111

(45 x 11.111 = 500   1 / x = 0.002)

(45 x .666 = 30 [29.999] = 2x Depth of water in the Molten Sea but it is an inverted sphere 1/x = 0.0333)
(266.666/30 = 8.888 1/x = 0.1125 the area of the surface of the molten sea water is [Area = pie r (15) squared) = 706.86 square hands   706.86/ 16 = 44.17875 / 22.5 = 1.9635 x 400 = 785.4 RFEU

(22.5 x 11.111 = 250 1/x = 0.004 x 266.666 = 1.0666) (22.5 x 0.666 = 15 hands is the Depth of water in the molten sea 15 /2 = 7.5 hands resonant air cavity (light travels 7.5 times around the earth in one second giving the Standard Light Earth Frequency (SLEF) above the water to the Rim)
(266.666 / 15 = 17.777   1/x = 0.05625,   266.666 / 7.5 = 35.555   1 / x = 0.28125)

360 x 11.111 = 4,000   1/x = 0.00025 [4,000/266.666 = 15/0.125 = 120.00000000000018 1 / x = .008333

180 x 11.111 = 2,000   1 / x = 0.0005

90 x 11.111 = 1,000   1 / x = 0.001 [1,000 / 266.666 = 3.75/ 0.125 = 30.00000000000075   1 / x = 0.001

(16 x 11.111 = 177.777   1 / x =) (400 x 11.111 = 4,444.444   1/x =)

(9 x 11.111 = 10 [99.9999]   1/x = 0.01])

(11.111 x 11.111 = 123.456790123457   1 / x = .0081)

NUMBERS FOR THE MOLTEN SEA IN CUBITS 10 and 5 TIMES 11.111

(5 x 11.111 = 55.55), (10 x 11.111 = 111.111), (30 x 11.111 = 333.333).

ARK OF THE COVENANT SACRED RULING NUMBERS

NUMBERS FOR THE ARK OF THE COVENANT IN CUBITS

(2.5 x 11.111 = 27.777) (1.25 x 11.111 = 13.888) (1.5 x 11.111 = 16.666)
(2.5 x 1.5 x 1.5 = 5.625 / 0.125 = 45)
(1.25 x 1.5 x 1.5 = 2.8125 / 0.125 = 22.5)
27.777 x 16.666 x 16.666 = 7,715.2160790120 / 0.125 = 61,721.728632096

NUMBERS FOR THE ARK OF THE COVENANT IN INCHES

(45 x 11.111 = 500), (22.5 x 11.111 = 250), (27 x 11.111 = 300)
(45 x 27 x 27 = 32,805 / 0.125 = 262,440)
(300 x 300 x 500 = 45,000 / 0.125 = 360,000,000)
(300 x 300 x 250 = 22,500,000 / 0.125 = 180,000,000)

NUMBERS FOR THE ARK OF THE COVENANT IN HANDS

(11.25 x 11.111 = 125), (5.625 x 11.111 = 62.5), (6.75 x 11.111 = 75)
(11.25 x 6.75 x 6.75 = 512.578125 / 0.125 = 4,100.625
(75 x 75 x 125 = 703,125 / 0.125 = 5,625,000)
(75 x 75 x 62.5 = 351,562.5 / 0.125 = 2,812,500)

THE HIGH PRIESTS EPHOD RULING NUMBERS AND FORM

NUMBERS FOR THE EPHOD IN HANDS

(2.25 x 11.111 = 25, 2.25 x 2.25 = 5.0625 / 0.125 = 40.5
(25 x 25 = 625 / 0.125 = 5,000)

MOLTEN SEA RULING NUMBERS

RULING NUMBER 1.1111

RULING NUMBER 1.3333

RULING NUMBER 5

RULING NUMBER 9

RULING NUMBER 10

RULING NUMBER 11.11

RULING NUMBER 12

RULING NUMBER 16

RULING NUMBER 22.5

4,500/

RULING NUMBER 27

RULING NUMBERS 37

RULING NUMBER 45

RULING SUN'S LIGHT NUMBER 144

RULING NUMBER 150 SUNS AND 150 MOONS ON OUTER MOLTEN SEA RIM

150/ 12 / 1.111 =11.25   150 / 360 / 1.111 = .375
150 x 1.333 = 199.999   150 / 16 = 150 / 12

RULING NUMBER 300 TOTAL NUMBER OF SUNS AND MOONS ON MOLTEN SEA RIM

300 x 1.333 x 1.111 = 444.444 IN MUSIC A
300 / 12 / 1.111 = 22.5
300 / 16 = 18.753 300 / 12 = 25
LILLIES 4,800 lilies x 1,800 petals = 8,640,000

RULING NUMBER 360

RULING NUMBER 432

RULING NUMBER 450

RULING NUMBER 720

720 / 16 = 45   720 / 1.333 = 540   720 / 1.111 = 432 [432+234 = 666]

RULING NUMBER 2,520

2,520 = 3 x 4 x 5 x 6 x 7
2,520 = 3 x 7 x 10 x 12 (all perfect numbers)
2,520 x 2 =5,040
2,520

RULING NUMBER 4,500 VOLUME OF RIM OF THE MOLTEN SEA

4,500 / 0.125 = 36,000

RULING NUMBER 14,400

14,400 = 1+2+3+4+5

MOLTEN SEA NUMBERS

VOLUME OF THE MOLTEN SEA IS 3,000 BATHS x 72 = 216,000 LOGS

216,000 Logs = 300 Homer = 54,000 Cad

VOLUME OF THE MOULTEN SEA IS 3,000 BATHS X 72 = HANDS

VOLUME OF THE MOULTEN SEA IS 2,000 BATHS x 72 = 144,000
1,44,000 / 11.11 EQUALS 129,600 RESONANT FREQUENCY ENERGY UNITS [RFEU]

VOLUME OF THE MOULTEN SEA IS 2,000 BATHS x 72 = 144,000 logs
144,000 logs = 36,000 Cad = 200 homer = 18,000 Him = 36,000 Kab =144,000 Logs
1,44,000 / 11.11 EQUALS 129,600 RESONANT FREQUENCY ENERGY UNITS [RFEU]

VOLUME OF THE MOULTEN SEA IS 2,000 BATHS = 150 RITUAL BATHS

VOLUME OF THE MOLTEN SEA INNER SPHERE = 47,712 / 2 = 23,856 / 0.666 = 35,784

VOLUME OF THE MOULTEN SEA IS 2,000 BATHS x 72 = 129,600 (RFEU)

VOLUME OF THE MOULTEN SEA IS 129,600 x 1.333 = 172,800

172,800 RFE [1728 + 8271 = 9,999]

VOLUME OF THE MOLTEN SEA OUTER SPHERE = 56,684 / 2 = 28,342 / 0.666 = 42,513

 MINUS VOLUME OF THE INNER SPHERE = 8,972 / 2 = 4,486 / 0.6666 = 6,729

AREA OF TOP OF THE MOLTEN SEA =

CIRCUMFERANCE OF THE MOLTEN SEA =

RULING NUMBERS 129,600, 9,999 AND 36
MADE FROM REFLECTIVE NUMBER 3456 AND

259,200 [EARTHS EQUINOX] = 3 X 4 X 5 X 6 X 6 X 5 X 4 X 3 X 2 / 2 = 129,600 / 0.125 = 1,036,800

129,600 / 36 = 3,600 [36 = 3+4+5+6+6+5+4+3]   9,999 = 3,456 + 6,543

129,600 / 1.111 = 116,640   129,600 X 1.111 = 144,000
129,600 / 1.333 = 97,200.0000000024   129,600 x 1.333 = 172,800

129,600 / 360 = 129,600/180= 129,600/ 90 = 129,600/45 = 129,600/22.5 =

129,600 / 9,999=12.9612961   9,999/129,600 = 0.077152777

9,999 = 3,456 + 6,543   9,999/36 = 277.75 /0.125 = 2,222

9,999 / 3,456 = 2.8932291666 = 1/X = 0.3456345634563456

9,999 / 6,543 = 1.52819807427785   1/x = 0.654365436543

6,543 / 9,999 = 0.654365436543

9,999 / 720 = 13.8875   9,999 / 360 = 27.775   9,999 / 180 = 55.55 9,999 / 90 = 111.111
9,999 / 45 =222.2   9,999 / 11.25 = 888.8

144,000 / 1.111 = 129,600

360 **X** 360 = **129,600**   180 x **720** = **129,600**

129,600 x **7.5** = **972,000** / **360** = **2,700** / **9** = **300** / **1.333** = **225** / **1.111** =

202.5000000000070875 = **1/x** = **0.00493827160494**

RULING NUMBER 1,776 = 999 + 777 AND 48
MADE FROM REFLECTIVE NUMBER 789

789 + 987 = 1,776   999+ 777 = 1,776

7 + 8 + 9 +9 + 8 +7 = 48

7 x 8 x 9 x 9 x 8 x 7 = 254.016

RULING NUMBER 144,000, 7,777 AND 28
MADE FROM REFLECTIVE NUMBER 2345

At Base 9 the harmonic speed of light is **144,000** minutes of Arc
Where normal **86,400** (**864,000 Sun 888**) seconds in a day based on harmonic 8 are expanded to into 9,720 (9,720 Earth 999) based on harmonic 9 splitting the earth into 27 grid hours the number of grid seconds then equals **97,200**. Therefore the grid speed of light is **144,000** minutes of Arc per grid second.

2 to the 7th + 3 SQUARED + 5 CUBED = 144,000

144,000 = 2 X 3 X 4 X 5 X 5 X 4 X 3 X 2 X 10   **144,000** / 0.125 = 1,152,000

144,000 / 28 = 5,142.85714285714 [28 = 2+3+4+5+5+4+3+2] 7,777 = 2,345 + 5,432

144,000 / 7,777= 18.5161373280185   7,777 / **144,000** = 0.054006944
144,000 x **144,000** =20,736,000,000 / 7,777 = 2,66,323.7752346663238

144,000 x **144,000** x **144,000** = 2,985,984,000,000,000 / 7,777 = 383,950,623,633.792

144,000 SQUARE ROOT = 379.473319220205 / 7,777 = 0.0487943061875

144,000 CUBE ROOT = 52.4148278841779 / 7,777 = 52.4148278841779

7,777 / 1.5241579027586648377 = 5,102.48970000042

7,777 = 2,345 + 5,432   7,777/28 = 277.75 /0.125 = 2,222

7,777 / 2,345 = 3.31641791044776 =   1/x = 0.30153015

7,777 / 5,432 = 1.43170103092784

5,432 / 7,777 = 0.698469846984

2,345 / 7,777 = 0.30153015301

144,000 /1.111 = 129,600

360 x 400 = 144,000   180 x 800 = 144,000   200 x 720 = 144,000

RULING NUMBER 1,728,000 [Suns Resonant Frequency Energy units]
9999 AND 36 and 48 MADE FROM REFLECTIVE NUMBER 1728

1,728 + 8,271 = 9,999   1+7+2+8+8+2+7+1= 36

1,728 / 36 = 48

1 x 7 x 2 x 8 x 8 x 2 x 7 x 1 = 12,544   / 36 = 348.444

SANSKRIT LONG 100 YR RULING NUMBER   3,110,400,000,000

311,040,000,000,000 / 25,920 = 12,000,000,000,000

3,110,400, 000,000 / 25,920 = 120,000,000

3,110,400,000,000 / 864,000 = 3,600,000 / 266.666 = 13,500 / 22.5 =

3,110,400,000,000 / 7.5 =

APPLICATIONS FOR A PLASMOIDS FORM AND FUNCTIONS

MOLTEN SEA GEOMETRY

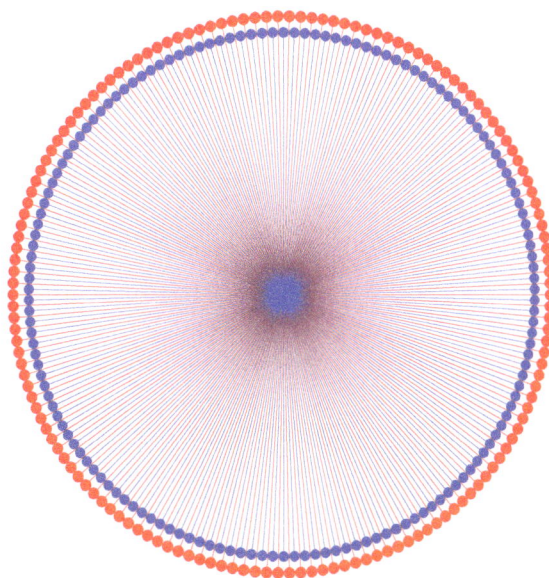

FIG 326 : -SOLOMONS MOLTEN SEA 150 SUNS AND 150 MOONS

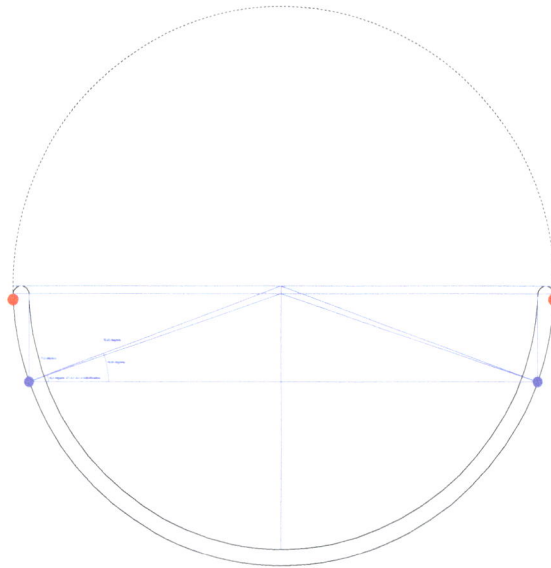

FIG 327 : -SOLOMONS MOLTEN SEA SCHEMATIC

PART 14

VESICA PISCIS

MSAART PATENT APPLICATION NOTES

APPENDIX 4: - VERSICA PISCIS

PART FOURTEEN OF TWENTY

DRAFT 518,400 B KMV

BY

MALCOLM V of SCOTLAND | MALCOLM BENDALL
THURSDAY 22ND SEPTEMBER 2022

APPLICATIONS FOR A PLASMOIDS FORM AND FUNCTIONS

DIAGRAM 328, 329 & 330 - VERSICA PISCIS

DIAGRAM 331 - VERSICA PISCIS

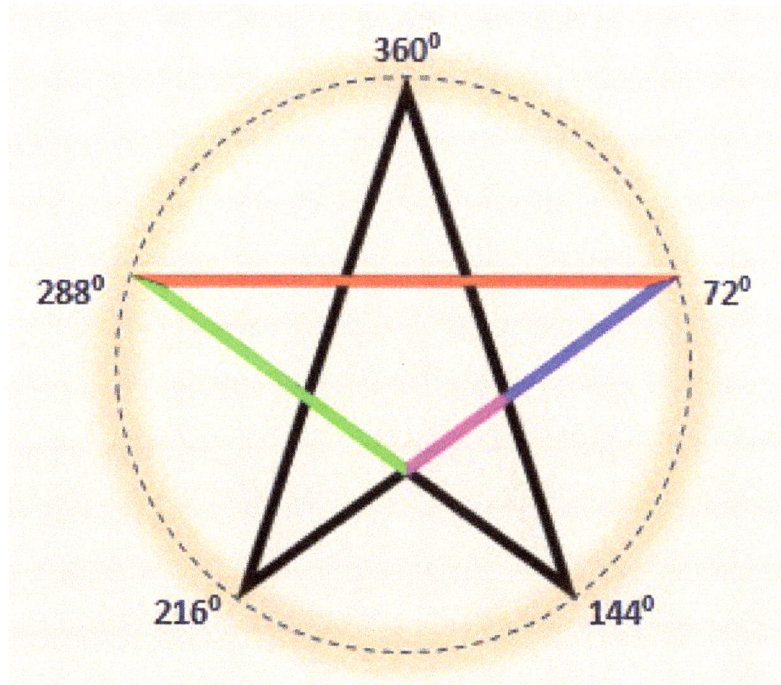

APPLICATIONS FOR A PLASMOIDS FORM AND FUNCTIONS

DIAGRAM 332 - VERSICA PISCIS

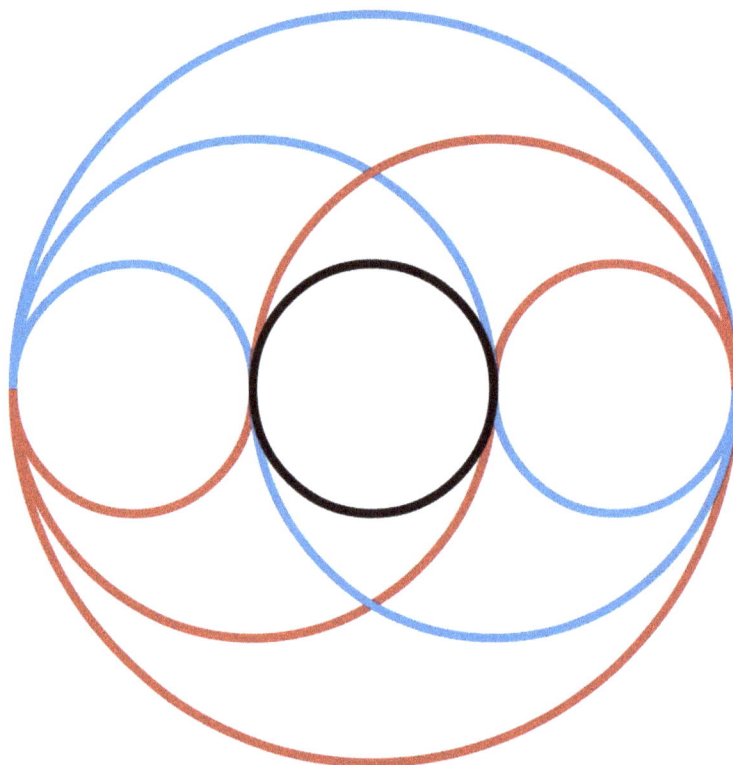

DIAGRAM 333 - VERSICA PISCIS

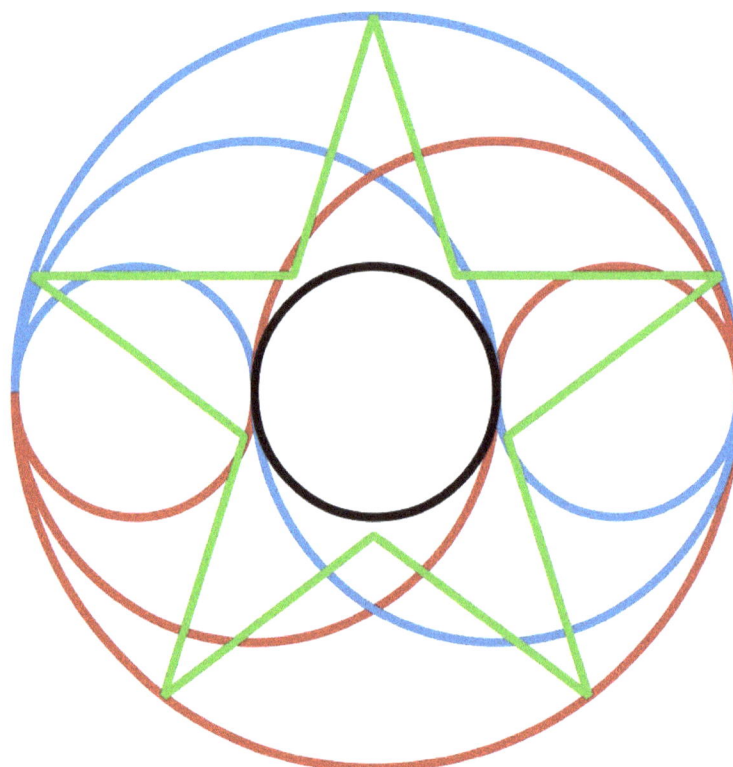

APPLICATIONS FOR A PLASMOIDS FORM AND FUNCTIONS

DIAGRAM 334 - VERSICA PISCIS

DIAGRAM 335 - VERSICA PISCIS

APPLICATIONS FOR A PLASMOIDS FORM AND FUNCTIONS

DIAGRAMS 336, 337 & 338 - VERSICA PISCIS

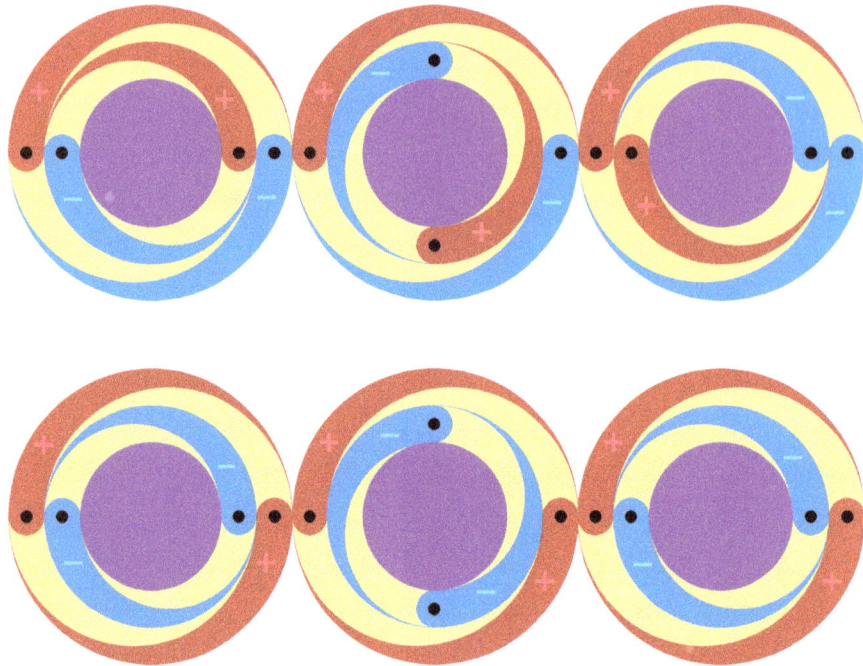

DIAGRAMS 339, 340 & 341 - VERSICA PISCIS

APPLICATIONS FOR A PLASMOIDS FORM AND FUNCTIONS

DIAGRAMS 342, 343 & 344 - VERSICA PISCIS

DIAGRAMS 345, 346 & 347 - VERSICA PISCIS

PART 15

ALIEN MATHS AND CHEMISTRY

APPLICATIONS FOR MSAART FORMS AND FUNCTIONS

ALIEN MATHS AND CHEMISTRY

PART FIFTEEN OF TWENTY – PP 375 - PP 406

DRAFT 518,400 B KMV

BY

MALCOLM V of SCOTLAND | MALCOLM BENDALL
THURSDAY 22ND SEPTEMBER 2022

APPLICATIONS FOR A PLASMOIDS FORM AND FUNCTIONS

ALIEN MATHS AND CHEMISTRY

HYDROGEN, NOBLE GASES, METHANE AND CARBON MONOXIDE

The melting point and boiling point of any element are the points at which enough energy has been put into the dynamic system for the resonant frequency intensity to be met. This is determined by a combination of dynamic factors. Primary of these factors is the sum of the internal angles of the elements or compounds natural crystal form manifesting in its solid state. These are adjusted to their atomic charge density and the Area / time / space that it occupies which are the secondary forces. The Area / Time / Space its self introduces the additional variable of the external positive and negative atmospheric and atomic pressure. This is in addition to the tertiary effect of implosive and explosive environmental and atomic forces. Contamination of other matter further varies the frequency /area / time / space equation and therefore must be measured and accounted for to determine a "group" mass resonance.

PROTIUM [PART OF THE HYDROGEN GROUP]

HYDROGEN MELTING POINT = − 259.2 so - 259.2 / 266.666 = 0.972

- 259.2 / 11.111 = 23.328

- 259.2 / 1.333 = 194.4 // 97.2//48.6//24.3

- 259.2 added sum = 2 + 5 + 9 + 2 [18] + 2 + 5 + 9 + 2 = 36

- 259.2 multiplied product = 2 x 5 x 9 x 2 (180) x 2 x 9 x 5 x 2 = 32,400

32,400 / 720 = 45 32,400 / 16 = 2,025 32,400 / 22.5 = 1,440 32,400 / 0.125 = 259,200

32,400 x 16 = 518,400   32,400 x 22.5 = 729,000   32,400 x .125 = 4,050

32,400 x 400 = 12,960,000 32,400 x 720 = 23,328,000 32,400 x 900 = 29,160,000

518,400 / 16 = 32,400   518,400 / 22.5 = 23,040   518,400 / 720 = 720

32,400 / 144 = 225 32,400 / 36 = 900 36 / 32,400 = 0.00111

144,000 / 32,400 = 4.4444   32,400 / 7.5 = 4,320 / 7.5 = 576 / 2 = 288 / 2 = 144

144 x 32,400 = 4,665,600 / 432 = 10,800 / 2,160 = 5

32,400 / 259.2 = 125
32,400 / 6.666 = 4,860   32,400 / 5.555 = 5,850   32,400 / 4.444 = 7,290
32,400 / 3.333 = 9,720   32,400 / 2.222 = 14,580   32,400 / 11.111 = 2,916

ALL SIX 864,000 COMBINATIONS

864,000	846,000	684,000	648,000	486,000	468,000
56,623,104*	55,443,456*	44,826,624*	42,467,328*	31,850,496*	30,670,848*
28,311,552*	27,721,728*	22,413,312*	21,233,664*	15,925,248*	15,335,424*
14,155,776*	13,860,864*	11,206,656*	10,616,832*	7,962,624*	7,667,712*
7,077,888*	6,930,432*	5,603,328*	5,308,416*	3,981,312*	3,833,856*
3,538,944*	3,465,216*	2,801,664*	2,654,208*	1,990,656*	1,916,928*
1,769,472*	1,732,608*	1,400,832*	1,327,104*	995,328*	958,464*
884,736*	866,304*	700,416*	663,552*	497,664*	479,232*
442,368*	433,152*	350,208*	331,776*	248,832*	239,616*
221,184*	216,576*	175,104*	165,888*	124,416	119,808*
110,592*	108,288*	87,552*	82,944*	62,208*	59,904*
55,296*	54,144*	43,776*	41,472*	31,104*	29,952*
27,684*	27,072*	21,888*	20,736*	15,552*	14,976*
13,824*	13,536*	10,944*	10,368*	7,776*	7,488*
6,912,000	6,768,000	5,472,000	5,184,000	3,888,000	3,744,000
3,456,000	3,384,000	2,736,000	2,592,000	1,944,000	1,872,000
1,728,000	1,692,000	1,368,000	1,296,000	972,000	936,000

864,000	846,000	684,000	648,000	486,000	468,000
432,000	423,000	342,000	H-324,000	243,000	234,000
216,000	211,500	171,000	162,000	121,500	117,000
108,000	105,000	85,500	81,000	60,750	58,500
54,000	52,875	42,750	40,500	30,375	29,250
27,000	26,437.5	21,375	20,250	15,187.5	14,625
13,500	13,218.75	10,687.5	10,125	7,593.75	7,312.5
6,750	6,609.375	5,343.75	5,062.5	3,796.875	
3,656.25					
3,375	3,304.6875	2,671.875	2,531.25	1,898.43	1,828.1
1,687.5	1,652.34375	1,335.9375	1,265.62	949.2187	914.06
843.75	826.171875	667.96875	632.812	474.6093	457.03
421.875	413.0859375	333.984375	316.406	237.3046	228.51
210.9375	206.54296875	166.99218	158.2031	118.6523	114.25
105.46875	103.271484375	83.496093	79.10156	59.32617	57.128
52.734375	51.6357421875	41.7480468	39.55078	29,66308	28.564
26.3671875	25.8187109375	20.8740234	19.77539	14.83154	14.282

864,000	846,000	684,000	648,000	486,000	468,000

PART 15 OF 20 – MSAART PATENT APPLICATION NOTES - DRAFT 518,400 – 22:22:22 THUR 22ND SEPT 2022
© STRIKE FOUNDATION GUARANTEE LIMITED | MALCOLM BENDALL 2022 | GRAPHICS - STEVE EARL

HYDROGEN RATIOS TO ALL OTHER NOBLE GASES

32,400 (Hydrogen -259.2) / 3,136 (Helium -272.2) = 10.33163265306122
3,136 (Helium -272.2) / 32,400 (Hydrogen -259.2) = 0.09679012345679

32,400 (Hydrogen -259.2) / 147,456 (Neon -248.6) = 0.2197265625
147,456 (Neon -248.6) / 32,400 (Hydrogen -259.2) = 4.55111

32,400 (Hydrogen -259.2) / 82,944 (Argon -189.4) = 0.390625
82,944 (Argon -189.4) / 32,400 (Hydrogen -259.2) = 2.56

32,400 (Hydrogen -259.2) / 19,600 (Krypton 157.4) = 1.653061224489796
19,600 (Krypton 157.4) / 32,400 (Hydrogen -259.2) = 0.604938271604938

32,400 (Hydrogen -259.2) / 64 (Xeon -111.8) = 506.25
64 (Xeon -111.8) / 32,400 (Hydrogen -259.2) = 0.00195308641975

32,400 (Hydrogen -259.2) / 1,225 (Radon -71.15) = 26.44897959183673
1,225 (Radon -71.15)/ 32,400 (Hydrogen -259.2) = 0.037808641975309 Argon/Helium

HYDROGEN RIGHT TRIANGLE MELTING POINT = - 259.2 C (Earths Equinox 25,920)

864,000 / 259.2 = 3,333.3   864,000 / 345.6 = 2,500   864,000/ 432 = 2,000
2,160 / 259.2 = 8.333   259.2 / 2,160 = 0.972

Triangles Area / Time / Space calculation

259.2 x -345.6 x 432 = 38,698,352.6
259.2 x -345.6 = - 89,579.5 (Area of the square) / 2 = 44,789.76 (Area of the Right Triangle)

44,789.76 / 24 Hrs = 1,866.24   1,866.24 / 60 Min = 31.104   31.104 / 60 Sec = 0.5184
0.5184 / 60 Arc = 0.00864   0.00864 / 60 = 0.000144   0.000144 / 60 = 0.0000024
0.0000024 / 60 = 0.00000004   0.00000004 / 24 = 0.000000001666
0.00000004 / 60 = 0.0000000000666   0.0000000000666 / 60 = 0.00000000001111

0.5184 / 7.5 = 0.06912 / 0.266666 = 0.2592 x 400 = 103.68 / 16 = 6.48 / 22.5 = 0.288
0.288/ 0.125 = 2.304 x 400 = 921.9 / 1.333 = 691.2 / 1.333 = 518.4 /1.333 = 388.8 / 1.333 =
291.6

-259.2 / 3 = - 86.4

86.4 x 5 = 432

- 345.6 = - 86.4 x 4

HYDROGEN SQUARE MELTING POINT -259.2 (Earths Equinox) x 4 = 1,036.8

1,036.8/3,456 = 0.3
259.2 x 259.2 = 67,184.64 / 11.111 = 6,046.6   67,184.64 / 266.666 = 251.9424
67,184.64 / 144 = 466.56
67.184.64 / 7.5 = 8,957.952
67,184.64 / 2 = 33,592.32
33,592.32 / 3 = 11,197.44   11,197.44 x 4 = 44,789.76   11,197.44 x 5 = 55,987.2

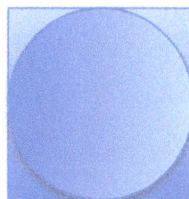

HYDROGEN CUBE MELTING POINT CUBED = 259.2 x 259.2 x 259.2 = 17,414,258.688
1 year, 360 day Time in seconds cube 17,414,258.688 / 360 = 48,372.9408
48,372.9408/24 = 2,016/60 = 33.6/60 = 0.56/ 720 = 0.000777/ 16 = 0.0000486111/ 22.5 =
0000002160493827

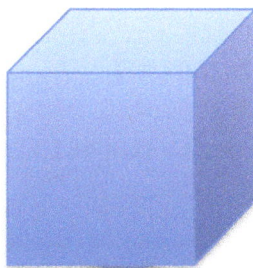

17,414,258.688 / 144,000 = 120.932352 / 7.5 = 16.1243136 / 16 = 1.0077696
 = 1.0077696 / 22.5 = 0.04478976
17,414,258.688 / 1.333 = 13,060,694.016
17,414,258.688 / 266.666 = 65,303.47008000002
17,414,258.688 / 6.666 = 2,612,138.8032
17,414,258.688 / 5.555 = 3,134,566.56384
17,414,258.688 / 4.444 = 3,918,208.2048
17,414,258.688 / 3.333 = 5,224,277.606400001
17,414,258.688 / 2.222 = 7,836,416.409600001
17,414,258.688 / 1.111 = 15,672,832.8192 / 9 = 1,741,425.8688

HYDROGEN RIGHT ANGLE TRIANGLE BOILING POINT = - 252.8 C

HYDROGEN Boiling point = - 252.87 so -252.87 / 266.666 = .9482625
- 252.87 so -252.87 / 11.111 = 22.7583
- 252.87 so -252.87 / 1.333 = 189.699925

252.87 = 2 + 5 + 2 + 8 + 7 (24) +7 + 8 + 2 + 5 + 2 = 48

252.87 = 2 x 5 x 2 x 8 x 7 (1,120) x 7 x 8 x 2 x 5 x 2 = 1,254,400 / 5.555 = 225,792

252.8 x 337.0666 = 85,210.4365 therefore triangle area 85,210.4365 / 2 = 42,605.2182
42,605.2182 / 24 Hours = 1,775.21743   1,775.21743 / 60 Minutes = 29.5869571
29.5869571 / 60 Seconds = 0.493115951
0.493115951 / 60 Arc seconds = 0.0821859918333

- 252.8 / 3 = 84.2666

421.333 = 84.2666 x 5

337.0666 = 84.2666 x 4

ARGON

ARGON RIGHT ANGLE TRIANGLE MELTING POINT – 189.4

- 189.4 (3 side) x 252.5333 (4 side) = 47,829.81333 (Area of Square)
47,829.81333 / 2 = 23,914.90666 (Area of the Right Triangle)
23,914.90666 / 24 Hours = 996.45444
996.45444 / 60 Minutes = 16.60757407407
16.60757407407 / 60 Seconds = 0.276792901234568
0.276792901234568 / 60 Arc Seconds = 0.004613215020576

189.4 / 3 = 63.1333

315.666 = 5 x 63.1333

252.5333 = 4 x 63.1333

1 + 8 + 9 + 4 (22) + 4 + 9 + 8 + 1 = 44   1
1 x 8 x 9 x 4 (288) x 4 x 9 x 8 x 1 = 82,944

82,944 (Argon - 189.4) / 44 = 1,885.0909
82,944 (Argon - 189.4) x 44 = 3,649,536
82,944 (Argon - 189.4) / 266.666 = 311.0400000000001
82,944 (Argon - 189.4) / 11.111 = 7,464.960000000001
82,944 (Argon - 189.4) / 1.333 = 62,208.00000000002
82,944 (Argon - 189.4) / 144 = 576 //288//144//72//36//16//8//4//2//1

ARGON RATIOS TO ALL OTHER NOBLE GASES INCLUDING HYDROGEN

82,944 (Argon- 189.4) / 32,400 (Hydrogen – 259.2) = 2.56
2.56// 1.28//0.64// 0.32//0.16//0.08
32,400 (Hydrogen – 259.2) / 82,944 (Argon - 189.4)= 0.390625

82,944 (Argon - 189.4) / 3,136 (Helium - 272.2) = 26.44897959183673
3,136 (Helium - 272.2) / 82,944 (Argon - 189.4) = 0.037808641975309
          0.037808641975309 //+8 = 9.679012345679012

82,944 (Argon - 189.4) / 147,456 (-248.6 Neon) = 0.000140625//+5 = 0.144
147,456 (-248.6 Neon) / 82,944 (Argon - 189.4) = 1.777
1.777//0.888/0.444/0.222/0.111/0.0555

82,944 (Argon - 189.4) / 19,600 (Krypton 157.4) = 4.231836734693878
19,600 (Krypton 157.4) / 82,944 (Argon - 189.4) = 0.236304012345679 //+5 =3.7808

82,944 (Argon - 189.4) / 64 (Xeon 111.8) = 1,296
64 (Xeon 111.8) / 82,944 (Argon - 189.4) = 0.000771604938272

82,944 (Argon - 189.4) / 1,225 (Radon -71.15) = 67.70938775510204
1,225 (Radon -71.15) / 82,944 (Argon - 189.4) = 0.0147690007716

ARGON

ARGON RATIOS TO ALL OTHER NOBLE GASES INCLUDING HYDROGEN

82,944 (Argon – 189.4) / 32,400 (Hydrogen -259.2) = 2.56
2.56 // 1.28//0.64// 0.32//0.16//0.08 //0.04//0.02//
32,400 (Hydrogen -259.2) / 82,944 (Argon – 189.4)= 0.390625

82,944 (Argon – 189.4) / 3,136 (Helium -272.2) = 26.44897959183673
3,136 (Helium -272.2) / 82,944 (Argon – 189.4) = 0.037808641975309
          0.037808641975309 //+8 = 9.679012345679012

82,944 (Argon – 189.4) / 147,456 (-248.6 Neon) = 0.5625
147,456 (-248.6 Neon) / 82,944 (Argon – 189.4) = 1.777
1.777//0.888//0.444//0.222/0.111/0.0555

82,944 (Argon – 189.4) / 19,600 (Krypton 157.4) = 4.231836734693878
19,600 (Krypton 157.4) / 82,944 (Argon – 189.4) = 0.236304012345679 //+5 =3.7808

82,944 (Argon – 189.4) / 64 (Xeon 111.8) = 1,296
64 (Xeon 111.8) / 82,944 (Argon – 189.4) = 0.000771604938272

82,944 (Argon – 189.4) / 1,225 (Radon -71.15) = 67.70938775510204
1,225 (Radon -71.15) / 82,944 (Argon – 189.4) = 0.0147690007716

CARBON'S RESONANT FORMS

Carbon has 5 states, each state has a unique melting point or phase change

1) - DIAMOND

MELTING POINT = 3,700K or 3,426.85C   The Sum = 56 and Product = 33,177,600 (5,760 Sq)
33,177,600 / 144,000 = 230.4   33,177,600 / 518,400 = 64   33,177,600 / 31,104,000 = 1.0666 /// 0.2666
33,177,600 / 24 = 1,382,400 / 60 = 23,040 / 60 = 384 / 60 = 6.4 / 7.5 = 0.85333 \\\\\ 0.02666
33,177,600 / 864 = 38.4 / 144 = 0.2666   33,177,600 / 2,160 = 15,360   33,177,600 / 129,600 = 256\\\\16

MELTING POINT = 5,000K or 4,726.85C The Sum = 32 and Product = 180,633,600 (13,440 Sq)
180,633,600 / 144,000 = 1,254.4   180,633,600/518,400 = 384.444 180,633,600/31,104,000 = 5.80740
180,633,600 / 7.5 = 24,084,480 / 24 = 1,003,520 / 60 = 16,725.333 / 60 = 278.7555 / 60 = 4.64592592
180,633,600 / 25,920 = 6,968.888   180,633,600 /864,000 = 209.0666 180,633,600 / 266.666 = 677,376

2) - GRAPHENE

MELTING POINT = 4,510K or 4,236.85C The Sum= 28 and Product= 33,177,600 SAME AS DIAMOND
33,177,600 / 144,000 = 230.4   33,177,600 / 518,400 = 64   33,177,600 / 31,104,000 = 1.0666 \\\0.2666

MELTING POINT = 4,900K or 4,626.85C The Sum = 31 and Product = 132,710,400 (11,520 Sq)
132,710,400 / 24 = 5,529,600 / 60 = 92,160 / 60 = 1,536 / 60 = 25.6 / 7.5 = 3.41333\\\\\\\0.02666
132,710,400 / 144,000 = 921.6 / 266.666 = 3.456 / 11.111 = 0.31104000031104

3) - GRAPHITE

MELTING POINT = 3,800K or 3,526.85C Sum = 58 and Product = 51,840,000 (7,200 Squared)
51,840,000 / 24 Hours = 2,160,000 / 60 min = 36,000 / 60 Sec = 600 / 60 Arc = 10 / 7.5 = 1.333 \266.666
51,840,000/864,000=60 51,840,000/144,000=360 51,840,000/129,600=400 51,840,000/266.666=1944
51,840,000 / 25,920 (EQUINOX 25,920 Years) = 2,000 (2,000 BATHS IN THE 2/3 FULLMOLTEN SEA)

4) - BUCKEY BALLS CARBON FULLERENE C60

MELTING POINT = 1523K or 1,249.85C The Sum = 58 and Product = 8,294,400 (2,880 Sq)
8,294,400/ 25,920 = 320 8,294,400/864,000=9.6 8,294,400/144= 57,600 8,294,400/21,600=384\\\\24
8,294,400 / 24 = 345,600 / 60 = 5,760 / 60 = 96 / 60 = 1.6 / 7.5 = 0.21333\\\0.02666
MELTING POINT = 278.15K or 5C The Sum = 10 and Product = 25

MELTING POINT = 800K or 526.85C The Sum = 52 and Product = 5,760,000 (2,400 Sq)
5,760,000 / 25,920= 222.222 5,760,000/864,000= 6.666 5,760,000/144= 40 5,760,000/21,600= 266.666
5,760,000 / 24 = 240,000 / 60 = 4,000 / 60 = 6.666 / 60 = 1.111 / 7.5 = 0.148148148

MELTING POINT = 707K or 433.85C The Sum = 46 and Product = 2,073,600 (1,440 Sq)
2,073,600 / 24 = 86,400 / 60 = 1,440 / 60 = 24 / 60 = 0.4 / 0.00075 = 266.666 (C music elemental Octave)
MELTING POINT = 739K or 465.85C The Sum = 56 and Product = 23,040,000 (4,800)
23,040,000 / 24 = 960,000 / 60 = 16,000 / 60 = 266.666 / 60 = 4.444   23,040,000 / 25,920 = 888.888

5) - NANOTUBULES

MELTING POINT = 5,200K or 4,926.85C The Sum = 68 and Product = 298,598,400 (17,280 Sq)
298,598,400 /24 = 12,441,600 / 60 = 207,360 / 60 = 3,456 / 60 = 57.6 (14.4 x 3)/ 7.5 = 7.68 \\\\\0.24
298,598,400 /144 = 2,073.6\\\259.2(H)   298,598,400/25,920 = 11,520   298,598,400/= 129,600 = 2,304
MELTING POINT VACUME = 3,073 K or 2,800C The Sum = 20 and Product = 256 (16 Sq)
MELTING POINT AIR = 1,023K or 750C The Sum = 13 and Product = 1,225 (35 Sq)

1). DIAMOND
33,1777,600

MELTING POINT = 3,700K or 3,426.85C Sum = 56 Product = 33,177,600 (5,760Sq)

33,177,600 / 144,000 = 230.4   33,177,600 / 518,400 = 64   33,177,600 / 31,104,000 = 1.0666 /// 0.2666

33,177,600 / 24 = 1,382,400 / 60 = 23,040 / 60 = 384 / 60 = 6.4 / 7.5 = 0.85333 \\\\\ 0.02666

33,177,600 / 864 = 38.4 / 144 = 0.2666 33,177,600 / 2,160 = 15,360 33,177,600 / 129,600 = 256\\\\16

MELTING POINT = 5,000K 4,726.85C The Sum = 32 Product = 180,633,600(13,440S q)

180,633,600 / 144,000 = 1,254.4   180,633,600/518,400 = 384.444 180,633,600/31,104,000 = 5.80740

180,633,600 / 7.5 = 24,084,480 / 24 = 1,003,520 / 60 = 16,725.333 / 60 = 278.7555 / 60 = 4.64592592

180,633,600 / 25,920 = 6,968.888 180,633,600 /864,000 = 209.0666 180,633,600 / 266.666 = 677,376

2). GRAPHENE
33,177,600

MELTING POINT = 4,510K or 4,236.85C Sum= 28 Product= 33,177,600 SAME AS DIAMOND

33,177,600 / 144,000 = 230.4   33,177,600 / 518,400 = 64   33,177,600 / 31,104,000 = 1.0666 \\\0.2666

MELTING POINT = 4,900K or 4,626.85C Sum = 31 Product = 132,710,400 (11,520 Sq)

132,710,400 / 24 = 5,529,600 / 60 = 92,160 / 60 = 1,536 / 60 = 25.6 / 7.5 = 3.41333\\\\\\\0.02666

132,710,400 / 144,000 = 921.6 / 266.666 = 3.456 / 11.111 = 0.31104000031104

3). GRAPHITE
51,840,000

MELTING POINT = 3,800K or 3,526.85C Sum = 58 Product = 51,840,000 (7,200 Squared)

51,840,000 / 24 Hours = 2,160,000 / 60 Min = 36,000 / 60 Sec = 600 / 60 Arc = 10 / 7.5 = 1.333\266.666

51,840,000/864,000=60 51,840,000/144,000=360 51,840,000/129,600=400 51,840,000/266.666=1944

51,840,000 / 25,920 (EQUINOX 25,920 Years) = 2,000 (2,000 BATHS IN THE 2/3 FULLMOLTEN SEA)

4). BUCKEY BALLS CARBON FULLERENE C60
23,040,000 – 2,073,600 - 8,294,400 - 5,760,000

MELTING POINT = 1523K or 1,249.85C Sum = 58 Product = 8,294,400 (2,880 Sq)

8,294,400 / 25,920 = 320   8,294,400 / 864,000 = 9.6

8,294,400 / 144 = 57,600   8,294,400 / 21,600 = 384\\\\24

8,294,400 / 24 = 345,600 / 60 = 5,760 / 60 = 96 / 60 = 1.6 / 7.5 = 0.21333 \\\0.02666

MELTING POINT = 278.15K or 5C The Sum = 10 and Product = 25

MELTING POINT = 800K or 526.85C Sum = 52 Product = 5,760,000 (2,400 Sq)

5,760,000 / 25,920 = 222.222

5,760,000/ 864,000 = 6.666   5,760,000/ 144 = 40   5,760,000 / 21,600= 266.666

5,760,000 / 24 = 240,000 / 60 = 4,000 / 60 = 6.666 / 60 = 1.111 / 7.5 = 0.148148148

MELTING POINT = 707K or 433.85C The Sum = 46 Product = 2,073,600 (1,440 Sq)

2,073,600 / 24 = 86,400 / 60 = 1,440 / 60 = 24 / 60 = 0.4 / 0.00075 = 266.666 (C music elemental Octave)

MELTING POINT = 739K or 465.85C The Sum = 56 and Product = 23,040,000 (4,800)

23,040,000 / 24 = 960,000 / 60 = 16,000 / 60 = 266.666 / 60 = 4.444   23,040,000 / 25,920 = 888.888

5) - CARBON NANOTUBULES

Melting Point 5,200 K or 4,926.85 C

MELTING POINT = 5,200K or 4,926.85C Sum = 68 Product = 298,598,400 (17,280 Sq)

298,598,400 /24 = 12,441,600 / 60 = 207,360 / 60 = 3,456 / 60 = 57.6 (14.4 x 3)/ 7.5 = 7.68 \\\\\0.24
298,598,400 /144 = 2,073.6\\\259.2(H)   298,598,400/25,920 = 11,520   298,598,400/= 129,600 = 2,304

MELTING POINT VACUME = 3,073 K or 2,800C The Sum = 20 and Product = 256 (16 Sq)

MELTING POINT AIR = 1,023K or 750C The Sum = 13 and Product = 1,225 (35 Sq)

4,926.85 / 1 year Time 360 = 13.68569
4,926.85 x 1 year Time 360 = 1,773,666

1,773,666/24 = 73,902.75 /60= 1,231.7125 /60= 20.528541666 /60 = 0.34214236111

4,926.85 / 24 = 205.28541666 / 60 = 3.4214236111 / 60 = 0.0570237226851852
0.0570237226851852 / 3.333 = 0.000950395447531

4,926.85 x 1 year Time 360 = 1,773,666

4,926.85 / 144,000 = 0.034214236111 / 7.5 = 0.00456189814814
4,926.85 x 144,000 x 7.5 = x 720 = x 16 = x 22.5 =

4,926.85 / 266.666 = 18.4756875
4,926.85 x 266.666 = 1,313,826.666

4,926.85 / 1.333 = 3,695.1375
4,926.85 x 1.333 = 6,569.1333

4,926.85 / 11.111 = 443.4165
4,926.85 x 11.111 = 54,742.777

4,926.85 / 144 = 34.214236111
4,926.85 x 144 = 709,466.4

4,926.85 / 7.5 = 656.91333
4,926.85 x 7.5 = 36,951.375

4,926.85 / 24 = 205.28541666
4,926.85 x 24 = 118,244.444

4,926.85 / 2,000 = 2.463425
4,926.85 x 2,000 = 9,853,700

5) - CARBON NANOTUBULES

4,926.85 / 864 = 5.702372685185185
4,926.85 x 864 = 4,256,798.4

4,926.85 / 432 = 11.40474537037037
4,926.85 x 432 = 2,128,399.2

4,926.85 / 216 = 22.0809490740740740
4,926.85 x 216 = 1.064,199.6

4,926.85 / 108 = 45.61898148148148
4,926.85 x 108 = 532,099.8

4,926.85 / 792 = 6.220770202020
4,926.85 x 792 = 3,902,65.2

4,926.85 / 396 = 12.44154040404
4,926.85 x 396 = 1,951,032.6

4,926.85 / 25.92 = 109.0790895061728
4,926.85 x 25.92 = 127,703.952

4,926.85 / 6.666 = 739.0275
4,926.85 x 6.666 = 32,845.666

4,926.85 / 5.555 = 886.833
4,926.85 x 5.555 = 27,371.3888

4,926.85 / 4.444 = 1,108.54125
4,926.85 x 4.444 = 21,897.111

4,926.85 / 3.333 = 1,478.055
4,926.85 x 3.333 = 16,422.8333

4,926.85 / 2.222 = 2,217.0825
4,926.85 x 2.222 = 10,948.555

4,926.85 / 1.111 = 4,434.165 / 9 = 492.685
4,926.85 x 1.111 = 44,341.65 x 9 = 399,074.85

5) - CARBON NANOTUBULES

CARBON RIGHT ANGLE TRIANGLE MELTING POINT 4,926.85 C

4,926.85 (3 side) x 6,569.1333 (4 side) = 32,365,134.56333 (Area of square)

32,365,134.56333 / 2 = 16,182,567.281666 (Area of the Right Triangle)

16,182,567.281666 / 24 Hours = 674,273.636736111

674,273.636736111 / 60 Minutes = 11,237.89394560185

11,237.89394560185 / 60 Seconds = 187.2982324266975

187.2982324266975 / 60 Arc Seconds = 3.121637207111625

4,926.85 / 3 = 1,642.28333

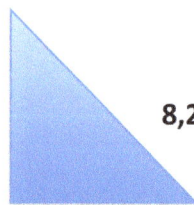

8,211.41666 = 5 x

1,642.28333

6,569.1333 = 4 x 1,642.28333

CARBON SQUARE MELTING POINT 4,926.85 x 4,926.85 = 24,273,850.9225

24,273,850.9225 / 518,400 = 46.82455810667438

24,273,850.9225 / 311,040,000 = 78.4093017779064

24,273,850.9225 / 266.666 = 91,026.94095937502

5) - CARBON NANOTUBULES

24,273,850.9225 / 11.111 = 2,184,646.583025
24,273,850.9225 / 1.333 = 18,205,388.191875
24,273,850.9225 / 144 = 168.568.4091840278
24,273,850.9225 / 7.5 = 3,236,513.456333

4,926.85 x 4,926.85 (Square) = 24,273,850.9225

24,273,850.9225 / 2 = 12,136,9256.46125
12,136,9256.46125 / 24 Hours = 505,705.2275520833
505,705.2275520833 / 60 Minutes = 8,428.420459201389
8,428.420459201389 / 60 Seconds = 140.4736743200231
140.4736743200231 / 60 Arc = 0.009528400205761

4,926.85

4,926.85

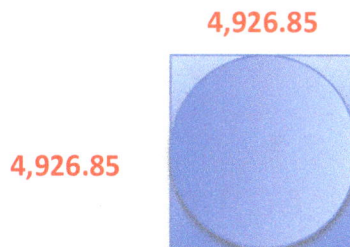

CARBON NANOTUBULE MELTING POINT 4,926.85 CUBED = 119,593,622,417.5191

119,593,622,417.5191 / 266.666 = 448,476,084.0656968

119,593,622,417.5191 / 11.111 = 10,763,426,017.57672

119,593,622,417.5191 / 1.333 = 89,695,216,813.13937

119,593,622,417.5191 / 144,000 = 830,511.2667883273

119,593,622,417.5191 / 7.5 = 15,945,816,322.33588

119,593,622,417.5191 / 24 = 4,983,067,600.729964

119,593,622,417.5191 / 864,000 = 138,418.5444647212

119,593,622,417.5191 / 432,000 = 276,837.0889294424

119,593,622,417.5191 / 2,160 = 55,367,417.78588848

119,593,622,417.5191 / 10,800 = 11,073,483.5571777

119,593,622,417.5191 / 6.666 = 17,939,043,362.62787

119,593,622,417.5191 / 5.555 = 21,526,852,035.15344

119,593,622,417.5191 / 4.444 = 26,908,565,043.94183

5) - CARBON NANOTUBULES

1 year, 360 day Time in seconds **cube 119,593,622,417.5191/ =**

/24/60/60 =

/ 3,456,000 =   /720/16/22.5 =

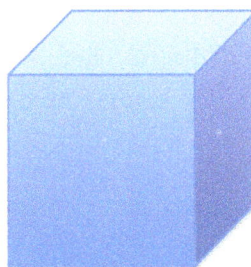

CARBON NANOTUBULES RATIOS TO HYDROGEN

CARBON NANOTUBULES 4,926.85 / HYDROGEN -259.2 = 19.007909

CARBON NANOTUBULES 4,926.85 x HYDROGEN -259.2 = 1,277,039.52

CARBON NANOTUBULES 4,926.85 + HYDROGEN -259.2 = 5,186.05

5 + 1 + 8 + 6 + 5 (25) 5 + 1 + 8 + 6 + 5 = 50
5 x 1 x 8 x 6 x 5 (1,200) x 5 x 1 x 8 x 6 x 5 = 1,440,000

1,440,000 / 50 = 28,800   1,440,000 x 50 = 72,000,000 / 2,160 = 33,333.333
72,000 / 7.5 = 9,600 / 24 = 400 / 60 = 6.666 /60 = 0.111 / 60 = 0.00185185185
72,000 / 24 = 3,000 / 60 = 50 / 60 = 0.8333 / 60 = 0.013888 x 400 = 5.555 / 0.125 = 44.444
72,000 x 720 = 51,840,000 / 864 = 60,000 / 144 = 416.666 / 720 = 0.57870370 / 16 = 0.036 / 22.5 = x
400 = /0.125 =

CARBON NANOTUBULES 4,926.85 - HYDROGEN -259.2 = 4,667.65

5) - CARBON NANOTUBULES

PRODUCT AND SUM OF MIRRORED MELTING POINT 4,926.85 C

4 + 9 + 2 + 6 + 8 + 5 (34) + 4 + 9 + 2 + 6 + 8 + 5 = 68

4 x 9 x 2 x 6 x 8 x 5 (17,280) x 5 x 8 x 6 x 9 x 4 = 298,598,400

298,598,400 / 24 Hours = 12,441,600 / 60 minutes = 207,360 / 60 seconds = 3,456 / 60 Arc Sec = 57.6
298,598,400 / 864 = 345,600 / 144 = 2,400 / 7.5 = 320 / 16 = 20 / 22.5 = 0.888 / 0.125 = 7.111

298,598,400 (Carbon 4,926.85) / 68 = 4,391,152.941176471
298,598,400 (Carbon 4,926.85) x 68 = 20,304,691,200

298,598,400 (Carbon 4,926.85) / 266.666 = 1,119,744
298,598,400 (Carbon 4,926.85) x 266.666 = 79,626,239,999.999

298,598,400 (Carbon 4,926.85) / 11.111 = 26,873,856//
298,598,400 (Carbon 4,926.85) x 11.111 = 3,317760,000

298,598,400 (Carbon 4,926.85) / 1.333 = 223,948,800
298,598,400 (Carbon 4,926.85) x 1.333 =398,131,200

298,598,400 (Carbon 4,926.85) / 288 = 1,036,800*/518,400/259,200/129,600/64,800/32,400
298,598,400 (Carbon 4,926.85) x 288 = 85,996,339,200 / 864 = 99,532,800/144 = 691,200

298,598,400 (Carbon 4,926.85) / 144 = 2,073,600*/1,036,800/518,400/259,200/129,600
/64,800/ 32,400/16,200/8,100/4,050/2,025/1,012.5/506.25
298,598,400 (Carbon 4,926.85) x 144 = 42,998,169,600

298,598,400 (Carbon 4,926.85) / 864 = 345,600
298,598,400 (Carbon 4,926.85) x 864 = 257,989,017,600

298,598,400 (Carbon 4,926.85) / 432 = 691,200
298,598,400 (Carbon 4,926.85) x 432 = 128,994,508,800

298,598,400 (Carbon 4,926.85) / 2,160 = 138,240 /69,120/34,560/17,280/8,640/4,320/2,160
298,598,400 (Carbon 4,926.85) x 2,160 =644,972,544,000

298,598,400 (Carbon 4,926.85) / 108 = 2,764,800/1,382,400/691,200/345,600/172,800/86,400
298,598,400 (Carbon 4,926.85) x 108 = 32,248,627,200

298,598,400 (Carbon 4,926.85) / 792 = 377,018.181818
298,598,400 (Carbon 4,926.85) x 792 = 236,489,932,800

298,598,400 (Carbon 4,926.85) / 396 = 809,209.756097561
298,598,400 (Carbon 4,926.85) x 396 = 118,244,966,400

298,598,400 (Carbon 4,926.85) / 3,110,400 = 96/48/24/12/6/3/1.5/0.75
298,598,400 (Carbon 4,926.85) x 3,110,400 = 928,760,463,360,000

298,598,400 (Carbon 4,926.85) / 518,400 = 576//288//144//72//36//18//9//4.5//2.25
298,598,400 (Carbon 4,926.85) x 518,400 = 154,793,410560,000

5) - CARBON NANOTUBULES

298,598,400 (Carbon 4,926.85) / 900 = 331,766/165,888/82,944/41,472/20,10,368/5,184/

/2,592/1,296/648/324/162/81/40.50/20.25

298,598,400 (Carbon 4,926.85) x 900 = 268,738,560,000

298,598,400 (Carbon 4,926.85) / 360 = 829,440/414,720/207,360/103,680 /51,840 /25,920
/12,960/6,480/3,240/1,620/810/405/202.5/101.25/50.625

298,598,400 (Carbon 4,926.85) x 360 = 107,495,424,000

298,598,400 (Carbon 4,926.85) / 720 = 414,720/207,360/103,680/51,840/25,920/ 12,960/
/6,480/3,240/1,620/810/405/202.5/101.25

298,598,400 (Carbon 4,926.85) x 720 = 214,990,848,000

298,598,400 (Carbon 4,926.85) / 16 = 18,662,400 / 22.5 =
298,598,400 (Carbon 4,926.85) x 16 = 4,777,574,400

298,598,400 (Carbon 4,926.85) / 22.5 = 13,271,040
298,598,400 (Carbon 4,926.85) x 22.5 = 6,718,464,000

298,598,400 (Carbon 4,926.85) / 400 = 746,496
298,598,400 (Carbon 4,926.85) x 400 = 119,439,360,000

CARBON RIGHT ANGLE TRIANGLE MELTING POINT PRODUCT 298,598,400

298,598,400 (3 side) x 398,131,200 (4 side) = 11,888,133,931,010,000 (Area of the square)

11,888,133,931,010,000 / 2 = 59,406,696,550,400,000 (Area of the Right Triangle)
59,406,696,550,400,000 / 24 Hours = 2,476,694,568,960,000
2,476,694,568,960,000 / 60 Minutes = 41,278,242,816,000
687,970,713,600 / 60 Seconds = 11,466,187,560
11,466,187,560 / 60 Arc Sec = 191,102,976 (191,102,976 / 7.5 = 25,480,396.8\\-13\\ = 3,110.4)

298,598,400 / 3 = 99,532,800

99,532,800

497,664,000 = 5 x

398,131,200 = 4 x 99,532,800

CARBON NANOTUBULE SQUARE MELTING POINT
298,598,400 x 298,598,400 = 89,161,004,482,560,000

89,161,004,482,560,000 / 518,400 = 171,992,678,400
89,161,004,482,560,000 / 311,040,000 = 286,654,464
89,161,004,482,560,000 / 129,600 = 687,970,713,600
89,161,004,482,560,000 / 266.666 = 334,353,766,809,600.1
89,161,004,482,560,000 / 11.111 = 8,024,490,403,430,403
89,161,004,482,560,000 / 1.333 = 66,870,753,361,920,000
89,161,004,482,560,000 / 144,000,000,000 = 619173.64224
89,161,004,482,560,000 / 2,160 = 41,278,242,816,000
89,161,004,482,560,000 / 25,920 = 3,439,853,568,000

5) - CARBON NANOTUBULES

298,598,400 x 298,598,400 (Square) = 89,161,004,482,560,000

89,161,004,482,560,000 / 24 Hours = **3,715,041,853,440,000**
3,715,041,853,440,000 / 60 Minutes = **6,197,364,224,000**
61,917,364,224,000 / 60 Seconds = **1,031,956,070,400**
1,031,956,070,400 / 60 Arc = **17,199,267,840 / 7.5 = 2,293,235,712**

298,598,400

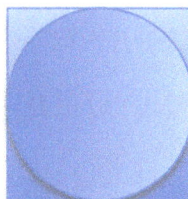

298,598,400

CARBON NANOTUBULE MELTING POINT PRODUCT 298,598,400 CUBED

298,598,400 x 298,598,400 x 298,598,400 = 26,623,333,280,890,000,000,000,000

26,623,333,280,890,000,000,000,000 / 266,666,666 = 99,837,499,803,320,000
26,623,333,280,890,000,000,000,000 / 111,111,111 = 239,699,999,767,600,000
26,623,333,280,890,000,000,000,000 / 1,333,333 = 199,674,999,606,600,000
26,623,333,280,890,000,000,000,000 / 144,000,000 = 184,884,258,895,000,000
26,623,333,280,890,000,000,000,000 / 75,000,000,000,000 = 354,977,777,078.4699
26,623,333,280,890,000,000,000,000 / 24,000,000,000,000 = 1,109,305,553,370.218
26,623,333,280,890,000,000,000,000 / 3,456,000,000,000 = 7,703510,787,293.184
26,623,333,280,890,000,000,000,000 / 864,000,000,000 = 30,814,043,149,172.74
26,623,333,280,890,000,000,000,000 / 432,000,000,000 = 61,628,086,298,345.47
26,623,333,280,890,000,000,000,000 / 2,160,000,000 = 12,325,617,259,690,000
26,623,333,280,890,000,000,000,000 / 108,000,000,000 = 246,512,345,193,381.9
26,623,333,280,890,000,000,000,000 / 8,884,000,000 (MARS D) = 2,996,773,219,370,244
26,623,333,280,890,000,000,000,000 / 4,222,000,000 (VENUS D) = 6,305,858,109,640,750

1 year, 360 day Time in seconds cube 26,623,333,280,890,000,000,000,000 / 360 =
73,953,703,558,010,000,000,000/24=3.081404324917/60=51356738581950000000
= 85,594,564,303,260,000/60 = 14,265,760,717,210,000 / 60 = 237,762,678,620,160
73,953,703,558,010,000,000,000 / 3,456,000 = / 720 = / 16 = / 22.5 =

298,598,400

298,598,400

298,598,400

5) - CARBON NANOTUBULES

298,598,400 / 144 = 2,073,600 / 7.5 = 276,480 (\\864)/720=384/16= 24/22.5= 1.0666 (266.6)
298,598,400 / 266.666 = 1,119,744 / 144 = 7,776 / 720 = 10.80 / 16 = 0.675 / 22.5 = 0.03 / 0.125 =
298,598,400 / 6.666 = 44,789,760/ 144 = 311,040 / 720 = 432 / 16 = 27 / 22.5 = 1.2 / 0.125 = 9.6
298,598,400 / 5.555 = 53,747,712/ 144 = 373,248 /720 = 518.4 /16 = 32.4 /22.5 = 1.44 / .125 = 11.52
298,598,400 / 4.444 = 67,184,640/ 144 = 466,560 / 720 = 684 /16 = 40.50 /22.5 = 1.8 (1/x = 0.555)
298,598,400 / 3.333 = 89,579,520/ 144 = 622,080 / 720 = 864 /16 = 54 / 22.5 = 2.4 / 0.125 = 19.2
298,598,400 / 2.222 = 134,369,280/ 144 = 933,120 / 720 = 1,296 /16 = 81 /22.5 = 3.6 /0.125 = 28.8
298,598,400 /1.111 = 268,738,560/ 144 = 1,866,240/ 720 = 2,592 /16 = 162 /22.5 = 7.2 /0.125 = 57.6
298,598,400 / 3,456 = 86,400 / 24 Hours = 3,600 / 60 Minutes = 60 / 60 Sec = 1 / 7.5 = 0.1333 \ 0.2666

CARBON NANOTUBULES RATIOS TO HYDROGEN

CARBON NANOTUBULES 4,926.85 / HYDROGEN -259.2 = 19.007909

CARBON NANOTUBULES 4,926.85 x HYDROGEN -259.2 = 1,277,039.52

CARBON NANOTUBULES 4,926.85 + HYDROGEN -259.2 = 5,186.05

5 + 1 + 8 + 6 + 5 (25) 5 + 1 + 8 + 6 + 5 = 50

5 x 1 x 8 x 6 x 5 (1,200) x 5 x 1 x 8 x 6 x 5 = 1,440,000

1,440,000 / 50 = 28,800   1,440,000 x 50 = 72,000,000 / 2,160 = 33,333.333

72,000 / 7.5 = 9,600 / 24 Hrs = 400 / 60 Min = 6.666 /60 Sec = 0.111 / 60 Arc Sec = 0.00185185185

72,000 / 24 Hrs = 3,000 / 60 Min = 50 / 60 Sec = 0.8333 / 60 Arc = 0.0138 x 400 = 5.555 /0.125 = 44.444

72,000 x 720 = 51,840,000 /864 = 60,000 /144 = 416.666 /720 = 0.57870370/16 =0.036 /22.5 = 0.16

CARBON NANOTUBULES 4,926.85 - HYDROGEN -259.2 = 4,667.65

CARBON MONOXIDE (Mp -205.2 C Bp -191.5 C)
AND HYDROGEN

CARBON RIGHT ANGLE TRIANGLE MELTING POINT C

(3 side) x (4 side) = (Area of square)

/ 2 = (Area of the Right Triangle)

/ 24 Hours =
/ 60 Minutes =
/ 60 Seconds =
/ 60 Arc Seconds =
/ 7.5 LCF =

/ 3 =

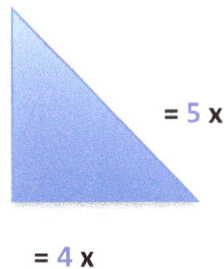

= 5 x

= 4 x

CARBON MONOXIDE SQUARE MELTING POINT x =

/ 518,400 =
/ 311,040,000 =
/ 266.666 =
/ 11.111 =
/ 1.333 =
/ 144 =
/ 7.5 =

x (Square) =

/ 2 =
/ 24 Hours =
/ 60 Minutes =
/ 60 Seconds =
/ 60 Arc =

CARBON MONOXIDE MELTING POINT CUBED =

/ 266.666 =
/ 11.111 =
/ 1.333 =
/ 144 =
/ 7.5 =
/ 24 =
/ 864 =
/ 432
/ 216
/ 108

1 year, 360 day Time in seconds cube / =
/24/60/60 =
/ 3,456,000 =   /720/16/22.5 =

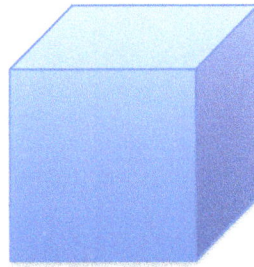

/ 1 year Time in Seconds cube =
/24 = /60 = /60 = / 3.333 =
/144,000 / 7.5 / 720 = / 16 = / 22.5 =
/ 266.666 =
/ 6.666 =
/ 5.555 =
/ 4.444 =
/ 3.333 =
/ 2.222 =
/ 1.111 = / 9 =

CARBON MONOXIDE RATIOS TO HYDROGEN

CARBON MONOXIDE / HYDROGEN -259.2 =
CARBON MONOXIDE x HYDROGEN -259.2 =
CARBON MONOXIDE + HYDROGEN -259.2 =
5 + 1 + 8 + 6 + 5 () 5 + 1 + 8 + 6 + 5 =
5 x 1 x 8 x 6 x 5 () x 5 x 1 x 8 x 6 x 5 =

/ 50 =   x 50 = / 2,160 =
/7.5 = / 24 = / 60 = /60 = / 60 =
/ 24 = / 60 = / 60 = / 60 = x 400 = / 0.125 =
= / 864 = / 144 = / 720 = / 16 = / 22.5 = x 400 = / 0.125 =

CARBON MONOXIDE - HYDROGEN -259.2 = 4,667.65

METHANE AND CARBON MONOXIDE

CARBON MONOXIDE AND METHANE RIGHT TRIANGLE MELTING POINT C

(3 side) x (4 side) = (Area of square)

/ 2 = (Area of the Right Triangle)

/ 24 Hours =
/ 60 Minutes =
/ 60 Seconds =
/ 60 Arc Seconds =

/ 3 =

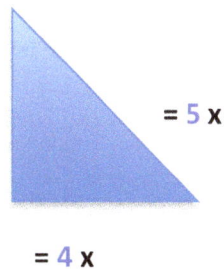

= 5 x

= 4 x

CARBON MONOXIDE AND METHANE SQUARE MELTING POINT x =

/ 518,400 =
/ 311,040,000 =
/ 266.666 =
/ 11.111 =
/ 1.333 =
/ 144 =
/ 7.5 =

x (Square) =

/ 2 =
/ 24 Hours =
/ 60 Minutes =
/ 60 Seconds =
/ 60 Arc =

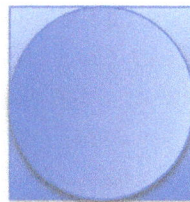

METHANE AND CARBON MONOXIDE MELTING POINT CUBED =

/ 266.666 =
/ 11.111 =
/ 1.333 =
/ 144 =
/ 7.5 =
/ 24 =
/ 864 =
/ 432
/ 216
/ 108

1 year, 360 day Time in seconds cube / =
/24 Hrs = / 60 Min = / 60 Sec = / 60 Arc Sec =
/ 3,456,000 = / 720 = / 16 = / 22.5 =

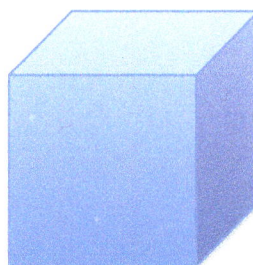

/ 1 year Time in Seconds cube =
/ 24 Hrs = / 60 Min = / 60 Sec = / 60 Arc Sec = / 3.333 =
/ 144,000 = / 7.5 = / 720 = / 16 = / 22.5 =
/ 266.666 =
/ 6.666 =
/ 5.555 =
/ 4.444 =
/ 3.333 =
/ 2.222 =
/ 1.111 = / 9 =

CARBON NANOTUBULES RATIOS TO HYDROGEN

CARBON NANOTUBULES 4,926.85 / HYDROGEN -259.2 = 19.007909
CARBON NANOTUBULES 4,926.85 x HYDROGEN -259.2 = 1,277,039.52
CARBON NANOTUBULES 4,926.85 + HYDROGEN -259.2 = 5,186.05
5 + 1 + 8 + 6 + 5 (25) 5 + 1 + 8 + 6 + 5 = 50
5 x 1 x 8 x 6 x 5 (1,200) x 5 x 1 x 8 x 6 x 5 = 1,440,000
1,440,000 / 50 = 28,800   1,440,000 x 50 = 72,000,000 / 2,160 = 33,333.333
72,000 / 7.5 = 9,600 / 24 = 400 / 60 = 6.666 / 60 = 0.111 / 60 = 0.00185185185
72,000 / 24 = 3,000 / 60 = 50 / 60 = 0.8333 / 60 = 0.013888 x 400 = 5.555 / 0.125 = 44.444
72,000 x 720 = 51,840,000 / 864 = 60,000 / 144 = 416.666 / 720 = 0.57870370 / 16 = 0.036 / 22.5
= x 400 = / 0.125 =

CARBON NANOTUBULES 4,926.85 - HYDROGEN -259.2 = 4,667.65

CARBON 3,500 C + OXYGEN -183 C
HYDROGEN -259.2

HYDROGEN PLUS PALLADIUM COLD FUSION INTERACTION STRIPPING THE ELECTRON FROM THE PROTON NON NUCLEAR REACTION, NO NUCLEUS, NO POLLUTION

2,963 (Palladium) x 252.87 (Hydrogen) = 749,253.81 x 1.333 = 999,005.078 / 11.111 / 7.5 / 16 / 22.5 =
2,963 (Palladium) / 252.87 = 11.71748329181   252.87 / 2,963 = 0.085342

252.87 + 2,963 (Palladium) = 3,215.87 is [3 + 2 + 1 + 5 + 8 + 7 (26) + 7 + 8 +5 + 1 + 2 + 3] = 52
3,215.87 = 3 x 2 x 1 x 5 x 8 x 7 (1,680) x 7 x 8 x 5 x 1 x 2 x 3 = 2,822,400 / 3.333 = 864,720

2,822,400 / 5.555 = 508,032   2,822,400 / 6.666 = 423,360
2,822,400/ 144 = 19,600   3,215.87/ 144,000 = / 7.5 = 2.61

CARBON RIGHT ANGLE TRIANGLE MELTING POINT 4,926.85 C

4,926.85 (3 side) x 6,569.1333 (4 side) = 32,365,134.56333 (Area of square)

32,365,134.56333 / 2 = 16,182,567.281666 (Area of the Right Triangle)

16,182,567.281666 / 24 Hours = 674,273.636736111
674,273.636736111 / 60 Minutes = 11,237.89394560185
11,237.89394560185 / 60 Seconds = 187.2982324266975
187.2982324266975   / 60 Arc Seconds = 3.121637207111625

4,926.85 / 3 = 1,642.28333

8,211.41666 = 5 x

1,642.28333

6,569.1333 = 4 x 1,642.28333

CARBON SQUARE MELTING POINT 4,926.85 x 4,926.85 = 24,273,850.9225

24,273,850.9225 / 518,400 = 46.82455810667438
24,273,850.9225 / 311,040,000 = 78.4093017779064
24,273,850.9225 / 266.666 = 91,026.94095937502
24,273,850.9225 / 11.111 = 2,184,646.583025
24,273,850.9225 / 1.333 = 18,205,388.191875
24,273,850.9225 / 144 = 168.568.4091840278
24,273,850.9225 / 7.5 = 3,236,513.456333

4,926.85 x 4,926.85 (Square) = 24,273,850.9225

24,273,850.9225 / 2 = 12,136,9256.46125
12,136,9256.46125 / 24 Hours = 505,705.2275520833
505,705.2275520833 / 60 Minutes = 8,428.420459201389
8,428.420459201389 / 60 Seconds = 140.4736743200231
140.4736743200231 / 60 Arc Sec = 0.009528400205761

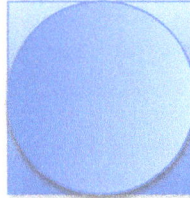

CARBON NANOTUBULE MELTING POINT 4,926.85 CUBED = 119,593,622,417.5191

119,593,622,417.5191 / 266.666 =
119,593,622,417.5191 / 11.111 =
119,593,622,417.5191 / 1.333 =
119,593,622,417.5191 / 144 =
119,593,622,417.5191 / 7.5 =
119,593,622,417.5191 / 24 =
119,593,622,417.5191 / 864 =
119,593,622,417.5191 / 432
119,593,622,417.5191 / 216
119,593,622,417.5191 / 108

1 year, 360 day Time in seconds cube 119,593,622,417.5191/ =
/ 24 = / 60 = / 60 = / 60 =
/ 3,456,000 = / 720 = / 16 = / 22.5 =

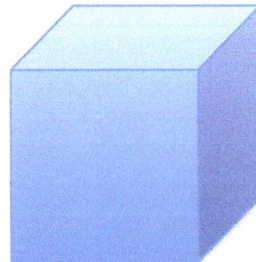

CARBON NANOTUBULES RATIOS TO HYDROGEN

CARBON NANOTUBULES 4,926.85 / HYDROGEN -259.2 = 19.007909
CARBON NANOTUBULES 4,926.85 x HYDROGEN -259.2 = 1,277,039.52
CARBON NANOTUBULES 4,926.85 + HYDROGEN -259.2 = 5,186.05
5 + 1 + 8 + 6 + 5 (25) 5 + 1 + 8 + 6 + 5 = 50
5 x 1 x 8 x 6 x 5 (1,200) x 5 x 1 x 8 x 6 x 5 = 1,440,000
1,440,000 / 50 = 28,800   1,440,000 x 50 = 72,000,000 / 2,160 = 33,333.333
72,000 / 7.5 = 9,600 / 24 = 400 / 60 = 6.666 /60 = 0.111 / 60 = 0.00185185185
72,000 / 24 = 3,000 / 60 = 50 / 60 = 0.8333 / 60 = 0.013888 x 400 = 5.555 / 0.125 = 44.444
72,000 x 720 = 51,840,000 / 864 = 60,000 / 144 = 416.666 / 720 = 0.57870370 / 16 = 0.036 / 22.5 = x
400 = /0.125 =

PALLADIUM MELTING POINT

MELTING POINT = 1,555.2 x 2 = 3,110.4 (Sanskrit long year TIME) 1,555.2 / 11.111 = 139.68
1 x 5 x 5 x 5 x 2 (250) x 2 x 5 x 5 x 5 x 1 = 62,500 (6.25 Pi square cubits 1/16 th sphere x 10,000)
62,500 + 250 + 250 = 63,000
The product of Palladium's BOILING POINT (2,963) is 46,656 divided by the melting point (1,555.2) is 62,500
equals 0.746496 and 62,500 / 46,656 = 1.333 x 2 = 2.666
1 + 5 + 5 + 5 + 2 (18) + 2 + 5 + 5 + 5 + 1 = 36

PALLADIUM RIGHT ANGLE TRIANGLE BOILING POINT 1,555.2

AVT = 3,224,862.72 /2 = 1,612,431.36*
*/24/60/60/60 = 18.6624
18.6624 /16 / 22.5 = 0.05184
*/720/16/22.5 = 6.2208 (648)

1,555.2 / 3 = 518.4   518.4 x 5 = 2,592

2,073.6 = 518.4 x 4

PALLADIUM SQUARE BOILING POINT 1,555.2 x 4 = 6,220.8 / 3,456 = 1.8 (1/x = 0.555)

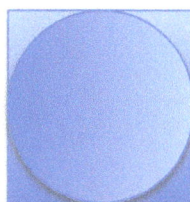

PALLADIUM CUBE BOILING POINT 1,555.2 CUBED = 3,761,479,876.608
1 yr = 360 days in Seconds = 31,104,000 Sec cubed = 30,091,839,012,860,000,000,000
30,091,839,012,860,000,000,000 / 3,761,479,876.6.608 = 8,000,000,000,000.
3,761,479,876.608 / 24 Hrs / 60 Min / 60 Sec / 60 Arc Sec / 7.5 = 96.7458816
3,761,479,876.608 / 3,456 = 1,088,391.168
8,000,000,000,000 / 7,200,000,000 = 1,111.111 / 16 = 69.444/ 22.5 =
3.08641975308641975 / 0.125 = 24.69135802469136 x 400 = 987.654320987654320

1 year Time in Seconds cubed /   3,761,479,876.608 =
8,000,000,000,000/24 = 333,333,333,333.333 /60 =   5,555,555,555.555 /60 =
92,592,592.592 /60 = 154,320.987654321   /   3.333 = 46,296.296296296

8,000,000,000,000 / 144,000 = 5,555,555.555/ 7.5 = 740,740.740 (1/x = 0.00000135)
740,740.740 / 720 = 1,028.80658436214 /16 = 64.3004115226337448559 /22.5 = 2.857796067672611
8,000,000,000,000 /129,600= 6,172,839,50617284/1.333/144,000=32.150205761316881/x=0.031104
8,000,000,000,000 / 266.666 = 3,000,000,000.
8,000,000,000,000 / 1.333 = 600,000,000,000
8,000,000,000,000 / 6.666 = 120,000,000,000
8,000,000,000,000 / 5.555 = 144,000,000,000
8,000,000,000,000 / 4.444 = 180,000,000,000
8,000,000,000,000 / 3.333 = 240,000,000,000
8,000,000,000,000 / 2.222 = 360,000,000,000
8,000,000,000,000 / 1.111 = 720,000,000,000 / 9 = 80,000,000,000

PALLADIUM BOILING POINT

BOILING POINT = 2,963 / 266.666 = 11.11125

2,963 (Palladium) so 2 + 9 + 6 + 3 (20) + 3 + 6 + 9 + 2 = 40
2,963 (Palladium) so 2 x 9 x 6 x 3 (216) 3 x 6 x 9 x 2 = 46,656 / 5.555 = 8,398.080

2,963 (Palladium) / 11.111 = 266.666 / 24 = 11.111
2,963 (Palladium) / 5.555 = 533.34   2,963 x 1.333 = 2,222.25   2,963 / 3.333 = 888.88

PALLADIUM RIGHT ANGLE TRIANGLE BOILING POINT 2,963

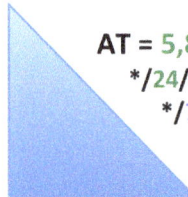

AT = 5,852,911.679*
*/24/60/60 = 67.74 / 16 = 4.23
*/720/16/22.5 / .125 = 180.064

2,963 / 3 = 987.666   987.666 x 5 = 4,938.333

3,950.666 = 987.666 x 4

PALLADIUM SQUARE BOILING POINT 2,963 x 4 = 11,852 / 3,456 = 3.429398148148

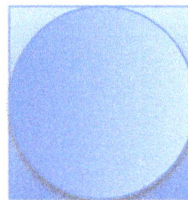

PALLADIUM CUBE BOILING POINT 2,963 CUBED = 26,013,270,347

1 year, 360 days in seconds 31,104,000 cubed 30,091,839,012,860,000,000,000
30,091,839,012,860,000,000,000 / 26,013,270,347 = 1,156,788,001,333.879
1,156,788,001,333.879/24/60/60 = 13,338,750.0154384
1,156,788,001,333.879/3,456,000 = 334,718.7503859605 /720/16/22.5 /.125 = 10.33

1 year Time in Seconds cube /   26,013,270,347 =
1,156,788,001,333.879
1,156,788,001,333.879 /24/60/60/60 =   223,145.8335906403
 223,145.8335906403 / 3.333 =   66,943.7500771921
1,156,788,001,333.879 / 144,000 / 7.5   =   1,071,100.001235073
/ 720 / 16 / 22.5 = 148.7638890604269

1,156,788,001,333.879 / 864,000 = 1,338,875.001543842
1,156,788,001,333.879 / 129,600 = 8,925,833.343625612
1,156,788,001,333.879 / 266.666 = 4,337,955,005.002049
1,156,788,001,333.879 / 6.666 = 173,518,200200.0819
1,156,788,001,333.879 / 5.555 = 208,221,840,240.0983
1,156,788,001,333.879 / 4.444 = 260,277,300,300.1229
1,156,788,001,333.879 / 3.333 = 347,036,400,400.1638
1,156,788,001,333.879 / 2.222 = 520,554,600,600.2458
1,156,788,001,333.879 / 1.111 = 1,041,109,201,200.492 / 9 = 115,678,800,133.3879

PLATINIUM - Boiling point = - 3,825

OXYGEN – Boiling point = -183
        - Melting point =

HYDROGEN – Boiling point = - 252.87
            - Melting point = -259.20

CARBON – MELTING POINT - CARBON SHEETS DIAMOND = 3,426.85
        - MELTING POINT - CARBON SHEETS GRAPHENE = 4,236.85
        - MELTING POINT - CARBON MATRIX GRAPHITE = 3,526.85
        - MELTING POINT CARBON FORMED SPHERES   = 1,249.85
        - MELTING POINT CARBON FORM NANOTUBULE = 4,926.85

HELIUM – Boiling point = -268.93
        - Melting point = -272.2

DEUTERIUM – Boiling point = 101.4

TUNGSTEN – Boiling point 5,555 (Sun Aether) / 11.111 = 500
        5,555 / 266.666 = 20.83125
        5 x 5 x 5 x 5 (625) x 5 x 5 x 5 x 5 = 390,625
        390,625 + 625 + 625 = 391,875
        5 + 5 + 5+ 5 (20) + 5 + 5 + 5 + 5 = 40

        The product of Tungsten's boiling point (5,555) product is 390,625 divided
        by the MELTING POINT (3,420) product 576 is 678.168
        MELTING POINT 3,420 (Sun's radius)

        3 x 4 x 2 (24) 2 x 4 x 3 = 576 / 24 = 24
        576 + 24 + 24 = 624 / 24 = 26
        3 + 4 + 2 (9) + 2 + 4 + 3 = 18

GOLD – Boiling point 2,970 (Earth)
        - Melting point 1,063

SILVER – Boiling point 2,162 (Moon)

COPPER – Boiling point 2,562 (Earths Equinox)

BRASS - Melting point 900

RED BRASS – Melting point 1,000

IRON – Boiling point – 2,861
        Melting point – 1,538

CARBON – Boiling point - 4,827
        – Melting point - 3,550

NITROGEN – Boiling point – 195.8

INTERNAL ANGLES OF A CUBE 2,160 OCTAVES UP AND DOWN TABLE

OCTAVES UP

2,160 / 1.111 = 1944

0.) 1,944 / 0.432 = 4,500

1.) 3,888 / 0.432 = 9,000

2.) 7,776 / 0.432 = 18,000

3.) 15,552 / 0.432 = 36,000

4.) 31,104 / 0.432 = 72,000 Torus

5.) 62,208 / 0.432 = 144,000 double Torus = Photon

6.) 124,416 / 0.432 = 288,000

7.) 248,832 / 0.432 = 576,000

8.) 497,664 / 0.432 = 1,152,000

9.) 995,328 / 0.432 = 2,304,000

OCTAVES DOWN

2,160 / 1.111 = 1944

0.) 1944 / 0.432 = 4,500

1.) 972 / 0.432 = 2,250

2.) 486 / 0.432 =1,125

3.) 243 / 0.432 = 562.5

4.) 121.5 / 0.432 = 281.25

5.) 607.5 / 0.432 = 140.625

6.) 303.75 / 0.432 = 70.3125

7.) 151.875 / 0.432 = 35.15625

8.) 075.9375 / 0.432 = 17.578125

9.) 037.96875 / 0.432 = 8.7890625

VAJRA SINGULARITY POINT AT VORTEX INVERSION

VORTEX INVERSION TABLE

REVERSAL LEFT HAND CLOCKWISE TO RIGHT HAND ANTICLOCKWISE SPIRAL SHIFT MIRROR NUMBER CREATION

1 x 8	+1 = 9
12 x 8	+2 = 98
123 x 8	+3 = 987
1234 x 8	+4 = 9876
12345 x 8	+5 = 98765
123456 x 8	+6 = 987654
1234567 x 8	+7 = 9876543
12345678 x 8	+8 = 98765432
123456789 x 8	+9 = 987654321

123456789 x 9 = 1,111,111,101

123456789 x 8 =   987,654,312

123456789 x 7 =   864,197,523

123456789 x 6 =   740,740,734

123456789 x 5 =   617,238,945

123456789 x 4 =   493,827,156

123456789 x 3 =   370,370,367

123456789 x 2 =   246,913,578

123456789 x 1 =   123,456,789

123456789 x 1.111 = 137,174,209.99999862826

123456789 x 11.111 = 1,371,742,099.9999862862

123456789 x 111 = 13,703,703,579

MILLENIUM TEMPLE

SOLAR SYSTEM YEAR 25,920, SUN CUBE 10,368,000
SUN 864,000 (888), EARTH 7,920 (999), MOON 2,160 (222)
RELATIONSHIP TO 2,160, 144, 108

EARTH'S EQUINOX CYCLE 25,920 x   7.5 = 194,400.....(POS)
EARTH'S EQUINOX CYCLE 25,920 /   7.5 = 3,456..........(POS)

EARTH'S EQUINOX CYCLE 25,920 x   1.333 = 34,560........(POS)
EARTH'S EQUINOX CYCLE 25,920 /   1.333 = 19,440........(POS)

EARTH'S EQUINOX CYCLE 25,920 x 11.111 = 288.............(LIGHT 144 x 2)
EARTH'S EQUINOX CYCLE 25,920 /   11.111 = 2,332.8......(COMPOSITE 864)

EARTH'S EQUINOX CYCLE 25,920 x 266.666 = 6,912,000..(RFEU)
EARTH'S EQUINOX CYCLE 25,920 / 266.666 = 97.2............(7,920 EARTH)

EARTH'S EQUINOX CYCLE 25,920 / 129,600 = 0.2 (1/x = 5)
EARTH'S EQUINOX CYCLE 25,920 x 129,600 = 3,359,232,000 / 144,000 = 23,328

EARTH'S EQUINOX CYCLE 25,920 /   108 = 240
EARTH'S EQUINOX CYCLE 25,920 x   108 = 2,799,360 / 144,000 = 19.44

EARTH'S EQUINOX CYCLE 25,920 /   144 = 180   (1/x = 0.00555555)
EARTH'S EQUINOX CYCLE 25,920 x   144 = 3,732,480 / 144,000 = 25.92

EARTH'S EQUINOX CYCLE 25,920 / 864,000 = 0.03   (1/x = 33.333)
EARTH'S EQUINOX CYCLE 25,920 x 864,000 = 22,394,880,000/144,000= 155,520/5= 31,104

EARTH'S EQUINOX CYCLE 25,920 / 432,000 = 0.06   (1/x = 16.666)
EARTH'S EQUINOX CYCLE 25,920 x 432,000 = 11,197,440,000 / 144,000 = 77,760

EARTH'S EQUINOX CYCLE 25,920 / MOON'S D 2,160 = 12.........(SUNS AETHER R)
EARTH'S EQUINOX CYCLE 25,920 x MOON'S D 2,160 = 55,987,200 / 144,000 = 388.8

EARTH'S EQUINOX CYCLE 25,920 / MOON'S R 1,080 = 24.........(SUNS AETHER D)
EARTH'S EQUINOX CYCLE 25,920 x MOON'S R 1,080 = 27,993,600 / 144,000 = 194.4

EARTH'S EQUINOX CYCLE 25,920 / EARTH'S D 7,920 = 3.272727 (1/x = 0.30555)
EARTH'S EQUINOX CYCLE 25,920 x EARTH'S D 7,920 = 205,286,400 / 144,000 = 1,425.6

EARTH'S EQUINOX CYCLE 25,920 / MAR'S D 4,222 = 6.13927 (1/x = 0.16288580)
EARTH'S EQUINOX CYCLE 25,920 x MAR'S D 4,222 = 109,434,240 / 144,000 = 759.96

EARTH'S EQUINOX CYCLE 25,920 / JUPITER'S D 88,800 = 0.29189 (1/x = 3.42592592)
EARTH'S EQUINOX CYCLE 25,920 x JUPITER'S D 88,800 = 2,301,696,000 / 144,000 = 15,984

EARTH'S EQUINOX CYCLE 25,920 / JUPITER'S R 44,400 = 0.58378378 (1/x = 1.712961296)
EARTH'S EQUINOX CYCLE 25,920 x JUPITER'S R 44,400 = 1,150,848,000/144,000 = 7,992 EARTH
180 / 12 = 15
180 / 22.5 = 8
25,920 / 266.666 = 97.20000000000243 1/x = 0.01028806584362
97.20000000000243 x 11.111 = 1,080.0000000000002 [Moons Radius]

SUN'S CUBE 10,368,000 / 266.666 = 38,880 / 266.666 = 145.8
SUN'S CUBE 10,368,000 /   108 = 96,000 / 266.666 = 360
SUN'S CUBE 10,368,000 /   144 = 72,000
SUN'S CUBE 10,368,000 /   2,160 = 4,800
                72,000 /   4,800 = 15

SUNS SQUARE 3,456,000 / 266.666 = 12,960 (RFEU)
SUNS SQUARE 3,456,000 /   108 = 32,000 / 266.666 = 120
SUNS SQUARE 3,456,000 /   144 = 24,000
SUNS SQUARE 3,456,000 / 2,160 = 1,600

24,000 / 22.5 = 1,066.666 \\266.666

3,456,000 / 266.666 = 12,960.000000000324   1/x = 0.00007716049383

3,456,000 x 11.111 = 384

SUN'S TORUS 2,160,000 (864,000 x 2.5) / 266.666 = 8,100
(SUN'S LIGHT POLARIZES, AS THE MOON IS THE SAME RESONANCE AS SUN'S TORUS 2,160)

8,100.0000000002025 //16,200.00000000405 // 32,400.0000000081 //64,800.0000000162
//129,600.0000000324 // 259,200.0000000648 //518,400.00000001296 [6 DAYS OF
CREATION]

SUN'S TORUS 2,160,000 (864,000 x 2.5) / 108 = 20,000 / 2.5 = 8,000
SUN'S TORUS 2,160,000 (864,000 x 2.5) / 144 = 15,000 (2 x 7.5)
SUN'S TORUS   2,160,000 / 2,160 = 1,000

SUN'S OCTAVE 1,728,000 / 108 = 16,000 16,000 / 22.5 =
SUN'S OCTAVE 1,728,000 / 144   = 12,000 12,000/22.5 = 533.333 / 2 = 266.666
SUN'S OCTAVE 1,728,000 / 2,160 = 800

SUN'S DIAMETER 864,000 /   108 = 8,000
SUN'S DIAMETER 864,000 /   144 = 6,000 / 22.5 = 266.666
SUN'S DIAMETER 864,000 / 2,160 = 400 / 22.5 = 1.5

                6,000 /   400 = 15
                  400 /   6,000 = 0.06666 (2/3 FULL MOLTEN SEA, RI 1.333, C 2.666)

6,000 / 22.5 = 266.666
864,000 / 266.666 = 3,240   1 / x = 0.00030864197531
3,240 x 11.111 = 36,000 x 1.333 = 48,000 / 7.5 = 6,400 / 16 = 400
311, 040,000,000,000 / 864,000 = 360,000 / 300 = 1,200,000 / 720 = 1,666.666 / 360 =
4.629 x 1.333 = 6.1728 / 1.111 = 5.555 x 1,000 = 5,555.555/ 16 = 347.222 / 22.5 =
15.432098765431944444/ 11.111 = 1.3888 x 8,000 = 11,111.111 x 7.5 = 83,333.333 /60 =
1,388.8 / 60 = 23.148148 / 37 = 0.625625 x 111 = 96.444

311,040,000,000,000 / 864,000 = 360,000,000

PART 16

OCTAVE MODEL OF THE
ELEMENTS AND ELEMENTAL
PLASMOID CALCULATIONS

APPLICATIONS FOR A PLASMOIDS FORM AND FUNCTIONS

OCTAVE MODEL OF THE ELEMENTS

AND

ELEMENTAL PLASMOID CALCULATIONS

PART SIXTEEN OF TWENTY

DRAFT 518,400 B KMV

BY

MALCOLM V of SCOTLAND | MALCOLM BENDALL
THURSDAY 22ND SEPTEMBER 2022

APPLICATIONS FOR A PLASMOIDS FORM AND FUNCTIONS

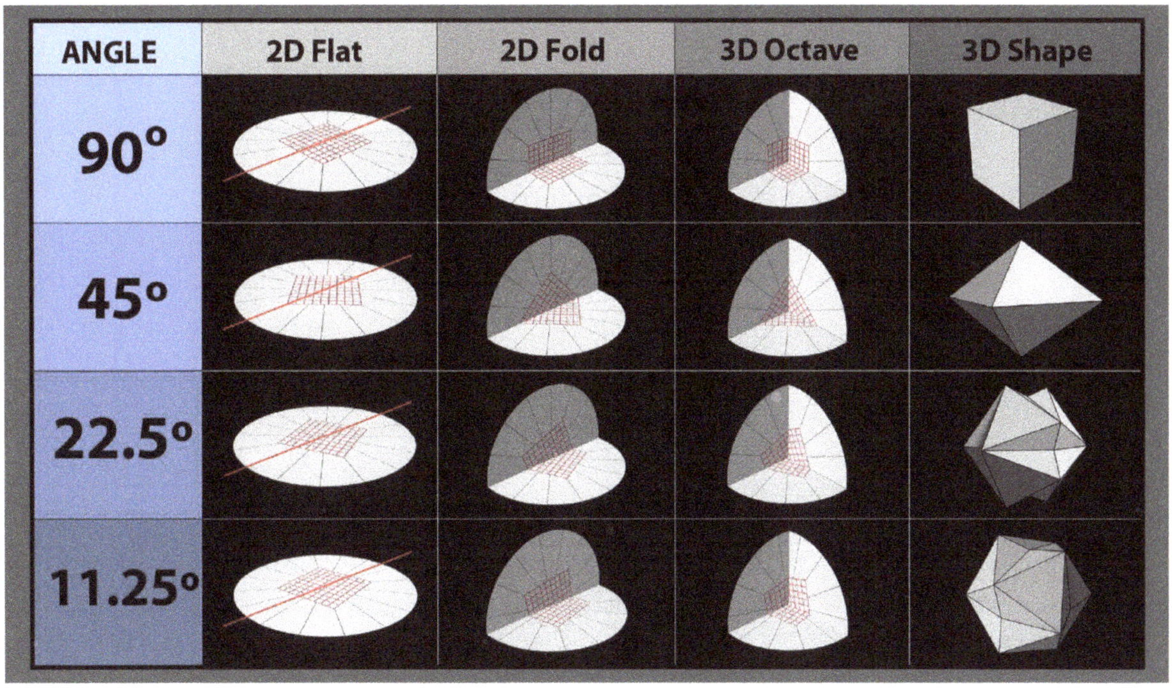

ANGLE
2D Flat
2D Fold
3D Octave
3D Shape
90°
45°
22.5°
11.25°

DIAGRAM 348

0
11.25
22.5
33.75
45
56.25
67.5
78.75
90

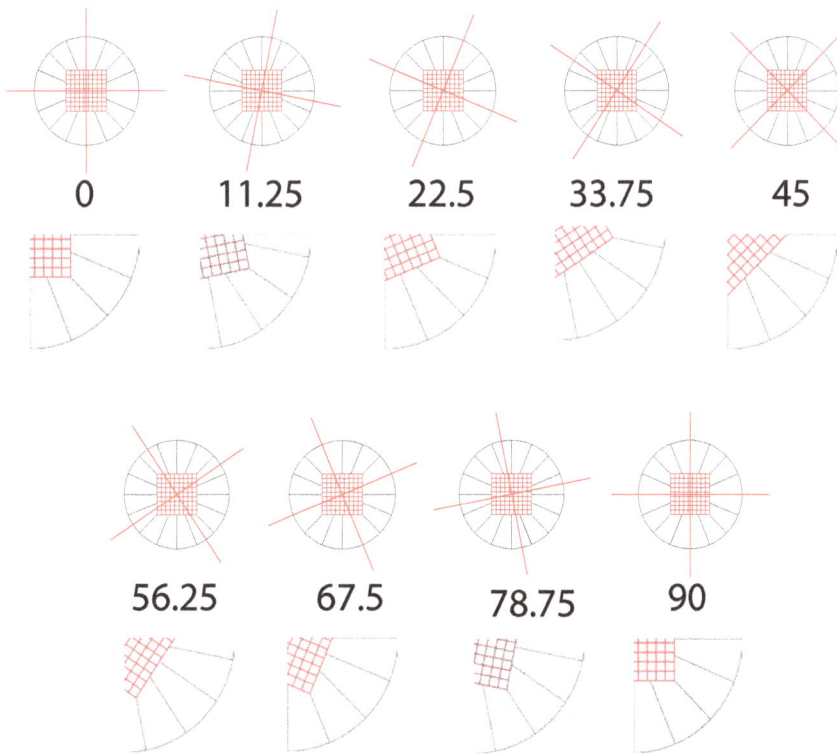

DIAGRAM 349

APPLICATIONS FOR A PLASMOIDS FORM AND FUNCTIONS

DIAGRAM 350

DIAGRAM 351

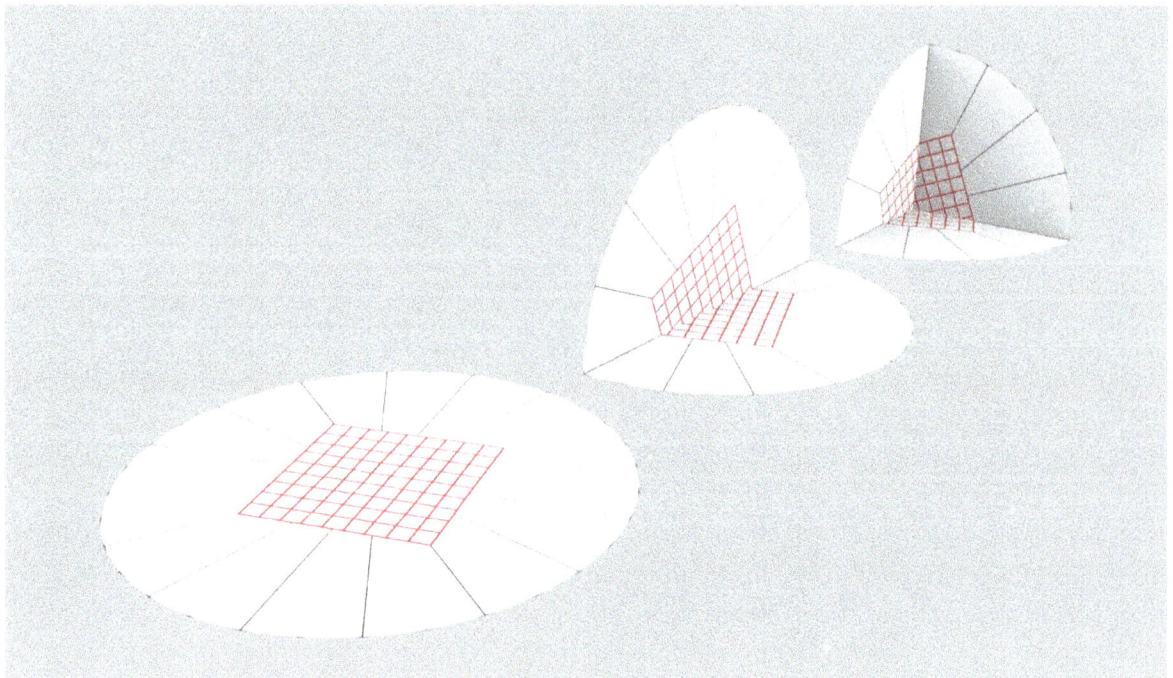

DIAGRAM 352

DIAGRAM 353

APPLICATIONS FOR A PLASMOIDS FORM AND FUNCTIONS

DIAGRAM 354

DIAGRAM 355

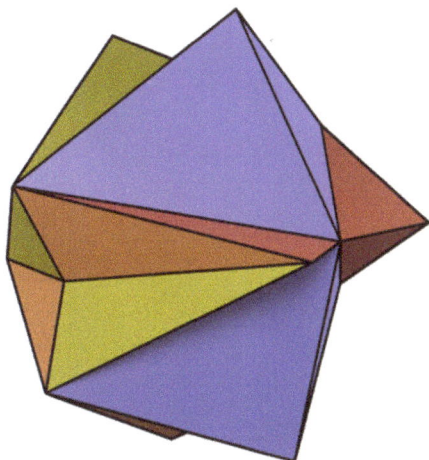
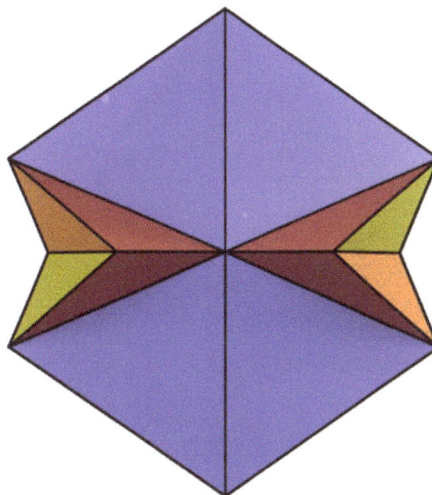

DIAGRAM 356

DIAGRAM 357

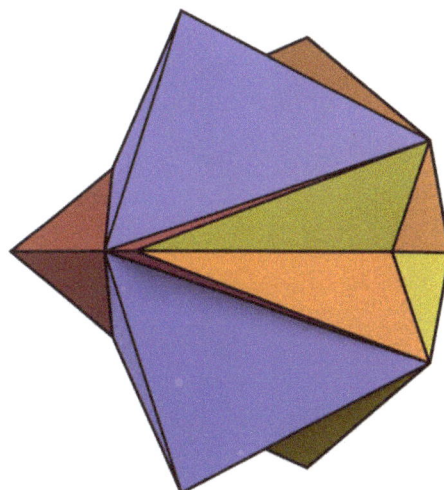

DIAGRAM 358

DIAGRAM 359

APPLICATIONS FOR A PLASMOIDS FORM AND FUNCTIONS

DIAGRAM 360

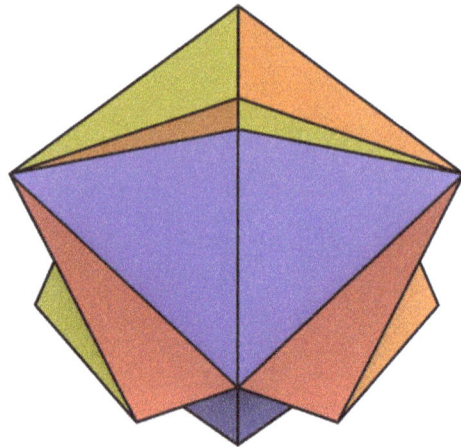

DIAGRAM 361

DIAGRAM 362

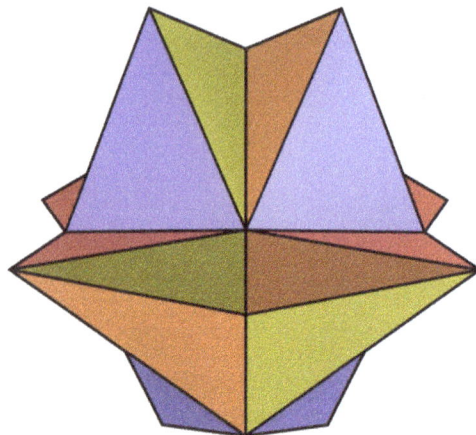

DIAGRAM 363

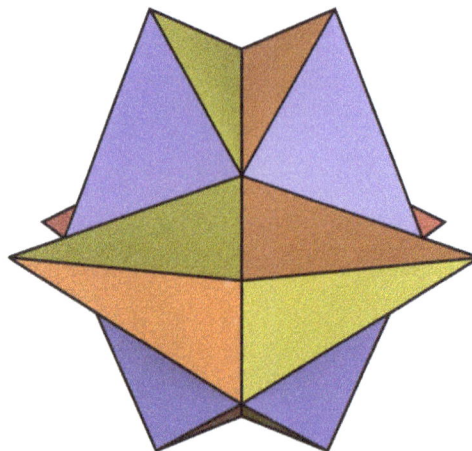

DIAGRAM 364

APPLICATIONS FOR A PLASMOIDS FORM AND FUNCTIONS

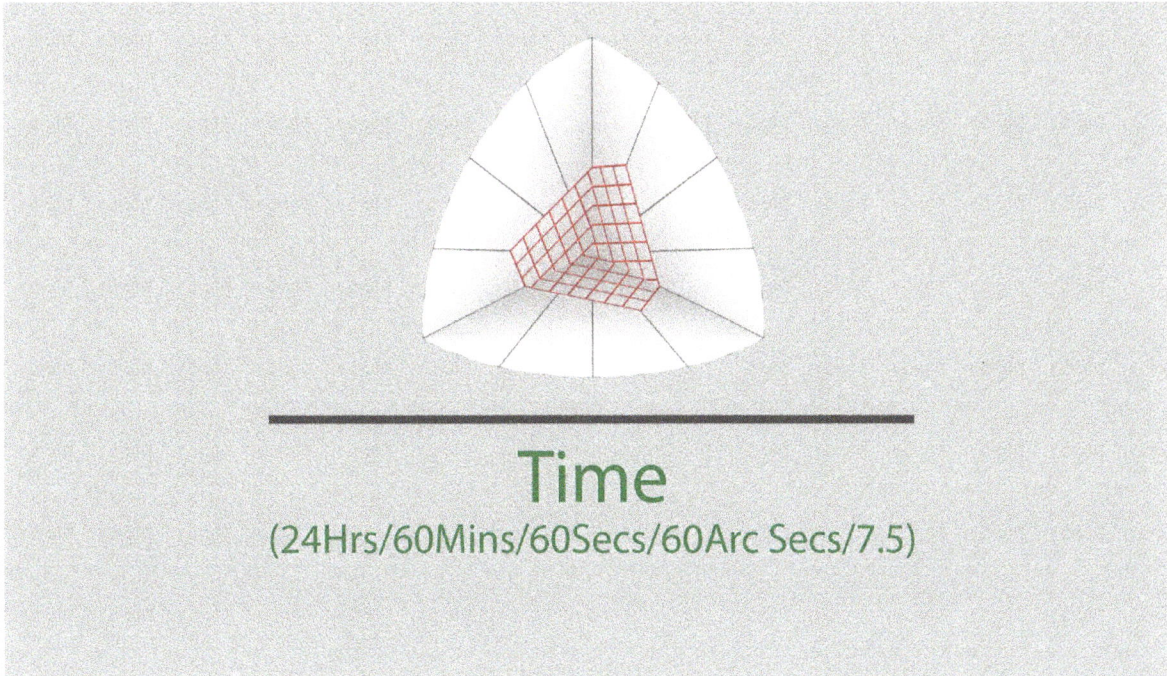

Time
(24Hrs/60Mins/60Secs/60Arc Secs/7.5)

DIAGRAM 365

APPLICATIONS FOR A PLASMOIDS FORM AND FUNCTIONS

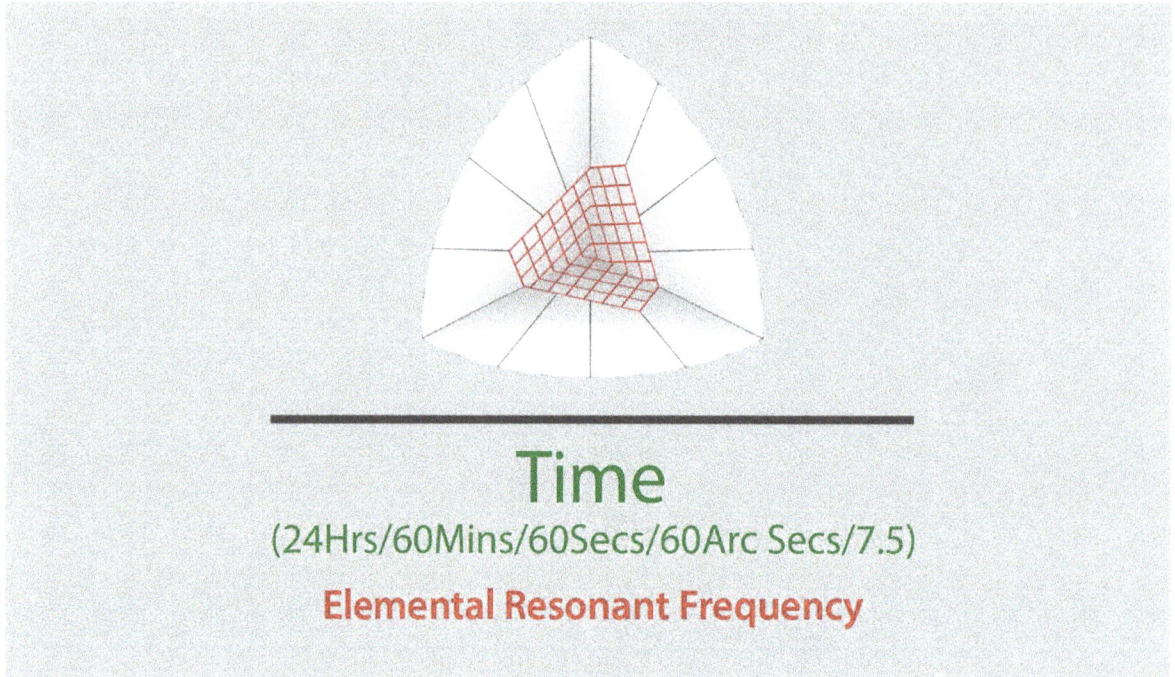

DIAGRAM 366

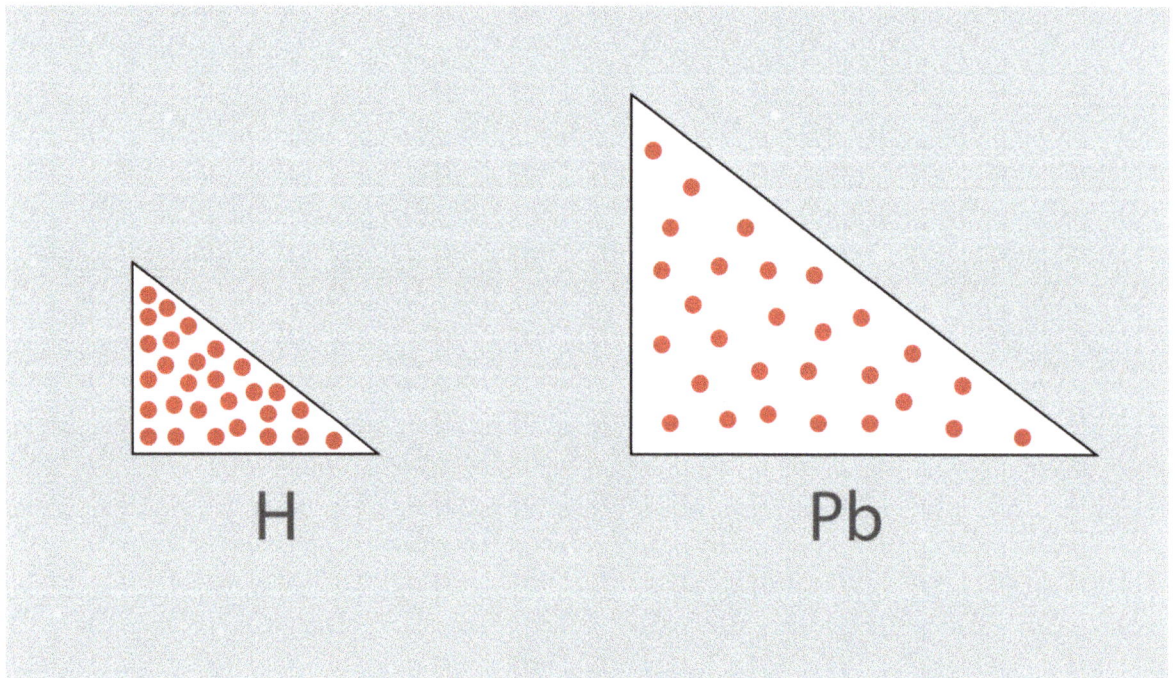

DIAGRAM 367

APPLICATIONS FOR A PLASMOIDS FORM AND FUNCTIONS

Classic Periodic Table plots vs Bendall's Model of the Elements plots

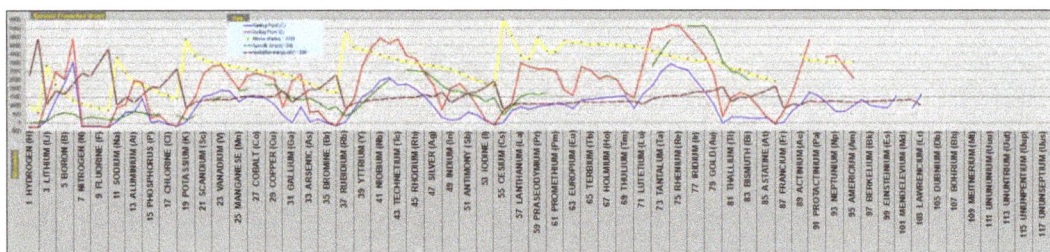

Classic Periodic Table plots

DIAGRAM 368 - STANDARD PERIODIC TABLE GRAPH

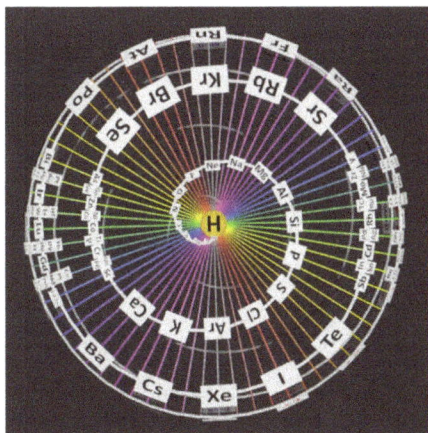

DIAGRAM 369

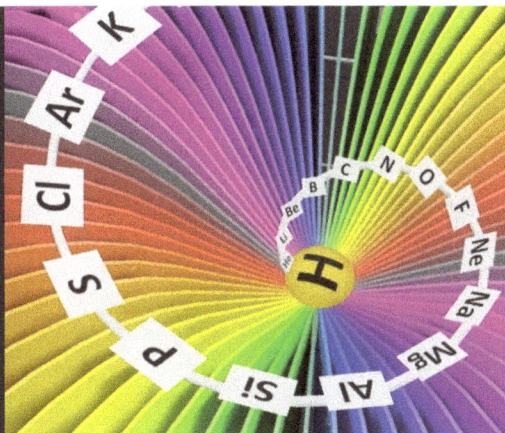

DIAGRAM 370

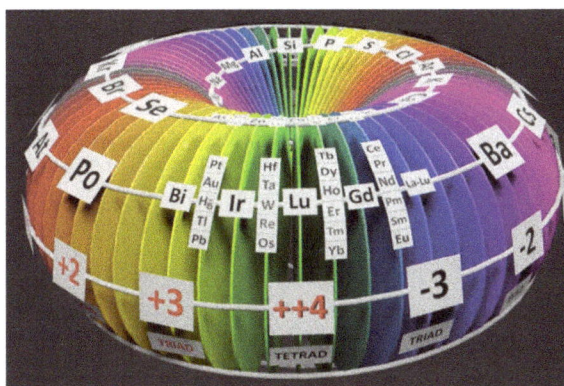

DIAGRAM 371

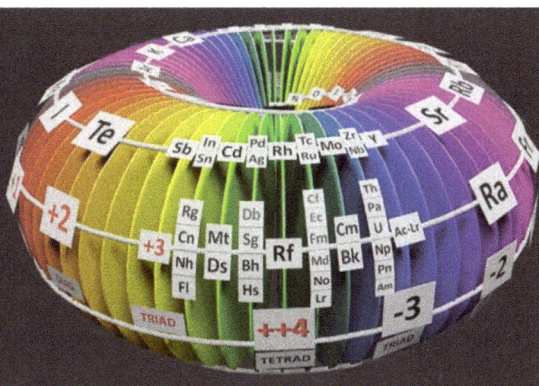

DIAGRAM 372

APPLICATIONS FOR A PLASMOIDS FORM AND FUNCTIONS

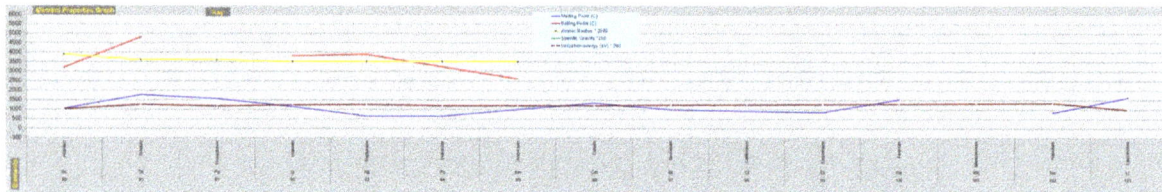

DIAGRAM 373 - ACTINIDE SERIES

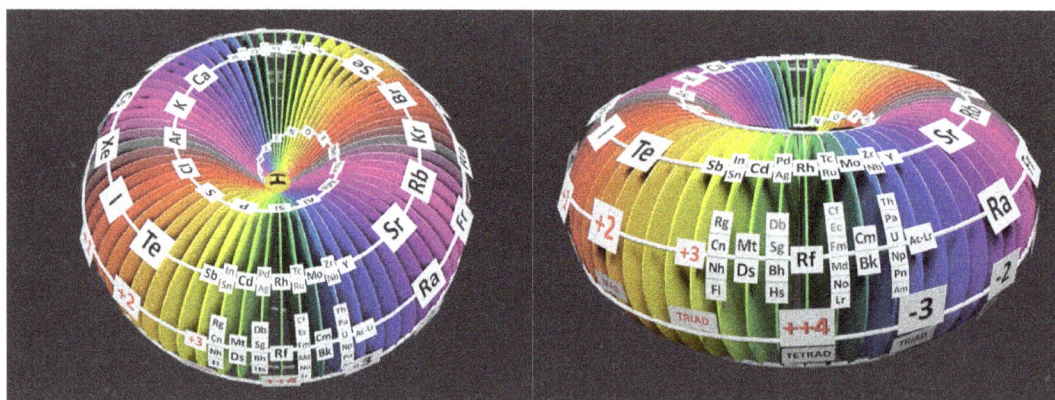

DIAGRAM 374

DIAGRAM 375

APPLICATIONS FOR A PLASMOIDS FORM AND FUNCTIONS

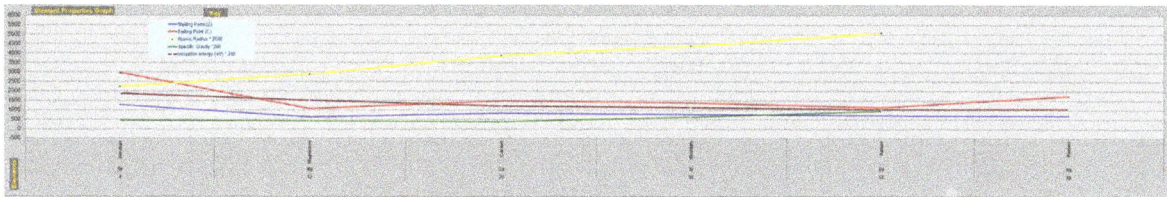

DIAGRAM 376 - BERYLLIUM SERIES

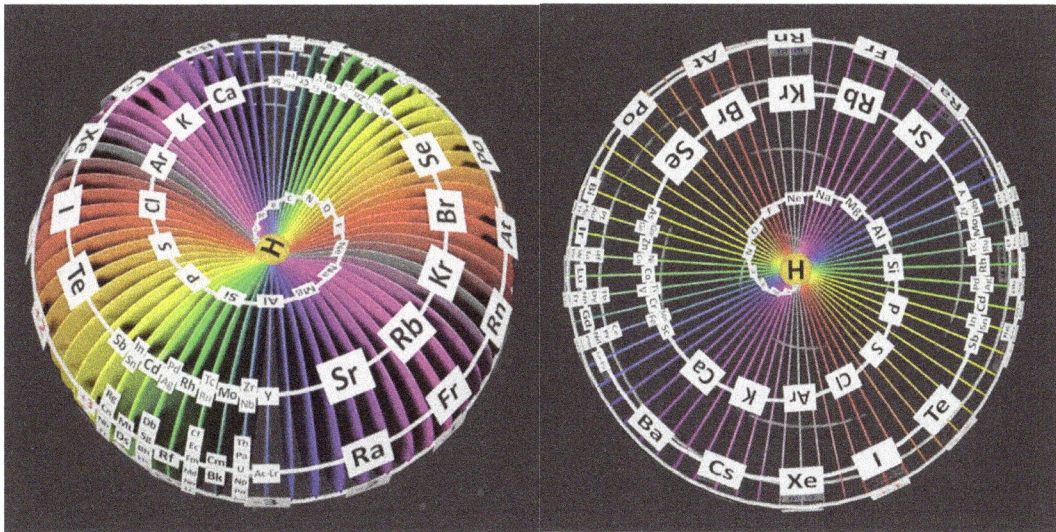

DIAGRAM 377

DIAGRAM 378

APPLICATIONS FOR A PLASMOIDS FORM AND FUNCTIONS

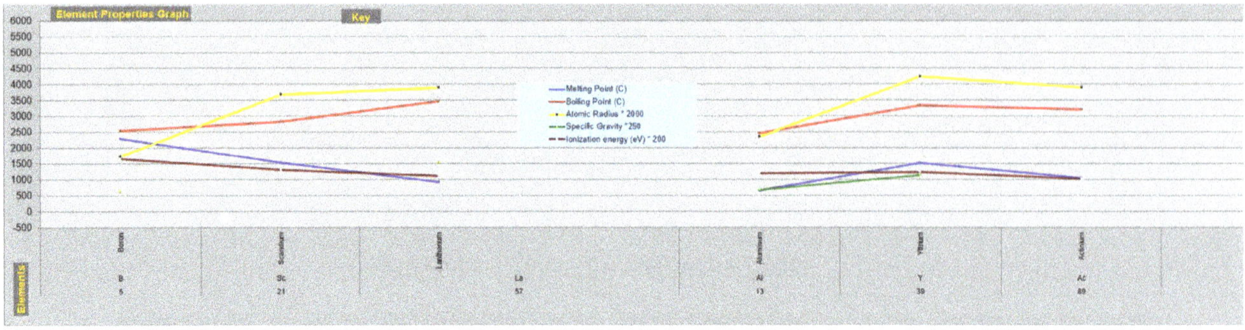

DIAGRAM 379 - BORON SERIES 2

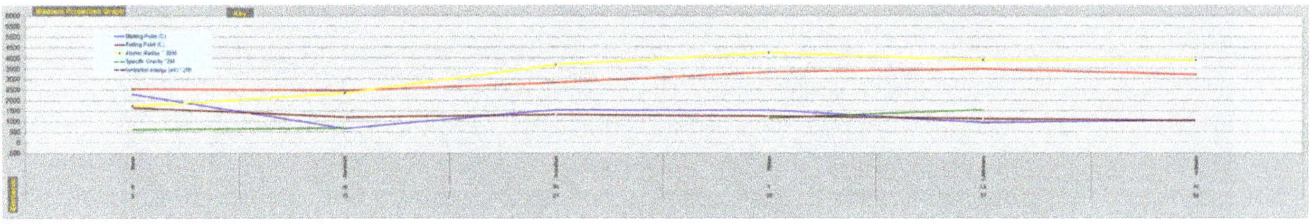

DIAGRAM 380 - BORON SERIES

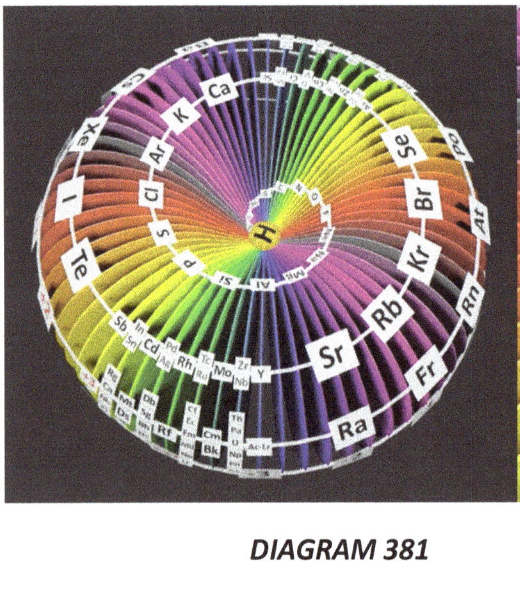

DIAGRAM 381 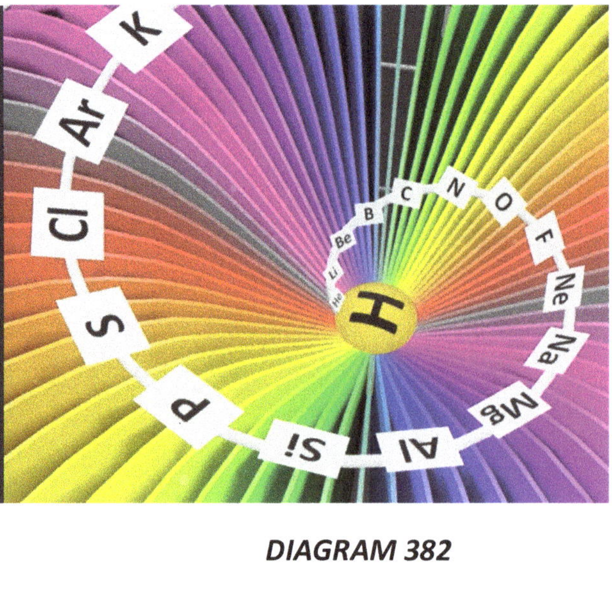   *DIAGRAM 382*

APPLICATIONS FOR A PLASMOIDS FORM AND FUNCTIONS

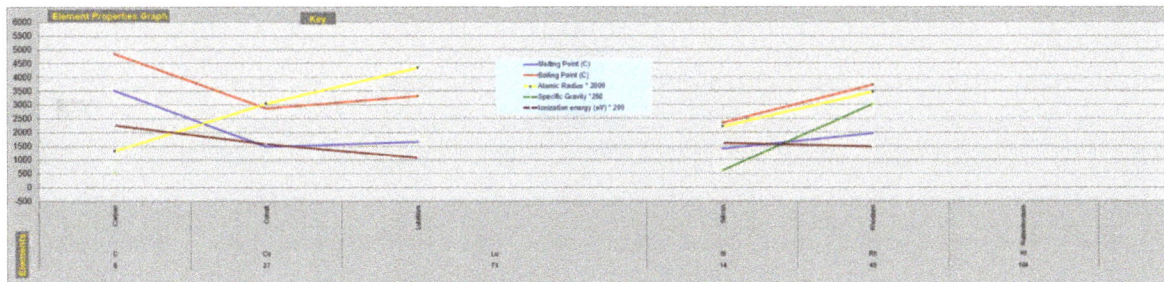

DIAGRAM 383 - CARBON SERIES 2

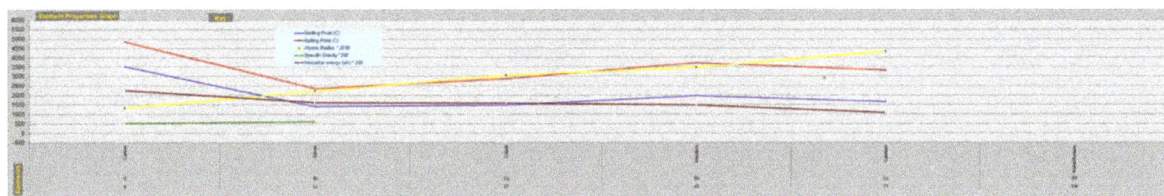

DIAGRAM 384 - CARBON SERIES

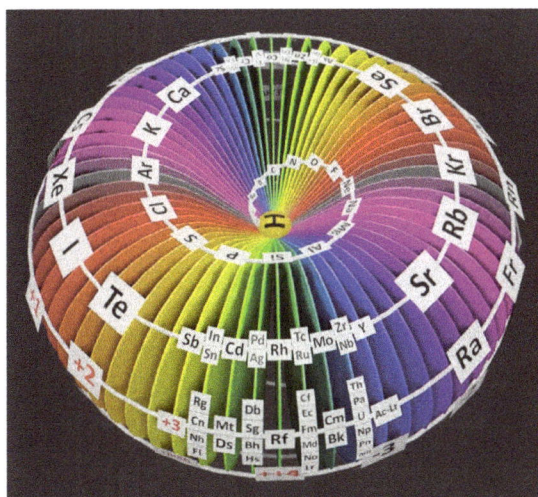

DIAGRAM 385　　　　　　　　　***DIAGRAM 386***

APPLICATIONS FOR A PLASMOIDS FORM AND FUNCTIONS

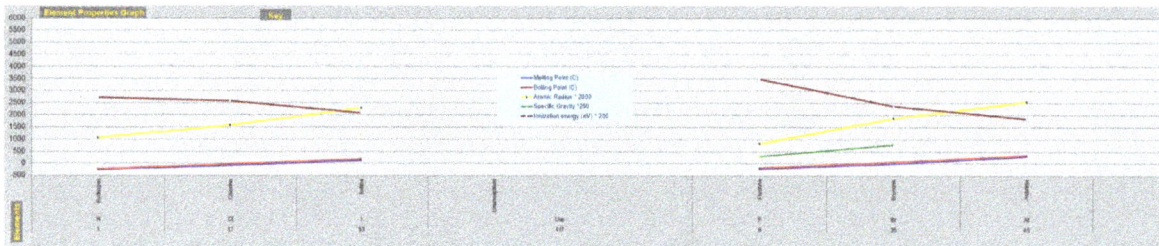

DIAGRAM 387 - FLORINE SERIES 2

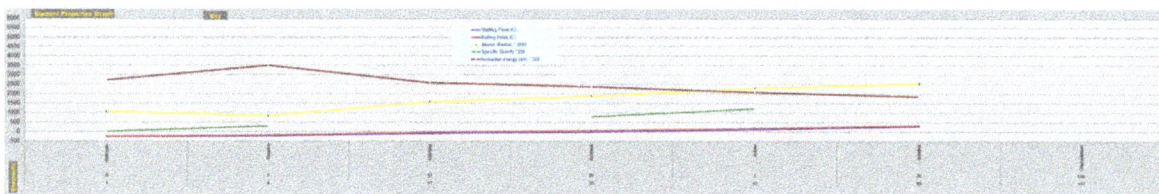

DIAGRAM 388 - FLORINE SERIES

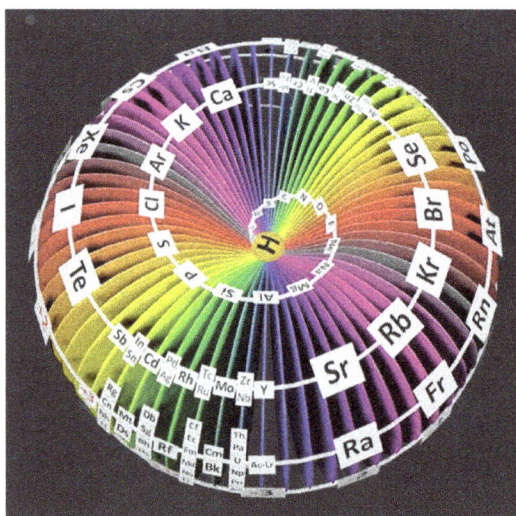

DIAGRAM 389

DIAGRAM 390

APPLICATIONS FOR A PLASMOIDS FORM AND FUNCTIONS

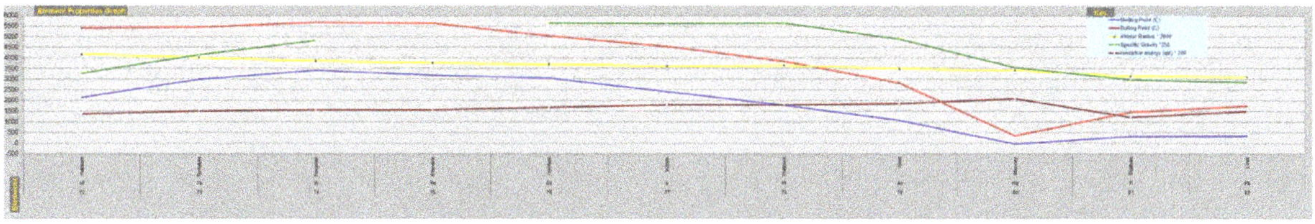

DIAGRAM 391 - HAFNIUM SERIES

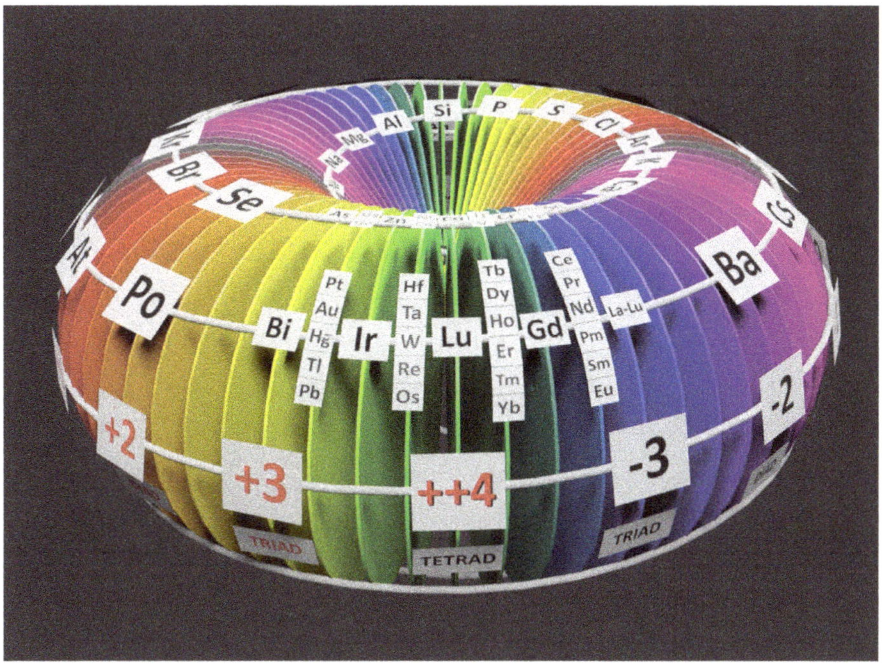

DIAGRAM 392

APPLICATIONS FOR A PLASMOIDS FORM AND FUNCTIONS

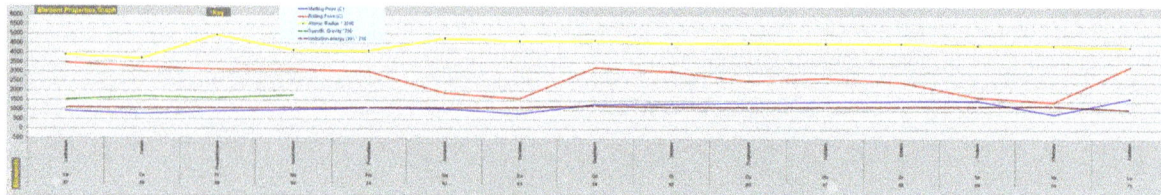

DIAGRAM 393 - LANTHANIDE SERIES

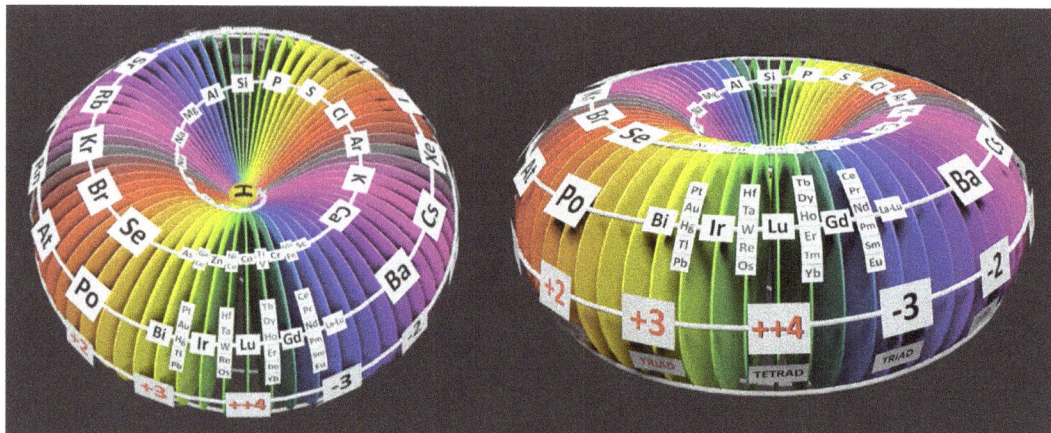

DIAGRAM 394

DIAGRAM 395

APPLICATIONS FOR A PLASMOIDS FORM AND FUNCTIONS

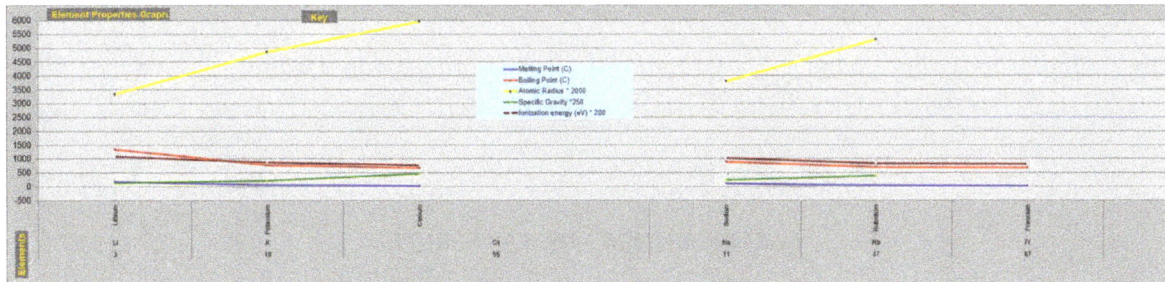

DIAGRAM 396 - LITHIUM SERIES 2

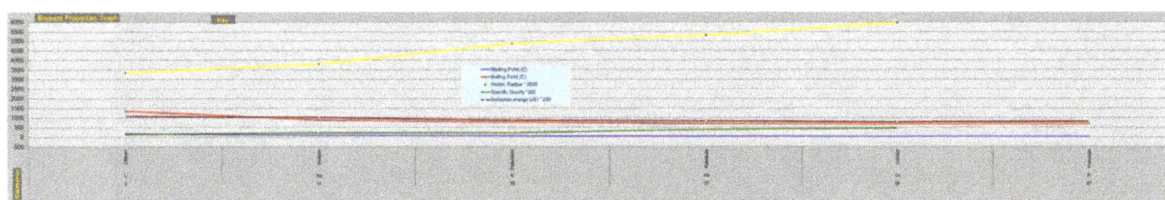

DIAGRAM 397 - LITHIUM SERIES

DIAGRAM 398

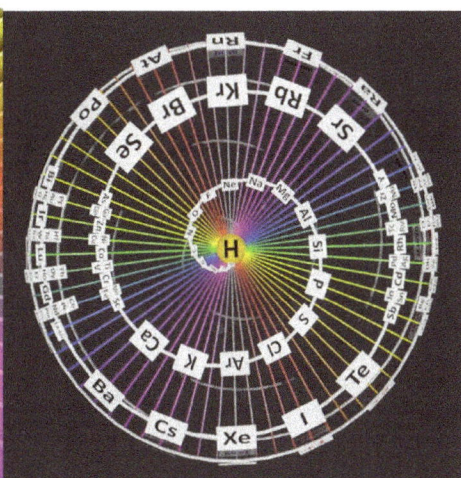

DIAGRAM 399

APPLICATIONS FOR A PLASMOIDS FORM AND FUNCTIONS

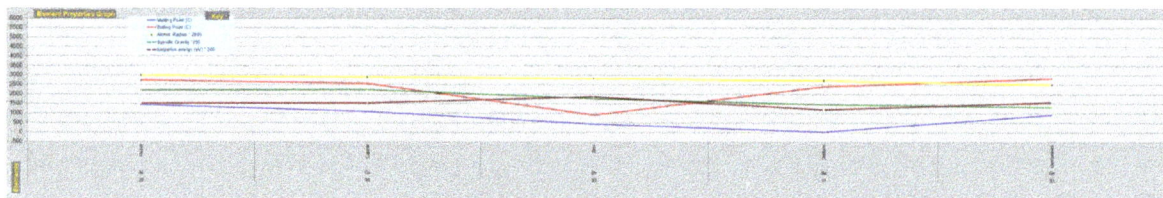

DIAGRAM 400 - NICKEL SERIES

DIAGRAM 401

DIAGRAM 402

APPLICATIONS FOR A PLASMOIDS FORM AND FUNCTIONS

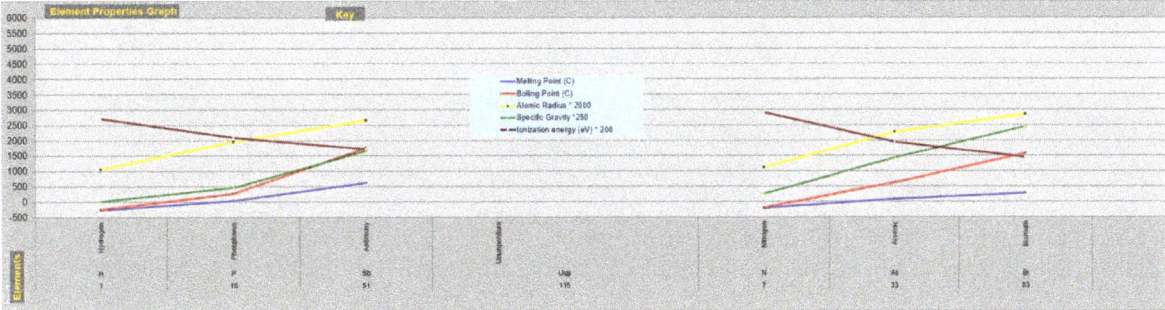

DIAGRAM 403 - NITROGEN SERIES 2

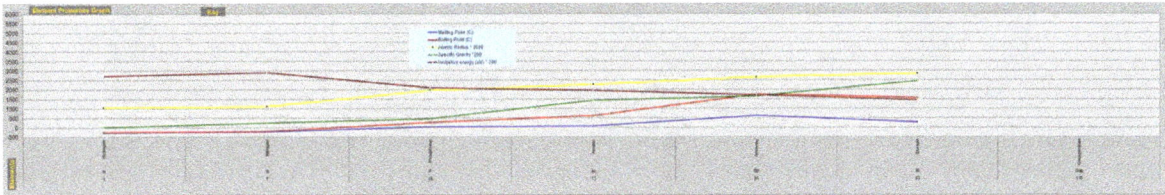

DIAGRAM 404 - NITROGEN SERIES

DIAGRAM 405

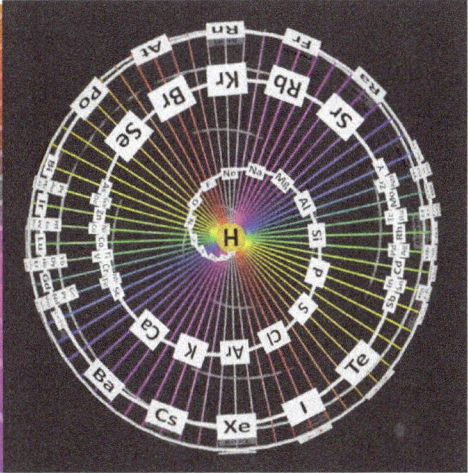

DIAGRAM 406

APPLICATIONS FOR A PLASMOIDS FORM AND FUNCTIONS

DIAGRAM 407 - NOBLE GASES 2

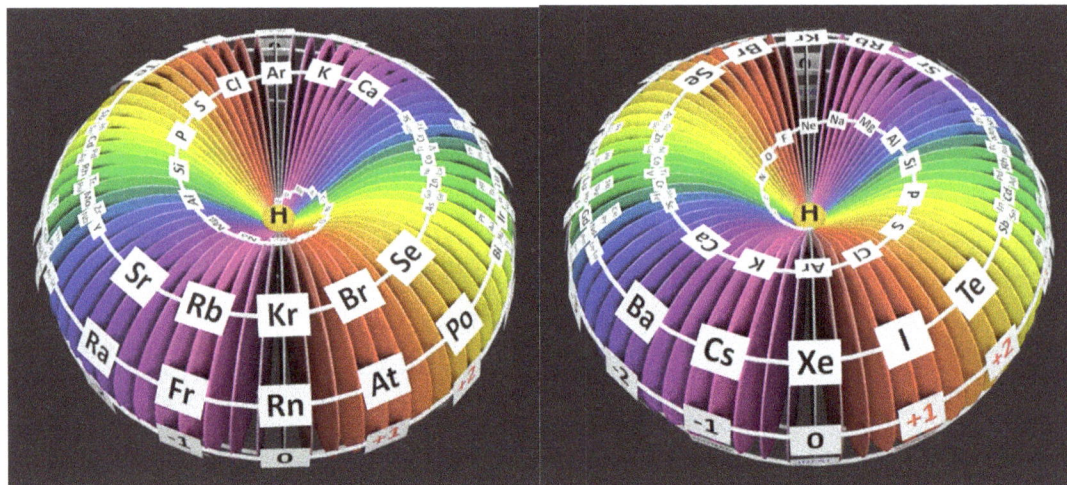

DIAGRAM 408

DIAGRAM 409

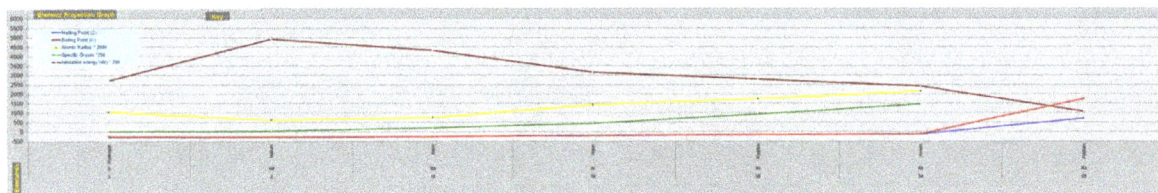

DIAGRAM 410 - NOBLE GASES

APPLICATIONS FOR A PLASMOIDS FORM AND FUNCTIONS

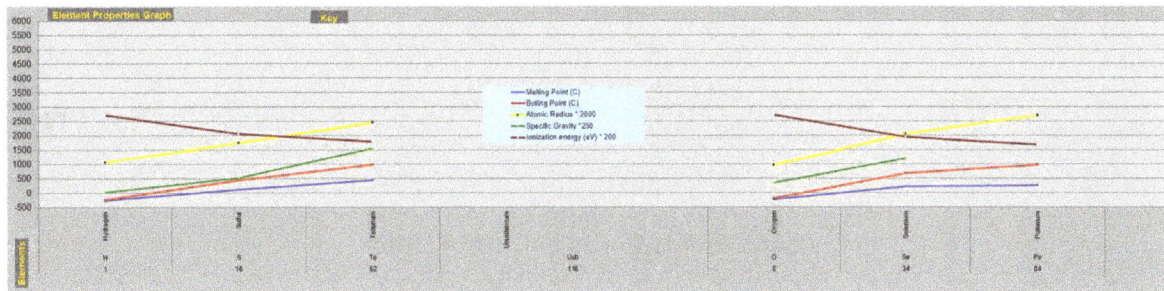

DIAGRAM 411 - OXYGEN SERIES 2

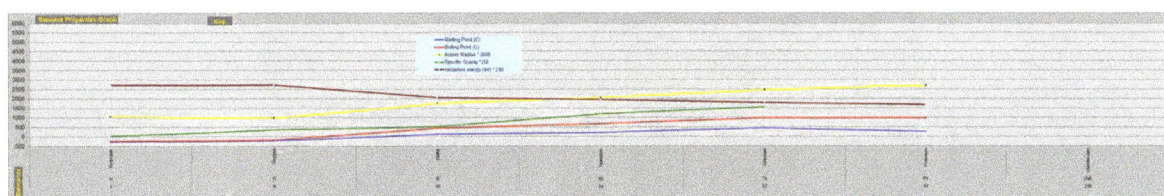

DIAGRAM 412 - OXYGEN SERIES

DIAGRAM 413

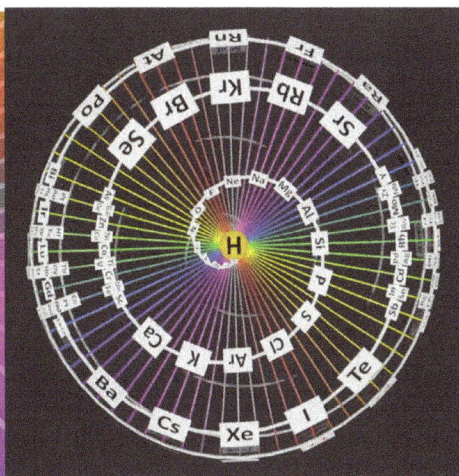

DIAGRAM 414

APPLICATIONS FOR A PLASMOIDS FORM AND FUNCTIONS

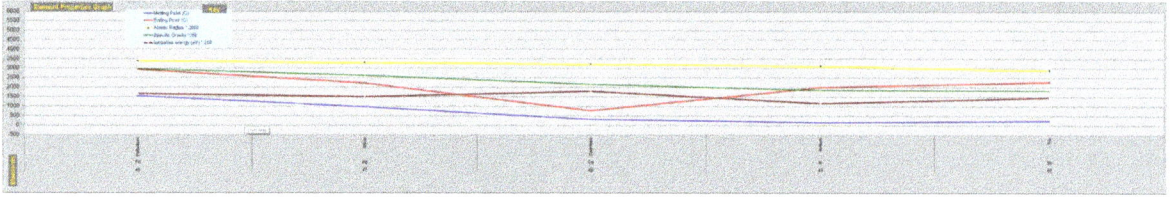

DIAGRAM 415 - PALLADIUM SERIES

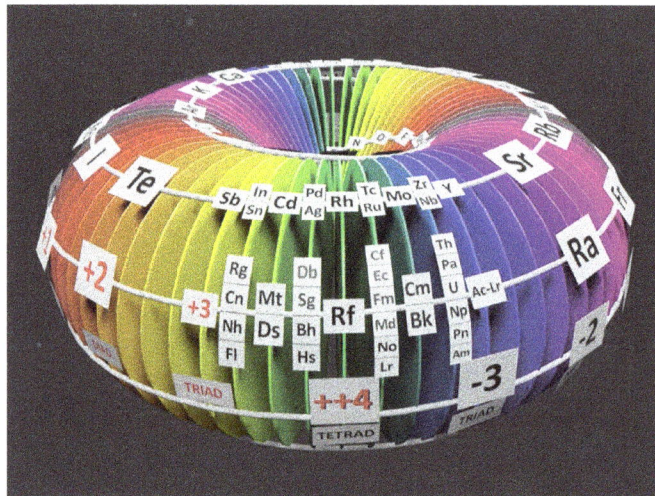

DIAGRAM 416

APPLICATIONS FOR A PLASMOIDS FORM AND FUNCTIONS

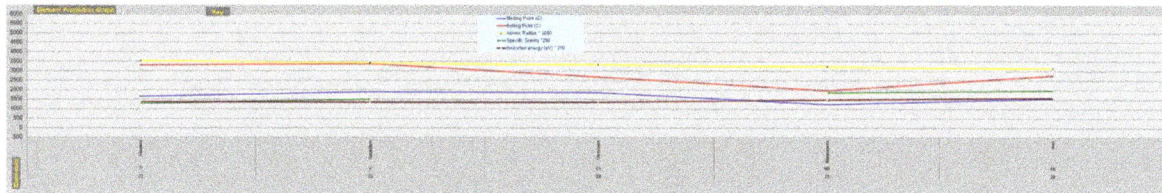

DIAGRAM 417 - TITANIUM SERIES

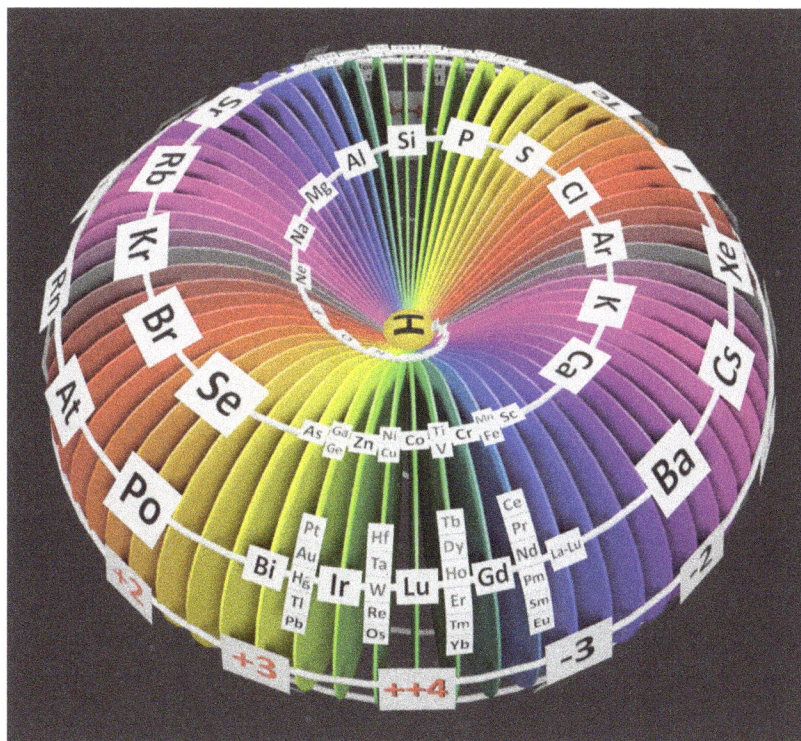

DIAGRAM 418

APPLICATIONS FOR A PLASMOIDS FORM AND FUNCTIONS

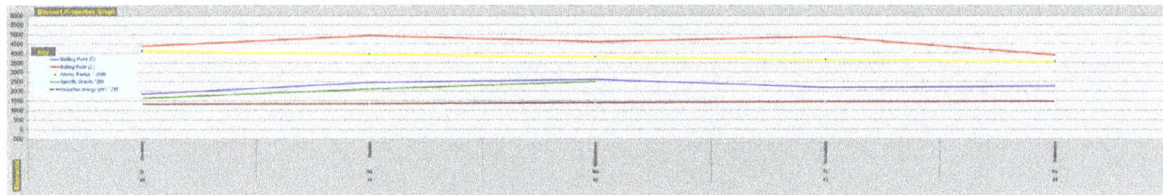

DIAGRAM 419 - ZIRCONIUM SERIES

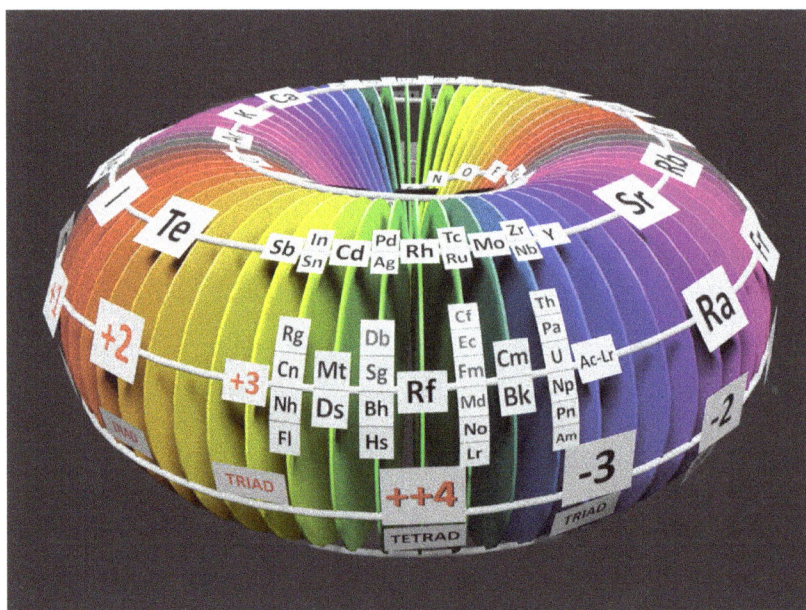

DIAGRAM 420

APPLICATIONS FOR A PLASMOIDS FORM AND FUNCTIONS

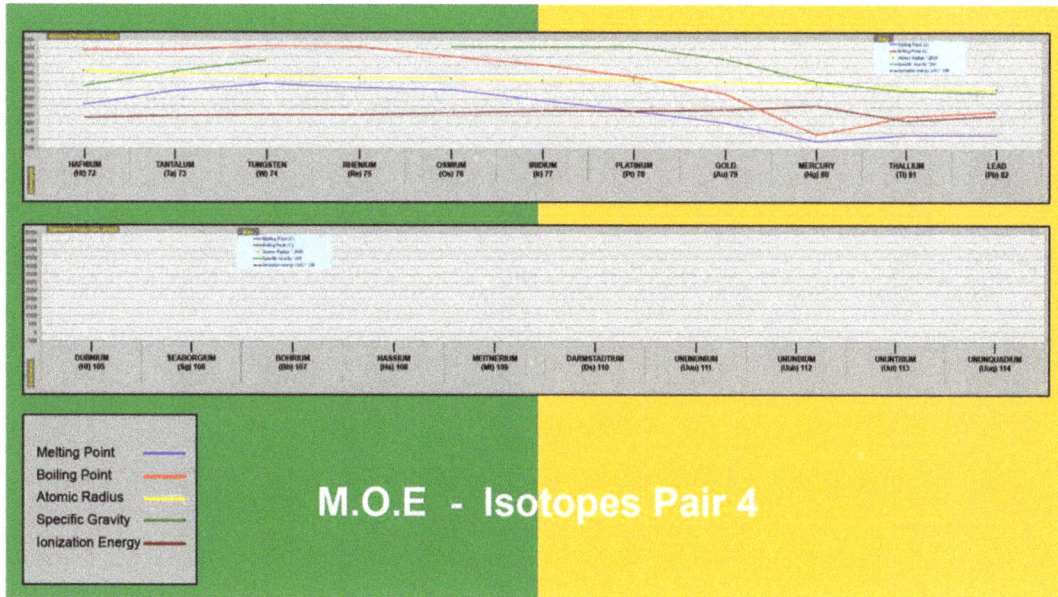

DIAGRAM 421

APPLICATIONS FOR A PLASMOIDS FORM AND FUNCTIONS

PART 17

SOLOMON'S MOLTEN SEA
ARK ATOMIC
RECONSTRUCTION
TECHNOLOGY USING THE
TORUS VAJRA MSAART

SOLOMON'S MOLTEN SEA ARK ATOMIC RECONSTRUCTION TECHNOLOGY USING THE TORUS VAJRA MSAART

ITS SCIENTIFIC PURPOSE

OPPOSITE CHARGED SPIRALS

TORUS 8 PLANE STRUCTURE

SINGULARITY ZERO POINT

MSAART EVO SACRED GEOMETRY

ALIEN DIMENSIONAL QUANTUM

VORTEX MATHS CHEMISTRY

PHYSICS AND ATOMIC

UNIFICATION MODEL

DRAFT 518,400 B KMV - SEVENTEEN OF TWENTY

MALCOLM V of SCOTLAND | MALCOLM BENDALL
THURSDAY 22ND SEPTEMBER 2022

INDEX

INDEX

INDEX

SECTION D – TORUS SINGULARITY POINT.

INDEX

INDEX

COLOUR CODING USED TO MARK MANIFESTATIONS

A.) DC AETHER

ROD

REED

STAFF

B.) SACRED TORUS DC TO AC CONVERSION AND FORM

THE SINGULARITY POINT

POTENTIAL DIFFERENCE AND GRAVITY

COMPASS ANGLES IN DEGREES AND MINUTES

AREA / TIME / VOLUME

Pi AND Phi

C.) MATTER (ELEMENTS) AC FREQUENCY MANIFESTATIONS

MUSIC

ELEMENTS

ENERGY

COLOR

LIGHT

TABLE OF NUMBER RELATIONSHIPS & MEANINGS COLOR CODED

...................[EARTH'S PROCESSION OF THE EQUINOX NUMBER 25,920]

[EARTH'S DIAMETER 7,920 x 2 = 15,840 x 2 = 31,680]

[EARTH'S DIAMETER 7,920]

[EARTH'S RADIUS 3,960]

[EARTH'S RADIUS 3,960 / 2 = 1,980 / 2 = 990]

....[SUN'S DAY 26.24 EARTH DAYS AT EQUATOR 31 EARTH DAYS AT ITS POLE]

[SUN'S DIAMETER 864,000 x 2 = 1,728,000 x 2 = 3,456,000 SUN SQUARE]

[SUN'S DIAMETER 864,000]

[SUN'S DIAMETER 864,000 MIRROR]

[SUN'S DIAMETER 864,000 SHUFFLE]

..................[SUN'S RADIUS 432,000]

..................[SUN'S RADIUS 432,000 SHUFFLE]

..................[SUN'S RADIUS MIRROR 432,000]

...[LIGHT SPEED DIAMETER NUMBER 144,000 x 2 = 288,000 LIGHT DIA EARTH 7.5]

...[SANSKRIT 100 YEARS 311,040 OCTAVES / 60 = 518,400 SECONDS IN SIX DAYS]

...[MOON'S DIAMETER 2,160 x 2 = 4,320 x 2 = 8,640]

...................[MOON'S DIAMETER 2,160]

...................[MOON'S DIAMETER 2,160 MIRROR]

...................[MOON'S DIAMETER 2,160 SHIFTED]

...................[MOON'S RADIUS 1,080]

...................[MOON'S RADIUS 1,080 AND EARTH'S + MOON'S DIA = 10,080]

...[RESONANT FREQUENCY RESONANT UNIT (RFEU) 129,600]

...[TIME OCTAVE 311,040 AND 518,400 SEQUENCE (TOS)]

...[VAJAR TORUS OCTAVE 16 SEQUENCE (VOS)]

...[PLANET OCTAVE 864 SEQUENCE (POS)]

...[MUSIC 266.666 OCTAVE 11.111 SEQUENCE (MOS)]

...[LIGHT PLASMOID OCTAVE 144 SEQUENCE (LOS)]

FOREWORD

We live in a Universe that is clearly the result of intelligent design. That design is in perfect octave harmonic resonance with its self so therefore all of creation resonates in unison with a collective conscience with a interconnected light chord based collective energy net whose webbing constructs, laws and governances are all based on the same 16 SECTOR precepts. The documentation of all manifestations along with their translates into one Torus model finally explains all things both seen and unseen. The Torus model is the HOLY GRAIL of science and philosophy the documentation and explanation of which is contained within the pages of this document for the comprehension, edification empowerment and enlightenment of mankind. Therefore both living things, permanence in motion, and elements and Aether that are permanence at rest harmonically nest within each other maintaining a dynamic homeostasis manifesting Life and Matter from the infinite energy of the Aether.

The Aether is best viewed as the infinite (high DC charge density) still space of space, the best parable for which is that Matter (AC) is represented by bubbles of low charge density within the Aether and the Aether is the water that the bubbles occur within. The Torus model demonstrates the forms and devices used as mechanisms to create Life and Matter from the Aether and also demonstrates the opposite mechanisms to take away Life and destroy Matter and the way they are returned to the Aether. These systems are self evident and beyond question as evidenced by empirical observations of the nature of life, space, suns, black holes, planets and all manifest forms of energy and light and sound. All that the Torus model does is to make these observed and measurable manifestations capable of being understood by the human mind. In doing so it becomes the base for both scientific and philosophical enlightenment the inevitable consequence of which will be a new fundamental industrial and spiritual revolution. The Torus base for these revolutions incorporates quantum physics, relativity, Aether, Resonance, Light, Sound, Crystal form, Valencies, Positive and Negative charge, Para-magnetic and Dia-magnetic character, Phi, Pi, and all elements both in their light matter and opposite and opposed dark anti matter forms.

If we do not stop moving ourselves, air, water and matter by centrifugally explosive destructive forces we will destroy all living things including our planet by the positive charge generated by these actions. We must to survive ban all these explosive devices. We must immediately replace, explosive anti – clockwise death inducing positive charge generating devices, with life enhancing implosive constructive clockwise vortex devices now to save the planet and ourselves. If we do not do this now we will loose the net positively charged ionosphere that is held in place by the net negative charge of our atmosphere. An atmosphere that we are turning to a net positive charge by our ignorant use of counter life forces we have been manifesting and propagating within our now doomed society. Once we loose the ionosphere we loose our atmosphere and all oceans will evaporate and drown us in an eternal darkness. We will and must act now together to use water not hydrocarbons as the future fuel for our world.

The Octave Table numbers 11.111, representing the conversion factor from direct current DC to alternating current AC into matter, and 5.555 representing the frequency and dimensional conversion factor for Aether into matter. These numbers 11.111 and 5.555 are used as a demonstration of the relationships and ratios between Aether, Time, Resonance, Crystalline Form, Charge, Light, magnetic forces, gravitational forces, spiral movement and Matter.

The Solution is the Torus Molten Sea Vajra Element model manifesting the Molten Sea Vajra Fusion, Aether to Light to Matter, reactor as a power generator, matter converter and thruster. The implosive turbine employs frequency and quadrature to effect to create a zero point Torus portal.

11.111 - AETHER TO MATTER FACTOR SEQUENCE

864,000 – SUN'S D TORUS PLASMOID DC TO AC CONVERTOR.

3,456,000 [SUN SQUARE] / 11.111 [(12 AETHER DC/AC MATTER 16) = 1.111 x 10] = 311,040 TIME
FINITE SUN SQUARE DIVIDED BY CONVERSION FACTOR OF INFINITE AETHER TO MATTER NUMBER EQUALS TIME.
THE SUN CREATES TIME, ITS CLOCK IS THE 11.111 YEARS IT TAKES TO FLIPS ITS POLES. THEREFORE 3,456,000 / 11.111
EQUALS 311,040 WHICH WHEN DIVIDED BY 60 EQUALS 5,184, TIME, WHICH IS MAPPED OUT BY THE MOON THAT
COVERS ITS OWN DIAMETER 2,160 MILES EVERY HOUR BEING 51,840 MILES A DAY AND THE SECONDS IN THE 6 DAYS OF
CREATION 518,400 SECONDS AND BEING THE ANGLE ON THE KHEOPS PYRAMID 51.84 DEGREES. TIME 518,400 DIVIDED
BY TWO EQUALS 259,200 THE PHASE CHANGE OF PROTIUM (H) -259.20 AND THE FREQUENCY OF OUR SOLAR SYSTEMS
EQUINOX PROCESSION BEING 25,920 YEARS SET BY PROTIUM (H) DUE TO ITS ABUNDANCE.

3 x 4 x 5 x 6 = (360 [INFINITE AETHER CIRCLE]) x 6 x 5 x 4 x 3 = 129,600 (RFEU FINITE SQUARE MATTER)
3 + 4 + 5 + 6 + 3 + 4 + 5 + 6 = 36   129,600/36 = 3,600   3,456,000/129,600 = 26.666 (24 x 11.111 = 266.666)
129,600 RFEU / 16 (SEGMENTS OF THE TORUS) = 8,100 POS / 22.5 (DEGREES OF EACH SEGMENT) = 360
360 x 400 = 144,000 / 0.125 = 1,152,000 / 266.666 = 4,320   864,000 / 16 = 400

3,456,000 SUN SQ / 266.666 = 12,960 RFEU   3,456,000 / 11.111 = 311,040 TIME   3,456,000 / 144,000 = 24
3,456,000 SUN SQ / 720 = 4,800 /16 = 300 /22.5 = 13.333 /0.125 = 106.666 x 400 = 42,666.666\\\\2,666.66
3,456,000 SUN SQ / 24 = 144,000 / 60 = 2,400 / 60 = 40 / 60 = 0.666 x 7.5 = 5 (MOLTEN SEA RADIUS IN CUBITS)
3,456,000 SUN SQ / 2.5 = 1,382,400 \-\ = 345,600   2.5 x 3,456,000 = 8,640,000   3,456,000 / 7.5 = 460,800
SO 3,456,000 THE SUN SQUARE, REPRESENTING FINITE AC MATTER CREATED FROM THE INFINITE DC
AETHER REPRESENTED BY THE SUN CIRCLE (DIAMETER IN MILES) OF 864,000 WHEN DIVIDED BY THE
RATIO OF CONVERSION FROM AETHER TO MATTER 11.111 REPRESENTING AETHER DIRECT CURRENT DC /
AC MATTER ALTERNATING CURRENT = 311,040 SANSKRIT TIME.

311,040 / 24 = 12,960 RFEU / 60 = 216 (MOON 2,160) /60 = 3.6 (360 DEGREES) / 60 = 0.06 / 7.5 = 0.008
311,040 / 720 = 432 /16 = 27 POS /22.5 = 1.2 AETHER / 0.125 = 9.6 x 400 = 3,840 / 144 = 26.666
311,040 / 518,400 = 0.6   518,400 / 311,040 = 1.666\\\\26.666   311,040 / 7.5 = 41,472 \-\\\\\ = 648

3 x 1 x 1 x 4 = (12 [INFINITE AETHER CIRCLE]) x 4 x 1 x 1 x 3 (12 [INFINITE AETHER CIRCLE]) = 144 (LIGHT
FROM FINITE SUN SQUARE TO MATTER).   144,000 LIGHT / 2,160 MATTER MOON DIA = 66.666 (ENERGY)
3 + 1 + 1 + 4 = (9) + 3 + 1 + 1 + 4 = 18 144 /18 = 8 (8 TORUS PLANES OF MATTER FROM LIGHT 16 SEGMENTS)

1,728,000 [129,600 x 13.333] / 11.111 [(AETHER DC / AC MATTER) = 1.111 x 10] = 155,520 TIME.
AETHER (TAPPED BY THE SUN) DIVIDED BY CONVERSION FACTOR OF AETHER TO MATTER NUMBER EQUALS TIME.
SO 1,728,000 THE SUN'S OCTAVE REPRESENTING FINITE MATTER FROM INFINITE AETHER REPRESENTED
BY THE CIRCLE AND THE PRODUCT OF 129,600 (RFEU) TIMES 13.333 (RATIO OF AETHER TO MATTER)
WHEN DIVIDED BY THE RATIO OF CONVERSION FROM AETHER TO MATTER 11.111 REPRESENTING
AETHER DIRECT CURRENT DC / AC MATTER ALTERNATING CURRENT = 155,520 OCTAVE OF ALL TIME 311,040.

864,000 [129,600 RFEU x 6.666 ENERGY] /11.111 [AETHER DC / AC MATTER = 1.111 x 10]= 77,760 TIME
AETHER (TAPPED BY THE SUN) DIVIDED BY CONVERSION TO MATTER NUMBER EQUALS TIME.
SO 864,000 THE SUN REPRESENTING INFINITE AETHER REPRESENTED BY THE CIRCLE WHEN DIVIDED BY
THE RATIO OF CONVERSION FACTOR FROM AETHER TO MATTER 11.111 REPRESENTING AETHER DIRECT
CURRENT DC / AC MATTER ALTERNATING CURRENT = 77,760 OCTAVE OF ALL TIME 311,040.

432,000 [129,600 x 3.333] / 11.111 [(AETHER DC / AC MATTER) = 1.111 x 10] = 38,880 TIME.
BASICALLY AETHER (TAPPED BY THE SUN) DIVIDED BY CONVERSION TO MATTER FACTOR EQUALS TIME.
SO 432,000 THE SUN OCTAVE REPRESENTING FINITE MATTER GENERATED FROM INFINITE AETHER
REPRESENTED BY THE CIRCLE WHEN DIVIDED BY THE RATIO OF CONVERSION FACTOR FROM AETHER TO
MATTER 11.111 REPRESENTING AETHER DIRECT CURRENT DC / AC MATTER EQUALS ALTERNATING
CURRENT = 38,880 OCTAVE OF ALL SANSKRIT TIME 311,040.

11.111 - 144,000 LIGHT NUMBERS

144,000 [864,000 / 144,000 = 6] x 11.111 [(AETHER DC / AC MATTER) = 1.111 x 10] = 1,600,000
1,600,000/129,600 = 12.34567901234568 518,400/1,600,000 = 0.324 311,040/1,600,000 = 0.1944 (POS)
BASICALLY LIGHT (GENERATED BY THE SUN) DIVIDED BY MATTERS CONVERSION FACTOR EQUALS TIME.
SO 144,000 LIGHT SPEED WHEN DIVIDED BY THE RATIO OF CONVERSION FACTOR FROM AETHER TO
MATTER 11.111 REPRESENTING AETHER DIRECT CURRENT DC / AC MATTER ALTERNATING CURRENT =
1,600,000 REPRESENTS THE 16 SEGMENTS DIVIDING MATTER INTO ITS ANGULAR CRYSTALLINE FORMS.

144,000 [864,000 / 144,000 = 6] / 11.111 [(AETHER DC / AC MATTER) = 1.111 x 10] = 129,600 RFEU

BASICALLY LIGHT (GENERATED BY THE SUN) DIVIDED BY DC TO AC CONVERSION FACTOR EQUALS 129,000
WHICH IS THE BASE RESONANT FREQUENCY ENERGY UNIT RFEU
SO 144,000 LIGHT SPEED WHEN DIVIDED BY THE RATIO OF CONVERSION FROM AETHER TO MATTER
11.111 REPRESENTING AETHER DIRECT CURRENT DC / AC MATTER ALTERNATING CURRENT = 129,600
THE RESONANT FREQUENCY ENERGY UNIT WHICH REPRESENT THE DIVIDING OF DIRECT CURRENT DC
AETHER INTO AC MATTER INTO ITS OCTAVE ANGULAR DETERMINED ALLOCATED CRYSTALLINE RESONANT
FORM.

11.111 - 7.5 - LIGHT MATTER CONVERSION NUMBERS

7.5 (CORRECTED TIME FACTOR FOR LIGHT 7.5 x AROUND THE EARTH IN 1 SECOND) x 11.111 = 83.333

BASICALLY LIGHTS CORRECTED TIME FACTOR TO THE EARTH 7.5 x AETHER TO MATTER FACTOR = MATTER
SO WHEN DIVIDED BY THE RATIO OF CONVERSION FROM AETHER TO MATTER 11.111 REPRESENTING
AETHER DIRECT CURRENT DC / AC MATTER ALTERNATING CURRENT = 83.333 \+\\\\2,666.666

11.111 / 7.5 LIGHTS CORRECTED TIME FACTOR TO THE EARTH 7.5 = 1.48148

BASICALLY AETHER TO MATTER FACTOR / LIGHTS CORRECTED TIME FACTOR TO THE EARTH 7.5 = MATTER
SO 11.111 THE RATIO OF CONVERSION FACTOR FROM AETHER TO MATTER REPRESENTING AETHER
DIRECT CURRENT DC (1) / (0.9) LIGHTS CORRECTED TIME FACTOR TO THE EARTH 7.5 = MATTER.

11.111 - 311,040 ARC SEC AND 518,400 SEC TIME IN SIX DAYS

311,040 (ALL SANSKRIT TIME) x 11.111 [(16 MATTER / 12 AETHER) = 1.111 x 10] = 3,456,000

BASICALLY 311,040 SANSKRIT ALL TIME MULTIPLIED BY THE AETHER TO MATTER CONVERSION FACTOR
11.111= MATTER WHICH IS REPRESENTED BY THE SUN SQUARE [864,000 x 4].
SO 311,040 WHEN MULTIPLIED BY THE RATIO OF CONVERSION FROM AETHER TO MATTER 11.111
REPRESENTING AETHER DIRECT CURRENT DC RATIO TO AC MATTER ALTERNATING CURRENT = 3,456,000
WHICH IS THE SUN SQUARE WHICH REPRESENTS FINITE MATTER.
3 x 4 x 5 x 6 = (360 [INFINITE AETHER CIRCLE]) x 6 x 5 x 4 x 3 = 129,600 (RFEU FINITE SQUARE MATTER)
3 + 4 + 5 + 6 = (18) + 3 + 4 + 5 + 6 = 36

311,040 (ALL SANSKRIT TIME) / 11.111 [(16 MATTER / 12 AETHER) = 1.111 x 10] = 27,993.6
27,993.6 = EARTH'S D 7,920 SHUFFEL AND EARTH'S RADIUS 3,960 SHUFFEL NUMBERS COMBINED.
BASICALLY 311,040 SANSKRIT ALL TIME DIVIDED BY THE AETHER TO MATTER CONVERSION FACTOR
11.111 = 27,993.6

SO 311,040 WHEN DIVIDED BY THE RATIO OF CONVERSION FROM AETHER TO MATTER 11.111
REPRESENTING AETHER DIRECT CURRENT DC / AC MATTER ALTERNATING CURRENT = 27,993.6

11.111 - 311,040 ARC SEC AND 518,400 SEC TIME IN SIX DAYS

518,400 [6 DAYS IN SECONDS] / 11.111 [(16 MATTER / 12 AETHER) = 1.111 x 10] = 46,656

46,656 /24 = 1,944 (POS) /60 = 32.4 (SUN'S RADIUS) /60 = 0.54 (360 DEGREES) /60 = 0.009 /7.5 = 0.0012
46,656 / 720 (DEGREES) = 6.48 /16 = 4.05 (POS) /22.5 = 0.18 / 0.125 = 1.44 x 400 = 576 / 144 = 3
46,656 / 518,400 = 0.6   518,400 / 46,656 = 11.111   46,656 / 311,040 = 0.15   311,040 / 46,656 = 6.666

BASICALLY 518,400 IS THE SECONDS IN THE 6 DAYS OF CREATION AS IT IS THE BASE MIRROR NUMBER FOR ALL MATTER THE YING AND YANG, MALE AND FEMALE, LEFT AND RIGHT HAND SPIRALS WHICH ARE THE CLOCKWISE IMPLOSIVE CREATION FORCE AS OPPOSED TO THE ANTI CLOCK WISE EXPLOSIVE DEATH AND DESTRUCTION FORCE DIVIDED BY THE AETHER TO MATTER CONVERSION FACTOR EQUALS TIME 46,656

518,400 [6 DAYS IN SECONDS] x 11.111 [(AC MATTER / DC AETHER) = 1.111 x 10] = 5,760,000

518,400 [6 DAYS IN SECONDS] / 11.111 [(16 MATTER / 12 AETHER) = 1.111 x 10] = 46,656

576\\\\144

BASICALLY 311,040 SANSKRIT ALL TIME DIVIDED BY THE AETHER TO MATTER CONVERSION FACTOR 11.111 EQUALS THE THIRD OCTAVE OF TIME.

576 / 24 Hrs = 24 / 60 = 0.4 / 60 = (360 DEGREES) / 60 = 0.00666 / 7.5 = 0.000888
576 / 720 (DEGREES) = 24 /16 = 1.5 (LCT) /22.5 = 0.0666 /0.125 = 0.5333 x 400 = 213.333 /144 = 1.48 POS
576 / 518,400 TIME = 0.00111 518,400 TIME / 576 = 900 576 /311,040 = 0.00185 311,040 /576 = 540

518,400 [6 DAYS IN SECONDS] x 1.333 [(16 MATTER / 12 AETHER) = 1.333 x 10] = 691,200 (RFEU)

518,400 [6 DAYS IN SECONDS] / 1.333 [(16 MATTER / 12 AETHER) = 1.333 x 10] = 388,800 (POS)

5.555 AETHER OCTAVE SEQUENCE

5.555 - 864,000 SUN'S DIAMETER NUMBERS TORUS PLASMOID DC TO AC CONVERTOR.

3,456,000 [SUN SQUARE FINITE MATTER] / 5.555 (AETHER) = 622,080 [311,040 x 2 = TIME]

THE SUN'S SQUARE REPRESENTING FINITE MATTER DIVIDED BY THE INFINITE AETHER DEFINES TIME.
AETHER TO MATTER DIVIDED BY CONVERSION TO MATTER NUMBER EQUALS TIME.
SO 3,456,000 THE SUN SQUARE, REPRESENTING FINITE MATTER CREATED FROM THE INFINITE AETHER
REPRESENTED BY THE SUN CIRCLE (DIAMETER IN MILES) OF 864,000 WHEN DIVIDED BY THE RATIO OF
CONVERSION FROM AETHER TO MATTER EQUALS 11.111 REPRESENTING AETHER DIRECT CURRENT DC /
AC MATTER ALTERNATING CURRENT = 311,040 SANSKRIT ALL TIME

3 x 1 x 1 x 4 = (12 [INFINITE AETHER CIRCLE]) x 4 x 1 x 1 x 3 (12 [INFINITE AETHER CIRCLE]) = 144 (LIGHT
FINITE SQUARE MATTER)
3 + 1 + 1 + 4 = (9) + 3 + 4 + 5 + 6 = 18   144 / 18 = 8 (8 PLANES OF MATTER LIGHT)

6 x 2 x 2 x 8 = 192 (192 [INFINITE AETHER CIRCLE]) x 8 x 2 x 2 x 6 = 36,864
6 + 2 + 2 + 8 = (18) + 3 + 4 + 5 + 6 = 36   36,864 / 36 = 1,024
36,864 / 266.666 (24 x 11.111 = 266.666) = 138.24\\\\8.64

1,728,000 [SUN EM DIAMETER] / 5.555(AETHER) = 311,040 [TIME]

BASICALLY THE SUN'S EM DIAMETER DIVIDED BY THE AETHER EQUALS TIME.
BASICALLY AETHER (TAPPED BY THE SUN) DIVIDED BY CONVERSION TO MATTER NUMBER EQUALS TIME.
SO 1,728,000 THE SUN OCTAVE REPRESENTING FINITE MATTER FROM INFINITE AETHER REPRESENTED
BY THE CIRCLE WHEN DIVIDED BY THE RATIO OF CONVERSION FROM AETHER TO MATTER 11.111
REPRESENTING AETHER DIRECT CURRENT DC / AC MATTER ALTERNATING CURRENT = 155,520 TIME

3 x 1 x 1 x 4 = (12 [INFINITE AETHER CIRCLE]) x 4 x 1 x 1 x 3 = 144
3 + 1 + 1 + 4 = (9) + 3 + 1 + 1 + 4 = 18   144 / 18 = 8   311,040 / 266.666 (24 x 11.111 = 266.666) = 1,166.4

864,000 [SUN VISUAL DIAMETER] / 5.555 (AETHER) = 155,520 [311,040 / 2 = TIME]

AETHER (TAPPED BY THE SUN) DIVIDED BY CONVERSION FROM AETHER FIFTH DIMENSIONAL NUMBER
EQUALS TIME.
SO 864,000 THE SUN OCTAVE REPRESENTING FINITE MATTER FROM INFINITE AETHER REPRESENTED BY
THE CIRCLE WHEN DIVIDED BY THE RATIO OF CONVERSION FROM AETHER TO MATTER 11.111
REPRESENTING AETHER DIRECT CURRENT DC / AC MATTER ALTERNATING CURRENT = 155,520 TIME

5 x 5 x 5 x 2 = (250 [INFINITE AETHER CIRCLE]) x 2 x 5 x 5 x 5 = 62,500
5 + 5 + 5 + 2 = (17) + 2 + 5 + 5 + 5 = 34   62,500 / 34 = 1,838.23
155,520 / 266.666 (24 x 11.111 = 266.666) = 583.2

432,000 [SUN'S VISUAL RADIUS] / 5.555 (AETHER) = 77,760 [TIME]

7 x 7 x 7 x 6 = (2,058 [INFINITE AETHER CIRCLE]) x 6 x 7 x 7 x 7 = 4,235,364
7 + 7 + 7 + 6 = (27) + 6 + 7 + 7 + 7 = 54   4,235,364 / 54 = 78,432.666
77,760 / 266.666 (24 x 11.111 = 266.666) = 291.60 [129,600 RFEU]

5.555 AETHER - 144,000 - LIGHT NUMBERS

144,000 [LIGHT] / 5.555 (AETHER) = 25,920 [25,920 YEARS EARTH'S EQUINOX CYCLE]
LIGHT 144 DIVIDED BY THE AETHER 5.555 EQUALS TIME 25,920 AS A PRODUCT OF HYDROGEN BEING THE MAJOR FORCE IN THE OUR SOLAR SYSTEM WHOSE RESONANCE,(AS DEFINED BY ITS MELTING POINT OF MINUS 259.2 DEGREES CELCIUS) CREATES THE 25,920 YEAR CYCLE.

25,920 / 3.333 = 7,776 x 1.333 = 10,368
10,368 / 24 = 432 / 60 = 7.2 / 60 = 0.12 / 60 = 0.002

144,000 [LIGHT] x 5.555 (AETHER) = 800,000 (MATTER)
800,000 (MATTER) / 266.666 (BASE RESONANT FREQUENCY FOR MATTER) = 3,000

LIGHT 144 MULTIPLIED BY THE AETHER 5.555 EQUALS 800,000 MATTER. HYDROGEN BEING THE MAJOR FORCE IN THE OUR SOLAR SYSTEM WHOSE RESONANCE, (AS DEFINED BY ITS MELTING POINT OF MINUS 259.2 DEGREES CELCIUS), CREATES THE 25,920 YEAR PROCESSION OF THE EQUINOX CYCLE WHEN DIVIDED BY 800,000 EQUALS 0.0324 (SUN'S MIRROR RADIUS 432,000 MILES).

800,000 / 24 = 33,333.333 / 60 = 555.555 / 60 = 9.259225922592
9.259225922592 / 60 = ***0.154320987654321***
1x5x4x3x2x9x8x7x6x5x4x3x2x1 = 43,545,600 x 43,545,600 = 1,896,219,279,360,000
1,896,219,279,360,000 / 720 = 2,633,637,888,000 / 360 = 7,315,660,800
7,315,660,800 / 16 = 457,228,800 / 22.5 = 20,321,280 / 0.125 = 162,570,240\

5.555 AETHER - 7.5 - LIGHT MATTER CONVERSION NUMBERS

7.5 LIGHTS CORRECTED TIME FACTOR TO THE EARTH / 5.555 (AETHER) = 1.35
7.5 LIGHTS CORRECTED TIME FACTOR TO THE EARTH DIVIDED BY THE AETHER EQUALS 135 DEGREES

5.555 (AETHER) x 7.5 LIGHTS CORRECTED TIME FACTOR TO THE EARTH = 41.666 (POS)

THE AETHER TIMES THE 7.5 LIGHTS CORRECTED TIME FACTOR TO THE EARTH EQUALS THE POS OCTAVE

5.555 AETHER - 311,040 TIME AND 518,400 SEC IN 6 DAYS

311,040 (TIME) / 5.555 (AETHER) = 55,987.2

311,040 SANSKRIT ALL TIME DIVIDED BY THE AETHER TO MATTER CONVERSION FACTOR 11.111 = 27,993.6
SO 311,040 WHEN DIVIDED BY THE RATIO OF CONVERSION FROM AETHER TO MATTER 11.111 REPRESENTING AETHER DIRECT CURRENT DC / AC MATTER ALTERNATING CURRENT = 27,993.6

311,040 (TIME) x 5.555 (AETHER) = 1,728,000 [SUN EM]

311,040 SANSKRIT ALL TIME MULTIPLIED BY THE AETHER TO MATTER CONVERSION FACTOR 11.111= MATTER WHICH IS REPRESENTED BY THE SUN SQUARE [864,000 x 4].

SO 311,040 WHEN MULTIPLIED BY THE RATIO OF CONVERSION FROM AETHER TO MATTER 11.111 REPRESENTING AETHER DIRECT CURRENT DC RATIO TO AC MATTER ALTERNATING CURRENT = 3,456,000 WHICH IS THE SUN SQUARE WHICH REPRESENTS FINITE MATTER.

3 x 4 x 5 x 6 = (360 [INFINITE AETHER CIRCLE]) x 6 x 5 x 4 x 3 = 129,600 (RFEU FINITE SQUARE MATTER)

3 + 4 + 5 + 6 = (18) + 3 + 4 + 5 + 6 = 36

5.555 - 311,040 TIME AND 518,400 SEC IN 6 DAYS

518,400 (TIME) / 5.555 (AETHER) = 93,312 [EARTH 7,920 = 93,312]

SO 518,400 IS THE SECONDS IN THE 6 DAYS OF CREATION AS IT IS THE BASE MIRROR NUMBER FOR ALL MATTER THE YING AND YANG, MALE AND FEMALE, LEFT AND RIGHT HAND SPIRALS WHICH ARE THE CLOCKWISE IMPLOSIVE CREATION FORCE AS OPPOSED TO THE ANTI CLOCK WISE EXPLOSIVE DEATH AND DESTRUCTION FORCE DIVIDED BY THE AETHER CONVERSION FACTOR EQUALS EARTHS DIAMETER SHIFTED COMPOSITE NUMBER. 9 x 3 x 3 x 2 = (162) x 9 x 3 x 3 x 2 = 26,244   9 + 3 + 3 + 2 = (17) + 9 + 3 + 3 + 2 = 34

518,400 (TIME) x 5.555 (AETHER)= 2,880,000 [144(LIGHT)x2= 288] 2,880,000/1.333= 21,600 POS

SO 518,400 TIME DIVIDED BY THE AETHER 5.555 EQUALS THE FIRST OCTAVE UP OF LIGHT 288.

2 x 8 x 8 = (128) x 8 x 8 x 2 = 16,384   2 + 8 + 8 = (18) + 8 + 8 + 2 = 36   16,384 / 36 = 455.111\\\\\\\\0.0555

HISTORICAL QUOTES FROM CREDIBLE FAMOUS SCIENTISTS BACKING THE CALCULATIONS, OBSERVATIONS AND RELATIONSHIPS (DIMENSIONAL AND MATERIAL) EXPLAINED ABOVE.

NICOLA TESLA (1856 – 1943) QUOTE RE – ENERGY, FREQUENCY AND VIBRATION
" IF YOU WANT TO FIND THE SECRETS OF THE UNIVERSE, THINK IN TERMS OF
ENERGY, FREQUENCY AND VIBRATION."

MARK TWAIN (1835 TO 1910) QUOTE RE - SCHOOLING
"I HAVE NEVER LET SCHOOLING INTERFERE WITH MY EDUCATION"

WALTER RUSSELL (1871 TO 1963)
" ALL MOTION AND ALL FORM SPRING FROM STILLNESS WAVES OF MOTION SPRING FROM THE CALM OCEAN THIS IS A UNIVERSE OF REST FROM WHICH MOTION SPRINGS TO EXPRESS THE POWER WHICH IS IN REST". " SEEK TO BE ALONE MUCH TO COMMUNE WITH NATURE BE THUS INSPIRED BY HER MIGHTY WHISPERINGS WITHIN YOUR CONSCIOUSNESS. NATURE IS A MOST JEALOUS GOD, FOR SHE WILL NOT WHISPER HER INSPIRING REVELATIONS TO YOU UNLESS YOU ARE ABSOLUTELY ALONE WITH HER." "INSPIRED THINKING IS THINKING IN TUNE WITH DIVINITY ALL POWER IS IN TRUE THINKING. THINKING TRUE MAY BE LIKENED UNTO A VIOLIN STRING SOUNDED BY A GREAT MASTER. TEN THOUSAND OTHER STRINGS TUNED TO THE SAME PITCH WOULD SING IN UNISON UNPLAYEDUPON. TEN THOUSAND OTHER STRINGS NOT SO ATTUNED WOULD NOT KNOW THE ECSTACY."

ALBERT EINSTIEN (1879 TO 1955)
" TO DENY THE AETHER IS ULTIMATELY TO ASSUME THAT EMPTY SPACE HAS NO PHYSICAL QUALITIES WHAT SO EVER THE FUNDERMENTAL FACTS OF MECHANICS DO NOT HARMONISE WITH THIS VIEW ACCORDING TO THE GENERAL THEORY OF RELATIVITY SPACE IS ENDOWED WITH PHYSICAL QUALITIES IN THIS SENSE THEREFORE THERE EXISTS AN AETHER ACCORDING TO THE GENERAL THEORY OF RELATIVITY SPACE WITHOUT AETHER IS UNTHINKABLE."

VICTOR SCHAULZBERGER (1885 TO 1958)
" YOU MUST LOOK AT THE PROCESSES OF MOTION IN THE MACROCOSMOS AND MICROCOSMOS ACCURATELY AND COPY THEM."
{OBSERVE AND MIMIC NATURE}

ROBERT OPPENHEIMER - VISHNU SANSKRIT BHAGVAD GITA TEXT ABOUT THE ANCIENT ATOMIC BOMB
" I HAVE BECOME DEATH THE DESTROYER OF WORLDS "

PYTHAGORAS (570 BC TO 495 BC)
" CRYSTALLINE MATTER ALWAYS RESONATES AT THE SUM OF ITS INTERNAL ANGLES "

SIR ISAAC NEWTON (1642 - 1727)
" TO EVERY ACTION THERE IS ALWAYS OPPOSED AN EQUAL REACTION "
KEPLER   SIR WILLIAM CROOKES,
LORD KELVIN - JOULE (THOMPSON) THUNDERSTORM MACHINE,
MAXWELL

11.111 AND 5.555 AETHER OCTAVE SEQUENCE

186,413,511.111
93,206,755.555
46,603,337.777
23,301,688.888
11,650,844.444
5,825,422.222
2,912,711.111
1,456,355.555
728,177.777
364,088.888
182,044.444
91,022.222
45,511.111
22,755.555
11.377.777
5,688.888
2,844.444
1,422.222
711.111
355.555
177.777
88.888
44.444
22.222

11.111

3,456,000/11.111= 311,040 1,728,000/11.111= 155,520 864,000/11.111= 77,760 432,000/11.111= 38,880
311,040 x 11.111 = 3,456,000 311,040 / 11.111 = 27,993.6 518,400 / 11.111= 46,656 518,400 x 11.111= 576
144,000 x 11.111 = 1,600,000 144,000 / 11.111 = 129,600   7.5 x 11.111 =83.333   11.111 / 7.5 = 1.48148

5.555

3,456,000 /5.555 = 622,080 1,728,000 /5.555 = 311,040 864,000/5.555 = 155,520 432,000/5.555 = 77,760
311,040/5.555= 55,987.2 311,040 x 5.555= 1,728,000 518,400/5.555= 93,312 518,400 x 5.555= 2,880,000
144,000 / 5.555 = 25,920 / 3.333 = 7,776 x 1.333 = 10,368/ 24 = 432/60 = 7.2/60 = 0.12/60 = 0.002
144,000 x 5.555 = 800,000 / 24 = 33,333.333 / 60 = 555.555 / 60 = 9.2592 / 60 = ***0.154320987654321***
1x5x4x3x2x9x8x7x6x5x4x3x2x1 = 43,545,600 x 43,545,600 = 1,896,219,279,360,000
7.5 / 5.555 = 1.35 5.555 x 7.5 = 41.666

2.777
1.3888
0.69444
0.347222
0.1736111
0.08680555
0.0434027888
0.0217013888
0.01085069444
0.005425347222
0.0027126736111

PART 17 OF 20 – MSAART PATENT APPLICATION NOTES DRAFT 518,400 – 22:22:22 THUR 22ND SEPT 2022
© STRIKE FOUNDATION GUARANTEE LIMITED | MALCOLM BENDALL 2022 | GRAPHICS - STEVE EARL

12, 24, 48 AETHER TO MATTER OCTAVE SEQUENCE

402,653,184
201,326,592
100,663,296
50,331,648
25,165,824
12,582,912
6,291,456
3,145,728
1,572,864
786,432
393,216
196,608
98,304
49,152
23,576
12,228
6,144
3,072
1,536
768
384
192
96
48
24

3,456,000/24= 144,000 1,728,000/24= 72,000 864,000/24 = 36,000 432,000/24= 18,000
31,104,000,000 / 24 = 12,96,000,000 518,400 / 24 = 21,600 518,400 x 24 = 12,441,600
144,000 /24= 6,000(MOLTEN SEA SPHERE VOLUME IN BATHS) 12/7.5 = 1.6 7.5/24 = 0.3125 24/7.5= 3.2
24 / 11.111 = 2.16 24 x 11.111 = 266.666 12 x 7.5 = 90 24 x 7.5 = 180 48 x 7.5 = 360

12
6
3
1.5
0.75 (LIGHT TRAVELS 7.5 x AROUND THE EARTH IN 1 SECOND)
0.375
0.1875
0.09375
0.046875
0.0234375
0.01171875
0.005859375
0.0029296875
0.00146484375
0.000732421875

PLANET OCTAVE 864,000 SEQUENCE (POS)

452,984,832,000
226,492,416,000
113,246,208,000
56,623,104,000
28,311,552,000
14,155,776,000
7,077,888,000
3,538,944,000
1,769,472,000
884,736,000
442,368,000
221,184,000
110,592,000
55,296,000
27,684,000
13,824,000
6,912,000 [SUNS DIAMETER x 8]
3,456,000 [SUNS DIAMETER x 4]
1,728,000 [SUN'S DIAMETER 864,000 x 2]

864,000 SUN AETHER OCTAVES

8 / 6 / 4 = 0.333
3,456,000/ 864 = 4,000 1,728,000/ 864 = 2,000 864,000/ 864 = 1,000 432,000/ 864 = 500
311,040/864= 360 311,040 x 864= 268,738,560 518,400/864= 600 5184x 864=4,478,976
144,000 / 864 = 166.666 x 1.333 = 222.222 / 16 = 13.888/22.5 = 0.61728/0.125 = 4.93827
144,000 x 864 = 124,416,000 /24 = 5,184,000/60 = 86,400 /60 = 1,440 /60 = 24 /7.5 = 3.2
864,000 x 7.5 = 6,480,000 x 7.5 = 48,600,000
432,000 [SUN'S RADIUS 432,000]
216,000 [MOON'S DIAMETER 2,160]
108,000 [MOON'S RADIUS 1,080]
54,000
27,000
13,500
6,750
3,375
1,687.5
843.75
421.875
210.9375
105.46875
52.734375
26.3671875
13.18359375
6.591796875
3.2958984375
1.64794921875

LIGHT PLASMOID OCTAVE 144 SEQUENCE (POS)

603,979,776,000
301,989,888,000
150,994,944,000
75,497,472,000
37,748,736,000
18,874,368,000
9,437,184,000
4,718,592,000
2,359,296,000
1,179,648,000
589,824,000
294,912,000
147,456,000
73,728,000
36,864,000
18,432,000
9,216,000
4,608,000
2,304,000
1,152,000
576,000
288,000

144,000 LIGHT OCTAVES

3,456,000 / 144,000 = 24   1,728,000 / 144,000 = 12   864,000 / 144,000 = 6   432,000 / 144,000 = 3
311,040 / 144,000 = 2.16   311,040 x 144,000 = 44,789,760,000   518,400 / 144,000 = 3.6
518,400 x 144,000 = 74,649,600,000 / 3.333 = 22,394,880,000 x 1.333 = 29,859,840,000
29,859,840,000 / 24 = 1,244,160,000 /60 = 20,736,000 /60 = 345,600 /60 = 5,760 \\\1,440
144,000 / 144 = 1,000 / 3.333 = 300 x 1.333 = 400 / 24 = 16.666 /60 = 0.2777 /60 = 0.00462962
144,000 x 144 = 20,736,000 / 24 = 864,000 / 60 = 14,400 / 60 = 240 / 60 = 4 / 7.5 = 0.5333 \0.2666
144,000 / 7.5 = 19,200   7.5 / 144,000 = 0.00005208333 \9\ 0.02666   144,000 x 7.5 = 1,080,000

72,000
36,000
18,000
9,000
4,500
2,250
1,125
5,625
281.25
140.625
70.3125
35.15625
17.578125
8.7890625
4.39453125
2.197265625
1.0986328125

TIME OCTAVE 311,040 AND 518,400 SEQUENCE (TOS)

5,218,385,264,640
2,609,192,632,320
1,304,596,316,160
652,298,158,080
326,149,076,040
163,074,539,520
81,537,269,760
40,768,634,880
20,384,317,440
10,192,158,720
5,096,079,360
2,548,039,680
1,274,019,840
637,009,920
318,504,960
159,252,480
79,626,240
39,813,120
19,906,560
9,953,280
4,976,640
2,488,320
1,244,160
622,080

311,040 TIME OCTAVES (311,040 / 864 = 360 Degrees)

3,456,000/311,040=11.1111,728,000/311,040=5.7291666864,000/311,040=2.777432,000/311,040=1.388
311,040 / 311,040 = 1   311,040 x 311,040 = 96,745,881,600   518,400 / 311,040 = 1.666
518,400 x 311,040= 161,234,136,000/24= 6,718,464,000/60= 111,974,400/60= 1,866,240/60= 31,104
31,104 / 60 = 518.4/60 = 86.4 / 60 = 1.44 / 60 = 0.024 / 60 = 0.0004 / 60 = 0.00000666 / 7.5 = 0.00000888
144,000 / 311,040 = 0.462962962 / 3.333 = 0.13888 x 1.333 = 0.185185 / 24 = 0.0077160/60 = 0.00012
144,000 x 311,040= 44,789,760,000/24=1,866,240/60= 31,104,000/60= 518,400 /60 = 8,640 x 7.5 = 64,800
44,789,760,000 / 7.5 = 5,971,968,000 / 24 = 248,832,000 / 60 = 4,147,200 / 60 = 69,120 / 60 = 1,152\144
7.5 / 311,040 = 0.000024112654321   311,040 x 7.5 = 2,332,800

311,040 x 7.5 (The one second light takes to travel 7.5 times around the Earths equator) = 2,332,800

155,520
77,760
38,880
19,440
9,720
4,860
2,430
1,215
607.5
303.75
151.875
75.9375
37.96975
18.984375
9.4921875

RESONANCE 129,600 RFEU OCTAVES

543,581,798,400
271,790,899,200
135,895,449,600
67,947,724,800
33,973,862,400
16,986,931,200
8,493,465,600
4,246,732,800
2,123,366,400
1,061,683,200
530,841,600
265,420,800
132,710,400
66,355,200
33,177,600
16,588,800
8,294,400
4,147,200
2,073,600
1,036,800
518,400
259,200

129,600 RFEU 129,600 /864 = 150 129,600 /31,104= 4.1666 129,600 /5,184= 25 129,600 /7.5 = 1,728
3,456,000/129,600= 26.666 1,728,000/129,600=13.333 864,000/129,600=6.666 432,000/129,600= 3.333
311,040 / 129,600 = 2.4   311,040 x 129,600 = 40,310,784,000   518,400 / 129,600 = 4
518,400 x 129,600= 67,184,640,000/24=2,799,360,000/60=46,656,000/60=777,600/60=12,960/7.5= 1,728
144,000 / 1,296 = 111.111 / 3.333 = 33.333 x 1.333 = 44.444 / 7.5 = 5.925922592
144,000 x 129,600 = 16,796,160,000 / 24 = 699,840,000 / 60 = 11,664,000 / 60 = 194,400 / 60 = 3,240
129,600 x 7.5 = 972,000 x 7.5 = 7,290,000, 129,600 / 7.5 = 17,280 / 7.5 = 2,304

64,800
32,400
16,200
8,100
4,050

2,025

1,012.5
506.25
253.125
126.5625
63.28125
31.640625
15.8203125
7.91015625
3.955078125
1.9775390625
0.98876953125

135 DEGREES SOLOMANS MOLTEN SEA

2,264,924,160
1,132,462,080
566,231,404
283,115,520
141,557,760
70,778,880
38,389,440
17,604,720
8,847,360
4,432,680
2,211,840
1,105,920
552,960
276,480
138,240
69,120
34,560
17,280
8,640
4,320
2,160
1,080
540
270

135 DEGREES OCTAVES S.E CORNER SOLOMON'S TEMPLE

864,000/135=6,400 864,000x135= 116,640,000 311,040/135=2,304 311,040x135=41,990,400/864= 48.6
864 x 9 = 7,776   864,000 / 900 = 960   31,104 x 9 = 279,936   311,040 / 900 = 345.6
3,456,000 / 135 = 25,600   1,728,000 / 135 = 12,800   864,000 / 135 = 6,400   432,000 / 135 = 3,200
311,040 / 135 = 2,304   311,040 x 135 = 41,990,400   518,400 / 135 = 3,840   518,400 x 135 = 69,984,000
144,000 / 135 = 1,066.666 x 3.333 = 3,555.555 x 1.333 = 4,740.740740740
4,740.740740740 / 24 = 197.530864 /60 = 3.292181 /60 = 0.054869684499314 /60 = 0.000914494
144,000 x 135 = 19,440,000 / 24 = 810,000 / 60 = 13,500 / 60 = 225 / 60 = 3.75 / 7.5 = 0.5 / 0.125 = 4
7.5 x 135 = 1,012.5   7.5 / 135 = 0.0555   135 x 7.5 = 1,012.5   135 / 7.5 = 18 / 7.5 = 2.4 / 7.5 = 0.32

67.5
33.75
16.875
8.4375
4.21875
2.109375
1.0546875
0.52734375
0.263671875
0.1318359375
0.06591796875
0.032958984375
0.0164794921875

VAJRA TORUS OCTAVE 16 SEQUENCE (VOS)

268,435,456
134,217,728
67,108,864
33,554,432
16,777,216
8,388,608
4,194,304
2,097,152
1,048,576
524,288
262,144
131,072
65,536
32,768
16,384
8,192
4,096
2,048
1,024
512
256
128
64
32

16 VAJRA SECTIONS OCTAVES 864 /16= 54 432 /16= 27 311,040 /16= 19,440 518,400 /16= 32,400
3,456,000 / 16 = 55,296,000 1,728,000 / 16 = 108,000 864,000 / 16 = 54,000 432,000 / 16 = 27,000
3,456,000 x16= 55,296,000 1,728,000 x16 = 27,684,000 864,000 x16 = 13,824,000 432,000 x16 = 6,912,000
311,040 / 16 = 19,440 311,040 x 16 = 4,976,640 518,400 / 16 = 32,400 518,400 x 16 = 8,294,400
144,000 / 16 = 9,000 x 3.333 = 30,000 x 1.333 = 40,000 / 24 = 1,666.666 /60 = 27.777 /60 = 0.4629
144,000 x 16 = 2,304,000 / 24 = 96,000 / 60 = 1,600 / 60 = 26.666 / 60 = 0.444 / 7.5 = 0.0592592
7.5 / 16 = 0.46875 7.5 x 16= 120 16 / 7.5 = 2.1333\\\0.2666
16 x 7.5 = 120 (AETHER TO LIGHT TO MATTER) x 7.5 = 900

8
4
2
1
0.5
0.25
0.125
0.0625
0.03125
0.015625
0.0078125
0.00390625
0.001953125
0.0009765625

MATTER MUSIC 266.666 OCTAVESEQUENCE (POS)

1,118,481,066.666
559,240,533.333
279,620,266.666
139,810,133.333
69,905,066.666
34,952,533.333
17,476,266.666
8,738,133.333
4,369,066.666
2,184,533.333
1,092,266.666
546,133.333
273,066.666
136,533.333
68,266.666
34,133.333
17,066.666
8,533.333
4,266.666
2,133.333
1,066.666
533.333

266.666 MATTER ELEMENTAL OCTAVES

3,456,000/266.666 = 12,960 1,728,000/266.666 = 6,480 864,000/266.666 = 3,240 432,000/266.666 = 1,620
311,040 /266.6 = 1,166.4 311,040x 266.6 = 82,944,000 518,400/266.666 = 1,944 518,400x266.6 = 138,240
144,000 /266.666 = 540/3.333=162 x1.333=216/24= 9 /60= 0.15/60=0.0025/60=0.00041666\5\0.002666
144,000 x 266.666 = 38,400,000 / 24 = 1,600,000 / 60 = 26,666.666 / 60 = 444.444 / 60 = 7.4074074
7.4074074 / 7.5 = 0.98765432098765432 / 1.111 = 0.888 x 1.333 = 1.8518518518
7.5 / 266.666 = 0.028125 266.666 x 7.5 = 2,000
266.666 x 7.5 = 2,000 x 7.5 = 15,000 x 7.5 = 112,500

133.333
66.666
33.333
16.666
8.333
4.1666
2.08333
1.041666
0.5208333
0.26041666
0.130208333
0.0651041666
0.03255208333
0.016276041666
0.0081380208333
0.00406901041666

ANGLES 720, 360, 22.5 OCTAVE SEQUENCE (AOS)

12,079,595,520
6,039,797,760
3,019,898,880
1,509,949,440
754,974,720
377,487,360
188,743,680
94,371,840
47,185,920
23,592,960
11,796,480
5,898,240
2,949,120
1,474,560
737,280
368,640
184,320
92,160
46,080
23,040
11,520
5,760
2,880
1,440
720

2,880 1,440, 720, 360,180, 90, 45, 22.5 DEGREES ANGLE OCTAVES

3,456,000/ 360 = 9,600   1,728,000/ 360 = 4,800   864,000/ 360 = 2,400 432,000/ 360 = 1,200
311,040/ 360 = 864   311,040 x 360 = 111,974,400   518,400/ 360 = 1,440   518,400 x 360 = 186,624,000
144,000 / 360 = 864 / 3.333 = 259.2 x 1.333 = 345.6 / 24 = 14.4 /60 = .24 /60 = 0.004 /60 = 0.0000666
144,000 x 360 = 400 / 24 = 16.666 / 60 = 0.2777 / 60 = 0.004629629 / 60 = 0.000077160493827
7.5 / 360 = 0.0208333   360 x 7.5 = 2,700
360 x 7.5 = 2,700 x 7.5 = 20,250 / 7.5 = 151,875   360 / 7.5 = 48 / 7.5 = 6.4 / 7.5 = 0.85333\\\\\0.02666

180
90
45
22.5
11.25
5.625
2.8125
1.40625
0.703125
0.3515625
0.17578125
0.087890625
0.0439453125
0.02197265625

7,920 EARTHS OCTAVE SEQUENCE

132,875,550,720

66,437,775,360

33,218,887,680

16,609443,840

8,304,721,920

4,152,360,960

2,076,180,480

1,038,090,240 [9,720 SPLICED COMPOSITE SHARED]

519,045,120 [518,400 SPLICED COMPOSITE SHARED]

259,522,560 [25,920 Years EARTHS EQUINOX]

129,761,280 [129,600 RFEU]

64,880,640 [648-864 MIRROR NUMBER SUNS DIAMETER]

32,440,320 [324-432 MIRROR NUMBER SUNS RADIUS]

16,220,160 [162-216 MIRROR NUMBER MOON'S D 2,160]

8,110,080 [162-216 MIRROR NUMBER MOON'S R 1,080]

4,055,040 [045-540 MIRROR NUMBER MOONS R/2= 540]

2,027,520 [297 SHUFFLED COMPOSITE 7,920]

1,013,760 [297 SHUFFLED COMPOSITE 7,920]

506,880 [468 SHUFFLED COMPOSITE SUN'S DIA 864]

253,440 [297 SHUFFLED COMPOSITE 7,920]

126,720 [972 SHUFFLED COMPOSITE 7,920]

63,360 [036-360 MIRROR CIRCLE NUMBER]

31,680 [468 MIRROR SUN'S DIAMETER 864]

15,840 [684 COMPOSITE SUN'S DIAMETER 864]

7,920 EARTH'S DIAMETER IN MILES

3,456,000 /7,920 = 436.3636 1,728,000 /7,920 = 225 864,000 /7,920 = 109.0909 432,000 /7,920 = 54.5454

311,040 / 7,920 = 39.2727   311,040 x 7,920 = 2,463,436,800

518,400 / 7,920 = 65.4545   518,400 x 7,920 = 4,105,728,000

144,000 / 7,920 = 18.181818 x 3.333 = 60.6060 x 1.333 = 80.8080 / 24 = 3.367003367/ 60 = 0.0561167227

144,000 x 7,920 = 1,140,480,000 / 24 = 47,520,000 / 60 = 7,92,000 / 60 = 13,200 / 60 = 220 / 7.5 = 29.333

7.5 / 7,920= 0.00094696969 7,920 x7.5= 1,056 7,920 /266.66= 29.70 /7.5= 3.96 7,920 x 266.66= 2,111,100

3,960

1,980

990

495

247.5

123.75

61.875

30.9375

15.46875

7.734375

3.8671875

1.93359375

0.966796875

0.4833984375

0.24169921875

0.120849609375

0.0604248046875

0.03021240234375

ALL SIX 864,000 COMBINATIONS

864,000	846,000	684,000	648,000	486,000	468,000
56,623,104*	55,443,456*	44,826,624*	42,467,328*	31,850,496*	30,670,848*
28,311,552*	27,721,728*	22,413,312*	21,233,664*	15,925,248*	15,335,424*
14,155,776*	13,860,864*	11,206,656*	10,616,832*	7,962,624*	7,667,712*
7,077,888*	6,930,432*	5,603,328*	5,308,416*	3,981,312*	3,833,856*
3,538,944*	3,465,216*	2,801,664*	2,654,208*	1,990,656*	1,916,928*
1,769,472*	1,732,608*	1,400,832*	1,327,104*	995,328*	958,464*
884,736*	866,304*	700,416*	663,552*	497,664*	479,232*
442,368*	433,152*	350,208*	331,776*	248,832*	239,616*
221,184*	216,576*	175,104*	165,888*	124,416	119,808*
110,592*	108,288*	87,552*	82,944*	62,208*	59,904*
55,296*	54,144*	43,776*	41,472*	31,104*	29,952*
27,684*	27,072*	21,888*	20,736*	15,552*	14,976*
13,824*	13,536*	10,944*	10,368*	7,776*	7,488*
6,912,000	6,768,000	5,472,000	5184,000	3,888,000	3,744,000
3,456,000	3,384,000	2,736,000	2,592,000	1,944,000	1,872,000
1,728,000	1,692,000	1,368,000	1,296,000	972,000	936,000

864,000	846,000	684,000	648,000	486,000	468,000
432,000	423,000	342,000	324,000	243,000	234,000
216,000	211,500	171,000	162,000	121,500	117,000
108,000	105,000	85,500	81,000	60,750	58,500
54,000	52,875	42,750	40,500	30,375	29,250
27,000	26,437.5	21,375	20,250	15,187.5	14,625
13,500	13,218.75	10,687.5	10,125	7,593.75	7,312.5
6,750	6,609.375	5,343.75	5,062.5	3,796.875	3,656.25
3,375	3,304.6875	2,671.875	2,531.25	1,898.43	1,828.1
1,687.5	1,652.34375	1,335.9375	1,265.62	949.2187	914.06
843.75	826.171875	667.96875	632.812	474.6093	457.03
421.875	413.0859375	333.984375	316.406	237.3046	228.51
210.9375	206.54296875	166.99218	158.2031	118.6523	114.25
105.46875	103.271484375	83.496093	79.10156	59.32617	57.128
52.734375	51.6357421875	41.7480468	39.55078	29,66308	28.564
26.3671875	25.8187109375	20.8740234	19.77539	14.83154	14.282

864,000	864,000	864,000	864,000	864,000	864,000
-864,000	- 846,000	- 684,000	- 648,000	- 486,000	- 468,000
000,000	= 18,000	= 180,000	= 216,000	= 378,000	= 396,000

PLANET OCTAVE 864,000 SEQUENCE (POS)

452,984,832,000
226,492,416,000
113,246,208,000
56,623,104,000
28,311,552,000
14,155,776,000
7,077,888,000
3,538,944,000
1,769,472,000
884,736,000
442,368,000
221,184,000
110,592,000
55,296,000
27,684,000
13,824,000
6,912,000
3,456,000
1,728,000

864,000 SUN AETHER OCTAVES

8 / 6 / 4 = 0.333

3,456,000/ 864 = 4,000 1,728,000/ 864 = 2,000 864,000/ 864 = 1,000 432,000/ 864 = 500
311,040/864= 360 311,040x 864= 268,738,560 518,400/864= 600 5184x 864=4,478,976
144,000 / 864 = 166.666 x 1.333 = 222.222 / 16 = 13.888/22.5 = 0.61728/0.125 = 4.93827
144,000 x 864 = 124,416,000 /24 = 5,184,000/60 = 86,400 /60 = 1,440 /60 = 24 /7.5 = 3.2
864,000 x 7.5 = 6,480,000 x 7.5 = 48,600,000

432,000
216,000
108,000
54,000
27,000
13,500
6,750
3,375
1,687.5
843.75
421.875
210.9375
105.46875
52.734375
26.3671875
13.18359375
6.591796875
3.2958984375
1.6479492187

PLANET OCTAVE 846,000 SEQUENCE (POS)

1,774,190,592,000
887,095,296,000
443,547,648,000
221,773,824,000
110,886,912,000
55,443,456,000
27,721,728,000
13,860,864,000
6,930,432,000
3,465,216,000
1,732,608,000
866,304,000
433,152,000
216,576,000
108,288,000
54,144,000
27,072,000
13,536,000
6,768,000
3,384,000
1,692,000

846,000

8 / 4 / 6 = 0.333

3,456,000 / 846 = 4,085.1 1,728,000 / 846 = 2,042.5 864,000 / 846 = 1,021.2 432,000 / 846 = 510.6
311,040/846 = 367.65 311,040 x 846 = 263,139,840 518,400/846= 612.76 518,400x846= 438,566,400
144,000 / 846 = 170.212 x 1.333 = 226.95/ 24 = 9.456 /60 = 0.1576 /60 = 0.02626 /60 = 0.00043779
144,000 x 846 = 121,824 / 24 = 5,076,000 / 60 = 84,600 / 60 = 1,410 / 60 = 23.5
7.5 / 846 = 0.00886524822695   846 x 7.5 = 6,345

423,000
211,500
105,750
52,875
26,437.5
13,218.75
6,609.375
3,304.6875
1,652.34375
826.171875
413.0859375
206.54296875

PLANET OCTAVE 684,000 SEQUENCE (POS)

1,434,451,968,000
717,225,984,000
358,612,992,000
179,306,496,000
89,653,248,000
44,826,624,000
22,413,312,000
11,206,656,000
5,603,328,000
2,801,664,000
1,400,832,000
700,416,000
350,208,000
175,104,000
87,552,000
43,776,000
21,888,000
10,944,000
5,472,000
2,736,000
1,368,000

684,000

6 / 8 / 4 = 0.1875

3,456,000/ 684 = 5,052.63 1,728,000/ 684 = 2,526.315 864,000/ 684 = 1,263.1578 432,000/ 684 = 631.57
311,040/ 684 = 454.73684 311,040 x 684 = 212,751,360 518,400/684 = 757.89 518,400 x 684= 354,585,600
144,000 / 684 = 210.526 x 1.333 = 280.70 / 24 = 11.6959 /60 = 0.1949/60 = 0.0032488 /60 = 0.0000541477
144,000 x 684 = 98,496,000 /24 = 4,104,000 /60 = 68,400 /60 = 1,140 /60 = 19 / 7.5 = 2.5333
7.5 / 684 = 0.01096491 684 x 7.5 = 5,130

342,000
171,000
85,500
42,750 (EARTH 792)
21,375
10,687.5
5,343.75
2,671.875
1,335.9375
667.96875
333.984375
166.9921875

PLANET OCTAVE 648,000 SEQUENCE (POS)

1,358,954,496,000

679,477,248,000

339,738,624,000

169,869,312,000

84,934,656,000

42,467,328,000

21,233,664,000

10,616,832,000

5,308,416,000

2,654,208,000

1,327,104,000

663,552,000

331,776,000

165,888,000

82,944,000

41,472,000

20,736,000

10,368,000

5,184,000

2,592,000 (EARTHS EQ)

1,296,000 (RFEU)

648,000

6 / 4 / 8 = 0.1875

3,456,000 /648 = 5,333.333 1,728,000/648 = 2,666.666 864,000/648 = 1,333.333 432,000/648 = 666.666

311,040 / 648 = 480   311,040 x 648 = 201,553,920   518,400 / 648 = 800   518,400 x 648 = 335,923,200

144,000/648=222.222 x 1.333=296.296/24=12.34567901234568/60=0.2057/60=0.00342/60= 0.0000571

144,000 x 648 = 93,312,000 / 24 = 3,888,000 / 60 = 64,800/ 60 = 1,080 / 60 = 18 / 7.5 = 2.4

7.5 / 648 = 0.0115740740740   648 x 7.5 = 4,860

324,000

162,000

81,000

40,500

20,250

10,125

5,062.5

2,531.25

1,265.625

632.8125

316.40625

158.203125

PLANET OCTAVE 486,000 SEQUENCE (POS)

1,019,215,872,000
509,607,936,000
254,803,968,000
127,401,984,000
63,700,992,000
31,850,496,000
15,925,248,000
7,962,624,000
3,981,312,000
1,990,656,000
995,328,000
497,664,000
248,832,000
124,416,000
62,208,000
31,104,000
15,552,000
7,776,000
3,888,000
1,944,000
972,000 (EARTH)

486,000

4 / 8 / 6 = 0.08333

3,456,000/486 = 7,111.111   1,728,000/486 = 3,555.555 864,000/486= 1,77.777 432,000/486= 888.888
311,040/486 = 640 311,040 x 486 = 151,165,440 518,400/486= 1,066.666 518,400 x 486 = 251,942,400
144,000 / 486 = 296.296 x 1.333 = 395/ 24 = 16.46 /60 = 0.265 /60 = 0.0044 /60 = 0.0000737
144,000 x 486 = 69,984 / 24 = 2,916,000 /60= 48,600/60= 810 / 60 = 13.5 / 7.5 = 1.8
7.5 / 486= 0.0154320987654321   486 x 7.5 = 3,645

243,000
121,500
60,750
30,375
15,187.5
7,593.75
3,796.875
1,898.4375
949.21875
474.609375
237.3046875
118.65234375

PLANET OCTAVE 468,000 SEQUENCE (POS)

1,962,934,272,000
981,467,136,000
490,733,568,000
245,366,784,000
122,683,392,000
61,341,696,000
30,670,848,000
15,335,424,000
7,667,712,000
3,833,856,000
1,916,928,000
958,464,000
479,232,000
239,616,000
119,808,000
59,904,000
29,952,000
14,976,000
7,488,000
3,744,000
1,872,000
936,000

468,000

4 / 6 / 8 = 0.08333
3,456,000/468= 7,384.6 1,728,000/468= 3,692.30 864,000/468= 1,846.153846 432,000/468= 923.07692
311,040/468= 664.61 311,040 x 468= 145,566,720 518,400/468= 1,107.692 518,400x468 = 242,611,200
144,000 / 468 = 307.692 x 1.333 = 410.25610/24= 17.094017/60 = 0.2849 /60 = 0.004748 /60 = 0.00079
144,000 x 468 = 67,392,000 / 24 = 2,808,000/60 = 46,800 /60 = 780 /60 = 13 / 7.5 = 1.7333
7.5 / 468 =0.0160256410   468 x 7.5 = 3,510

234,000
117,000
58,500
29,250
14,625
7,312.5
3656.25
1828.125
914.0625
457.03125
228.515625

THE TORUS VAJRA AND ITS NATURE

DRAFT 266.666 29 – 8 - 2022

ABSTRACT

Nicola Tesla invented the 20th Century and only now in the 21st Century are the implications and possibilities of his work being rediscovered and realized. Tesla based his theories on the study of the properties of the Sun with its alternating and direct current helped substantially by William Crooks empirical observations Cathode ray tube results. Tesla believed the Sun to be a DC (anti – matter) to AC (matter) converter. The Sun by taking energy at rest DC (5D) and spinning it out creates a frequency AC (3D) which can be measured in (4D) time units hence quantifying AC and its vortex spin qualifies it's effects. As there are no straight lines in the Universe AC matter can only take a clockwise imploding vortex form or an anti – clockwise exploding vortex form. If AC looses its movement and is not spinning or resonating it reverts to its DC anti-matter form loosing 10 % of its original energy this also applies in the reverse scenario. The left hand clockwise imploding centripetal spin manifests as a negative ion charge and the right hand anti-clock wise explosive centrifugal spin manifests as a positive ion charge. Due to the concentrating of charge density implicit in the clockwise implosive vortex movement and form the "gravity" of the negative ions appear lighter. Conversely the reduction of charge density implicit in the dissipation of the anti - clockwise explosive vortex movement the "gravity" of the positive ion appear heavier.

So therefore the pure energy of direct current (DC) is subject to mechanisms that create from permanence at rest DC to permanence in motion AC and visa-versa. From the still DC to the dynamic vortex form pushing and pulling ions that creates frequency (AC) in doing so the Rod unit [R (Universal Constant)] was born. Due to the energy loss implicit in that push pull and vortex action (direct current DC anti-matter = 1) / (Alternating current matter .9 = 10% energy loss as a result of the push pull action creating vortex spin resonant Matter) that is the Rod unit. The Rod unit is the old universal constant 1.111 recurring. This Rod unit converts to matters resonance by multiples of 11.111. This unit I will refer to in this text as R, from the DC, from which the Sun transfers, creates and then transmits Resonant Frequency Energy Units (RFEU) AC, one unit of which is 1.111 the Rod (R).

The Molten Sea in the SE corner of the Temple Courtyard (at 135 Degrees) surrounding Solomon's Temple is a parable of the relationship between Dimensions, Aether, Light, Area / Time / Space, Degrees and Music, Matter, Energy and Colour is the first functional hot fusion device, based on Ancient Wisdom, recorded in modern times. As it is an inverted Sacred sphere (its surface area being Pi and radius 5 Cubits) creating a mono-polar effect (North and South poles opposite spins joined) at midday, when the Sun was directly overhead, because of the refractive index of water of 1.333 (16/12 the Matter to Aether ratio) and the half sphere being 66.666 percent full (Sun 864,000 / 66.666 x 10 = 129,600 RFEU) of water focuses the Suns rays below the surface tension creating plasma bubbles that upon their collapse form Torus singularity points. The volume of the full Molten Sea sphere is 6,000 Baths being 36,000 Hin or 432,000 Logs.

When you divide 36,000 Hin by the 16 Torus sectors you end up with 2,250 Hin per sector as there are 22.5 degrees per sector then 2,250 / 22.5 = 100 Hin per degree equals, 3600 seconds of arc in a degree, there are 36 seconds for each 1 Hin in volume.

The 150 Suns on the outside of the Molten Sea represents the 43,200 seconds (720 minutes) in a day and dividing 150 into 43,200 gives you 288 the first octave or diameter for the radius of 144 the speed of light (144,000 miles per hour) per arc second on the Earth. There are 21,600 minutes of arc on the Earth sphere, (the Earth 7,920 Miles Diameter), if you divide that by 7.5, (which is both the number of times light travels around the Earth in one second and the intervals of 7.5 degrees of arc N-S and E-W of the Earth), you get 2,880. That meaning a fundamental truth is exposed that is profound that the Earth is tuned to twice, that is the first octave, up of the speed of light per arc second.

The 150 Moons represent the same 43,200 seconds or 720 minutes in the night. Therefore the parable of the Torus being 720 degrees plus the center's right angle intersection of 180 degrees equals exactly 900 degrees (9 being the key number the zero's simply amplifications). Everything we perceive, and some things we can't, can be mapped on the 100 Pi imploded Sphere Torus. The Torus when divided into 16 sections of 22.5 degrees each demonstrates all manifestations and properties of both matter and anti matter. The manifestations of matter that are plotted are sound, elements, energy and color along with their Monad, Diad, Triad and Tetrad crystalline resonant properties. Also plotted are all the valence's and due to natural geometry and the Overlaid Right triangles all Area/Time /Space elemental resonances can be calculated. These Right Triangles centered on the zero point also define from the equatorial event horizon of the Torus the location of the elemental knops and nodes. These elemental knops and nodes when connected form a Parabolic Fibonacci curve on the 100 Pi imploded sphere surface. This joining of all the Elemental resonant knops and nodes is the way of simply locating and defining the position in time space on the 100 Pi surface of the exact location of any particular element. The implosive clockwise convergent creative constructive negative vortex spin and explosive anti-clockwise divergent destructive positive vortex spin and their paramagnetic and diamagnetic dispositions are mapped on the Torus.

The concilliant Sacred Imploded Sphere, with 100 Pi surface area, Torus law, defining ruling force and Pi scaffolding for both Energy and Matter is defined the Torus geometry and dimensions and it has 900 degrees internal angles for each of it's 8 core planes intersecting the singularity point this dictates that all 16 cardinal compass angles for the construction of form whether degrees or shapes or diameters or the radius of planetary spheres all add up to 9. Planetary spheres must be naturally pre - determined by the Torus Law and Spiral Alien Math governance as all perceptible manifestations in our reality are ruled by vibrations and their frequencies both resonate and conform with the sum of the internal angles of any form that dictate both its mass, boiling point, resonant frequency and form. The 8 harmonic cardinal Toroidal planes on which all concilliant Music, Elements, Energy, Colour and Light construct on are therefore bound by 9 and 16 and 1.111 by Law and Governance. Conversely any sound, matter or energy not conforming to the Toroidal law and governance will be dis - harmonious and destructive and self-extinguishing.

The addition of 9th places of even numbers and their mirror numbers (2 + 4 + 6 + 8 + 10 + 12 + 14 + 16 + 18 + 18 +16 + 14 + 12 + 10 + 8 + 6 + 4 + 2 = 180 (Octave of Light 144) [as in 180 degrees half of 360 degrees] equals 180 whereas the addition of the 9th place odd numbers and their mirror numbers (1 + 3 + 5 + 7 + 9 + 11 + 13 + 15 + 17 + 15 + 13 + 11 + 9 + 7 + 5 + 3 + 1 = 162) equals 162 (Octave of Sun and 266.666 Matter). When you divide the addition of the even numbers (180) by the addition of odd numbers (162) the result is 1.111 the Rod (R). This is because all resonant nodes, (manifesting to us as apparent matter), spiral into either constructive negative entropy left hand female implosive vortices (known as a contracting Negative charge) or destructive positive entropy right hand male explosive vortices (known as an expanding positive charge) as the negative exceeds the positive the universe expands.

As Aether is ruled by 12 (Gematria for Perfect Government) and Matter is ruled by 16 [360 Degrees / 16 = 22.5 Degrees], 16 / 12 = 1.333 giving us the conversion rate from DC Energy (Aether) to AC resonant manifest Matter of 1.333 (1.333 Radius [permanence in motion] of diameter 2.666 [permanence at rest]. 1.333 is also the width in Hands of the Molten Sea sphere wall in the Temple of Solomon. 266.666 (diameter of 2.666 x 100) is therefore the unreactive resonant key that rules all Matter that is why C in music is 266.666 (24 x 11.111) and the gap between C and C Sharp is also 266.666 (24 x 11.111). Musical A is 40 x 11.111 = 444 and Red in colour is 444 times one billion and Tungsten the Element with the highest boiling point (Phase change) of 5,555 Degrees Celsius is 5,555 = 500 x 11.111) and further 5,555 divided by 266.666.

Pythagoras taught us that all shapes resonate at the sum of their internal angles adjusted to certain variables, this fact is how in Ancient times all large blocks of stone were moved as when they are carved into a Tetrad (square) form, a fact ignored by all ignorant observers, they will resonate and then levitate at 2,160 vibrations per second [as an electric shaver will move over a hard surface]. 2,160 Degrees is the sum of the rectangles internal angles being 8 corners at 270 degrees per corner of a Tetrad block. This is how the large pyramid stones and Lebanon Baalbek stones were moved. For more efficiency a stone can be quarried at close to a natural resonant frequency of 266.666 (C in Pythagorean Music, 24 x 11.111) at a Specific gravity density of SG 2.666, which are the granites, dolerites and basaltic rocks such as in the Egyptian Pyramids Kings chamber, Bolivian and at Baalbek in Lebanon. Through Alchemy alkaline (Caustic soda as the reagent with salt as a catalyst, usually naturally occurring in Flamingo salt lakes, know by the Ancients at Natron) geo-polymer liquids can be created as in the Ancient Sanskrit texts designed to precipitate alkaline fluid rock with acids as solids with an inherent resonance to effect a response effectively amplifying the levitation in effect creating anti gravity resonant platforms.

As there are no straight lines in Nature, or indeed the Universe, therefore every thing is either spiraling to the left or right your thumbs palms down determine the direction of spiral being left hand clockwise (Female) and right hand anti-clockwise (Male). These directions naturally combine into a Fractal Torus at all levels and scales of the universe in every dimension as a product of Resonant Frequency except where they collide at the singularity point at the Centre of a Torus creating a DC Aether portal and therefore tapping the Aether. The emanations from this singularity point create the

manifestations that we observe as the Sun, unbalanced singularity point in motion (radius = r), or the planets, near balance singularity points at near rest (diameter = D). As the conversion of DC to AC basic units (ROD) is 1.111 hence Hydrogen which has the highest charge density of any element, detectable in our solar system and dimension, converts at a rate of 5 making 5.555 (DC Aether's conversion rate) when manifest as water, combining two Hydrogen with one Oxygen making 11.111 to 88.8999.
The origins of the 360 degrees of a compass are lost in time however the concept at the core of a compass is that 360 Degrees and Zero occur at the same point, everything and nothing, the first manifestation of the computer binary code 1 and 0.

100 x Pi EQUALS MOLTEN SEA SACRED TORUS IMPLODED SPHERE AREA

If you divide the 16 sector Solomon's Molten Sea, (22.5 degrees per sector), Sacred Torus imploded sphere surface area, which has a radius of 5 Cubits, the area is 314.159 square Pi Cubits (4 x Pi x r squared) into 100 sectors then 100 parts divided by 16 sectors = 6.25 (Pi Cubit units per 1 /16 sector). This number 6.25 can also be obtained by dividing the sum of the Sun circle's 12 angles (30 degrees per sector) which equals 2,340 degrees, (in 16 sectors angles total 3,060), by 144 which equals 16.25. Following Alien Vortex Math Laws (AVML) 16.25 is 6.25 the 1 is an order of magnitude not the defining product, 6.25 / .125 (octaves) = 50 which is 5 / 8 ratio of Pi Area-Time to Matter.

The surface area in square Hands of Solomon's Molten Sea Sacred Torus Sphere with a radius of 22.5 hands, 22.5 sector degrees and sector volume in Hin of 2,250 is calculated by 4 x Pi x (r squared) where r equals 22.5 therefore = 12.566370614359175158 x 506.25 = 6,361.7251235193324288. 6,361.7251235193324288/ 16 = 397.6078

6,361.7251235193324 / Pi = 2,025 / 16 = 126.5625 / 22.5 = 5.625 / .125 = 45

6,361.7251235193324 / 720 = 8.83557293382213

6,361.7251235193324 / 144 = 44.1786466911064752

6,361.7251235193324 / 11.111 = 576.5552611167464414

6,361.7251235193324 / 266.666 = 23.85646921319809302

6,361.7251235193324 / 360 = 17.67145867644259008

6,361.7251235193324 / 16 = 397.6078202199582768

6,361.7251235193324 / 7.5 = 848.23001646924432384

6,361.7251235193324 / 1.333 = 4,771.2938426396186039

6,361.7251235193324 / 50 = 127.23450247038664858

6,361.7251235193324288 / 16 = 397.607802199582768 / 22.5 = 17.6714586764426

17.67145867644259008 / 400 = 0.0441786466911064752 x 0.125 = 3534291735288518016

The sum total of all the cardinal angles in a 16 sector circle (3.060 Matter) divided by the sum total of all the cardinal angles in a 12 sector circle (2,340 Aether) is calculated by dividing 3,060 / 2,340 = 1.30769230769231.
1.30769230769231 / Pi = 0.4162513896249569587 [octave 6.66 and 266].
When reversed is 2,340 / 3,060 = 0.76470588235294 / Pi = 0.2434134423758398824.
As 129,000 (RFEU) is, 144,000 (LIGHT DC) / 1.111 (ROD), then Pi 3.1415 / 129,000 = 0.00002424068400554768 RFEU
and 2424 + 4242 = 666 and 684 + 486 = 1152 +2511 = 3663 and 3663 / 37 = 99
3663 / 36 = 101.75   3663 / 144 = 25.4375   3663 / 266.666 = 13.73625
2,340 / 37 = 63.24322432   2,340 /36 = 65   2,340 / 144 = 16.25   2,340 / 266.666 = 8.775

So the 1 / 16 th sector area is 6.25 square Pi Cubits that is 2.5 Pi Cubits x 2.5 Pi Cubits (Torus and Ephod). 0.00002424068400554768 x 6.25 = 0.000151502753466734
And 1515 + 5151 = 3663 the mirror Matter number as is the mirror Sun number for the sum of degrees in a 16-segment 360 degree Matter circle which is 3060, 3060 + 0603 = 3,663. The sum of the degrees in the 12 segment 360 degree Sun circle is 2,340 [(Suns radius number reversed as Matter (clockwise female thread consolidating force) spins opposite to Sun (anticlockwise male thread dissipating force)] so 234 + 432 = 666.
3,663 / 666 = 5 THE SUN TORUS + AETHER TO MATTER CONVERSION NUMBER.

In square Hands the 1 /16 th sector area of 6.25 square Pi Cubits equals 4 Pi r squared (11.25 Hands x 11.25 Hands) = 126.5625 Pi square Hands, 126.5625 square Hands / 6.25 Pi square Cubits = 202,500 (22.5 degrees) octaves up are // 405,000 // 810,000 // 1,620,000 (Moon 2,160)// 3,240,000 (Sun's radius 432,000)// 6,840,000 (Sun's Diameter 684,000) // 12,960,000 (Resonant Frequency Energy Units RFEU) // 25,920,000 (Earth,s equinox 25,920 years) // 51,840,000 (matter mirror number 518,400 seconds in 6 days x 60 = 31,104,000)

126.5625 / Pi 3.141592 = 40.286094970135999768
129,600 RFEU / Pi 3.1415 = 41,252.9612494193 x 6.25 = 257,831.0078039852

126.5625/ 6.25 = 20.25 octaves. Octaves up = 20.25 // 40.5// 81// 162 // 324 // 648 // 1,296 // 2,592 // 5,184 // 10,386 (Suns cube strings).

Pi squared = 9.8690440108936

Pi cubed = 31.0062766802998

Pi x 1,000,000,000 = 3,141,592,653.58979
3,141,592,653.58979 /24 / 60 / 60 = 36,361.026083215205926

* / 266.666 = 136.35384781206043107
* / 144 = 252.507125577883
* / 36 = 1,010.02850231153
* / 36.363636 = 999.9282172884181
* / 12 = 3,030.0855069346
* / 2.5 = 14,544.41043328608
* / 7.5 = 4,848.13681109536
* / 720 = 50.50142511557667
* / 16 = 2,272.56413020095 #
# / 22.5 = 101.0028502311533 / .125 = 808.0228018492266

THE 10 LAWS OF ALIEN VORTEX MATHEMATICS

LAW 1.)

LAW 2.)

LAW 3.)

LAW 4.)

LAW 5.)

LAW 6.)

LAW 7.)

LAW 8.)

LAW 9.)

LAW 10.)

GOVERNANCE OF RELATIONSHIPS BY THE BENDALL TRANSLATES

LANGUAGE

TORUS

POTENTIAL DIFFERENCE AND GRAVITY

THE SINGULARITY POINT

The 32 + 24 + 16 = 72 sides the Bendall Translater shape Aether to Matter converter

COMPASS CARDINAL MIRROR ANGLES

27 Letter in the Greek language down 1 - 9 = 45   10 – 90 = 450   100 – 900 = 4,500

27 Letter Greek language across 111, 222, 333, 444, 555, 666, 777, 888 and 999

MUSIC

11.111 x 24 = 266.666

11.111 x 40 = 444.444

To translate Music to Elements

To translate Music to Colour

To translate Music to Angles

ELEMENTS

Atom = 2 Hydrogen 11.111 / 2 = 5.555   Oxygen = 88.888 (11.111 x 8)

Molecule two Hydrogen plus Oxygen (water) =

ENERGY

864,000 / 11.111 = 77,760

3,456,000 (864,000 x 4) square / 11.111 = 311,040

644,972,540,000,000,000(864,000cubed)/ 31,104,000,000,000 Time = 20,736 / 144 = 144

COLOUR AND LIGHT

INTRODUCTION

DRAFT 518,400 MONDAY 29 – 08 -2022

CREATION IS AN IMPLOSIVE FORCE MANIFESTING FROM SOLOMON'S MOLTEN SEA TORUS VAJRA ORDER

"LET THERE BE LIGHT" AND THE SUN TORUS APPEARED (864,000 Miles in diameter) TO LIGHT THE DAYS CONVERTING DC ANTI-MATTER TO AC MATTER, CAME INTO BEING WHICH HAS SINCE SUSTAINED ALL LIFE ON EARTH (Earth 7,920 Miles in diameter). THE MOON (2,160 Miles in diameter [400 x 2,160 = 864,000]) WAS PLACED TO LIGHT THE NIGHTS WITH POLARIZED LIGHT REVERSING THE SPIN BY REFLECTION FROM POSITIVE IN TO NEGATIVE OUT. THE SUNS DIAMETER IS 864,000 MILES. EARTH DAYS WERE GENERATED DUE TO THE RATE OF ITS ROTATION AND AS A PRODUCT OF THE SUN, WHICH ADVANCES OVER THE FACE OF THE EARTH AT 582-MILES PER HOUR, 5,820 BEING THE NUMBER OF FEET IN A MILE. IF YOU DIVIDE THE SUNS DIAMETER, IN MILES, BY A DAY IT EQUALS 864,000 / 24 HOURS = 36,000 / 60 MINUTES = 600 / 60 SECONDS = 10 / 7.5 = 1.333 / 1.333 = 0 WHICH IS 1 AND 0, EVERYTHING AND NOTHING, 360 DEGREES AND 0 DEGREES THE PARADOX OF CREATION. 10 SECONDS / 60 ARC SECONDS = 0.1666 // + 0.333 // 0.666 // 1.333 // 266.666 (24 x 11.111 = 266.666) THE MILE IS USED, AS IT IS 660 FURLONGS, 7,920 INCHES, and THE EARTH'S DIAMETER IN MILES) AS 5,280 FEET EQUATES TO 528 MILES AN HOUR THE SUNS SHADOW ADVANCES ON THE PLANET.

THE EARTH WAS CREATED IN SIX DAYS AND ON THE SEVENTH DAY GOD RESTED.

6 DAYS TIMES 24 HOURS = 144 (1/1,000 the Speed of LIGHT per arc second) TIMES 60 MINUTES = 8,640 [Sun's diameter 864,000 / 100] TIMES 60 SECONDS = 518,400 [Key resonant MATTER number 129,600 (square) x 4 = 518,400 and 518,400 / 108 = 4,800] 518,400 x 60 ARC SECONDS = 31,104,000 (Sanskrit 100 long years number 6 days in seconds x 60 = 360)

(518,400 / 54 = 9,600 / 22.5 = 426.666 / 16 = 26.666 x 7.5 x 1.333 = 266.666)
(518,400 = 24 Hours x 60 Minutes x 60 Seconds x 6 Days of Creation and *518,400* / 60 = 8,643.3)
(518,400 / 2 = *259,200* / 2 = *129,600* (*RFEU*) / 2 = *64,800* (Sun's Diameter Time shifted / 10) / 2 = *32,400* (Sun's Radius Time shifted / 10) / 2 = *16,200* (Moon's diameter Time shifted) / 2 = 8,100 (Earth D + 180) / 2 = 4,050 / 2 = 2,025 // 1,012.5

518,400 / 7.5 = 69,120 (RFEU 129,600) / 2 = 34,560 (Sun's Square) / 2 = 17,280 / 2 = 8,640 (Sun D) / 2 = 4,320 (Sun R / 100) / 2 = 2,160 (Moon D)/ 2 = 1,080 (Sun D + Earth D) /2 = 540 // 270 // 135 // 67.5 // 33.75 // 16.875 // 8.4375 // 4.21875 // 2.109375 // 1.0546875

LIGHT (144,000 MILES PER ARC SECOND) SUSTAINS ALL LIFE ON EARTH, LIGHT COMES FROM THE SUN, THE SUN IS 93,000,000 MILES AWAY AND 864,000 MILES IN DIAMETER, (at that point its surface temperature is 6,000 Degrees Centigrade) AND CREATES DAYS ON THE EARTH, (86,400 [SUN'S D] SECONDS = 1 DAY = 1,440 MINUTES) DUE TO THE EARTHS SPEED OF ROTATION, ONE DAY IS 24 HOURS, LIGHT TAKES ONE SECOND TO TRAVEL 7.5 TIMES AROUND THE EARTH GIVING A STANDARD CONVERSION OF BOTH ENERGY AND DIMENSIONAL SHIFTS. DIAMETERS = DC [DIRECT CURRENT] PERMANENCE AT REST, THE RADIUS = AC [ALTERNATING CURRENT] PERMANENCE IN MOTION, THE RADIUS BEING FIRST OCTAVE RESONANCE OF THE DIAMETER.

IF YOU TAKE 360 DAYS IN THE SACRED YEAR x 24 HOURS x 60 MINUTES x 60 SECONDS = 31,104,000 THE SANDSKRIT LONG YEAR / 100,000. 31,104,000 / 144,000 x 7.5 x 1.333 = 2,160 (Moon's diameter, a Cubes sum of its internal angles and 21,600 is the sum of arc minutes around the Earth. 31,104,000 x 60 ARC SECONDS = 1,866,240,000 1,866,240,000 / 864,000 = 2,160 / 7.5 = 288 / 1.333 = 216 / 11.111 = 19.44 (POS)

31,104,000 (36 x SUN 864,000) // 15,552,000 (18 x SUN 864,000) // 7,776,000 (9 x SUN 864,000) // 3,888,000 // 1,944,000 // 972,000 (EARTH D) // 486,000 (SUN D) // 243,000 (SUN R) // 121,500 // 60,750 // 30,375 // 15,187.5 // 7,593.75.

31,104,000 / 7.5 (TIMES LIGHT TRAVEL AROUND THE EARTH IN ONE SECOND) = 4,147,200 // 2,073,600 // 1,036,800 // 518,400 (6 DAYS OF CREATION) // 259,200 (Earths Equinox) // 129,600 (RFEU) // 64,800 (SUN D) // 32,400 (SUN R) // 16,200 (MOON D) // 8,100 // 4,050 // 2,025 // 1,012.5 // 506.25 // 253.125 // 126.5625 (Pi Hands area of a Sacred Torus) // 63.28125 // 31.640625 // 15.8203125 // 7.91015625 // 3.955078125.

31,104,000 / 108 = 288,000 (2 TIMES 144,000)

IF YOU TAKE THE 6 DAYS OF CREATION INTO SECONDS IT IS 6 DAYS x 24 HOURS = 144 x 60 MINUTES = 8,640 (SUNS DIAMETER/100) x 60 SECONDS = 518,400 SECONDS x 60 ARC SECONDS = 31,104,000

518,400// 259,200 (EARTHS EQU 25,920) // 129,600 (RFEU 129,600) // 64,800 (SUN D) // 32,400 (SUN R) // 16,200 (MOON D) // 8,100// 4,050// 2,025// 1,012.5// 506.25// 253.125// 126.5625 (Pi Hands area of a Sacred Torus) // 63.28125// 31.640625// 15.8203125.

518,400/ 7.5 = 69,120.69,120 (RFEU 129,600) // 34,560 (SUN SQUARE) // 17,280 // 8,640 (SUN D) // 4,320 (SUN R) // 2,160 (MOON D) // 1,080 (MOON R) // 540 // 270 // 135 // 67.5 // 33.75 //16.875

IF YOU THEN TAKE THE 6 DAYS OF CREATION PLUS THE 1 DAY OF REST = 7 DAYS 7 x 24 HOURS = 168 x 60 MINUTES = 10,080 x 60 SECONDS = 604,800 SEC 604,800 x 60 ARC SECONDS = 36,288,000 (COMPOSITE NUMBER 360 DEGREES AND 288 LIGHT)

36,288,000/864,000 = 42   36,288,000/266.666 = 136,080   36,288,000/11.111 = 3,265,920

604,800 / 7.5 = 80,640   80,640 x 60 arc degrees = 4,838,400
80,640 / 2 = 40,320//20,160//10,080//5,040//2,520//1,260//630//315

604,800/ 1.111 = 544,320 x 1.333 = 725,760 x 11.111 = 8,064,000 x 1.111 = 896,000 x 1.111 = 8,960,000 x 1.111 = 9,955,555.555 / 266.666 = 37,333.333

604,800 // 302,400//151,200//75,600//37,800//18,900//9,450//4,725//2,362.5

604,800 / 7.5 = 80,640// 40,320//20,160//10,080//5,040//2,520//1,260//630//315

THE SUN IS DIRECT CURRET AETHER (DC = 1) WHEN THAT IS CONVERTED INTO ALTERNATING CURRENT MATTER (AC = .9) IT SUFFERS A 10 PERCENT LOSS, WHICH EQUATES TO 1 / .9 = 1.111 RECURING. WHEN THIS NUMBER 1.111 IS THEN APPLIED TO 864,000 (SUNS DIAMETER) / 6 OR 432,000 (SUNS RADIUS) / 3 THE PRODUCT IS 144,000.

SOLOMON'S MOLTEN SEA PARABLE OF AETHER AND THE CREATION OF MATTER

144,000 is also the result of multiplying 2,000 Baths (Volume of water in the Solomon's Temple Molten Sea half inverted Sphere, mono-polar half sphere) by 72 to get the volume in logs which is 144,000 Logs. 144,000 Logs / 1.111 (R) = 129,600 RFEU (Resonant Frequency Energy Units). The volume in logs, is also equal to the volume in ritual Baths is 150 (1 Bath = 72 Logs) of the Molten Sea. The Ephod, the Ark of the Covenant, Temple and Molten Sea all the dimensions are the keys to the Torus and Singularity point. A Torus that manifests the appearance of Matter is described with a constant H, V and G but with a variable, octave node spacing and vortex spin direction, charge density, time and light speed. Nicola Tesla identified, from the Bible and Sanskrit texts, and used, the Torus as a magnifying transmitter among other things and used alternating current (AC) to evoke sympathetic resonant frequency responses on knops at nodes to effect. Tesla collaborated with Sir William Crookes who introduced him to the concept of a spiral model of the elements based on his cathode ray tube work. By simply converting cubits to hands and with hands divided by 0.125 all vectors and dimensions become translated to a common language. As was said of the tree of knowledge it's meaning was hidden by him by using different languages, in this paper that which was hidden is now manifest. This unraveling also reveals the real root of Pi and Phi, which are demonstrably also based on 9 and Pi simply is the imploded Sacred Torus surface that the Phi curve connecting all the Music, Element, Energy and Colour.

An imploding vortex design based on the Ezekiel Chapter 1 description has been constructed using the information in this text to use nodes to effect to tap the Aether and reduce elements to produce above unity energy in our dimension. An Implosive Vortex Torus (IVT) model has been constructed which describes the relationships between all things it is the new wheel. This invokes ancient knowledge of implosive force, to replace the blundering in the dark of so called "modern" destructive primitive technologies based on a child's primary school understanding, leading us now into the age of enlightenment and a new industrial revolution without the axil and inevitable Centrifugal forces and toxic destructive forces of the old wheel.

Key sacred numbers common to the Tibetan Sanskrit, Indians, Chinese, Tibetans, Sumerians, Babylonians, Egyptians, Israelites, Mayans, Sanskrit Indians and lastly Greeks were all based on diameters in Miles of the Sun, which were known to them from sources undisclosed, Sun (864,000), Moon (2,160) and Earth (7,620). The numbers 0.125, 0.886, 1.111, 1.333, 7.5, 11.111, 108, 144, 432, [3, 6 and 9 made famous by Nicola Tesla], 12, 18, 24, 27, 36, 37, 45, 50, 48, 70, 150, 180, 360, 16, 22.5, 150, 266.666, 300, 400, 450, 720, 1,260, 1,440, 1,776, 8,640, 129,600, 144,000, 194,400, 216,000, 1,728,000, 518,400 and 311,040,000,000,000. The sequence 111 to 999 along with 1.111 to 9.999 relate to all Aether, Dimensions, Light, Matter, Music, Area / Time/ Volume, Angles and Energy. The Gematria for the Tower of Babel is 111 where 72 languages were spoken and confusion and scattering dispersed the truth this paper recombines the dispersed threads of a riven ancient tapestry of Sacred once revered knowledge. The ancient Sanskrit texts tell us that 100 long years is equal to 311,040,000,000,000 it also states that "when the cause of all causes becomes known then everything knowable becomes known and nothing remains unknown." this is also known as the Key of David and the Wisdom of Solomon, harmonics and frequencies understood and used to effect, "As above so below".
311,040,000,000,000 ALL TIME / 25,920 YEARS SOLAR EQUINOX = 12,000,000,000

311,040,000,000,000 ALL TIME / 144,000 LIGHT = 2,160,000,000 / 518,400 = 416.666
311,040,000,000,000 ALL TIME / 266.666 = 11,664,000,000,000 / 720
16,200,000,000 / 16 = 1,012,500,000 / 129,600 = 7,812.5 x 400 = 3,125,000 / 22.5 = 1,111,111.111
311,040,000,000,000 / 129,600 = 2,400,000,000 311,040,000,000,000 / 518,400 = 600,000,000
600,000,000 / 129,600 = 4,629.6296296296 / 266.666 = 17.36111 // 34.7222 / 69.444 / 138.888 /
277.777 / 555.555 / 1,111.111 / 2,222.222 / 4,444.444 / 8,888.888

Edgar Casey gave a reading on 11:11 "the first lesson for six months should be one, one one, one; oneness of God, oneness of mans relation, oneness of force, oneness of time, oneness of purpose, oneness in every effort oneness – oneness.

Reflective numbers are the key as seen in a compass dial the numbers from 0 to 180 reflective or mirror the numbers from 180 to 360. An example from the Sun's radius of 432,000 is such a number as 432 plus its mirror 234 equals 666. The speed of the planet earth around the Sun is 66,600 miles per hour. So 66,600 mph / 7.5 = 8,880 / 24 Hours = 370 370 / 60 Minutes = 60 60 / 60 Seconds = 1 1 / 60 ARC SECONDS = 0.1666 0.1666 // + 0.333 // 0.666 // 0.1333 // 0.2666 // 0.5333 // 1.0666 // 2.1333 // 4.2666 // 8.5333 // 17.0666 // 34.1333 // 68.2666 // 136.5333 // 273.0666 // 546.1333 Dimensional numbers based on 1 / .9 = 1.111 (AC) of 1.111 (1D), 2.222 (2D flat land), 3.333 (3D (AC = alternating current) Matter), 4.444 (4 D (AC) Time) and 5.555 (5D (DC = direct current) = Aether) 6.666 (Rod energy, the speed the earth moves in its orbit 66,600 miles per hour), 7.777 (Reed measurement), 8.888 (Staff), 9.999 are used to describe the base for Aether to Matter and Alchemy Matter to Matter Fusion Energy. Aethereal Pythagorean Music is based on the number 266.666 (C) and the C to C# span which is 266.666. The root numbers 24 (C), 27 (D), 30 (E), 32 (F), 36 (G), 40 (A), 44 (B), and 48 (C#) for notes are manifest by multiplying the root numbers by 11.111 creating musical sound, ultrasonic sound, elements, energy and colour all report to angles and both generate shapes and also resonate in shapes both geometric and crystalline to focus Energy.

108

108 Degrees [Earth (792) plus Moon (216) = 1,008] 1080 x 2 =2,160(Moon)

DIVIDED LINE PYTHAGORAS TRIANGLE

36 Degrees   36 Degrees

Plato's Divided Line of Pythagoras puts Phi, 1, 1, 1 / Phi on the bottom of a triangle that has 36 degrees 36 degrees and top angle of 108 degrees therefore the 108 is the internal octave angle defining within a circle a pentagon.

Phi 1.618 = 16 and 180 is part of a double octave sequence shown by the following 3.2360 = 32 octave and 360 Degrees / Phi = 1.6180 = 16 octave and 180 Degrees // .809 = 8 octave and 90 Degrees // .4045 = 4 and 45 degrees // .20225 = 2 Octaves and 22.5 degrees // .1011.25 Therefore giving the Octave series start at 2 of 22.5 degrees one 16th of 360.

KHUFU PYRAMID GIZA EGYPT
HEIGHT 486.777 FEET IS 5,841.324 INCHES

4 x 8 x 6 = 192 x 192 = 36,864 7 x 7 x 7 = 343 x 343 = 117,649   36,864 x 117,649 = 4,337,012,736
4,337,012,736 / 24 Hrs / 60 Min / 60 Sec / 60 Arc Sec = 836.615   836.615 / 7.5 = 111.54868148148

76.32 Degrees (792 Earths Diameter)

Phi   Phi

51.84 Degrees = 51min 50 sec 40   51 min 50 sec 40 = 51.84 Degrees

Two times 51.84 = 103.68 (864)

This triangle has a face angle of 51.84° (51°50'40) of the Khufu Giza pyramid.
Phi is the side of the Egyptian pyramids with 1 + 1 base with 1 unit height it has 51
Degrees and 51 minutes as the base angles. The right angle Triangle defined can
match the above triangles base of Phi and therefore crystal form as internal octave
angles create a point on the circle defining pie. 51.84 x 3 = 1,555.2, which is the first,
Octave of down from 3,110.4 and is in the planet and Sun diameter series.
864 cubed = 644,972,544
644,972,544 / 108 = 5,971,968 / 36 = 165,888 / 36 = 4,608
644,972,544 / 90 = 7,166,361.6 / 36.9 = 194,210.3414634146
194,210.3414634146 / 53.1 = 3,657.445225299711

644,972,544 / 24 = 26,873,856 26,873,856 / 60 = 447,897.6 / 60 = 7,464.96 /60 = 124.416
124.416 // 62.208 // 31.104 // 15.552 // 7.776 // 3.888 // 1.944 // 0.972 // 0.486 // 0.243 //
0.1215 // 0.06075 // 0.030375 // 0.0151875

7,464.96 / 7.5 = 995.328
995.328 // 497.664 // 248.832 // 124.416 // 62.208 // 31.104 // 15.552 // 7.776 // 3.888 //
1.944 // 0.972 // 0.486 // 0.243 // 0.1215

864 cubed = 644,972,544
644,972,544 / 24 / 60 / 60 / 7.5 = 995.328 / 720 = 1.3824 / 16 = 0.0864 / 22.5 = 0.00384 x
400 = 1.536 / 0.125 = 1.2288 x 6.25 = 76.8 / 11.111 = 6.912 (RFEU)

Then Taking 311,040 cubed = 3.0091839012864e+16 and dividing it by 108 / 36 / 36
then by 24 hours / 60 min / 60 sec / .75 / 720 / 16 = 288 (144 x 2)
288 / 22.5 = 12.8 / .125 = 102.4 x 400 =

Then taking 311,040 cubed which is equal to 3.0091839012864e+16
and then dividing it by 90 / 36.9 / 35.1 = 257,989,017,600
then by 24 / 60 / 60 / .75 = 3,981,312
257,989,017,600 / 720 / 16 / 22.5 = 995,328 / 288 = 3,456
3,456 [NOTE - 3 x 4 x 5 x 6 = (360) x 6 x 5 x 4 x 3 = 129,600]
3,456 / 11.111 = 311.040
31.1040 x 400 x 6.25 = 7,776
31.1040 / 6.25 = 1,990.656

Taking the Phi Triangle divided line angles of 36 / 36 / 108
518,400 / 36 / 36 / 108 = 3.7037037

Taking the Sacred Right Angle Triangle angles of 90 / 36.9 / 53.1
518,400 / 90 / 36.9 / 53.1 = 2.939690413853291
311,040 / 90 = 3,456 (3,456 / 11.111 = 311.04) 3,456 / 36.9 = 93.6585366
93.6585366 / 53.1 = 1.763814248311975 / 7.5 = 0.235175233108263

25,920 / 90 = 288 / 36.9 = 7.8 / 53.1 x 7.5 = 1.102383905194984

311,040 x 155,520 = 48,372,940,800 / 518,400 = 93,312

93,312 / 144 = 648 (SUN'S DIAMETER) 518,400 x 25,920 / 31,104 = 432,000 (SUN'S RADIUS)
This is important the octameter that the two octave series multiplied together divided by time 311,040 **equals the area equaling the Suns radius.**

864,000 x 432,000 = 373,248,000,000 / 311,040 (Time) = 1,200,000
1,200,000 / 266.666 = 4,500 / 11.111 = 405.0 **of the Planetary Octave series.**
864,000 x 864,000 = 746,496,000,000 / 311,040 (Time)= 2,400,000
2,400,000 / 11.111 = 216,000 **Arc seconds of the Earth.**

51.840 = .144 **of a turn**
51.840 (6 days = 518,400 seconds) Degrees = 51 Degrees 51 Minutes **pyramid angle.**

To understand the significance of the 108 temples of Anchor Wat one can simply multiply the Suns and Moons diameters by 108 and that respectively will be the distances to the planet Earth.

Sun's **diameter is** 864,000 **miles x** 108 = 93,312,000 (7,920 **D Earth) which is the distance to the** Earth.
93,312,000 / 24 = 3,888,000 / 60 = 64,800 / 60 = 1,080 / 60 **arc sec = 18**
93,312,000 / 144,000 = 684   93,312,000 / 24 hr / 60 min / 60 sec = 18 x 16 = 288
288 x 22.5 = 6,480 x 400 = 900 [Degrees in one Plasmoid sector] x .125 = 11.25

311,040,000,000,000 x 155,520 (864 x 18 **and** 108 x 144) = 48,372,940,800,000,000,000
/ 518,400 = 93,312,000,000,000 / 144,000 = 648 [Sun 864 **shuffle**]

(9[331]2 = Earth 7,920 **miles diameter)**
[93,312,000 / 144,000 = 684]

Moon's **diameter is** 2,160 **miles x** 108 = 233,280 **which is the distance from the Moon to the Earth.**
233,280 / 144 = 1,620   233,280 / 24 = 9,720 **E /** 60 = 162 **M /** 60 = 2.7 / 60 **arc sec = 0.045**

Mars 4,222 **miles D x 108 =** 455,976 / 24 = 18,999 / 60 = 316.65 / 60 = 5.2775 / 60 **arc sec = 0.087958333**

Jupiter 8,884 **miles D x 108 =** 959,475 / 24 = 39,978.125 / 60 = 666.30208333 / 60 = 11.1050347222 / 60 **arc sec = 0.185083912037037**

108 x 144 = 15,552 (SUN'S 864 x 18)

108 x 2 = 216 (MOON'S DIAMETER 2,160)

108 x 8 = 864 (SUN'S DIAMETER 864,000)

108 x 9 = 972 (972 x 7.5 = 729) (EARTH'S DIAMETER 7,920)

108 x 12 = 1,296 (RFEU 129,600)

108 x 16 = 1,728 (SUN'S 864 x 2)

108 x 20 = 2,160 (MOON'S DIAMETER 2,160)

108 x 24 = 2,592 (EARTH EQUINOX 25,920)= C in Music   3 x 8 = 24 x 11.111 = 266.66 C

108 x 27 = 2,916 (RFEU 129,600)   = D in Music   3 x 9 = 27 x 11.111 = 300 D

108 x 28 = 3,024 (SUN'S RADIUS 432,000)

108 x 30 = 3,240 (SUN'S RADIUS 432,000) = E in Music 3 x 10 = 30 x 11.111 = 333.333 E

108 x 32 = 3,456 (SUN SQUARE)   = F in Music   4 x 8 = 32 x 11.111 = 355.555 F

108 x 36 = 3,888 POS   = G in Music   4 x 9 = 36 x 11.111 = 400 G

108 x 37 = 3,996

108 x 38 = 4,104

108 x 40 = 4,320 (SUN'S RADIUS 432,000) = A in Music   4 x 10 = 40 x 11.111 = 444.444 A

108 x 44 = 4,752   = B in Music   4 x 11 = 44 x 11.111 = 488.888 B

108 x 48 = 5,184 (SIX DAYS IN SECONDS) = C# in Music 4 x 12 = 48 x 11.111 = 533.333 C
108 x 50 = 5,400 (PLANET OCTAVE SEQUENCE POS)

108 x 80 = 8,640 (SUN'S DIAMETER 864,000)

108 x 90 = 9,720 (EARTH'S DIAMETER 7,920)

108 x 72 = 7,776 (PLANET OCTAVE SEQUENCE POS)

108 x 108 = 11,664   11,664 / 3,110,400 = 0.00375 and 3,110,400 / 11,664 = 266.666

108 x 111 = 11,988

108 x 123 = 13,284

108 x 124 = 13,392

108 x 144 = 15,552

108 x 222 = 23,976

108 x 266.666 = 28,800...[LIGHT SPEED DIAMETER NUMBER 144,000 x 2 = 288,000]

108 x 288 = 31,104...[SANSKRIT 100 YEARS 311,040 OCTAVES]

108 x 216 = 23,328

108 x 432 = 46,656

108 x 864 = 93,312

108 x 888 = 95,904 (95,904 / 864 = 111) (95,904 / 144 = 666) (95,904 / 1.333 = 71,928) (95,904 / 11.111 = 8,631.36)

108 x 108 = 11,664   108 x 108 x 108 = 1,259,712   108 x 108 x 108 x 108 = 136,048,896

864 x 864 x 864 = 644,972,544
644,972,544 / 1,259,712 [108 cubed]= 512 *512 is the ninth octave of one and the root of Pi*
644,972,544 / 9 = 71,663,616
644,972,544 x 1,259,712 = 812,479,653,347,328

31,104 x 31,104 x 31,104 = 30,091,839,012,864 (3114 cubed)
3,114 cubed / 108 cubed = 23,887,872
30,091,839,012,864 (31,104 Cubed) / 864 Cubed = 46,656

864 cube x 108 cube / 31,104 Cubed = 27   1/x = 0.037037037
812,479,653,347,328
108 squared = 11,664
864 squared = 746,496

31,104 cube / 108 square = 2,579,890,176
31,104 cube / 864 square = 40,310,784
108 x 108 x 108 = 1,259,712
   1,259,712 / 972 = 1,296 RFEU 129,600)
   1,259,712 / 216 = 5,832
   1,259,712 / 432 = 2,916 (RFEU 129,600)
   1,259,712 / 864 = 1,458
   1,259,712 / 5,184 = 243 (SUN'S RADIUS 432,000)
   1,259,712 / 3,456 = 364.5 SUN SQUARE)
   1,259,712 / 31,104 = 4050
The Torus Planes internal angles = 900 Degrees x 8 planes = 7,200 / 108 = 66.666
[The Molten Sea was 66.666 % full and 864,000 / 66.666 x 10 = 129,600 RFEU]

108 x 216   = 23,328

108 x 432   = 46,656

108 x 864   = 93,312

108 x 31,1040   = 33,592,320

108 x 266.666   = 28,800 / 2 = 14,400

108 / 266.666   = 0.4050 POS

108 x 1.111   = 120 GOVERNMENT

108 / 1.111   = 97.2 EARTH

108 x 11.111   = 1200

108 / 11.111   = 9.72 EARTH

108 x 1.333   = 144 LIGHT

108 / 1.333   = 81 POS

108 / 1.23456   = 87.4805598755832

108 x 1.23456 = 133.33248

108 x 1.23456654321 = 133.3330666668

108 / 1.29600   = 83.333

108 / 1.44   = 75 Time Light T

108 / 2.5920 = 41.666 POS

108 / 3.11040 = 34.7222 POS

108 / 1.616161 = 66.825 / .125 =

108 / 12.1212 = 8.910000006682502

108 / 11.111 = 97.2   EARTH

108 / 9.999 = 10.80 (EARTH + MOON) 10.80 / 0.125 = 86.4 / 0.125 = 691.2 / 0.125 = 5,5296

108 / 8.888 = 12.15 / 0.125 = 97.2   97.2 x 400 = 38,880   38,880 / 720 = 54

108 / 7.777 = 13.88571428571429 / 0.125 = 111.0857142857143

108 x 7.777 = 840 / 0.125 = 6,720 / 16 = 420 = 18.666   840 / 5.555 = 151.2

108 / 6.666 = 16.20 (MOON'S DIAMETER 2,160)

108 / 5.555 = 19.44 (POS)

108 / 4.444 = 24.3   (SUN'S RADIUS 432,000)

108 / 3.333 = 32.4   (SUN'S RADIUS 432,000)

108 / 2.222 = 48.6   (SUN'S DIAMETER 864,000)

108 / 1.111 = 97.2   (EARTH'S DIAMETER 7,920)

108 / 5.18400 = 68.18181818181818

108 / 5.4 = 20

108 x 7.5 = 810

108 / 7.5 = 14.4

108 / 1.6 = 67.5

108 / 2.25 = 48

108 / 1.125 = 96

108 / 6.26 = 17.28 [SUN'S DIAMETER 864,000 x 2 = 1,728,000]

518,400 / 108 = 4,800 / 108 = 44.444

518,400 x 108 = 55,987,200 / 144,000 = 388.8
          24.3 // 48.6 // 97.2 // 194.4 // - 388.8 // + 777.6 // 1,555.2 // 3,110.4

518,400 / 6.25 = 82,944 x 82,944 x 4 x Pi equals 1,042,305.044237407
51.840 Degrees on a circle is 311.040 minutes of Arc 186.624 seconds of Arc speed of light at
360 days it is also 864,000 squared the Area / Time / Volume unit (ATV).

518,400 Degrees on a circle is 31,104,000 minutes of Arc
360 Degrees times 3,600 arc seconds in a degree = 1,296,000 arc seconds in 360 Degrees.

129,600 x 129,600 x 4 x Pi = 2.1106677145807e+11

518,400 meters per second equals 1,866,240 kilometers per hour

518,400 kilometers per hour equals 144,000 meters per second

518,400 x 518,400 x 4 x Pi = 3.3770683433292e+12 Area of sphere divided by 16 equals
2.1106677145807e+11 = 129,600 x 129,600 x 4 x Pi = 2.1106677145807e+11
518,400 / 2 = 259,200 x 259,200 x 4 x Pi = 846 x 10 to 11th so the area of the Earth's Equinox
time in years is the Suns Diameter In miles shifted.
25,920 / 4 = 6,480 / 10 = 648 x 648 x 648 x 648 [4 Ages] = 512,249,392,656
144,000 degrees is 400 turns.
There are 144 degrees being 3,600 a Degree in 518,400

144,000 Degrees is 518,400,000 Seconds of arc / 3,600 = 144,000

144,000 Degrees is 8,640,000 Minutes of arc.

144,000 x 144,000 x 4 Pi = surface area of the Light Sphere = 2.61 x 10 to 11th.

864,000 Degrees equals Seconds of Arc 3,110,400,000 Divided by 144,000 = Hours Arc 21,600

864,000 (Degrees equals Minutes of arc) 51,840,000 divided by 360 = 144,000

SECTION A

VAJRA AETHER TO MATTER CONVERSION

THE MOLTEN SEA AND EPHOD PARABLES OF AETHER TO MATTER BY TORUS VAJRA

The Molten Sea vectors and dimensions also go to the digital root of 9 when reduced from the unit of length in the bible of a cubit to that of the hand. Then to go into the atomic level, one must then ensure that, all dimensions are divided by one eighth or 0.125. Hence 45 hands divided by 0.125 = 360 hands and 22.5 divided by 0.125 = 180, thus effectively normalizing Time, Degrees and Dimensions along with Aether units with materialized elemental units through the shell (16 / 12 = 1.333 Hands is 1 / 7.5 times 10) of the Molten Sea. The number 7.5 being how many times the speed of Light travels around the earth in a second, creating a conversion factor from Light energy to electron energy. This is important for the translation of light to matter in that the photons in water slow down to equal to ionic speed of the electrons.

The inverted inside sphere has a diameter of 10 cubits that is converted from Cubits to Hands by a factor of 4.5 to 45 Hands (4 + 5 = 9) and radius of 22.5 Hands (2 + 2 + 5 = 9) equals the 22.5 degrees arcs achieving 360 degrees in 16 (1 + 6 = 7) segments in quadrature (90 degrees to each other) with the outer shell of the half sphere which is 50 Pi in area, which is 1.333 [1.333 = 1, 4 / 3, 4 + 3 = 7] Hands thick. As the inner, inverted sphere - creation, is ruled by 16, the outer, with the 12 bulls (12 BULL'S TAILS, 12 HEADS, 24 HORNS, 24 EYES, 24 BALLS - left and right spirals, 48 LEGS, 32 x 12 TEETH = 384 TEETH / 16 = 24, 12 SUN DISKS BETWEEN THE BULL'S HORNS = 12 x 864,000 = 10,368,000 / 144,000 = 72, 10,368,000 / 266.666 = 38,880, 10,368,000 / 11.111 = 933,120 / 16 = 58,320 / 22.5 = 2,592 / .125 = 20,736 / 72 = 288 20,736 / 900 = 23.04 x 3 = 69.12), is ruled by 12, which is literally the foundation of the Molten Sea. Hence 360 / 12 = 30 and 360 / 16 = 22.5 therefore 30 – 22.5 = 7.5 is the Bendall Translate that is Light photons travel around the earth 7.5 times in one second so therefore 7.5 is a factor in time along with the Aetherial number 144,000. THE DEPTH OF THE MOLTEN SEA WATER IS 2 x 7.5 HANDS (15 HANDS) AND THE AIR CAVITY ABOVE 7.5 DEEP AND 7.5 RADIUS FLAT LAND 7.5 x 7.5 = 56.25 / 0.125 = 450 7.5 x 7.5 x 7.5 = 421.875/0.125 = 3,375/266.666 = 12.65625 x 11.111 = 140.625 / 0.125 = 1,125 x 1.333 = 1,500 / 16 = 93.75 x 400 = 37,500 / 30 = 1,250

It has 300 knobs with lilies (300 x 6 petals =1,800), on the outer circumference of the Molten Sea shell on the outside in two rows, equals 150 per row one row representing 150 suns (photons) and the other row representing 150 moons (secondary polarized photon luminescence). A day (Sun) has 43,200 seconds divided by 150 = 288 [144 x 2] and a night has 43,200 seconds divided by 150 = 288. The volume of the Molten Sea in logs is 432,000, (where the surface area of the sphere with a radius of 5 is 314.15926535899100 Pi) therefore confirming the 10 x (7.5 x 1.333) conversion factor from Aether (Area / Time / Volume (ATV)) to Matter. Taking the 300 hands circumference and 300 knops [150 Suns = Day 150 Moons = Night] is one knop per hand, 360 [3 + 6 = 9] divided by 150 equals, 2.4 degrees, per sun or per moon, divided by Rod factor of 1.111 equals 2.16 Resonant Frequency Energy Units (RFEU). 129,600 / 2.16 = 60,000. Taking the 300 hands circumference and 150 Sun (Day 43,200 sec) or 150 moon (Night 43,200 seconds) knops equals two knops one knop per 0.5 of a hand, 360 degrees [3 + 6 = 9] divided by 300 equals, 1.2 degrees, per Sun and per Moon from the inside, divided by the ROD factor of 1.111 equals 1.08 RFEU.

There are 6 petals per Lilly times 300 lilies, 1,800 [1 + 8 = 9] petals is 5 petals per degree and each petal represents one ROD, that is 2,000 Baths divided by one ROD (1.111 = 1,800)

The Molten Sea, an inverted half sphere, that is two thirds full of water (0.66666) the Molten Sea contains 2,000 baths of water which is 150 (2,000 / 150 = 13.333 / 10 = Rim width 1.333 Hands) Ritual Baths (2,000 Baths times 72 = 144,000 Logs) which is a volume of 144,000 Logs (1 + 4 + 4 = 9) divided by the 16 segments is 9,000 logs per segment divided by 360 degrees equals 25 Logs per degree divided by 1.111 is 22.5 RFEU, the radius in hands of the inverted sphere. 9,000 logs per segment divided by 22.5 degrees is 400 Log resonant units per Degree.

Then divide 144,000 logs by 360 equals 400 logs divided by 16 segments equals 25 divided by the ROD resonant frequency ratio to energy of 1.111 equals 22.5 the Molten Sea's radius in Hands and the degrees of the 1 / 16 Sector. 144,000 logs volume divided by the AC resonant frequency ratio to DC energy of 1.111 equals 129,600 (RFEU) resonant log volumes or resonant potential photon energy units, the ROD.

144,000 divided by the ROD factor of 1.111 equals 129,600 (1+2+9+6=1+8=9) divided by 360 degrees (outer inverted sphere) equals 360 divided by 360 degrees (inner inverted sphere) equals one log volume resonating at 1. Therefore it must be so that 360 degrees squared equals 129,600 degrees, the number of energy log octave volume metric units per degree, hence giving a direct degree energy ratio to volume.

Full the Molten Sea half Sphere contains a volume of 3,000 Baths of water times 72 equals 216,000 logs (2 + 1 + 6 = 9) divided by 360 degrees equals 600 logs per degree divided by 16 equals 37.5 divided by 1.111 equals 33.75 divided by 22.5 equals 1.5 times 100 = 150 Ritual Baths.

216,000 logs divided by 16 segments equals 13,500 logs per segment divided by 360 equals 37.5 divided by 1.111 = 33.75. 13,500 logs per segment divided by 22.5 degrees = 1.5 logs per segment. 216,000 logs divided by 1.111 equals 194,400 log volumes of resonant potential photon energy units. 194,400 divided by 16 = 12,150 divided by 22.5 = 540. 194,400 divided by 360 equals 540 divided by 360 equals 1.5. 194,400 divided by 16 equals 12,150 divided by 22.5 equals 540 divided by 360 = 1.5 540 divided by 720 equals 0.75, 1 / x = 1.333. Then take 540 divided by equals

The Specific Gravity SG of water is one; all elemental weights therefore are units of waters SG, which in turn are also units of relative charge density. Hydrogen has the largest charge density and therefore the smallest osmotic Gravity pull. As Gravity is the osmotic charge attraction between low charge density AC Matter and high charge density DC Aether and should be 1.111 how many ones equals 0.9. The Molten Sea, as it is 2/3 full of water, has a refractive index (the amount by which light slows and bends in a given material is described by the refractive index) of 1.333, which is a Hand in bible measurement, and the 4 / 3 in Pi r cubed for the calculation of the volume of the sphere that is (4 / 3 Pi r cubed). It is critical to note that electrons in water now travel faster than the slowed speed of light. The 2 circles meeting at their centers describe a line between their centers, sharing a single radius that the two circles of the Torus sit; this relationship introduces the square root of 3 (1.7320) that is important in the formation of the structure and function of the Sacred Torus.

PART 17 OF 20 – MSAART PATENT APPLICATION NOTES DRAFT 518,400 – 22:22:22 THUR 22ND SEPT 2022
© STRIKE FOUNDATION GUARANTEE LIMITED | MALCOLM BENDALL 2022 | GRAPHICS - STEVE EARL

SOLOMON'S MOLTEN SEA IS THE TRANSLATE FROM THE AETHER INTO ELEMENTS USING COSMIC HARMONICS TO ENHANSE LIFE GIVING IMPLOSIVE FORCES INTO THE TEMPLE WATER

VOLUME OF THE FULL MOLTEN SEA SPHERE IS 6,000 BATHS x 72 = 432,000 LOGS

VOLUME OF HALF THE MOLTEN SEA SPHERE IS 3,000 BATHS x 72 = 216,000 LOGS

VOLUME OF HALF THE MOLTEN SEA SPHERE IS 2,000 BATHS x 72 = 144,000 LOGS

216,000 (Moon x 100) Logs = 300 (Molten Sea Circle) Homer = 54,000 CAD

VOLUME OF THE MOLTEN SEA HALF SPHERE IS 3,000 BATHS x 72 = 216,000 HANDS

VOLUME OF THE MOLTEN SEA (1/3 full whole Sphere) IS 2,000 BATHS x 72 = 144,000 L

144,000 Logs / 11.111 EQUALS 129,600 RESONANT FREQUENCY ENERGY UNITS (RFEU)

VOLUME OF THE 2/3 FULL MOLTEN SEA IS 2,000 BATHS x 72 = 144,000 logs

144,000 logs = 36,000 Cad = 200 homer = 18,000 Him = 36,000 Kab =144,000 Logs

144,000 / 11.111 EQUALS 129,600 RESONANT FREQUENCY ENERGY UNITS

VOLUME OF THE MOLTEN SEA IS 2,000 BATHS = 150 RITUAL BATHS

On the outside 300 Hand diameter of the Molten Sea are 150 Moons and 150 Suns

360 / 300 = 1.2 (12 Aether ruling number / 10) 360 / 150 = 2.4 (24 HOURS)

VOLUME OF THE MOLTEN SEA INNER SPHERE
4/3 PIE R (22.5 Hands) cubed = 47,712 cubic hands (6,000 Baths and 432,000 logs)
= 47,712 / 2 = 23,856 x .6666 = 15,904 = 2/3 water in half sphere
15,904 / 266.666 = 59.640   15,904 / 144 = 110.444 / 1.111 = 99.4
15,904 x 11.111 = 176,711

15,904 x 1.111 = 17,671.111 x 1.333 = 23,561.481 x 7.5 = 176,711.111 x 11.111 =
1,963,456.7901233684346 / 360 = 5,454 (9999RF) 0.0466392315789849 / 16 =
340.877914952 / 22.5 = 15.1501295534 x 8 = 121.201036427 x 400 = 48,4804145709
/ 2 = 24,240.2072855 / 2 = 12,120.1036427 / 2 = 6,060.05182137 / 2 = 3,030.0259106

15,904 / 2 = 7,952 / 360 = 22.0888
22.0888 / 16 = 1.380555 / 22.5 = 0.06135802469 x 8 = 0.49086419753 x 400 =
196.3456790123456790 x 11.111 = 2,181.61865569271 x 1.333 =
2,908.8248742568712135 x 1.111 = 3232.0276380630787908 (3232 + 2323 = 5555)

7,952 / 266.666 = 29.820074550186375466   7,952 / 22.5 = 353.4222

VOLUME OF THE MOLTEN SEA IS 2,000 BATHS x 72 = 144,000 / 1.111 = 129,600
Resonant Frequency Energy units (RFEU)

VOLUME OF THE MOLTEN SEA IS 129,600 x 1.333 = 172,800

172,800 RFEU [1,728 + 8,271 = 9,999]

VOLUME OF THE MOLTEN SEA OUTER SPHERE = 56,684 / 2 = 28,342 x .666 = 18,894.6666

MINUS VOLUME OF THE INNER SPHERE = 8,972 / 2 = 4,486 / .6666 = 6,729

AREA OF TOP OF THE MOLTEN SEA =

CIRCUMFERANCE OF THE MOLTEN SEA =

DESCRIPTION OF THE TORUS FORM AND FUNCTION

The Tesla Clockwise Implosive Vortex Torus is an octave, Pi and Phi based doughnut shape, where Pi defines the surface the Phi curve intersects the 8 cardinal planes octave nodes (insert Torus formula) centered on a singularity point G.

At G, time does not exist, as on an atomic scale the slowing of time to D max and the later acceleration of time returning to D min G must have a zero net effect on time being that they are acted upon by equal but opposite forces. I am therefore defining G as a zero time and zero entropy point, with V the node size as a constant and the torque induced changes in charge density as variable along with time, speed of light, node spin direction of both spin charge and entropy. As entropy is the fabric of Time both are by definition clearly subject to varied states and are not constants as the previous observers have declared. The fabric of Time entropy and at 90 degrees centropy, clearly and demonstrably acts on different states at variable rates and requires different energy inputs to resist falling apart and to maintain their operational designed function, but with time all creation is demonstrably all affected by death and destruction and ultimately, in time, fail. The connection of clockwise Female imploding centripetal negative entropy, to a slowing and or reversal of manifest male counter clockwise destructive exploding centrifugal vortexes can be demonstrated when counteracted by applying in force centripetal imploding forces and by maintaining stasis the reversal of time is also manifest when reversal of Entropy is achieved that is negative Entropy is applied with effect.

The node has a volumetric constant but its singularity point G is the focus of its own inward implosive forces interacting with the weakening counterclockwise explosive outer shell positive charges. The weakening outer positive charge allows the clockwise negative charge previously attracted towards the outer shell to congregate back toward the center G creating an internally stronger net negative charge density precipitating the zero point quadrature phase change seen at point D max. The concentration back to the center of the node effectively switches the polarity at the outer shells potential difference with the SN decreases. Conversely as the outer positive charge increases the internal node negative charge is attracted out, charge moving out decreases the charge density, as the torque induced counter spin slows the spin the moving out of the charge also slows the spin in the same way that the increasing negative outer charge increases the charge density at the center and therefore naturally increases the rate of spin as the outer shells net positive charge is overcome with the infusing negative charge.

The Torus has eight inner layers, four with a decreasing angle and effect of female spiral and four with a decreasing angle and effect of male spiral. It is divided on the outer skin into 16 primary points of origin with 32 secondary and 64 tertiary origins. These describe both a positively charged male spiral around the Torus, crossing at predetermined and geometrically defined nodes manifesting divergent centrifugal forces following both the long Fibonacci curves moving away from the singularity point but with a latitudinal straight series of discharge points perpendicular to and intersecting all the nodes, with the charge directed over the curve of the Torus to the singularity point S. until Dmax central torus longitudinal outside equator Dmax point is reached. There at Dmax the minimum positive charge density within the constant volume V is obtained with the corresponding proportional response of the maximum mass and also, and directly proportional to it, the slowest spin and slowest progression of time are obtained. The outward velocity of the node being overcome by the pull of the singularity point G. it first neutralizes its susceptibility by rotating through 90 degrees therefore initiating the first step of reverses its poles after it has briefly spun at 90 degrees to both forces then after progressing past the point of quadrature at D max with no pole or interference then flips poles by reversing its apparent spin by inverting the node and starts its arc back to the line of least resistance and maximum infusing torsional negative charge route along the long Fibonacci curve to G.

TORUS MOVING MODEL (INSERT)

Beyond the inflection point at Dmax the charge density is increased thru the infusion of negatively charged male centripetal, imploding, clockwise spiral energy flowing out through the latitudinal straight lines, torque on the node is also applied by the centripetal converging forces returning along the longitudinal Fibonacci curves to G where time is at its zero and inflection point.

The Tesla Torus model also describes through the ROD and Bendall Translate every manifest measurable characteristic of gods creation a person is capable of perceiving, and some not, all moving up or down in circulate left and right spiral octave steps from knop to knop, to and from G identifiable to us as elemental crystal forms, as sounds, as elements, as high energy states and color. This progression continues right up to the second and third order colors of the light spectrum three visible, first second and third order colors we see, and four we do not.

The progression of the entropy male counterclockwise spirals, emitting their moving out and away, dissipating positive charge effusions in exploding destructive phase is countered by centropy, clockwise, moving in, imploding, Female aggregating negative infusion in constructive imploding stage. Paramagnetic and Diamagnetic character, Monad, Diad, Triad and Tetrad dispositions are predicted, along with the octave scaling of the charge density, centrifugal energy, Osmotic charge attraction manifest as Mass and their relationships to Aether, Light, Dimensions, Music, Matter, Energy, color, Area / Time / Space and Motion.

FORMULA FOR THE DEMONSTRATION OF RELATIONSHIPS DESCRIBED ABOVE (INSERT)

TORUS INSIDE THE MOLTEN SEA

THE MOLTEN SEA AND TESLA'S WARDENCLIFFE TOWER

THE TORUS CAN BE IMAGINED AS HAVING 16 SPHERES TRAVELLING AROUND INSIDE IT

TORUS RESONANT FORM THE SACRED
VAJRA IMPLOSIVE VORTEX

The Implosive vortex can have 14 clockwise oriented semicircles of outer 240 + inner 120 degrees equaling 360 degrees times 14 = 5,040.

50,040 / 16 (Segments)= 315
5,040 / 22.5 (Degrees per segment) = 224
5,040 / 266.666 = 18.9 / 16 = 1.18125 1/x = 0.846560846560 / .125 = 6.77248677248677
5,040/ 11.111 = 453.6 x 22.5 = 10,206
7,776 /2/2/2/ = 972 /2 = 486 / 2 = 243

5,040 times two for the anticlockwise equals 10,080.

10,080 / 16 = 630
10,080 / 22.5 = 488
10,080 / 266.666 = 37.8
10,080 / 11.111 = 907.2 x 7.5 = 6,804

The Implosive vortex can also have 15 clockwise oriented semicircles of 240 degrees + inner 120 Degrees equaling 360 degrees times 15 = 5,400.

5,400 / 216 = 25
5,400 / 22.5 = 240
5,400 / 266.666 = 20.2500506251562816 / 16 = 1.2656281 1/x = 0.790121481 / .125 = 6.3
5,400 / 11.111 = 486 / 1.333 = 364.5
5,400 / 7.5 = 720
5,400 x 7.5 = 3,888,000
3,888,000 /2/2/2/ = 486,000 / 2 = 243

5,400 Degrees times two for the anticlockwise equals 10,800

10,800 / 2,160 = 5
10,800 / 16 = 675
10,800 / 22.5 = 480
10,800 / 266.666 = 40.50
10,800 / 11.111 = 972   972 x 7.5 = 129.600
10,800 / 7.5 = 1,440
10,800 x 7.5 = 81,000

The Implosive vortex can also have 16 clockwise oriented semicircles of 360 degrees equaling 5,760.

5,760 / 216 = 26.666
5,760 / 432 = 13.333
5,760 / 266.666 = 21.6
5,760 / 1.333 = 4,320
5,760 / 22.5 = 256
5,760 / 266.666 = 21.6 / 16 = 1.35 1/x =   / .125 =

5,760 / 11.111 = 518.4 x 22.5 = 11,664
5,760 / 2 = 2,880 / 2 = 1,440 / 2 = 720 / 2 = 360 / 2 = 180 / 2 = 90 / 2 = 45 / 2 = 22.5 / 2 = 11.25 / = 5.625
5.625 / 2 = 2.8125 C / 2 = 1.4625 C / 2 = 0.703125 C / 2 = 0.3515625 / 2 = 0.17578125 / 2 = 0.08789062

5,760 times two for the 16 anticlockwise equals 11,520.

11,520 / 2160 = 5.333
11,520 / 16 = 720
11,520 / 22.5 = 512
11,520 / 266.666 = 43.2
11,520 / 11.111 = 1,036.8 x 22.5 = 23,328
.125 or one eighth is half the 2.5 of the torus

The Implosive vortex can also have 16 clockwise oriented semicircles of 240 degrees equaling.

3,840 / 720 = 5.333
3,840 / 16 = 240
3,840 / 22.5 = 170.6666666666667 octave of 266.666
3,840 / 266.666 = 14.4 / 16 = .9 1/x = 1.111 14.4 / .125 =
3,840 / 11.111 = 345.6 x 22.5 = 7,776 x 2 x 2 = 31,104
7,776 /2/2/2/ = 972/2 = 486 / 2 = 243

3,840 times two for the 16 anticlockwise equals 7,680.

7,680 / 22.5 = 341.333 octave of 266.666 and .666
7,680 / 266.666 = 28.800000072
7,680 / 11.111 = 691.2 x 22.5 = 15,552

.125 or one eighth is half the 2.5 of the torus

Torus has the sum of internal angles, Pythagorean resonance being the sum of the internal angles, on one of its eight planes of 360 plus joiners 90 + 90 = 180 plus 360 = 900 degrees times eight = 7,200 / 22.5 = 320.

7,200 / 16 = 450 (molten sea 45 hands)
7,200 / .125 = 57,600
7,200 / 11.111 = 648.0000000000001
7,200 / 266.666 = 27
27 / 16 = 1.6875
1.6875 the up octaves of which are all the planetary diameters
3.375000000000001 // 6.75 // 13.5 // 27 // 54 // 108 // 216 Moon // 432 Sun radius // 864 Suns diameter //
1,728 // 3,456 SUN SQUARE // 6,912 RFEU // 13,824 // 27,648

900 vertical plus 360 horizontal =

1,260 / 11.1111 = 113.4
1,260 / 266.666 = 4.725

1,260 // 2,520 // 5,040 // 10,080 // 20,160 // 40,320 // 80,640 // 161,280 // 322,560

518,400 x 2.5 = 1,296,000 RFEU

**Tests for Light / 6.666 for Aether / 5.555 for Time / 4.444 for Matter 3.333 and 266.666
Divide by 37 AND 5.555 for ENERGY and Divide by 36 AND 3.333 for MATTER**

ALIEN MATHS AND CHEMISTRY

HYDROGEN, NOBLE GASES, METHANE AND CARBON MONOXIDE

The melting point and boiling point of any element are the points at which enough energy has been put into the dynamic system for the resonant frequency intensity to be met. This is determined by a combination of dynamic factors. Primary of these factors is the sum of the internal angles of the elements or compounds natural crystal form manifesting in its solid state. These are adjusted to their atomic charge density and the Area / time / space that it occupies which are the secondary forces. The Area / Time / Space its self introduces the additional variable of the external positive and negative atmospheric and atomic pressure. This is in addition to the tertiary effect of implosive and explosive environmental and atomic forces. Contamination of other matter further varies the frequency /area / time / space equation and therefore must be measured and accounted for to determine a "group " mass resonance.

HYDROGEN

HYDROGEN MELTING POINT = − 259.2 so - 259.2 / 266.666 = 0.972
- 259.2 / 11.111 = 23.328
- 259.2 / 1.333 = 194.4 // 97.2//48.6//24.3

- 259.2 added sum = 2 + 5 + 9 + 2 [18] + 2 + 5 + 9 + 2 = 36

- 259.2 multiplied product = 2 x 5 x 9 x 2 (180) x 2 x 9 x 5 x 2 = 32,400

32,400 / 720 = 45 32,400 / 16 = 2,025 32,400 / 22.5 = 1,440 32,400 / 0.125 = 259,200

32,400 x 16 = 518,400   32,400 x 22.5 = 729,000   32,400 x .125 = 4,050

32,400 x 400 = 12,960,000   32,400 x 720 = 23,328,000   32,400 x 900 = 29,160,000

518,400 / 16 = 32,400   518,400 / 22.5 = 23,040   518,400 / 720 = 720

32,400 / 144 = 225   32,400 / 36 = 900   36 / 32,400 = 0.00111

144,000 / 32,400 = 4.4444   32,400 / 7.5 = 4,320 / 7.5 = 576 / 2 = 288 / 2 = 144

144 x 32,400 = 4,665,600 / 432 = 10,800 / 2,160 = 5

32,400 / 259.2 = 125
32,400 / 6.666 = 4,860   32,400 / 5.555 = 5,850   32,400 / 4.444 = 7,290
32,400 / 3.333 = 9,720   32,400 / 2.222 = 14,580   32,400 / 11.111 = 2,916

ALL SIX 864,000 COMBINATIONS

864,000	846,000	684,000	648,000	486,000	468,000
56,623,104*	55,443,456*	44,826,624*	42,467,328*	31,850,496*	30,670,848*
28,311,552*	27,721,728*	22,413,312*	21,233,664*	15,925,248*	15,335,424*
14,155,776*	13,860,864*	11,206,656*	10,616,832*	7,962,624*	7,667,712*
7,077,888*	6,930,432*	5,603,328*	5,308,416*	3,981,312*	3,833,856*
3,538,944*	3,465,216*	2,801,664*	2,654,208*	1,990,656*	1,916,928*
1,769,472*	1,732,608*	1,400,832*	1,327,104*	995,328*	958,464*
884,736*	866,304*	700,416*	663,552*	497,664*	479,232*
442,368*	433,152*	350,208*	331,776*	248,832*	239,616*
221,184*	216,576*	175,104*	165,888*	124,416	119,808*
110,592*	108,288*	87,552*	82,944*	62,208*	59,904*
55,296*	54,144*	43,776*	41,472*	31,104*	29,952*
27,684*	27,072*	21,888*	20,736*	15,552*	14,976*
13,824*	13,536*	10,944*	10,368*	7,776*	7,488*
6,912,000	6,768,000	5,472,000	5,184,000	3,888,000	3,744,000
3,456,000	3,384,000	2,736,000	2,592,000	1,944,000	1,872,000
1,728,000	1,692,000	1,368,000	1,296,000	972,000	936,000

864,000	846,000	684,000	648,000	486,000	468,000
432,000	423,000	342,000	H-324,000	243,000	234,000
216,000	211,500	171,000	162,000	121,500	117,000
108,000	105,000	85,500	81,000	60,750	58,500
54,000	52,875	42,750	40,500	30,375	29,250
27,000	26,437.5	21,375	20,250	15,187.5	14,625
13,500	13,218.75	10,687.5	10,125	7,593.75	7,312.5
6,750	6,609.375	5,343.75	5,062.5	3,796.875	3,656.25
3,375	3,304.6875	2,671.875	2,531.25	1,898.43	1,828.1
1,687.5	1,652.34375	1,335.9375	1,265.62	949.2187	914.06
843.75	826.171875	667.96875	632.812	474.6093	457.03
421.875	413.0859375	333.984375	316.406	237.3046	228.51
210.9375	206.54296875	166.99218	158.2031	118.6523	114.25
105.46875	103.271484375	83.496093	79.10156	59.32617	57.128
52.734375	51.6357421875	41.7480468	39.55078	29,66308	28.564
26.3671875	25.8187109375	20.8740234	19.77539	14.83154	14.282

864,000	846,000	684,000	648,000	486,000	468,000

HYDROGEN RATIOS TO ALL OTHER NOBLE GASES

32,400 (Hydrogen -259.2) / 3,136 (Helium -272.2) = 10.33163265306122
3,136 (Helium -272.2) / 32,400 (Hydrogen -259.2) = 0.09679012345679

32,400 (Hydrogen -259.2) / 147,456 (Neon -248.6) = 0.2197265625
147,456 (Neon -248.6) / 32,400 (Hydrogen -259.2) = 4.55111

32,400 (Hydrogen -259.2) / 82,944 (Argon -189.4) = 0.390625
82,944 (Argon -189.4) / 32,400 (Hydrogen -259.2) = 2.56

32,400 (Hydrogen -259.2) / 19,600 (Krypton 157.4) = 1.653061224489796
19,600 (Krypton 157.4) / 32,400 (Hydrogen -259.2) = 0.604938271604938

32,400 (Hydrogen -259.2) / 64 (Xeon -111.8) = 506.25
64 (Xeon -111.8) / 32,400 (Hydrogen -259.2) = 0.00195308641975

32,400 (Hydrogen -259.2) / 1,225 (Radon -71.15) = 26.44897959183673
1,225 (Radon -71.15)/ 32,400 (Hydrogen -259.2) = 0.037808641975309 Argon/Helium

HYDROGEN RIGHT TRIANGLE MELTING POINT = - 259.2 C (Earths Equinox 25,920)

864,000 / 259.2 = 3,333.3   864,000 / 345.6 = 2,500   864,000/ 432 = 2,000
2,160 / 259.2 = 8.333   259.2 / 2,160 = 0.972

Triangles Area / Time / Space calculation

259.2 x -345.6 x 432 = 38,698,352.6
259.2 x -345.6 = - 89,579.5 (Area of the square) / 2 = 44,789.76 (Area of the Right Triangle)

44,789.76 / 24 Hrs = 1,866.24   1,866.24 / 60 Min = 31.104   31.104 / 60 Sec = 0.5184
0.5184 / 60 Arc = 0.00864   0.00864 / 60 = 0.000144   0.000144 / 60 = 0.0000024
0.0000024 / 60 = 0.00000004   0.00000004 / 24 = 0.000000001666
0.00000004 / 60 = 0.0000000000666   0.0000000000666 / 60 = 0.00000000001111

0.5184 / 7.5 = 0.06912 / 0.266666 = 0.2592 x 400 = 103.68 / 16 = 6.48 / 22.5 = 0.288
0.288/ 0.125 = 2.304 x 400 = 921.9 / 1.333 = 691.2 / 1.333 = 518.4 /1.333 = 388.8 / 1.333 =
291.6

-259.2 / 3 = - 86.4

86.4 x 5 = 432

- 345.6 = - 86.4 x 4

HYDROGEN SQUARE MELTING POINT -259.2 (Earths Equinox) x 4 = 1,036.8

1,036.8/3,456 = 0.3
259.2 x 259.2 = 67,184.64 / 11.111 = 6,046.6   67,184.64 / 266.666 = 251.9424
67,184.64 / 144 = 466.56
67.184.64 / 7.5 = 8,957.952
67,184.64 / 2 = 33,592.32
33,592.32 / 3 = 11,197.44   11,197.44 x 4 = 44,789.76   11,197.44 x 5 = 55,987.2

HYDROGEN CUBE MELTING POINT CUBED   = 259.2 x 259.2 x 259.2 = 17,414,258.688
1 year, 360 day Time in seconds cube 17,414,258.688 / 360 = 48,372.9408
48,372.9408/24 = 2,016/60 = 33.6/60 = 0.56/ 720 = 0.000777/ 16 = 0.0000486111/
22.5 = 0000002160493827

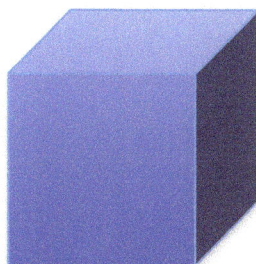

17,414,258.688 / 144,000 = 120.932352 / 7.5 = 16.1243136 / 16 = 1.0077696
 = 1.0077696 / 22.5 = 0.04478976
17,414,258.688 / 1.333 = 13,060,694.016
17,414,258.688 / 266.666 = 65,303.47008000002
17,414,258.688 / 6.666 = 2,612,138.8032
17,414,258.688 / 5.555 = 3,134,566.56384
17,414,258.688 / 4.444 = 3,918,208.2048
17,414,258.688 / 3.333 = 5,224,277.606400001
17,414,258.688 / 2.222 = 7,836,416.409600001
17,414,258.688 / 1.111 = 15,672,832.8192 / 9 = 1,741,425.8688

HYDROGEN RIGHT ANGLE TRIANGLE BOILING POINT = - 252.8 C

HYDROGEN Boiling point = - 252.87 so -252.87 / 266.666 = .9482625
- 252.87 so -252.87 / 11.111 = 22.7583
- 252.87 so -252.87 / 1.333 = 189.699925

252.87 = 2 + 5 + 2 + 8 + 7 (24) +7 + 8 + 2 + 5 + 2 = 48

252.87 = 2 x 5 x 2 x 8 x 7 (1,120) x 7 x 8 x 2 x 5 x 2 = 1,254,400 / 5.555 = 225,792

252.8 x 337.0666 = 85,210.4365 therefore triangle area 85,210.4365 / 2 = 42,605.2182
42,605.2182 / 24 Hours = 1,775.21743   1,775.21743 / 60 Minutes = 29.5869571
29.5869571 / 60 Seconds = 0.493115951
0.493115951 / 60 Arc seconds = 0.0821859918333

- 252.8 / 3 = 84.2666

421.333 = 84.2666 x 5

337.0666 = 84.2666 x 4

HELIUM

HELIUM MELTING POINT = -272.2

HELIUM RIGHT ANGLE TRIANGLE MELTING POINT – 272.2

272.2 x 362.9333 = 98,790.45333
98,790.45333 / 2 = 49,395.22666

49,395.22666 / 24 Hours = 2,058.13444   2,058.13444 / 60 Minutes = 34.30224074
34.30224074 / 60 Seconds = 0.571704012345679
0.571704012345679 / 60 Arc Seconds = 0.009528400205761

- 272.2 / 3 = 90.7333

453.666 = 5 x 90.7333

362.9333 = 4 x 90.7333

2 + 7 + 2 + 2 (13) + 2 + 2 + 7 + 2 = 26
2 x 7 x 2 x 2 (56) x 2 x 2 x 7 x 2 = 3,136

3,136 / 5.555 = 564.48   3,136 / 266.666 = 11.76   3,136 / 144 = 21.777

3,136 / -272.2 = 11.5209405 (1/x = 0.086798469387755)
3,136 / 5.555 = 564.48 / 11.111 = 50.80320000000001
32,400 (Hydrogen -259.2) / 3,136 (Helium -272.2) = 10.33163265306122

HELIUM RATIOS TO ALL OTHER NOBLE GASES INCLUDING HYDROGEN

3,136 (Helium - 272.2) / 32,400 (Hydrogen -259.2) = 0.09679012345679
   0.09679012345679 //- 9 = 0.000378086419753 = Helium to Argon 8th Octave
32,400 (Hydrogen -259.2) / 3,136 (Helium -272.2) = 10.33163265306122

3,136 (Helium - 272.2) / 147,456 (-248.6 Neon) = 0.02126736111
147,456 (-248.6 Neon) / 3,136 (Helium - 272.2) = 47.02040816326531

3,136 (Helium - 272.2) / 82,944 (Argon) = 0.037808641975309
         0.037808641975309 //+8 = 9.679012345679012
82,944 (Argon) / 3,136 (Helium - 272.2) = 26.44897959183673

3,136 (Helium - 272.2) / 19,600 (Krypton 157.4) = 0.16
19,600 (Krypton 157.4) / 3,136 (Helium - 272.2) = 6.25

3,136 (Helium - 272.2) / 64 (Xeon 111.8) = 49
64 (Xeon 111.8) / 3,136 (Helium - 272.2) = 0.020408163265306

3,136 (Helium - 272.2) / 1,225 (Radon -71.15) = 2.56
1,225 (Radon -71.15) / 3,136 (Helium - 272.2) = 0.390625 //+ 12 = 1,600 – 25,600

NEON

NEON RIGHT ANGLE TRIANGLE MELTING POINT -248.6

-248.6 (3 side) x 331.4666 (4 side) = 82,402.61333 (Area of Square)
82,402.61333 / 2 = 41,201.30666 (Area of the Right Triangle)
41,201.30666 / 24 Hours = 1,716.72111
1,716.72111 / 60 Minutes = 28.6120185185185
 28.6120185185185 / 60 Seconds = 0.476866975308642

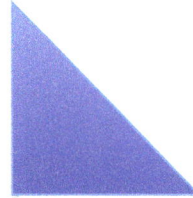

-248.6/ 3 = 82.8666

414.333 = 5 x 82.8666

331.4666 = 4 x 82.8666

2 + 4 + 8 + 6 (20) + 6 + 8 + 4 + 2 = 40
2 x 4 x 8 x 6 (384) x 6 x 8 x 4 x 2 = 147,456

147,456 / 1.111 = 132,710.4
147,456 (Neon 248.6) / 40 = 3,686.4
147,456 (Neon 248.6) x 40 = 5,898,240

147,456 (Neon 248.6) / 266.666 = 552.9600000000001
147,456 (Neon 248.6) / 11.111 = 13,271.04
147,456 (Neon 248.6) / 1.333 = 110,592
147,456 (Neon 248.6) / 144 = 1,024

NEON RATIOS TO ALL OTHER NOBLE GASES INCLUDING HYDROGEN

147,456 (Neon -248.6) / 32,400 (Hydrogen -259.2) = 4.5111
32,400 (Hydrogen -259.2) / 147,456 (Neon 248.6)= 0.2197265625

147,456 (Neon -248.6) / 3,136 (Helium - 272.2) = 47.02040816326531
3,136 (Helium - 272.2) / 147,456 (Neon 248.6) = 0.02126736111

147,456 (Neon -248.6) / 147,456 (-248.6 Neon) = 1
147,456 (-248.6 Neon) / 147,456 (Neon 248.6) = 1

147,456 (Neon - 248.6) / 82,944 (Argon -189.4) = 1.777
 82,944 (Argon 189.4) / 147,456 (Neon - 248.6) = 0.5625

147,456 (Neon - 248.6) / 19,600 (Krypton 157.4)   = 7.523265306122449
19,600 (Krypton 157.4) / 147,456 (Neon - 248.6) = 0.1329210069444

 147,456 (Neon 248.6) / 64 (Xeon 111.8) = 2,304
64 (Xeon 111.8) / 147,456 (Neon 248.6) = 0.00043402777

147,456 (Neon -248.6) / 1,225 (Radon -71.15) = 120.3722448979592
1,225 (Radon -71.15) / 147,456 (Neon 248.6) = 0.008307562934028

ARGON

ARGON RIGHT ANGLE TRIANGLE MELTING POINT – 189.4

- 189.4 (3 side) x 252.5333 (4 side) = 47,829.81333 (Area of Square)
47,829.81333 / 2 = 23,914.90666 (Area of the Right Triangle)
23,914.90666 / 24 Hours = 996.45444
996.45444 / 60 Minutes = 16.60757407407
16.60757407407 / 60 Seconds = 0.276792901234568
0.276792901234568 / 60 Arc Seconds = 0.004613215020576

189.4 / 3 = 63.1333

315.666 = 5 x 63.1333

252.5333 = 4 x 63.1333
1

1 + 8 + 9 + 4 (22) + 4 + 9 + 8 + 1 = 44

1 x 8 x 9 x 4 (288) x 4 x 9 x 8 x 1 = 82,944

82,944 (Argon - 189.4) / 44 = 1,885.0909
82,944 (Argon - 189.4) x 44 = 3,649,536
82,944 (Argon - 189.4) / 266.666 = 311.0400000000001
82,944 (Argon - 189.4) / 11.111 = 7,464.960000000001
82,944 (Argon - 189.4) / 1.333 = 62,208.00000000002
82,944 (Argon - 189.4) / 144 = 576 //288//144//72//36//16//8//4//2//1

ARGON RATIOS TO ALL OTHER NOBLE GASES INCLUDING HYDROGEN

82,944 (Argon- 189.4) / 32,400 (Hydrogen – 259.2) = 2.56
2.56// 1.28//0.64// 0.32//0.16//0.08
32,400 (Hydrogen – 259.2) / 82,944 (Argon - 189.4)= 0.390625

82,944 (Argon - 189.4) / 3,136 (Helium - 272.2) = 26.44897959183673
3,136 (Helium - 272.2) / 82,944 (Argon - 189.4) = 0.037808641975309
　　　　　0.037808641975309 //+8 = 9.679012345679012

82,944 (Argon - 189.4) / 147,456 (-248.6 Neon) = 0.000140625//+5 = 0.144
147,456 (-248.6 Neon) / 82,944 (Argon - 189.4) = 1.777
1.777//0.888/0.444/0.222/0.111/0.0555

82,944 (Argon - 189.4) / 19,600 (Krypton 157.4)　　= 4.231836734693878
19,600 (Krypton 157.4)　 / 82,944 (Argon - 189.4) = 0.236304012345679 //+5 =3.7808

82,944 (Argon - 189.4) / 64 (Xeon 111.8) = 1,296
64 (Xeon 111.8) / 82,944 (Argon - 189.4) = 0.000771604938272

82,944 (Argon - 189.4) / 1,225 (Radon -71.15) = 67.70938775510204
1,225 (Radon -71.15) / 82,944 (Argon - 189.4) = 0.0147690007716

KRYPTON

KRYPTON RIGHT ANGLE TRIANGLE MELTING POINT – 157.4 C

– 157.4 C (3 side) x 209.8666 (4 side) = 33,033.01333 (Area of Square)
33,033.01333 / 2 = 16,516.50666 (Area / Time / Space of the Right Triangle)

16,516.50666 / 24 Hrs = 688.18777
688.18777/ 60 Min = 11.469796296296296
11.469796296296296 / 60 Sec = 0.191163271604938
0.191163271604938 / 60 Arc Sec = 0.003186054526749

-157.4 / 3 = 52.4666

262.333 = 5 x 52.4666

209.8666 = 4 x 52.4666

1 + 5 + 7 + 4 (17) + 4 + 7 + 5 + 1 = 34

1 x 5 x 7 x 4 (140) x 4 x 7 x 5 x 1 = 19,600

19,600 (Krypton -157.4) / 34 = 576.4705882352941
19,600 (Krypton -157.4) x 34 = 666,400

19,600 (Krypton -157.4) / 266.666 = 73.50000000000002
19,600 (Krypton -157.4) / 11.111 = 1,764
19,600 (Krypton -157.4) / 1.333 = 14,700
19,600 (Krypton -157.4) / 144 = 136.111

KRYPTON RATIOS TO ALL OTHER NOBLE GASES INCLUDING HYDROGEN

19,600 (Krypton -157.4) / 32,400 (Hydrogen -259.6) = 0.604938271604938
32,400 (Hydrogen – 259.6) / 19,600 (Krypton -157.4) = 1.653061224489796

19,600 (Krypton -157.4) / 3,136 (Helium – 272.2) = 6.25
3,136 (Helium – 272.2) / 19,600 (Krypton -157.4) = 0.16

19,600 (Krypton -157.4) / 147,456 (-248.6 Neon) = 0.1329210069444
147,456 (-248.6 Neon) / 19,600 (Krypton -157.4) = 7.523265306122449

19,600 (Krypton -157.4) / 19,600 (Krypton - 157.4)   = 1
19,600 (Krypton - 157.4)   / 19,600 (Krypton -157.4) = 1

19,600 (Krypton -157.4) / 64 (Xeon - 111.8) = 306.25
64 (Xeon 111.8) / 19,600 (Krypton -157.4) = 0.003265306122449

19,600 (Krypton -157.4) / 1,225 (Radon -71.15) = 16
1,225 (Radon -71.15) / 19,600 (Krypton -157.4) = 0.0625

XENON

XEON RIGHT ANGLE TRIANGLE MELTING POINT -111.8

-111.8 (3 side) x 149.0666(4 side) = 16,665.65333 (Area of Square)
 16,665.65333 / 2 = 8,332.82666 (Area of the Right Triangle)

8,332.82666 / 24 Hours = 347.20111
347.20111 / 60 Minutes = 5.786685185185183
5.786685185185183 / 60 Seconds = 0.09644475308642
0.09644475308642 / 60 Arc Seconds = 0.00160741255144

-111.8/ 3 = 37.2666

186.333 = 5 x 37.2666

149.0666 = 4 x 37.2666

1 + 1 + 1 + 8 (11) + 8 + 1 + 1 + 1 = 22

1 x 1 x 1 x 8 (8) x 8 x 1 x 1 x 1 = 64

64 (Xeon -111.8) / 22 = 2.909090
64 (Xeon -111.8) x 22 = 1,408

64 (Xeon -111.8) / 266.666 = 0.24   (1/x = 4.1666)
64 (Xeon -111.8) / 11.111 = 5.76   (1/x = 0.1736111)
64 (Xeon -111.8) / 1.333 = 48   (1/x = 0.0208333)
64 (Xeon -111.8) / 144 = 0.444   (1/x = 2.25)

XENON RATIOS TO ALL OTHER NOBLE GASES INCLUDING HYDROGEN

64 (Xeon -111.8) / 32,400 (Hydrogen) = 0.001975308641975
32,400 (Hydrogen) / 64 (Xeon -111.8) = 506.25

64 (Xeon -111.8) / 3,136 (Helium – 272.2) = 0.02408163265306
3,136 (Helium – 272.2) / 64 (Xeon -111.8) = 49

64 (Xeon -111.8) / 147,456 (-248.6 Neon) = 0.00043402777
147,456 (-248.6 Neon) / 64 (Xeon -111.8) = 2,304

64 (Xeon -111.8) / 19,600 (Krypton 157.4)   = 0.003265306122449
19,600 (Krypton 157.4)   / 64 (Xeon -111.8) = 306.25

 64 (Xeon -111.8) / 64 (Xeon 111.8) = 1
64 (Xeon 111.8) /64 (Xeon -111.8) = 1

64 (Xeon -111.8) / 1,225 (Radon -71.15) = 0.052244897959184
1,225 (Radon -71.15) / 64 (Xeon -111.8) = 19.140625

RADON

RADON RIGHT ANGLE TRIANGLE MELTING POINT -71.15

-71.15 (3 side) x (4 side) = 6,749.76333 (Area of Square)
6,749.76333 / 2 = 3,374.881666 (Area of the – 71.15 Right Triangle)
3,374.881666 / 24 Hours = 140.620069444
140.620069444 / 60 Minutes = 2.343667824074074
2.343667824074074 / 60 Seconds = 0.039061130401235
0.039061130401235 / 60 Arc Seconds = 0.000651018840021

-71.15 / 3 = 23.71666

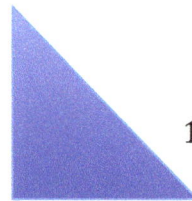

118.58333 = 5 x 23.1666

94.8666= 4 x 23.71666

7 + 1 + 1 + 5 (14) + 5 + 1 + 1 + 7 = 28

7 x 1 x 1 x 5 (35) x 5 x 1 x 1 x 7 = 1,225

1,225 (Radon -71.15) / 28 = 43.75
1,225 (Radon -71.15) x 28 = 34,300

1,225 (Radon -71.15) / 266.666 = 4.593750000000001
1,225 (Radon -71.15) / 11.111 = 110.25
1,225 (Radon -71.15) / 1.333 = 918.75
1,225 (Radon -71.15) / 144 = 8.5069444

RADON RATIOS TO ALL OTHER NOBLE GASES INCLUDING HYDROGEN

1,225 (Radon -71.15) / 32,400 (Hydrogen) = 0.037808641975309
32,400 (Hydrogen) / 1,225 (Radon -71.15)= 26.44897959183673

1,225 (Radon -71.15) / 3,136 (Helium – 272.2) = 0.390625
3,136 (Helium – 272.2) / 1,225 (Radon -71.15) = 2.56

1,225 (Radon -71.15) / 147,456 (-248.6 Neon) = 0.008307562934028
147,456 (-248.6 Neon) / 1,225 (Radon -71.15) = 120.3722448979592

1,225 (Radon -71.15) / 19,600 (Krypton 157.4) = 0.0625
19,600 (Krypton 157.4) / 1,225 (Radon -71.15) = 16

1,225 (Radon -71.15) / 64 (Xeon 111.8) = 19.140625
64 (Xeon 111.8) / 1,225 (Radon -71.15) = 0.052244897959184

1,225 (Radon -71.15) / 1,225 (Radon -71.15) = 1
1,225 (Radon -71.15) / 1,225 (Radon -71.15) = 1

NOBLE GASES FULL COMBINED TABLES

3,136 (Helium – 272.2) / 64 (Xeon 111.8) = 49
3,136 (Helium – 272.2) / 19,600 (Krypton 157.4) = 0.16 [1/x = 6.25]

32,400 (Hydrogen) / 64 (Xeon 111.8) = 506.25

3,136 (Helium – 272.2) / 64 (Xeon 111.8) = 49
64 (Xeon 111.8) / 3,136 (Helium – 272.2) = 0.020408163265306

3,136 (Helium – 272.2) / 82,944 (Argon) = 0.037808641975309
                  0.037808641975309 //+8 = 9.679012345679012
82,944 (Argon) / 3,136 (Helium – 272.2) = 26.44897959183673

3,136 (Helium – 272.2) / 19,600 (Krypton 157.4) = 0.16
19,600 (Krypton 157.4) / 3,136 (Helium – 272.2) = 6.25

3,136 (Helium – 272.2) / 147,456 (-248.6 Neon) = 0.02126736111
                                    [1/x = 47.020816326531]
147,456 (-248.6 Neon) / 3,136 (Helium – 272.2) = 47.02040816326531
3,136 (Helium – 272.2) / 82,944 (Argon) = 0.037808641975309
0.037808641975309 //+8 = 9.679012345679012
82,944 (Argon) / 3,136 (Helium – 272.2) = 26.44897959183673

3,136 (Helium – 272.2) / 32,400 (Hydrogen) = 0.09679012345679
0.09679012345679 //- 9 = 0.000378086419753 = Helium to Argon 8 th Octave

32,400 (Hydrogen) / 3,136 (Helium -272.2) = 10.33163265306122
3,136 (Helium -272.2) / 32,400 (Hydrogen) = 0.09679012345679

32,400 (Hydrogen) / 147,456 (Neon -248.6) = 0.2197265625
147,456 (Neon -248.6) / 32,400 (Hydrogen) = 4.55111

32,400 (Hydrogen) / 82,944 (Argon -189.4) = 0.390625
82,944 (Argon -189.4) / 32,400 (Hydrogen) = 2.56

32,400 (Hydrogen) / 19,600 (Krypton 157.4) = 1.653061224489796
19,600 (Krypton 157.4) / 32,400 (Hydrogen) = 0.604938271604938

32,400 (Hydrogen) / 64 (Xeon -111.8) = 506.25
64 (Xeon -111.8) / 32,400 (Hydrogen) = 0.00195308641975

32,400 (Hydrogen) / 1,225 (Radon -71.15) = 26.44897959183673
1,225 (Radon -71.15) / 32,400 (Hydrogen) = 0.037808641975309 Argon/Helium ratio

HYDROGEN

HYDROGEN RATIOS TO ALL OTHER NOBLE GASES

32,400 (Hydrogen -259.2) / 3,136 (Helium -272.2) = 10.33163265306122
3,136 (Helium -272.2) / 32,400 (Hydrogen -259.2) = 0.09679012345679

32,400 (Hydrogen -259.2) / 147,456 (Neon -248.6) = 0.2197265625
147,456 (Neon -248.6) / 32,400 (Hydrogen -259.2) = 4.55111

32,400 (Hydrogen -259.2) / 82,944 (Argon -189.4) = 0.390625
82,944 (Argon -189.4) / 32,400 (Hydrogen -259.2) = 2.56

32,400 (Hydrogen -259.2) / 19,600 (Krypton 157.4) = 1.653061224489796
19,600 (Krypton 157.4) / 32,400 (Hydrogen -259.2) = 0.604938271604938

32,400 (Hydrogen -259.2) / 64 (Xeon -111.8) = 506.25
64 (Xeon -111.8) / 32,400 (Hydrogen -259.2) = 0.00195308641975

32,400 (Hydrogen -259.2) / 1,225 (Radon -71.15) = 26.44897959183673
1,225 (Radon -71.15) / 32,400 (Hydrogen -259.2) = 0.037808641975309
Argon/Helium ratio

HELIUM

HELIUM RATIOS TO ALL OTHER NOBLE GASES INCLUDING HYDROGEN

3,136 (Helium – 272.2) / 32,400 (Hydrogen -259.2) = 0.09679012345679
0.09679012345679 //- 9 = 0.000378086419753 = Helium to Argon 8th Octave
32,400 (Hydrogen -259.2) / 3,136 (Helium -272.2) = 10.33163265306122

3,136 (Helium – 272.2) / 147,456 (-248.6 Neon) = 0.02126736111
147,456 (-248.6 Neon) / 3,136 (Helium – 272.2) = 47.02040816326531

3,136 (Helium – 272.2) / 82,944 (Argon) = 0.037808641975309
0.037808641975309 //+8 = 9.679012345679012
82,944 (Argon) / 3,136 (Helium – 272.2) = 26.44897959183673

3,136 (Helium – 272.2) / 19,600 (Krypton 157.4) = 0.16
19,600 (Krypton 157.4) / 3,136 (Helium – 272.2) = 6.25

3,136 (Helium – 272.2) / 64 (Xeon 111.8) = 49
64 (Xeon 111.8) / 3,136 (Helium – 272.2) = 0.020408163265306

3,136 (Helium – 272.2) / 1,225 (Radon -71.15) = 2.56
1,225 (Radon -71.15) / 3,136 (Helium – 272.2) = 0.390625 //+ 12 = 1,600 – 25,600

NEON

NEON RATIOS TO ALL OTHER NOBLE GASES INCLUDING HYDROGEN

147,456 (Neon -248.6) / 32,400 (Hydrogen -259.2) = 4.5111
32,400 (Hydrogen -259.2) / 147,456 (Neon 248.6)= 0.2197265625

147,456 (Neon – 248.6) / 3,136 (Helium – 272.2) = 47.02040816326531
3,136 (Helium – 272.2) / 147,456 (Neon - 248.6) = 0.02126736111

147,456 (Neon - 248.6) / 147,456 (-248.6 Neon) = 1
147,456 (-248.6 Neon) / 147,456 (Neon -248.6) = 1

147,456 (Neon - 248.6) / 82,944 (Argon -189.4) = 1.777
 82,944 (Argon - 189.4) / 147,456 (Neon - 248.6) = 0.5625

147,456 (Neon - 248.6) / 19,600 (Krypton 157.4) = 7.523265306122449
19,600 (Krypton - 157.4) / 147,456 (Neon - 248.6) = 0.1329210069444

 147,456 (Neon - 248.6) / 64 (Xeon - 111.8) = 2,304
64 (Xeon - 111.8) / 147,456 (Neon - 248.6) = 0.00043402777

147,456 (Neon -248.6) / 1,225 (Radon -71.15) = 120.3722448979592
1,225 (Radon -71.15) / 147,456 (Neon - 248.6) = 0.008307562934028

ARGON

ARGON RATIOS TO ALL OTHER NOBLE GASES INCLUDING HYDROGEN

82,944 (Argon – 189.4) / 32,400 (Hydrogen -259.2) = 2.56
2.56 // 1.28//0.64// 0.32//0.16//0.08 //0.04//0.02//
32,400 (Hydrogen -259.2) / 82,944 (Argon – 189.4)= 0.390625

82,944 (Argon – 189.4) / 3,136 (Helium -272.2) = 26.44897959183673
3,136 (Helium -272.2) / 82,944 (Argon – 189.4) = 0.037808641975309
          0.037808641975309 //+8 = 9.679012345679012

82,944 (Argon – 189.4) / 147,456 (-248.6 Neon) = 0.5625
147,456 (-248.6 Neon) / 82,944 (Argon – 189.4) = 1.777
1.777//0.888//0.444//0.222/0.111/0.0555

82,944 (Argon – 189.4) / 19,600 (Krypton 157.4)   = 4.231836734693878
19,600 (Krypton 157.4) / 82,944 (Argon – 189.4) = 0.236304012345679 //+5 =3.7808

82,944 (Argon – 189.4) / 64 (Xeon 111.8) = 1,296
64 (Xeon 111.8) / 82,944 (Argon – 189.4) = 0.000771604938272

82,944 (Argon – 189.4) / 1,225 (Radon -71.15) = 67.70938775510204
1,225 (Radon -71.15) / 82,944 (Argon – 189.4) = 0.0147690007716

KRYPTON

KRYPTON RATIOS TO ALL OTHER NOBLE GASES INCLUDING HYDROGEN

19,600 (Krypton) / 32,400 (Hydrogen -259.2) = 0.604938271604938
32,400 (Hydrogen -259.2) / 19,600 (Krypton) = 1.653061224489796

19,600 (Krypton - 157.4) / 3,136 (Helium -272.2) = 6.25
3,136 (Helium -272.2) / 19,600 (Krypton - 157.4) = 0.16

19,600 (Krypton - 157.4) / 147,456 (-248.6 Neon) = 0.1329210069444
147,456 (-248.6 Neon) / 19,600 (Krypton - 157.4) = 7.523265306122449

19,600 (Krypton - 157.4) / 19,600 (Krypton -157.4) = 1
19,600 (Krypton - 157.4)   / 19,600 (Krypton - 157.4) = 1

19,600 (Krypton - 157.4) / 64 (Xeon - 111.8) = 306.25
64 (Xeon - 111.8) / 19,600 (Krypton - 157.4) = 0.003265306122449

19,600 (Krypton - 157.4) / 1,225 (Radon -71.15) = 16
1,225 (Radon -71.15) / 19,600 (Krypton - 157.4) = 0.0625

XENON

XEON RATIOS TO ALL OTHER NOBLE GASES INCLUDING HYDROGEN

64 (Xeon -111.8) / 32,400 (Hydrogen -259.2) = 0.001975308641975
32,400 (Hydrogen -259.2) / 64 (Xeon -111.8) = 506.25

64 (Xeon -111.8) / 3,136 (Helium -272.2) = 0.02408163265306
3,136 (Helium -272.2) / 64 (Xeon -111.8) = 49

64 (Xeon -111.8) / 147,456 (-248.6 Neon) = 0.00043402777
147,456 (-248.6 Neon) / 64 (Xeon -111.8) = 2,304

64 (Xeon -111.8) / 19,600 (Krypton - 157.4) = 0.003265306122449
19,600 (Krypton - 157.4)   / 64 (Xeon -111.8) = 306.25

64 (Xeon -111.8) / 64 (Xeon 111.8) = 1
64 (Xeon 111.8) /64 (Xeon -111.8) = 1

64 (Xeon -111.8) / 1,225 (Radon -71.15) = 0.052244897959184
1,225 (Radon -71.15) / 64 (Xeon -111.8) = 19.140625

RADON

RADON RATIOS TO ALL OTHER NOBLE GASES INCLUDING HYDROGEN

1,225 (Radon -71.15) / 32,400 (Hydrogen -259.2) = 0.037808641975309
32,400 (Hydrogen -259.2) / 1,225 (Radon -71.15)= 26.44897959183673

1,225 (Radon -71.15) / 3,136 (Helium – 272.2) = 0.390625
3,136 (Helium – 272.2) / 1,225 (Radon -71.15) = 2.56

1,225 (Radon -71.15) / 147,456 (-248.6 Neon) = 0.008307562934028
147,456 (-248.6 Neon) / 1,225 (Radon -71.15) = 120.3722448979592

1,225 (Radon -71.15) / 19,600 (Krypton 157.4) = 0.0625
19,600 (Krypton 157.4)   / 1,225 (Radon -71.15) = 16

1,225 (Radon -71.15) / 64 (Xeon 111.8) = 19.140625
64 (Xeon 111.8) / 1,225 (Radon -71.15) = 0.052244897959184

1,225 (Radon -71.15) / 1,225 (Radon -71.15) = 1
1,225 (Radon -71.15) / 1,225 (Radon -71.15) = 1

OXYGEN

OXYGEN Melting point = -218.8 / 266.666 = 0.8205   [1/x = 1.218769043266301]
            and = -218.8 / 11.111 = 19.692   [1/x = 0.050782043469429]
            and = -218.8 / 1.333 = 164.1   [1/x = 0.006093845216332]
            and = -218.8 / 144 = 1.519444   [1/x = 0.658135283363803]

2 + 1 + 8 + 8 = (19) + 8 + 8 + 1 + 2 = 38

2 x 1 x 8 x 8 = (128) x 8 x 8 x 1 x 2 = 16,384

16,384 (Oxygen 218.8) / 22 = 2.909090
16,384 (Oxygen 218.8) x 22 = 1,408

16,384 (Oxygen 218.8) / 266.666 = 61.44   [1/x = 0.16276041666]
16,384 (Oxygen 218.8) / 11.111 = 1,474.56   [1/x = 0.000678168402778]
16,384 (Oxygen 218.8) / 1.333 = 122,288   [1/x = 0.000081380208333]
16,384 (Oxygen 218.8) / 144 = 113.777   [1/x = 0.0087890625]

OXYGEN RATIOS TO ALL OTHER NOBLE GASES INCLUDING HYDROGEN

OXYGEN RIGHT ANGLE TRIANGLE MELTING POINT – 218.8

-218.8 x 291.7333 = 63,831.25333
63,831.25333 / 2 = 31,915.62666
31,915.62666 / 24 Hours = 1,329.81 / 60 Minutes = 22.1636296296296
22.1636296296296 / 60 Seconds = 0.369393827160494
0.369393827160494 / 60 Arc Seconds = 0.006156563786008

-218.8 / 3 = 72.9333

364.666 = 5 x 72.9333

291.7333 = 4 x 72.9333

OXYGEN SQUARE MELTING POINT -218.8 x 4 = 875.2 / 3,456 = 0.25324074074074

-218.8

-218.8

OXYGEN

OXYGEN CUBE MELTING POINT -218.8 CUBED   = -10,474,708.672

1 year, 360 day Time in seconds cube 10,474,708.672 / 360 = 29,096.4129777
29,096.4129777 / 24 Hours = 1212.35054074074 / 60 Minutes = 20.20584234567901
20.20584234567901 / 60 Seconds = 0.33676403909465
0.33676403909465 / 60 Arc Seconds = 0.005612733984911
0.005612733984911 / 16 = 0.0003507958740570 / 22.5 = 0.000015509027736

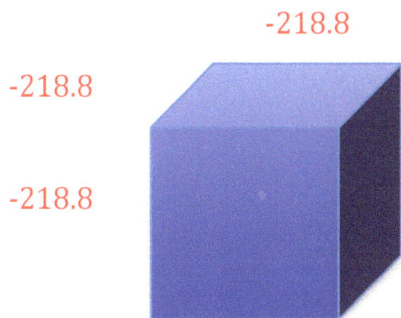

-218.8

-218.8

-218.8

10,474,708.672 / 360 (1 year) Time in Seconds cube = 29,096.4129777
10,474,708.672 / 24 = 436,446.194666 /60 = 7,274.1-32444 /60= 121.235 /60= 2.00205842345679
10,474,708.672 / 144,000 = 72.7410324   / 24 = 3.03087635185 / 60 =   / 60 = / 60 =
10,474,708.672 / 266.666 = 39,280.15   / 24 = / 60 = / 60 = / 60 =
10,474,708.672 / 6.666 = 1,571,206.3008   / 24 = / 60 =   / 60 = / 60 =
10,474,708.672 / 5.555 = 1,885,447.56096 / 24 = / 60 =   / 60 = / 60 =
10,474,708.672 / 4.444 = 2,356,8094512   / 24 = / 60 =   / 60 = / 60 =
10,474,708.672 / 3.333 = 3,142,412.6016   / 24 = / 60 =   / 60 = / 60 =
10,474,708.672 / 2.222 = 4,713,618.9024   / 24 = / 60 =   / 60 = / 60 =
10,474,708.672 / 1.111 = 9,427,237.8048   / 24 = / 60 =   / 60 = / 60 =

CARBON'S RESONANT FORMS

Carbon has 5 states, each state has a unique melting point or phase change

1) - DIAMOND

MELTING POINT = 3,700K or 3,426.85C   The Sum = 56 and Product = 33,177,600 (5,760 Sq)
33,177,600 / 144,000 = 230.4   33,177,600 / 518,400 = 64   33,177,600 / 31,104,000 = 1.0666 /// 0.2666
33,177,600 / 24 = 1,382,400 / 60 = 23,040 / 60 = 384 / 60 = 6.4 / 7.5 = 0.85333 \\\\\ 0.02666
33,177,600 / 864 = 38.4 / 144 = 0.2666 33,177,600 / 2,160 = 15,360 33,177,600 / 129,600 = 256\\\\16

MELTING POINT = 5,000K or 4,726.85C The Sum = 32 and Product = 180,633,600 (13,440 Sq)
180,633,600 / 144,000 = 1,254.4 180,633,600/518,400 = 384.444 180,633,600/31,104,000 = 5.80740
180,633,600 / 7.5 = 24,084,480 / 24 = 1,003,520 / 60 = 16,725.333 / 60 = 278.7555 / 60 = 4.64592592
180,633,600 / 25,920 = 6,968.888 180,633,600 /864,000 = 209.0666 180,633,600 / 266.666 = 677,376

2) - GRAPHENE

MELTING POINT = 4,510K or 4,236.85C The Sum= 28 and Product= 33,177,600 SAME AS DIAMOND
33,177,600 / 144,000 = 230.4   33,177,600 / 518,400 = 64   33,177,600 / 31,104,000 = 1.0666 \\\0.2666

MELTING POINT = 4,900K or 4,626.85C The Sum = 31 and Product = 132,710,400 (11,520 Sq)
132,710,400 / 24 = 5,529,600 / 60 = 92,160 / 60 = 1,536 / 60 = 25.6 / 7.5 = 3.41333\\\\\\\0.02666
132,710,400 / 144,000 = 921.6 / 266.666 = 3.456 / 11.111 = 0.31104000031104

3) - GRAPHITE

MELTING POINT = 3,800K or 3,526.85C Sum = 58 and Product = 51,840,000 (7,200 Squared)
51,840,000 / 24 Hours = 2,160,000 / 60 min = 36,000 / 60 Sec = 600 / 60 Arc = 10 / 7.5 = 1.333 \266.666
51,840,000/864,000=60 51,840,000/144,000=360 51,840,000/129,600=400 51,840,000/266.666=1944
51,840,000 / 25,920 (EQUINOX 25,920 Years) = 2,000 (2,000 BATHS IN THE 2/3 FULLMOLTEN SEA)

4) - BUCKEY BALLS CARBON FULLERENE C60

MELTING POINT = 1523K or 1,249.85C The Sum = 58 and Product = 8,294,400 (2,880 Sq)
8,294,400/ 25,920= 320 8,294,400/864,000= 9.6 8,294,400/144= 57,600 8,294,400/21,600=384\\\\24
8,294,400 / 24 = 345,600 / 60 = 5,760 / 60 = 96 / 60 = 1.6 / 7.5 = 0.21333\\\0.02666
MELTING POINT = 278.15K or 5C The Sum = 10 and Product = 25

MELTING POINT = 800K or 526.85C The Sum = 52 and Product = 5,760,000 (2,400 Sq)
5,760,000 / 25,920 = 222.222 5,760,000/864,000= 6.666 5,760,000/144= 40 5,760,000/21,600= 266.666
5,760,000 / 24 = 240,000 / 60 = 4,000 / 60 = 6.666 / 60 = 1.111 / 7.5 = 0.148148148

MELTING POINT = 707K or 433.85C The Sum = 46 and Product = 2,073,600 (1,440 Sq)
2,073,600 / 24 = 86,400 / 60 = 1,440 / 60 = 24 / 60 = 0.4 / 0.00075 = 266.666 (C music elemental Octave)
MELTING POINT = 739K or 465.85C The Sum = 56 and Product = 23,040,000 (4,800)
23,040,000 / 24 = 960,000 / 60 = 16,000 / 60 = 266.666 / 60 = 4.444   23,040,000 / 25,920 = 888.888

5) - NANOTUBULES

MELTING POINT = 5,200K or 4,926.85C The Sum = 68 and Product = 298,598,400 (17,280 Sq)
298,598,400 /24 = 12,441,600 / 60 = 207,360 / 60 = 3,456 / 60 = 57.6 (14.4 x 3)/ 7.5 = 7.68 \\\\\0.24
298,598,400 /144 = 2,073.6\\\259.2(H) 298,598,400/25,920 = 11,520 298,598,400/= 129,600 = 2,304

MELTING POINT VACUME = 3,073 K or 2,800C The Sum = 20 and Product = 256 (16 Sq)
MELTING POINT AIR = 1,023K or 750C The Sum = 13 and Product = 1,225 (35 Sq)

ALL SIX 864,000 COMBINATIONS

864,000	846,000	684,000	648,000	486,000	468,000
56,623,104*	55,443,456*	44,826,624*	42,467,328*	31,850,496*	30,670,848*
28,311,552*	27,721,728*	22,413,312*	21,233,664*	15,925,248*	15,335,424*
14,155,776*	13,860,864*	11,206,656*	10,616,832*	7,962,624*	7,667,712*
7,077,888*	6,930,432*	5,603,328*	5,308,416*	3,981,312*	3,833,856*
3,538,944*	3,465,216*	2,801,664*	2,654,208*	1,990,656*	1,916,928*
1,769,472*	1,732,608*	1,400,832*	1,327,104*	995,328*	958,464*
884,736*	866,304*	700,416*	663,552*	497,664*	479,232*
442,368*	433,152*	350,208*	331,776*	248,832*	239,616*
221,184*	216,576*	175,104*	165,888*	124,416	119,808*
110,592*	108,288*	87,552*	82,944*	62,208*	59,904*
55,296*	54,144*	43,776*	41,472*	31,104*	29,952*
27,684*	27,072*	21,888*	20,736*	15,552*	14,976*
13,824*	13,536*	10,944*	10,368*	7,776*	7,488*
6,912,000	6,768,000	5,472,000	5184,000	3,888,000	3,744,000
3,456,000	3,384,000	2,736,000	2,592,000	1,944,000	1,872,000
1,728,000	1,692,000	1,368,000	1,296,000	972,000	936,000

864,000	846,000	684,000	648,000	486,000	468,000
432,000	423,000	342,000	324,000	243,000	234,000
216,00	211,500	171,000	162,000	121,500	117,000
108,000	105,000	85,500	81,000	60,750	58,500
54,000	52,875	42,750	40,500	30,375	29,250
27,000	26,437.5	21,375	20,250	15,187.5	14,625
13,500	13,218.75	10,687.5	10,125	7,593.75	7,312.5
6,750	6,609.375	5,343.75	5,062.5	3,796.875	3,656.25
3,375	3,304.6875	2,671.875	2,531.25	1,898.43	1,828.1
1,687.5	1,652.34375	1,335.9375	1,265.62	949.2187	914.06
843.75	826.171875	667.96875	632.812	474.6093	457.03
421.875	413.0859375	333.984375	316.406	237.3046	228.51
210.9375	206.54296875	166.99218	158.2031	118.6523	114.25
105.46875	103.271484375	83.496093	79.10156	59.32617	57.128
52.734375	51.6357421875	41.7480468	39.55078	29,66308	28.564
26.3671875	25.8187109375	20.8740234	19.77539	14.83154	14.282

864,000	846,000	684,000	648,000	486,000	468,000

MATTER-MUSIC 266.666 OCTAVE SEQUENCE

1,118,481,066.666
559,240,533.333
279,620,266.666
139,810,133.333
69,905,066.666
34,952,533.333
17,476,266.666
8,738,133.333
4,369,066.666
2,184,533.333
1,092,266.666
546,133.333
273,066.666
136,533.333
68,266.666
34,133.333
17,066.666
==8,533.333==
4,266.666
==2,133.333== BUCKEY BALLS FULLERINE
==1,066.666== DIAMOND AND GRAPHENE
533.333

==266.666== MATTER ELEMENTAL OCTAVES
3,456,000 1,728,000 864,000 432,000
31,104,000,000 518,400
144,000 7.5
266.666 x 7.5 = 2,000 x 7.5 = 15,000
==133.333== GRAPHITE
66.666
33.333
16.666
8.333
4.1666
2.08333
1.041666
0.5208333
0.26041666
0.130208333
0.0651041666
0.03255208333
0.016276041666
0.0081380208333
0.00406901041666
0.002034505208333

PART 17 OF 20 – MSAART PATENT APPLICATION NOTES DRAFT 518,400 – 22:22:22 THUR 22ND SEPT 2022
© STRIKE FOUNDATION GUARANTEE LIMITED | MALCOLM BENDALL 2022 | GRAPHICS - STEVE EARL

ANGLE 1,440, 720, 360, 22.5 OCTAVE SEQUENCE (AOS)

12,079,595,520
6,039,797,760
3,019,898,880
1,509,949,440
754,974,720
377,487,360
188,743,680
94,371,840
47,185,920
23,592,960
11,796,480
5,898,240
2,949,120
1,474,560
737,280
<mark>368,640</mark> DIAMOND AND GRAPHENE
184,320
<mark>92,160</mark> DIAMOND AND GRAPHINE
<mark>46,080</mark> DIAMOND AND GRAPHINE
<mark>23,040</mark> NANOTUBULES
<mark>11,520</mark> NANOTUBULES
<mark>5,760</mark> BUCKEY BALLS
2,880
<mark>1,440</mark> BUCKEY BALLS AND GRAPHITE
720
360

2,880 1,440, 720, 360,180, 90, 45, 22.5 DEGREES ANGLE OCTAVES

360 x 7.5 = 2,700 x 7.5 = 20,250   360 / 7.5 = 48 / 7.5 = 6.4

180
90
45
22.5
11.25
5.625
2.8125
1.40625
0.703125
0.3515625
0.17578125
0.087890625
0.0439453125
0.02197265625
0.010986328125
0.0054931640625
0.00274658203125

PART 17 OF 20 – MSAART PATENT APPLICATION NOTES DRAFT 518,400 – 22:22:22 THUR 22ND SEPT 2022
© STRIKE FOUNDATION GUARANTEE LIMITED | MALCOLM BENDALL 2022 | GRAPHICS - STEVE EARL

RESONANCE 129,600 RFEU OCTAVES

543,581,798,400
271,790,899,200
135,895,449,600
67,947,724,800
33,973,862,400
16,986,931,200
8,493,465,600
4,246,732,800
2,123,366,400
1,061,683,200
530,841,600
265,420,800
132,710,400 / 4 = 33,177,600 = DIAMOND AND GRAPHENE
66,355,200
33,177,600 = DIAMOND AND GRAPHENE
16,588,800
8,294,400 BUCKY BALLS AND NANOTUBULES
4,147,200
2,073,600 BUCKY BALLS AND NANOTUBULES
1,036,800
518,400 GRAPHITE
259,200

129,600 RFEU (129,600 / 864 = 150)
(129,600 x 7.5 = 972,000 x 7.5 = 7,290,000, 129,600 / 7.5 = 17,280 / 7.5 = 2,304)

64,800
32,400
16,200
8,100
4,050
2,025
1,012.5
506.25
253.125
126.5625
63.28125
31.640625
15.8203125
7.91015625
3.955078125
1.9775390625
0.98876953125
0.494384765625
0.2471923828125

11.111 AND 5.555 AETHER OCTAVE SEQUENCE

186,413,511.111
93,206,755.555
46,603,337.777
23,301,688.888
11,650,844.444
5,825,422.222
2,912,711.111
1,456,355.555
728,177.777
364,088.888
182,044.444
91,022.222
45,511.111
22,755.555
11.377.777
5,688.888
2,844.444
1,422.222
711.111
355.555
177.777
88.888
44.444
22.222

11.111

3,456,000/11.111= 311,040 1,728,000/11.111= 155,520 864,000/11.111= 77,760 432,000/11.111= 38,880
311,040 x 11.111 = 3,456,000 311,040/11.111 = 27,993.6 518,400 /11.111 = 46,656 518,400 x 11.111= 576
144,000 x 11.111 = 1,600,000 144,000 / 11.111 = 129,600   7.5 x 11.111 =83.333   11.111 / 7.5 = 1.48148

5.555

3,456,000/5.555 = 622,080   1,728,000/5.555 = 311,040   864,000/5.555 = 155,520   432,000/5.555 = 77,760
311,040/5.555= 55,987.2 311,040 x 5.555= 1,728,000 518,400/5.555 = 93,312 518,400 x 5.555 = 2,880,000
144,000 / 5.555 = 25,920 / 3.333 = 7,776 x 1.333 = 10,368/ 24 = 432/60 = 7.2/60 = 0.12/60 = 0.002
144,000 x 5.555 = 800,000 / 24 = 33,333.333 / 60 = 555.555 / 60 = 9.2592 / 60 = ***0.154320987654321***
1x5x4x3x2x9x8x7x6x5x4x3x2x1 = 43,545,600 x 43,545,600 = 1,896,219,279,360,000
7.5 / 5.555 = 1.35   5.555 x 7.5 = 41.666

2.777
1.3888
0.69444
0.347222
0.1736111
0.08680555
0.0434027888
0.0217013888
0.01085069444
0.005425347222
0.0027126736111

12, 24, 48 AETHER TO MATTER OCTAVE SEQUENCE

402,653,184
201,326,592
100,663,296
50,331,648
25,165,824
12,582,912
6,291,456
3,145,728
1,572,<mark>864</mark>
786,<mark>432</mark>
393,<mark>216</mark>
196,608
98,304
49,152
23,576
12,228
6,144
3,072
1,536
768
384
192
96
48
24

3,456,000/24= 144,000 1,728,000/24= 72,000 864,000/24 = 36,000 432,000/24= 18,000
31,104,000,000 / 24 = 12,96,000,000 518,400 / 24 = 21,600 518,400 x 24 = 12,441,600
144,000 /24= 6,000(MOLTEN SEA SPHERE VOLUME IN BATHS) 12/7.5 = 1.6 7.5/24 = 0.3125 24/7.5= 3.2
24 / 11.111 = 2.16 24 x 11.111 = 266.666 12 x 7.5 = 90 24 x 7.5 = 180 48 x 7.5 = 360

12
6
3
1.5
0.75 (LIGHT TRAVELS 7.5 x AROUND THE EARTH IN 1 SECOND)
0.375
0.1875
0.09375
0.046875
0.0234375
0.01171875
0.005859375
0.0029296875
0.00146484375
0.000732421875

1). DIAMOND
33,1777,600

MELTING POINT = 3,700K or 3,426.85C Sum = 56 Product = 33,177,600 (5,760Sq)

33,177,600 / 144,000 = 230.4   33,177,600 / 518,400 = 64   33,177,600 / 31,104,000 = 1.0666 /// 0.2666

33,177,600 / 24 = 1,382,400 / 60 = 23,040 / 60 = 384 / 60 = 6.4 / 7.5 = 0.85333 \\\\\ 0.02666

33,177,600 / 864 = 38.4 / 144 = 0.2666 33,177,600 / 2,160 = 15,360 33,177,600 / 129,600 = 256\\\\16

MELTING POINT = 5,000K 4,726.85C The Sum = 32 Product = 180,633,600(13,440S q)

180,633,600 / 144,000 = 1,254.4 180,633,600/518,400 = 384.444 180,633,600/31,104,000 = 5.80740

180,633,600 / 7.5 = 24,084,480 / 24 = 1,003,520 / 60 = 16,725.333 / 60 = 278.7555 / 60 = 4.64592592

180,633,600 / 25,920 = 6,968.888 180,633,600 /864,000 = 209.0666 180,633,600 / 266.666 = 677,376

2). GRAPHENE
33,177,600

MELTING POINT = 4,510K or 4,236.85C Sum= 28 Product= 33,177,600 SAME AS DIAMOND

33,177,600 / 144,000 = 230.4   33,177,600 / 518,400 = 64   33,177,600 / 31,104,000 = 1.0666 \\\0.2666

MELTING POINT = 4,900K or 4,626.85C Sum = 31 Product = 132,710,400 (11,520 Sq)

132,710,400 / 24 = 5,529,600 / 60 = 92,160 / 60 = 1,536 / 60 = 25.6 / 7.5 = 3.41333\\\\\\\0.02666

132,710,400 / 144,000 = 921.6 / 266.666 = 3.456 / 11.111 = 0.31104000031104

3). GRAPHITE
51,840,000

MELTING POINT = 3,800K or 3,526.85C Sum = 58 Product = 51,840,000 (7,200 Squared)

51,840,000 / 24 Hours = 2,160,000 / 60 Min = 36,000 / 60 Sec = 600 / 60 Arc = 10 / 7.5 = 1.333\266.666

51,840,000/864,000=60 51,840,000/144,000=360 51,840,000/129,600=400 51,840,000/266.666=1944

51,840,000 / 25,920 (EQUINOX 25,920 Years) = 2,000 (2,000 BATHS IN THE 2/3 FULLMOLTEN SEA)

4). BUCKEY BALLS CARBON FULLERENE C60
23,040,000 – 2,073,600 - 8,294,400 - 5,760,000

MELTING POINT = 1523K or 1,249.85C Sum = 58 Product = 8,294,400 (2,880 Sq)

8,294,400 / 25,920 = 320   8,294,400 / 864,000 = 9.6

8,294,400 / 144 = 57,600   8,294,400 / 21,600 = 384\\\\24

8,294,400 / 24 = 345,600 / 60 = 5,760 / 60 = 96 / 60 = 1.6 / 7.5 = 0.21333 \\\0.02666

MELTING POINT = 278.15K or 5C The Sum = 10 and Product = 25

MELTING POINT = 800K or 526.85C Sum = 52 Product = 5,760,000 (2,400 Sq)

5,760,000 / 25,920 = 222.222

5,760,000 / 864,000 = 6.666   5,760,000 / 144 = 40   5,760,000 / 21,600= 266.666

5,760,000 / 24 = 240,000 / 60 = 4,000 / 60 = 6.666 / 60 = 1.111 / 7.5 = 0.148148148

MELTING POINT = 707K or 433.85C The Sum = 46 Product = 2,073,600 (1,440 Sq)

2,073,600 / 24 = 86,400 / 60 = 1,440 / 60 = 24 / 60 = 0.4 / 0.00075 = 266.666 (C music elemental Octave)

MELTING POINT = 739K or 465.85C The Sum = 56 and Product = 23,040,000 (4,800)

23,040,000 / 24 = 960,000 / 60 = 16,000 / 60 = 266.666 / 60 = 4.444   23,040,000 / 25,920 = 888.888

5) - CARBON NANOTUBULES

Melting Point 5,200 K or 4,926.85 C

MELTING POINT = 5,200K or 4,926.85C Sum = 68 Product = 298,598,400 (17,280 Sq)

298,598,400 /24 = 12,441,600 / 60 = 207,360 / 60 = 3,456 / 60 = 57.6 (14.4 x 3)/ 7.5 = 7.68 \\\\\0.24
298,598,400 /144 = 2,073.6\\\259.2(H)   298,598,400/25,920 = 11,520   298,598,400/= 129,600 = 2,304

MELTING POINT VACUME = 3,073 K or 2,800C The Sum = 20 and Product = 256 (16 Sq)

MELTING POINT AIR = 1,023K or 750C The Sum = 13 and Product = 1,225 (35 Sq)

4,926.85 / 1 year Time 360 = 13.68569
4,926.85 x 1 year Time 360 = 1,773,666

1,773,666/24 = 73,902.75 /60= 1,231.7125 /60= 20.528541666 /60 = 0.34214236111

4,926.85 / 24 = 205.28541666 / 60 = 3.4214236111 / 60 = 0.0570237226851852
0.0570237226851852 / 3.333 = 0.000950395447531

4,926.85 x 1 year Time 360 = 1,773,666

4,926.85 / 144,000 = 0.034214236111 / 7.5 = 0.00456189814814
4,926.85 x 144,000 x 7.5 =   x 720 = x 16 = x 22.5 =

4,926.85 / 266.666 = 18.4756875
4,926.85 x 266.666 = 1,313,826.666

4,926.85 / 1.333 = 3,695.1375
4,926.85 x 1.333 = 6,569.1333

4,926.85 / 11.111 = 443.4165
4,926.85 x 11.111 = 54,742.777

4,926.85 / 144 = 34.214236111
4,926.85 x 144 = 709,466.4

4,926.85 / 7.5 = 656.91333
4,926.85 x 7.5 = 36,951.375

4,926.85 / 24 = 205.28541666
4,926.85 x 24 = 118,244.444

4,926.85 / 2,000 = 2.463425
4,926.85 x 2,000 = 9,853,700

5) - CARBON NANOTUBULES

4,926.85 / 864 = 5.702372685185185
4,926.85 x 864 = 4,256,798.4

4,926.85 / 432 = 11.40474537037037
4,926.85 x 432 = 2,128,399.2

4,926.85 / 216 = 22.0809490740740740
4,926.85 x 216 = 1.064,199.6

4,926.85 / 108 = 45.61898148148148
4,926.85 x 108 = 532,099.8

4,926.85 / 792 = 6.220770202020
4,926.85 x 792 = 3,902,65.2

4,926.85 / 396 = 12.44154040404
4,926.85 x 396 = 1,951,032.6

4,926.85 / 25.92 = 109.0790895061728
4,926.85 x 25.92 = 127,703.952

4,926.85 / 6.666 = 739.0275
4,926.85 x 6.666 = 32,845.666

4,926.85 / 5.555 = 886.833
4,926.85 x 5.555 = 27,371.3888

4,926.85 / 4.444 = 1,108.54125
4,926.85 x 4.444 = 21,897.111

4,926.85 / 3.333 = 1,478.055
4,926.85 x 3.333 = 16,422.8333

4,926.85 / 2.222 = 2,217.0825
4,926.85 x 2.222 = 10,948.555

4,926.85 / 1.111 = 4,434.165 / 9 = 492.685
4,926.85 x 1.111 = 44,341.65 x 9 = 399,074.85

5) - CARBON NANOTUBULES

CARBON RIGHT ANGLE TRIANGLE MELTING POINT 4,926.85 C

4,926.85 (3 side) x 6,569.1333 (4 side) = 32,365,134.56333 (Area of square)

32,365,134.56333 / 2 = 16,182,567.281666 (Area of the Right Triangle)

16,182,567.281666 / 24 Hours = 674,273.636736111

674,273.636736111 / 60 Minutes = 11,237.89394560185

11,237.89394560185 / 60 Seconds = 187.2982324266975

187.2982324266975 / 60 Arc Seconds = 3.121637207111625

4,926.85 / 3 = 1,642.28333

8,211.41666 = 5 x 1,642.28333

6,569.1333 = 4 x 1,642.28333

CARBON SQUARE MELTING POINT 4,926.85 x 4,926.85 = 24,273,850.9225

24,273,850.9225 / 518,400 = 46.82455810667438

24,273,850.9225 / 311,040,000 = 78.4093017779064

24,273,850.9225 / 266.666 = 91,026.94095937502

5) - CARBON NANOTUBULES

24,273,850.9225 / 11.111 = 2,184,646.583025
24,273,850.9225 / 1.333 = 18,205,388.191875
24,273,850.9225 / 144 = 168.568.4091840278
24,273,850.9225 / 7.5 = 3,236,513.456333

4,926.85 x 4,926.85 (Square) = 24,273,850.9225

24,273,850.9225 / 2 = 12,136,9256.46125
12,136,9256.46125 / 24 Hours = 505,705.2275520833
505,705.2275520833 / 60 Minutes = 8,428.420459201389
8,428.420459201389 / 60 Seconds = 140.4736743200231
140.4736743200231 / 60 Arc = 0.009528400205761

4,926.85

4,926.85

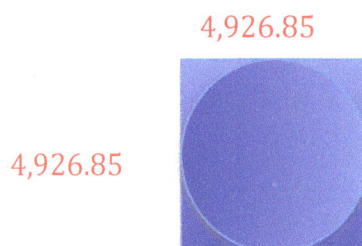

CARBON NANOTUBULE MELTING POINT 4,926.85 CUBED = 119,593,622,417.5191

119,593,622,417.5191 / 266.666 = 448,476,084.0656968

119,593,622,417.5191 / 11.111 = 10,763,426,017.57672

119,593,622,417.5191 / 1.333 = 89,695,216,813.13937

119,593,622,417.5191 / 144,000 = 830,511.2667883273

119,593,622,417.5191 / 7.5 = 15,945,816,322.33588

119,593,622,417.5191 / 24 = 4,983,067,600.729964

119,593,622,417.5191 / 864,000 = 138,418.5444647212

119,593,622,417.5191 / 432,000 = 276,837.0889294424

119,593,622,417.5191 / 2,160 = 55,367,417.78588848

119,593,622,417.5191 / 10,800 = 11,073,483.5571777

119,593,622,417.5191 / 6.666 = 17,939,043,362.62787

119,593,622,417.5191 / 5.555 = 21,526,852,035.15344

119,593,622,417.5191 / 4.444 = 26,908,565,043.94183

5) - CARBON NANOTUBULES

1 year, 360 day Time in seconds cube 119,593,622,417.5191/ =
/24/60/60 =
/ 3,456,000 =   /720/16/22.5 =

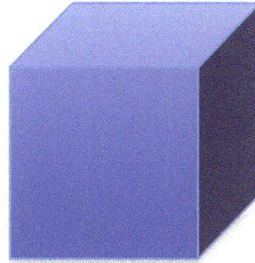

CARBON NANOTUBULES RATIOS TO HYDROGEN

CARBON NANOTUBULES 4,926.85 / HYDROGEN -259.2 = 19.007909
CARBON NANOTUBULES 4,926.85 x HYDROGEN -259.2 = 1,277,039.52
CARBON NANOTUBULES 4,926.85 + HYDROGEN -259.2 = 5,186.05

5 + 1 + 8 + 6 + 5 (25) 5 + 1 + 8 + 6 + 5 = 50
5 x 1 x 8 x 6 x 5 (1,200) x 5 x 1 x 8 x 6 x 5 = 1,440,000

1,440,000 / 50 = 28,800   1,440,000 x 50 = 72,000,000 / 2,160 = 33,333.333
72,000 / 7.5 = 9,600 / 24 = 400 / 60 = 6.666 /60 = 0.111 / 60 = 0.00185185185
72,000 / 24 = 3,000 / 60 = 50 / 60 = 0.8333 / 60 = 0.013888 x 400 = 5.555 / 0.125 = 44.444
72,000 x 720 = 51,840,000 / 864 = 60,000 / 144 = 416.666 / 720 = 0.57870370 / 16 = 0.036 / 22.5 =
x 400 = /0.125 =

CARBON NANOTUBULES 4,926.85 - HYDROGEN -259.2 = 4,667.65

5) - CARBON NANOTUBULES

PRODUCT AND SUM OF MIRRORED MELTING POINT 4,926.85 C

4 + 9 + 2 + 6 + 8 + 5 (34) + 4 + 9 + 2 + 6 + 8 + 5 = 68

4 x 9 x 2 x 6 x 8 x 5 (17,280) x 5 x 8 x 6 x 9 x 4 = 298,598,400

298,598,400 / 24 Hours = 12,441,600 / 60 minutes = 207,360 / 60 seconds = 3,456 / 60 Arc Sec = 57.6
298,598,400 / 864 = 345,600 / 144 = 2,400 / 7.5 = 320 / 16 = 20 / 22.5 = 0.888 / 0.125 = 7.111

298,598,400 (Carbon 4,926.85) / 68 = 4,391,152.941176471
298,598,400 (Carbon 4,926.85) x 68 = 20,304,691,200

298,598,400 (Carbon 4,926.85) / 266.666 = 1,119,744
298,598,400 (Carbon 4,926.85) x 266.666 = 79,626,239,999.999

298,598,400 (Carbon 4,926.85) / 11.111 = 26,873,856//
298,598,400 (Carbon 4,926.85) x 11.111 = 3,317760,000

298,598,400 (Carbon 4,926.85) / 1.333 = 223,948,800
298,598,400 (Carbon 4,926.85) x 1.333 =398,131,200

298,598,400 (Carbon 4,926.85) / 288 = 1,036,800*/518,400/259,200/129,600/64,800/32,400
298,598,400 (Carbon 4,926.85) x 288 = 85,996,339,200 / 864 = 99,532,800/144 = 691,200

298,598,400 (Carbon 4,926.85) / 144 = 2,073,600*/1,036,800/518,400/259,200/129,600
/64,800/ 32,400/16,200/8,100/4,050/2,025/1,012.5/506.25
298,598,400 (Carbon 4,926.85) x 144 = 42,998,169,600

298,598,400 (Carbon 4,926.85) / 864 = 345,600
298,598,400 (Carbon 4,926.85) x 864 = 257,989,017,600

298,598,400 (Carbon 4,926.85) / 432 = 691,200
298,598,400 (Carbon 4,926.85) x 432 = 128,994,508,800

298,598,400 (Carbon 4,926.85) / 2,160 = 138,240 /69,120/34,560/17,280/8,640/4,320/2,160
298,598,400 (Carbon 4,926.85) x 2,160 =644,972,544,000

298,598,400 (Carbon 4,926.85) / 108 = 2,764,800/1,382,400/691,200/345,600/172,800/86,400
298,598,400 (Carbon 4,926.85) x 108 = 32,248,627,200

298,598,400 (Carbon 4,926.85) / 792 = 377,018.181818
298,598,400 (Carbon 4,926.85) x 792 = 236,489,932,800

298,598,400 (Carbon 4,926.85) / 396 = 809,209.756097561
298,598,400 (Carbon 4,926.85) x 396 = 118,244,966,400

298,598,400 (Carbon 4,926.85) / 3,110,400 = 96/48/24/12/6/3/1.5/0.75
298,598,400 (Carbon 4,926.85) x 3,110,400 = 928,760,463,360,000

298,598,400 (Carbon 4,926.85) / 518,400 = 576//288//144//72//36//18//9//4.5//2.25
298,598,400 (Carbon 4,926.85) x 518,400 = 154,793,410560,000

5) - CARBON NANOTUBULES

298,598,400 (Carbon 4,926.85) / 900 = 331,766/165,888/82,944/41,472/20,10,368/5,184/
/2,592/1,296/648/324/162/81/40.50/20.25

298,598,400 (Carbon 4,926.85) x 900 = 268,738,560,000

298,598,400 (Carbon 4,926.85) / 360 = 829,440/414,720/207,360/103,680 /51,840 /25,920
/12,960/6,480/3,240/1,620/810/405/202.5/101.25/50.625

298,598,400 (Carbon 4,926.85) x 360 = 107,495,424,000

298,598,400 (Carbon 4,926.85) / 720 = 414,720/207,360/103,680/51,840/25,920/ 12,960/
/6,480/3,240/1,620/810/405/202.5/101.25

298,598,400 (Carbon 4,926.85) x 720 = 214,990,848,000

298,598,400 (Carbon 4,926.85) / 16 = 18,662,400 / 22.5 =
298,598,400 (Carbon 4,926.85) x 16 = 4,777,574,400

298,598,400 (Carbon 4,926.85) / 22.5 = 13,271,040
298,598,400 (Carbon 4,926.85) x 22.5 = 6,718,464,000

298,598,400 (Carbon 4,926.85) / 400 = 746,496
298,598,400 (Carbon 4,926.85) x 400 = 119,439,360,000

CARBON RIGHT ANGLE TRIANGLE MELTING POINT PRODUCT 298,598,400

298,598,400 (3 side) x 398,131,200 (4 side) = 11,888,133,931,010,000 (Area of the square)

11,888,133,931,010,000 / 2 = 59,406,696,550,400,000 (Area of the Right Triangle)

59,406,696,550,400,000 / 24 Hours = 2,476,694,568,960,000
2,476,694,568,960,000 / 60 Minutes = 41,278,242,816,000
687,970,713,600 / 60 Seconds = 11,466,187,560
11,466,187,560 / 60 Arc Sec = 191,102,976 (191,102,976 / 7.5 = 25,480,396.8\\-13\\ = 3,110.4)

298,598,400 / 3 = 99,532,800

497,664,000 = 5 x 99,532,800

398,131,200 = 4 x 99,532,800

CARBON NANOTUBULE SQUARE MELTING POINT
298,598,400 x 298,598,400 = 89,161,004,482,560,000

89,161,004,482,560,000 / 518,400 = 171,992,678,400
89,161,004,482,560,000 / 311,040,000 = 286,654,464
89,161,004,482,560,000 / 129,600 = 687,970,713,600
89,161,004,482,560,000 / 266.666 = 334,353,766,809,600.1
89,161,004,482,560,000 / 11.111 = 8,024,490,403,430,403
89,161,004,482,560,000 / 1.333 = 66,870,753,361,920,000
89,161,004,482,560,000 / 144,000,000,000 = 619173.64224
89,161,004,482,560,000 / 2,160 = 41,278,242,816,000
89,161,004,482,560,000 / 25,920 = 3,439,853,568,000

5) - CARBON NANOTUBULES

298,598,400 x 298,598,400 (Square) = 89,161,004,482,560,000

89,161,004,482,560,000 / 24 Hours = 3,715,041,853,440,000
3,715,041,853,440,000 / 60 Minutes = 6,197,364,224,000
61,917,364,224,000 / 60 Seconds = 1,031,956,070,400
1,031,956,070,400 / 60 Arc = 17,199,267,840 / 7.5 = 2,293,235,712

298,598,400

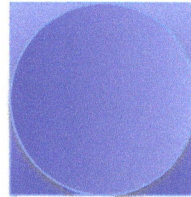

298,598,400

CARBON NANOTUBULE MELTING POINT PRODUCT 298,598,400 CUBED

298,598,400 x 298,598,400 x 298,598,400 = 26,623,333,280,890,000,000,000,000

26,623,333,280,890,000,000,000,000 / 266,666,666 = 99,837,499,803,320,000
26,623,333,280,890,000,000,000,000 / 111,111,111 = 239,699,999,767,600,000
26,623,333,280,890,000,000,000,000 / 1,333,333 = 199,674,999,606,600,000
26,623,333,280,890,000,000,000,000 / 144,000,000 = 184,884,258,895,000,000
26,623,333,280,890,000,000,000,000 / 75,000,000,000,000 = 354,977,777,078.4699
26,623,333,280,890,000,000,000,000 / 24,000,000,000,000 = 1,109,305,553,370.218
26,623,333,280,890,000,000,000,000 / 3,456,000,000,000 = 7,703510,787,293.184
26,623,333,280,890,000,000,000,000 / 864,000,000,000 = 30,814,043,149,172.74
26,623,333,280,890,000,000,000,000 / 432,000,000,000 = 61,628,086,298,345.47
26,623,333,280,890,000,000,000,000 / 2,160,000,000 = 12,325,617,259,690,000
26,623,333,280,890,000,000,000,000 / 108,000,000,000 = 246,512,345,193,381.9
26,623,333,280,890,000,000,000,000 / 8,884,000,000 (MARS D) = 2,996,773,219,370,244
26,623,333,280,890,000,000,000,000 / 4,222,000,000 (VENUS D) = 6,305,858,109,640,750

1 year, 360 day Time in seconds cube 26,623,333,280,890,000,000,000,000 / 360 =
73,953,703,558,010,000,000,000/24=3.081404324917/60=51356738581950000000
= 85,594,564,303,260,000/60 = 14,265,760,717,210,000 / 60 = 237,762,678,620,160
73,953,703,558,010,000,000,000 / 3,456,000 = / 720 = / 16 = / 22.5 =

298,598,400

298,598,400

298,598,400

5) - CARBON NANOTUBULES

298,598,400 / 144 = 2,073,600 / 7.5 = 276,480 (\\864)/720=384/16= 24/22.5= 1.0666 (266.6)
298,598,400 / 266.666 = 1,119,744 / 144 = 7,776 / 720 = 10.80 / 16 = 0.675 / 22.5 = 0.03 / 0.125 =
298,598,400 / 6.666 = 44,789,760/ 144 = 311,040 / 720 = 432 / 16 = 27 / 22.5 = 1.2 / 0.125 = 9.6
298,598,400 / 5.555 = 53,747,712/ 144 = 373,248 /720 = 518.4 /16 = 32.4 /22.5 = 1.44 / .125 = 11.52
298,598,400 / 4.444 = 67,184,640/ 144 = 466,560 / 720 = 684 /16 = 40.50 /22.5 = 1.8 (1/x = 0.555)
298,598,400 / 3.333 = 89,579,520/ 144 = 622,080 / 720 = 864 /16 = 54 / 22.5 = 2.4 / 0.125 = 19.2
298,598,400 / 2.222 = 134,369,280/ 144 = 933,120 / 720 = 1,296 /16 = 81 /22.5 = 3.6 /0.125 = 28.8
298,598,400 /1.111 = 268,738,560/ 144 = 1,866,240/ 720 = 2,592 /16 = 162 /22.5 = 7.2 /0.125 = 57.6
298,598,400 / 3,456 = 86,400 / 24 Hours = 3,600 / 60 Minutes = 60 / 60 Sec = 1 / 7.5 = 0.1333 \ 0.2666

CARBON NANOTUBULES RATIOS TO HYDROGEN

CARBON NANOTUBULES 4,926.85 / HYDROGEN -259.2 = 19.007909

CARBON NANOTUBULES 4,926.85 x HYDROGEN -259.2 = 1,277,039.52

CARBON NANOTUBULES 4,926.85 + HYDROGEN -259.2 = 5,186.05

5 + 1 + 8 + 6 + 5 (25) 5 + 1 + 8 + 6 + 5 = 50

5 x 1 x 8 x 6 x 5 (1,200) x 5 x 1 x 8 x 6 x 5 = 1,440,000

1,440,000 / 50 = 28,800   1,440,000 x 50 = 72,000,000 / 2,160 = 33,333.333

72,000 / 7.5 = 9,600 / 24 Hrs = 400 / 60 Min = 6.666 /60 Sec = 0.111 / 60 Arc Sec = 0.00185185185

72,000 / 24 Hrs = 3,000 / 60 Min = 50 / 60 Sec = 0.8333 / 60 Arc = 0.0138 x 400 = 5.555 /0.125 = 44.444

72,000 x 720 = 51,840,000 /864 = 60,000 /144 = 416.666 /720 = 0.57870370/16 =0.036 /22.5 = 0.16

CARBON NANOTUBULES 4,926.85 - HYDROGEN -259.2 = 4,667.65

EXPLANATORY NOTES ON DIAMOND AND GRAPHENE

CARBON MONOXIDE (Mp -205.2 C Bp -191.5 C) AND HYDROGEN

CARBON RIGHT ANGLE TRIANGLE MELTING POINT C

(3 side) x (4 side) = (Area of square)

/ 2 = (Area of the Right Triangle)

/ 24 Hours =
/ 60 Minutes =
/ 60 Seconds =
/ 60 Arc Seconds =
/ 7.5 LCF =

/ 3 =

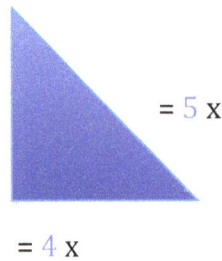

= 5 x

= 4 x

CARBON MONOXIDE SQUARE MELTING POINT   x =

/ 518,400 =
/ 311,040,000 =
/ 266.666 =
/ 11.111 =
/ 1.333 =
/ 144 =
/ 7.5 =

x (Square) =

/ 2 =
/ 24 Hours =
/ 60 Minutes =
/ 60 Seconds =
/ 60 Arc =

CARBON MONOXIDE MELTING POINT CUBED =

/ 266.666 =
/ 11.111 =
/ 1.333 =
/ 144 =
/ 7.5 =
/ 24 =
/ 864 =
/ 432
/ 216
/ 108

1 year, 360 day Time in seconds cube / =
/24/60/60 =
/ 3,456,000 =   /720/16/22.5 =

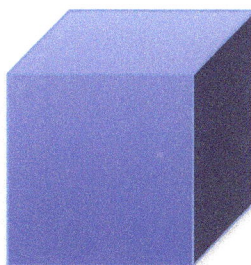

/ 1 year Time in Seconds cube =
/24 = /60 = /60 =   / 3.333 =
/144,000 / 7.5 / 720 = / 16 = / 22.5 =
/ 266.666 =
/ 6.666 =
/ 5.555 =
/ 4.444 =
/ 3.333 =
/ 2.222 =
/ 1.111 = / 9 =

CARBON MONOXIDE RATIOS TO HYDROGEN

CARBON MONOXIDE / HYDROGEN -259.2 =
CARBON MONOXIDE x HYDROGEN -259.2 =
CARBON MONOXIDE + HYDROGEN -259.2 =
$$5 + 1 + 8 + 6 + 5 \, () \, 5 + 1 + 8 + 6 + 5 =$$
$$5 \times 1 \times 8 \times 6 \times 5 \, () \times 5 \times 1 \times 8 \times 6 \times 5 =$$

/ 50 =   x 50 = / 2,160 =
/7.5 = / 24 = / 60 = /60 = / 60 =
/ 24 = / 60 = / 60 = / 60 = x 400 =   / 0.125 =
 = / 864 = / 144 = / 720 = / 16 = / 22.5 = x 400 = / 0.125 =

CARBON MONOXIDE - HYDROGEN -259.2 = 4,667.65

METHANE AND CARBON MONOXIDE

CARBON MONOXIDE AND METHANE RIGHT TRIANGLE MELTING POINT C

(3 side) x (4 side) = (Area of square)

/ 2 = (Area of the Right Triangle)

/ 24 Hours =
/ 60 Minutes =
/ 60 Seconds =
/ 60 Arc Seconds =

/ 3 =

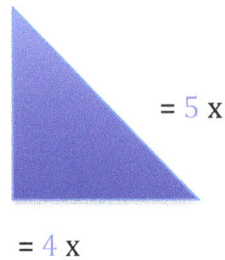

= 5 x

= 4 x

CARBON MONOXIDE AND METHANE SQUARE MELTING POINT x =

/ 518,400 =
/ 311,040,000 =
/ 266.666 =
/ 11.111 =
/ 1.333 =
/ 144 =
/ 7.5 =

x (Square) =

/ 2 =
/ 24 Hours =
/ 60 Minutes =
/ 60 Seconds =
/ 60 Arc =

METHANE AND CARBON MONOXIDE MELTING POINT CUBED =

/ 266.666 =
/ 11.111 =
/ 1.333 =
/ 144 =
/ 7.5 =
/ 24 =
/ 864 =
/ 432
/ 216
/ 108

1 year, 360 day Time in seconds cube / =
/24 Hrs = / 60 Min = / 60 Sec = / 60 Arc Sec =
/ 3,456,000 = / 720 = / 16 = / 22.5 =

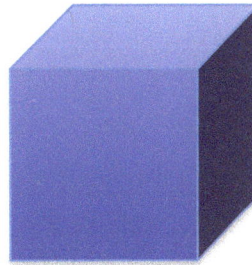

/ 1 year Time in Seconds cube =
/ 24 Hrs = / 60 Min = / 60 Sec = / 60 Arc Sec = / 3.333 =
/ 144,000 = / 7.5 = / 720 = / 16 = / 22.5 =
/ 266.666 =
/ 6.666 =
/ 5.555 =
/ 4.444 =
/ 3.333 =
/ 2.222 =
/ 1.111 = / 9 =

CARBON NANOTUBULES RATIOS TO HYDROGEN

CARBON NANOTUBULES 4,926.85 / HYDROGEN -259.2 = 19.007909
CARBON NANOTUBULES 4,926.85 x HYDROGEN -259.2 = 1,277,039.52
CARBON NANOTUBULES 4,926.85 + HYDROGEN -259.2 = 5,186.05
5 + 1 + 8 + 6 + 5 (25) 5 + 1 + 8 + 6 + 5 = 50
5 x 1 x 8 x 6 x 5 (1,200) x 5 x 1 x 8 x 6 x 5 = 1,440,000
1,440,000 / 50 = 28,800   1,440,000 x 50 = 72,000,000 / 2,160 = 33,333.333
72,000 / 7.5 = 9,600 / 24 = 400 / 60 = 6.666 / 60 = 0.111 / 60 = 0.00185185185
72,000 / 24 = 3,000 / 60 = 50 / 60 = 0.8333 / 60 = 0.013888 x 400 = 5.555 / 0.125 = 44.444
72,000 x 720 = 51,840,000 / 864 = 60,000 / 144 = 416.666 / 720 = 0.57870370 / 16 = 0.036 / 22.5
= x 400 = / 0.125 =
CARBON NANOTUBULES 4,926.85 - HYDROGEN -259.2 = 4,667.65

CARBON 3,500 C + OXYGEN -183 C
HYDROGEN -259.2

This table below needs editing for the combined numbers
HYDROGEN PLUS PALLADIUM COLD FUSION INTERACTION STRIPPING THE ELECTRON
FROM THE PROTON NON NUCLEAR REACTION, NO NUCLEUS, NO POLLUTION

2,963 (Palladium) x 252.87 (Hydrogen) = 749,253.81 x 1.333 = 999,005.078 / 11.111 / 7.5 / 16 / 22.5 =
2,963 (Palladium) / 252.87 = 11.71748329181   252.87 / 2,963 = 0.085342

252.87 + 2,963 (Palladium) = 3,215.87 is [3 + 2 + 1 + 5 + 8 + 7 (26) + 7 + 8 +5 + 1 + 2 + 3] = 52
3,215.87 = 3 x 2 x 1 x 5 x 8 x 7 (1,680) x 7 x 8 x 5 x 1 x 2 x 3 = 2,822,400 / 3.333 = 864,720

2,822,400 / 5.555 = 508,032   2,822,400 / 6.666 = 423,360
2,822,400/ 144 = 19,600   3,215.87/ 144,000 = / 7.5 = 2.61

CARBON RIGHT ANGLE TRIANGLE MELTING POINT 4,926.85 C

4,926.85 (3 side) x 6,569.1333 (4 side) = 32,365,134.56333 (Area of square)

32,365,134.56333 / 2 = 16,182,567.281666 (Area of the Right Triangle)

16,182,567.281666 / 24 Hours = 674,273.636736111
674,273.636736111 / 60 Minutes = 11,237.89394560185
11,237.89394560185 / 60 Seconds = 187.2982324266975
187.2982324266975   / 60 Arc Seconds = 3.121637207111625

4,926.85 / 3 = 1,642.28333

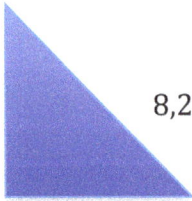

8,211.41666 = 5 x 1,642.28333

6,569.1333 = 4 x 1,642.28333

CARBON SQUARE MELTING POINT 4,926.85 x 4,926.85 = 24,273,850.9225

24,273,850.9225 / 518,400 = 46.82455810667438
24,273,850.9225 / 311,040,000 = 78.4093017779064
24,273,850.9225 / 266.666 = 91,026.94095937502
24,273,850.9225 / 11.111 = 2,184,646.583025
24,273,850.9225 / 1.333 = 18,205,388.191875
24,273,850.9225 / 144 = 168.568.4091840278
24,273,850.9225 / 7.5 = 3,236,513.456333

4,926.85 x 4,926.85 (Square) = 24,273,850.9225

24,273,850.9225 / 2 = 12,136,9256.46125
12,136,9256.46125 / 24 Hours = 505,705.2275520833
505,705.2275520833 / 60 Minutes = 8,428.420459201389
8,428.420459201389 / 60 Seconds = 140.4736743200231
140.4736743200231 / 60 Arc Sec = 0.009528400205761

CARBON NANOTUBULE MELTING POINT 4,926.85 CUBED = 119,593,622,417.5191

119,593,622,417.5191 / 266.666 =
119,593,622,417.5191 / 11.111 =
119,593,622,417.5191 / 1.333 =
119,593,622,417.5191 / 144 =
119,593,622,417.5191 / 7.5 =
119,593,622,417.5191 / 24 =
119,593,622,417.5191 / 864 =
119,593,622,417.5191 / 432
119,593,622,417.5191 / 216
119,593,622,417.5191 / 108

1 year, 360 day Time in seconds cube 119,593,622,417.5191/ =
/ 24 = / 60 = / 60 = / 60 =
/ 3,456,000 = / 720 = / 16 = / 22.5 =

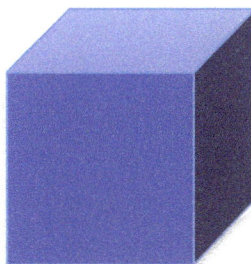

CARBON NANOTUBULES RATIOS TO HYDROGEN

CARBON NANOTUBULES 4,926.85 / HYDROGEN -259.2 = 19.007909
CARBON NANOTUBULES 4,926.85 x HYDROGEN -259.2 = 1,277,039.52
CARBON NANOTUBULES 4,926.85 + HYDROGEN -259.2 = 5,186.05
5 + 1 + 8 + 6 + 5 (25) 5 + 1 + 8 + 6 + 5 = 50
5 x 1 x 8 x 6 x 5 (1,200) x 5 x 1 x 8 x 6 x 5 = 1,440,000
1,440,000 / 50 = 28,800   1,440,000 x 50 = 72,000,000 / 2,160 = 33,333.333
72,000 / 7.5 = 9,600 / 24 = 400 / 60 = 6.666 /60 = 0.111 / 60 = 0.00185185185
72,000 / 24 = 3,000 / 60 = 50 / 60 = 0.8333 / 60 = 0.013888 x 400 = 5.555 / 0.125 = 44.444
72,000 x 720 = 51,840,000 / 864 = 60,000 / 144 = 416.666 / 720 = 0.57870370 / 16 = 0.036 / 22.5 =
x 400 = /0.125 =

PALLADIUM MELTING POINT

MELTING POINT = 1,555.2 x 2 = 3,110.4 (Sanskrit long year TIME) 1,555.2 / 11.111 = 139.68
1 x 5 x 5 x 5 x 2 (250) x 2 x 5 x 5 x 5 x 1 = 62,500 (6.25 Pi square cubits 1/16 th sphere x 10,000)
62,500 + 250 + 250 = 63,000
The product of Palladium's BOILING POINT (2,963) is 46,656 divided by the melting point (1,555.2) is
62,500 equals 0.746496 and 62,500 / 46,656 = 1.333 x 2 = 2.666
1 + 5 + 5 + 5 + 2 (18) + 2 + 5 + 5 + 5 + 1 = 36

PALLADIUM RIGHT ANGLE TRIANGLE BOILING POINT 1,555.2

AVT = 3,224,862.72 /2 = 1,612,431.36*
*/24/60/60/60 = 18.6624
18.6624 /16 / 22.5 = 0.05184
*/720/16/22.5 = 6.2208 (648)

1,555.2 / 3 = 518.4   518.4 x 5 = 2,592

2,073.6 = 518.4 x 4

PALLADIUM SQUARE BOILING POINT 1,555.2 x 4 = 6,220.8 / 3,456 = 1.8 (1/x = 0.555)

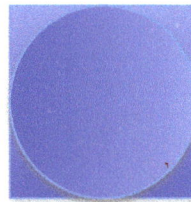

PALLADIUM CUBE BOILING POINT 1,555.2 CUBED = 3,761,479,876.608

1 yr = 360 days in Seconds = 31,104,000 Sec cubed = 30,091,839,012,860,000,000,000
30,091,839,012,860,000,000,000 / 3,761,479,876.6.608 = 8,000,000,000,000.
3,761,479,876.608 / 24 Hrs / 60 Min / 60 Sec / 60 Arc Sec / 7.5 = 96.7458816
3,761,479,876.608 / 3,456 = 1,088,391.168
8,000,000,000,000 / 7,200,000,000 = 1,111.111 / 16 = 69.444 / 22.5 =
3.08641975308641975 / 0.125 = 24.69135802469136 x 400 = 987.654320987654320

1 year Time in Seconds cubed /   3,761,479,876.608 =
8,000,000,000,000/24 = 333,333,333,333.333   /60 = 5,555,555,555.555 /60 =
92,592,592.592 /60 = 154,320.987654321   / 3.333 = 46,296.296296296

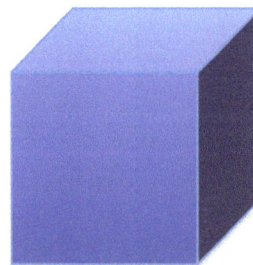

8,000,000,000,000 / 144,000 = 5,555,555.555/ 7.5 = 740,740.740 (1/x = 0.00000135)
740,740.740 / 720 = 1,028.80658436214 /16 = 64.3004115226337448559 /22.5 = 2.857796067672611
8,000,000,000,000 /129,600= 6,172,839,50617284/1.333/144,000=32.150205761316881/x=0.031104
8,000,000,000,000 / 266.666 = 3,000,000,000.
8,000,000,000,000 / 1.333 = 600,000,000,000
8,000,000,000,000 / 6.666 = 120,000,000,000
8,000,000,000,000 / 5.555 = 144,000,000,000
8,000,000,000,000 / 4.444 = 180,000,000,000
8,000,000,000,000 / 3.333 = 240,000,000,000
8,000,000,000,000 / 2.222 = 360,000,000,000
8,000,000,000,000 / 1.111 = 720,000,000,000 / 9 = 80,000,000,000

PALLADIUM BOILING POINT

BOILING POINT = 2,963 / 266.666 = 11.11125

2,963 (Palladium) so 2 + 9 + 6 + 3 (20) + 3 + 6 + 9 + 2 = 40
2,963 (Palladium) so 2 x 9 x 6 x 3 (216) 3 x 6 x 9 x 2 = 46,656 / 5.555 = 8,398.080

2,963 (Palladium) / 11.111 = 266.666 / 24 = 11.111
2,963 (Palladium) / 5.555 = 533.34   2,963 x 1.333 = 2,222.25   2,963 / 3.333 = 888.88

PALLADIUM RIGHT ANGLE TRIANGLE BOILING POINT 2,963

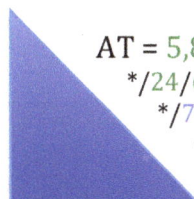

AT = 5,852,911.679*
*/24/60/60 = 67.74 / 16 = 4.23
*/720/16/22.5/.125 = 180.064
2,963 / 3 = 987.666   987.666 x 5 = 4,938.333

3,950.666 = 987.666 x 4

PALLADIUM SQUARE BOILING POINT 2,963 x 4 = 11,852 / 3,456 = 3.429398148148

/
PALLADIUM CUBE BOILING POINT 2,963 CUBED = 26,013,270,347
1 year, 360 days in seconds 31,104,000 cubed 30,091,839,012,860,000,000,000
30,091,839,012,860,000,000,000 / 26,013,270,347 = 1,156,788,001,333.879
1,156,788,001,333.879/24/60/60 = 13,338,750.0154384
1,156,788,001,333.879/3,456,000 = 334,718.7503859605 /720/16/22.5 /.125 = 10.33

1 year Time in Seconds cube /   26,013,270,347 =
1,156,788,001,333.879
1,156,788,001,333.879 /24/60/60/60   = 223,145.8335906403
 223,145.8335906403 / 3.333 =   66,943.7500771921
1,156,788,001,333.879 / 144,000 / 7.5   = 1,071,100.001235073
/ 720 / 16 / 22.5 =   148.7638890604269
1,156,788,001,333.879 / 864,000 =   1,338,875.001543842
1,156,788,001,333.879 / 129,600 = 8,925,833.343625612
1,156,788,001,333.879 / 266.666 = 4,337,955,005.002049
1,156,788,001,333.879 / 6.666 = 173,518,200200.0819
1,156,788,001,333.879 / 5.555 = 208,221,840,240.0983
1,156,788,001,333.879 / 4.444 = 260,277,300,300.1229
1,156,788,001,333.879 / 3.333 = 347,036,400,400.1638
1,156,788,001,333.879 / 2.222 = 520,554,600,600.2458
1,156,788,001,333.879 / 1.111 = 1,041,109,201,200.492 / 9 = 115,678,800,133.3879

PLATINIUM - Boiling point = - 3,825

OXYGEN – Boiling point = -183
         - Melting point =

HYDROGEN – Boiling point = - 252.87
            - Melting point = -259.20

CARBON – MELTING POINT - CARBON SHEETS DIAMOND = 3,426.85
        - MELTING POINT - CARBON SHEETS GRAPHENE = 4,236.85
        - MELTING POINT - CARBON MATRIX GRAPHITE = 3,526.85
        - MELTING POINT CARBON FORMED SPHERES   = 1,249.85
        - MELTING POINT CARBON FORM NANOTUBULE = 4,926.85

HELIUM – Boiling point = -268.93
        - Melting point = -272.2

DEUTERIUM – Boiling point = 101.4

TUNGSTEN – Boiling point 5,555 (Sun Aether) / 11.111 = 500
            5,555 / 266.666 = 20.83125
            5 x 5 x 5 x 5 (625) x 5 x 5 x 5 x 5 = 390,625
            390,625 + 625 + 625 = 391,875
            5 + 5 + 5+ 5 (20) + 5 + 5 + 5 + 5 = 40

            The product of Tungsten's boiling point (5,555) product is 390,625 divided
            by the MELTING POINT (3,420) product 576 is 678.168
            MELTING POINT 3,420 (Sun's radius)

            3 x 4 x 2 (24) 2 x 4 x 3 = 576 / 24 = 24
            576 + 24 + 24 = 624 / 24 = 26
            3 + 4 + 2 (9) + 2 + 4 + 3 = 18

GOLD – Boiling point 2,970 (Earth)
       - Melting point 1,063

SILVER – Boiling point 2,162 (Moon)

COPPER – Boiling point 2,562 (Earths Equinox)

BRASS - Melting point 900

RED BRASS – Melting point 1,000

IRON – Boiling point – 2,861
       Melting point – 1,538

CARBON – Boiling point - 4,827
         – Melting point - 3,550

NITROGEN – Boiling point – 195.8

INTERNAL ANGLES OF A CUBE 2,160 OCTAVES UP AND DOWN TABLE

OCTAVES UP

2,160 / 1.111 = 1944

0.) 1,944 / 0.432 = 4,500

1.) 3,888 / 0.432 = 9,000

2.) 7,776 / 0.432 = 18,000

3.) 15,552 / 0.432 = 36,000

4.) 31,104 / 0.432 = 72,000 Torus

5.) 62,208 / 0.432 = 144,000 double Torus = Photon

6.) 124,416 / 0.432 = 288,000

7.) 248,832 / 0.432 = 576,000

8.) 497,664 / 0.432 = 1,152,000

9.) 995,328 / 0.432 = 2,304,000

OCTAVES DOWN

2,160 / 1.111 = 1944

0.) 1944 / 0.432 = 4,500

1.) 972 / 0.432 = 2,250

2.) 486 / 0.432 =1,125

3.) 243 / 0.432 = 562.5

4.) 121.5 / 0.432 = 281.25

5.) 607.5 / 0.432 = 140.625

6.) 303.75 / 0.432 = 70.3125

7.) 151.875 / 0.432 = 35.15625

8.) 075.9375 / 0.432 = 17.578125

9.) 037.96875 / 0.432 = 8.7890625

VAJRA SINGULARITY POINT AT VORTEX INVERSION

VORTEX INVERSION TABLE

REVERSAL LEFT HAND CLOCKWISE TO RIGHT HAND ANTICLOCKWISE SPIRAL SHIFT
MIRROR NUMBER CREATION

1 x 8	+1 = 9
12 x 8	+2 = 98
123 x 8	+3 = 987
1234 x 8	+4 = 9876
12345 x 8	+5 = 98765
123456 x 8	+6 = 987654
1234567 x 8	+7 = 9876543
12345678 x 8	+8 = 98765432
123456789 x 8	+9 = 987654321

123456789 x 9 = 1,111,111,101

123456789 x 8 =   987,654,312

123456789 x 7 =   864,197,523

123456789 x 6 =   740,740,734

123456789 x 5 =   617,238,945

123456789 x 4   =   493,827,156

123456789 x 3 =   370,370,367

123456789 x 2   =   246,913,578

123456789 x 1 =   123,456,789

123456789 x 1.111 = 137,174,209.99999862826

123456789 x 11.111 = 1,371,742,099.9999862862

123456789 x 111 = 13,703,703,579

MILLENIUM TEMPLE

SOLAR SYSTEM YEAR 25,920, SUN CUBE 10,368,000
SUN 864,000 (888), EARTH 7,920 (999), MOON 2,160 (222)
RELATIONSHIP TO 2,160, 144, 108

EARTH'S EQUINOX CYCLE 25,920 x   7.5 = 194,400.....(POS)
EARTH'S EQUINOX CYCLE 25,920 /   7.5 = 3,456..........(POS)

EARTH'S EQUINOX CYCLE 25,920 x 1.333 = 34,560........(POS)
EARTH'S EQUINOX CYCLE 25,920 / 1.333 = 19,440........(POS)

EARTH'S EQUINOX CYCLE 25,920 x 11.111 = 288.............(LIGHT 144 x 2)
EARTH'S EQUINOX CYCLE 25,920 / 11.111 = 2,332.8......(COMPOSITE 864)

EARTH'S EQUINOX CYCLE 25,920 x 266.666 = 6,912,000..(RFEU)
EARTH'S EQUINOX CYCLE 25,920 / 266.666 = 97.2.............(7,920 EARTH)

EARTH'S EQUINOX CYCLE 25,920 / 129,600 = 0.2 (1/x = 5)
EARTH'S EQUINOX CYCLE 25,920 x 129,600 = 3,359,232,000 / 144,000 = 23,328

EARTH'S EQUINOX CYCLE 25,920 /   108 = 240
EARTH'S EQUINOX CYCLE 25,920 x   108 = 2,799,360 / 144,000 = 19.44

EARTH'S EQUINOX CYCLE 25,920 /   144 = 180   (1/x = 0.00555555)
EARTH'S EQUINOX CYCLE 25,920 x   144 = 3,732,480 / 144,000 = 25.92

EARTH'S EQUINOX CYCLE 25,920 / 864,000 = 0.03   (1/x = 33.333)
EARTH'S EQUINOX CYCLE 25,920 x 864,000 = 22,394,880,000/144,000= 155,520/5= 31,104

EARTH'S EQUINOX CYCLE 25,920 / 432,000 = 0.06   (1/x = 16.666)
EARTH'S EQUINOX CYCLE 25,920 x 432,000 = 11,197,440,000 / 144,000 = 77,760

EARTH'S EQUINOX CYCLE 25,920 / MOON'S D 2,160 = 12.........(SUNS AETHER R)
EARTH'S EQUINOX CYCLE 25,920 x MOON'S D 2,160 = 55,987,200 / 144,000 = 388.8

EARTH'S EQUINOX CYCLE 25,920 / MOON'S R 1,080 = 24.........(SUNS AETHER D)
EARTH'S EQUINOX CYCLE 25,920 x MOON'S R 1,080 = 27,993,600 / 144,000 = 194.4

EARTH'S EQUINOX CYCLE 25,920 / EARTH'S D 7,920 = 3.272727   (1/x = 0.30555)
EARTH'S EQUINOX CYCLE 25,920 x EARTH'S D 7,920 = 205,286,400 / 144,000 = 1,425.6

EARTH'S EQUINOX CYCLE 25,920 / MAR'S D 4,222 = 6.13927   (1/x = 0.16288580)
EARTH'S EQUINOX CYCLE 25,920 x MAR'S D 4,222 = 109,434,240 / 144,000 = 759.96

EARTH'S EQUINOX CYCLE 25,920 / JUPITER'S D 88,800 = 0.29189   (1/x = 3.42592592)
EARTH'S EQUINOX CYCLE 25,920 x JUPITER'S D 88,800 = 2,301,696,000 / 144,000 = 15,984

EARTH'S EQUINOX CYCLE 25,920 / JUPITER'S R 44,400 = 0.58378378   (1/x = 1.712961296)
EARTH'S EQUINOX CYCLE 25,920 x JUPITER'S R 44,400 = 1,150,848,000/144,000 = 7,992 EARTH

180 / 12 = 15
180 / 22.5 = 8
25,920 / 266.666 = 97.20000000000243   1/x = 0.01028806584362
97.20000000000243 x 11.111 = 1,080.0000000000002 [Moons Radius]

SUN'S CUBE 10,368,000 / 266.666 = 38,880 / 266.666 = 145.8
SUN'S CUBE 10,368,000 /   108 = 96,000 / 266.666 = 360
SUN'S CUBE 10,368,000 /   144 = 72,000
SUN'S CUBE 10,368,000 /   2,160 = 4,800
                    72,000 /   4,800 = 15

SUNS SQUARE 3,456,000 / 266.666 = 12,960 (RFEU)
SUNS SQUARE 3,456,000 /   108 = 32,000 / 266.666 = 120
SUNS SQUARE 3,456,000 /   144 = 24,000
SUNS SQUARE 3,456,000 / 2,160 = 1,600

24,000 / 22.5 = 1,066.666 \\266.666

3,456,000 / 266.666 = 12,960.000000000324   1/x = 0.00007716049383

3,456,000 x 11.111 = 384

SUN'S TORUS 2,160,000 (864,000 x 2.5) / 266.666 = 8,100
(SUN'S LIGHT POLARIZES, AS THE MOON IS THE SAME RESONANCE AS SUN'S TORUS 2,160)

8,100.0000000002025 //16,200.00000000405 // 32,400.0000000081
//64,800.0000000162 //129,600.0000000324 // 259,200.0000000648
//518,400.00000001296 [6 DAYS OF CREATION]

SUN'S TORUS   2,160,000 (864,000 x 2.5) / 108 = 20,000 / 2.5 = 8,000
SUN'S TORUS   2,160,000 (864,000 x 2.5) / 144 = 15,000 (2 x 7.5)
SUN'S TORUS   2,160,000 / 2,160 = 1,000

SUN'S OCTAVE   1,728,000 / 108 = 16,000 16,000 / 22.5 =
SUN'S OCTAVE   1,728,000 / 144   = 12,000 12,000/22.5 = 533.333 / 2 = 266.666
SUN'S OCTAVE   1,728,000 / 2,160 = 800

SUN'S DIAMETER 864,000 /   108 = 8,000
SUN'S DIAMETER 864,000 /   144 = 6,000 / 22.5 = 266.666
SUN'S DIAMETER 864,000 / 2,160 = 400 / 22.5 = 1.5
                    6,000 /   400 = 15
                      400 /   6,000 = 0.06666 (2/3 FULL MOLTEN SEA, RI 1.333, C 2.666)

6,000 / 22.5 = 266.666
864,000 / 266.666 = 3,240   1 / x = 0.00030864197531
3,240 x 11.111 = 36,000 x 1.333 = 48,000 / 7.5 = 6,400 / 16 = 400

311, 040,000,000,000 / 864,000 = 360,000 / 300 = 1,200,000 / 720 = 1,666.666 / 360 =
4.629 x 1.333 = 6.1728 / 1.111 = 5.555 x 1,000   = 5,555.555/ 16 = 347.222 / 22.5 =
15.432098765431944444/ 11.111 = 1.3888 x 8,000 = 11,111.111 x 7.5 = 83,333.333
/60 = 1,388.8 / 60 = 23.148148 / 37 = 0.625625 x 111 = 96.444

311,040,000,000,000 / 864,000 = 360,000,000

EARTHS CUBE   95,040 / 144 = 660
                     95,040 / 2,160 = 44
                     660 / 44 = 15   44 / 660 = .6666

EARTH'S SQUARE   31,680 / 144 = 220
                     31,680 / 2,160 = 14.666

EARTH'S TORUS 19,800 / 144 = 137.5
                     19,800 / 2,160 = 9.1666
                     137.5 / 9.1666 = 15

EARTH'S OCTAVE 15,840 / 144 = 110
                     15,840/ 2,160 = 7.333

EARTH'S DIAMETER 7,920 / 144 = 55

55 / 22.5 = 2.4 x 1,000 = 2,400 x 400 = 960,000 / 8 = 120,000
7,920/ 266.666 = 29.7   1/x = 0.03367003367

29.7 x 11.111 = 330 x 1.333 = 440/7.5 =58.666/16 = 3.666
311, 040,000,000,000 / 7,920 = 39,272727272 / 300 = 130,909,090.9090 / 720 =
181,818.1818181 / 360 = 505.0505050 x 1.333 = 673.400672400657 / 1.111 =
606.0606060 x 400 = 242,424.242 / 16 = 15,151.515 / 22.5 =
673.40067340066329966 / 11.111 = 60.606060 x 8,000 = 484,848.48484 x 7.5 =
3,636,363.6363636 / 60 = 60,606.060606 / 60 = 1,010.10101010050505 / 37 =
27.3000027300027163527 x 111 = 3,030.303030
311,040,000,000,000 / 7,920 = 39,272,727,272.727272

MOON'S CUBE 25,920 (EQUINOX 25,920) / 144 = 180
                     25,920 / 2,160 = 12
                     25,920 x 12 = 311,040

MOON'S SQUARE 8,640 / 144 = 60

MOON'S TORUS 5,400 / 144 = 37.5

MOON 2,160 / 144   = 15

55/ 22.5 = .666 x 1,000 = 666.666 x 400 = 266,666.666 / 8 = 33.333 / 11.111 = 2.99

2,160 / 266.666 = 8.1   1/x = 0.12345679012345
8.1//16.2//32.4//64.8//129.6//259.2//518.4//1,036.8//2,073.6//4,147.2//8,294.4//
16,588.8//33,177.6//66,355.2//132,710.4
    2,160 = A CUBES TOTAL INTERNAL ANGLES EQUALS ITS RESONANT FREQUENCY
311, 040,000,000,000 / 2,160 = 144,000,000,000 / 300 = 480,000,000 / 720 =
666,666.666 / 360 = 1,851.851 x 1.333 = 2,469.13580247 / 1.111 = 2,22.222 x 400 =
888,888.888 / 16 = 55,555.555 / 22.5 = 2,469.13580246987654 / 11.111 =
2,22.2244444666635552 x 8,000 = 1,777,795.5557333084442 x 7.5 =
13,333,466.667999813331 / 60 = 222,224.44446666355552 / 60
=3,703.7407411110592587 / 37 = 100.1011011111097097 x 111 = 11,111.222233321

666,666.666 x 1.333 = 888,88.888 / 1.111 = 800,000 x 400 = 320,000,000 / 7.5 =
42,666,666.666/ 360 = 118,518.518 /16 = 7407.40740/ 22.5 = 329.218106995885 x 8
= 2,633.7448559670781893 x 400 = 1,053,497.94238683127757

1 TO 36 EARTH MATTER ELEMENT NUMBERS

36 x	1	2	3	4	5	6	7	8	9	10	11	12	13	14	15	16	17	18	19	20	21	22	23	24	25	26	27	28
=	36	72	108	144	180	216	252	288	324	360	396	432	468	504	540	576	612	648	702	756	792	828	900	936	972	1008	1044	1080

=	37	74	111	148	185	222	259	296	333	370	407	444	481	518	555	592	629	666	703	740	777	814	851	888	925	962	999	1036
37 x	1	2	3	4	5	6	7	8	9	10	11	12	13	14	15	16	17	18	19	20	21	22	23	24	25	26	27	28

36 x	29	30	31	32	33	34	35	36	37	38	39	40	41	42	43	44	45	46	47	48
=	1044	1081	1118	1155	1192	1229	1266	1303	1340	1377	1414	1451	1488	1525	1562	1599	1636	1673	1710	1747

=	1073	1110	1147	1184	1221	1258	1295	1332	1369	1406	1443	1480	1517	1554	1591	1628	1665	1702	1739	1776
37 x	29	30	31	32	33	34	35	36	37	38	39	40	41	42	43	44	45	46	47	48

36x	49	50	51	52	53	54	55	56	57	58	59	60	61	62	63	64	65	66	67	68	69	70
=	1764	1800	1836	1872	1908	1944	1980	2016	2052	2088	2124	2160	2196	2232	2268	2304	2340	2376	2412	2448	2484	2520

=	1813	1850	1887	1924	1961	1998	2035	2072	2109	2146	2183	2220	2257	2294	2331	2368	2405	2442	2479	2516	2553	2516
37x	49	50	51	52	53	54	55	56	57	58	59	60	61	62	63	64	65	66	67	68	69	70

80 90 100 110 120 160 180 225 360

37 AETHER, TIME AND DIMENSIONAL NUMBERS

FIRST SOUND OCTAVES THEN THE ELEMENT OCTAVES

From the one ROD unit of 1.111 progressing in Octaves (simply doubling) first manifesting to us as Sound to the limit of human hearing of 20,000 vibrations per second then through the 144 known elements covering the ultrasonic range 20,000 to 60,000 vibrations per second to the high energy particles then through to First, Second and Third order light all governed by the 16 segments and ruled by the same spiral geometry using the molten sea as a standard. Aether and Time are ruled by 12, therefore the 7.5 degrees of separation, between 16 and 12 segments, also equals the second it takes light to circle the earth 7.5 times. As the Oxygen Isotopes progress 8, 9 and 10 so Music progresses as the table below demonstrates and the ratio of one Oxygen 88.999 to two Hydrogen 11.111 of living water.

TABLE OF ELEMENTAL OCTAVE PROGRESSION OF DAVID'S HARP MUSIC

C DOUBLE FLAT – 12   x 11.111   = 133.333
F DOUBLE FLAT - 16   x 11.111   = 177.777
A FLAT - 2 x 10 = 20   x 11.111   = 222.222
B FLAT - 2 x 11 = 22   x 11.111   = 244.444

C - 3 x 8 = 24   x 11.111 = 266.66 (C)
D - 3 x 9 = 27   x 11.111 = 300
E - 3 x 10 = 30   x 11.111 = 333.333
F - 4 x 8 = 32   x 11.111 = 355.555
G - 4 x 9 = 36   x 11.111 = 400
A - 4 x 10 = 40   x 11.111 = 444.444
B - 4 x 11 = 44   x 11.111 = 488.888
C - 4 x 12 = 48   x 11.111 = 533.333
533.333 (C#) – 266.666 C = 266.666 Interval

D SHARP - 6 x 9 = 54   x 11.111   = 600
E SHARP - 6 x 10 = 60   x 11.111   = 666.666
F SHARP - 6 x 11 = 66   x 11.111   = 711.11
G SHARP - 6 x 12 = 72   x 11.111   = 800
A SHARP - 8 x 10 = 80   x 11.111   = 888.888
B SHARP - 8 x 11 = 88   x 11.111   = 977.777
C SHARP - 8 x 12 = 96   x 11.111   = 1,066.64
D ##   - 9 x 12 = 108 x 11.111   = 1,200

MOLTEN SEA NUMBERS

NUMBERS FOR THE OUTSIDE OF THE MOLTEN SEA TIMES 11.111

Earth 7,920 x 11.111 = 87,999.99999999912 [88,000]   1 / x = 0.00001136363

88,000 / 266.666 = 330   1 / x = 0.00303030

Earth 7,920 / 266.666 = 29.7   1/x = 0.12345679012345

 7,920 x 266.666 = 2,112,000

Sun 864,000 x 11.111 = 9,599,999.999999904 [9,600,000]   1 / x = 0.0000001041667

9,600,000 / 266.666 = 36,000

Moon 2,160 x 11.111 = 23,999.99999999976 [24,000]   1/x =0.000041666

2,160 / 266.666 = 8.10000002025   1 / x = 0.123456789814815

(432,000 x 11.111 = (300 x 11.111 = 3,333.333) (150 x 11.111 = 1,666.666) (50 x 11.111 = 555.555)

(6 x 11.111 = 66.666) (3,333.333 / 0.125 = 26,666.666) (1,666.666 / 0.125 = 13333.33)

(555,555 / 0.125 = 4,444,440) (4,444,440 / 1.3333 = 3,333,330.000000083333)

(4,444,440 / 0.125 = 35,555,520 / 0.125 = 284,444,160 / 0.125 = 2,275,553,280 / 0.125 =)

NUMBERS FOR THE INSIDE OF THE MOLTEN SEA IN HANDS TIMES 11.111

(45 x 11.111 = 500   1 / x = 0.002)

(45 x .666 = 30 [29.999] = 2x Depth of water in the Molten Sea but it is an inverted sphere 1/x = 0.0333)
(266.666/30 = 8.888 1/x = 0.1125 the area of the surface of the molten sea water is [Area = pie r (15) squared) = 706.86 square hands   706.86/ 16 = 44.17875 / 22.5 = 1.9635 x 400 = 785.4 RFEU

(22.5 x 11.111 = 250 1/x = 0.004 x 266.666 = 1.0666) (22.5 x 0.666 = 15 hands is the Depth of water in the molten sea 15 /2 = 7.5 hands resonant air cavity (light travels 7.5 times around the earth in one second giving the Standard Light Earth Frequency (SLEF) above the water to the Rim)
(266.666 / 15 = 17.777   1/x = 0.05625,   266.666 / 7.5 = 35.555   1 / x = 0.28125)

360 x 11.111 = 4,000   1/x = 0.00025 [4,000/266.666 = 15/0.125 = 120.00000000000018 1 / x = .008333

180 x 11.111 = 2,000   1 / x = 0.0005

90 x 11.111 = 1,000   1 / x = 0.001 [1,000 / 266.666 = 3.75/ 0.125 = 30.00000000000075   1 / x = 0.001

(16 x 11.111 = 177.777   1 / x =) (400 x 11.111 = 4,444.444   1/x =)

(9 x 11.111 = 10 [99.9999]   1/x = 0.01])

(11.111 x 11.111 = 1==23.456790123457==   1 / x = .0081)

NUMBERS FOR THE MOLTEN SEA IN CUBITS 10 and 5 TIMES 11.111

(5 x 11.111 = 55.55), (10 x 11.111 = 111.111), (30 x 11.111 = 333.333).

ARK OF THE COVENANT SACRED RULING NUMBERS

NUMBERS FOR THE ARK OF THE COVENANT IN CUBITS
(2.5 x 11.111 = 27.777) (1.25 x 11.111 = 13.888) (1.5 x 11.111 = 16.666)
(2.5 x 1.5 x 1.5 = 5.625 / 0.125 = 45)
(1.25 x 1.5 x 1.5 = 2.8125 / 0.125 = 22.5)
27.777 x 16.666 x 16.666 = 7,715.2160790120 / 0.125 = 61,721.728632096

NUMBERS FOR THE ARK OF THE COVENANT IN INCHES

(45 x 11.111 = 500), (22.5 x 11.111 = 250), (27 x 11.111 = 300)
(45 x 27 x 27 = 32,805 / 0.125 = 262,440)
(300 x 300 x 500 = 45,000 / 0.125 = 360,000,000)
(300 x 300 x 250 = 22,500,000 / 0.125 = 180,000,000)

NUMBERS FOR THE ARK OF THE COVENANT IN HANDS

(11.25 x 11.111 = 125), (5.625 x 11.111 = 62.5), (6.75 x 11.111 = 75)
(11.25 x 6.75 x 6.75 = 512.578125 / 0.125 = 4,100.625
(75 x 75 x 125 = 703,125 / 0.125 = 5,625,000)
(75 x 75 x 62.5 = 351,562.5 / 0.125 = 2,812,500)

THE HIGH PRIESTS EPHOD RULING NUMBERS AND FORM

NUMBERS FOR THE EPHOD IN HANDS

(2.25 x 11.111 = 25, 2.25 x 2.25 = 5.0625 / 0.125 = 40.5
(25 x 25 = 625 / 0.125 = 5,000)

MOLTEN SEA ARK ATOMIC RECONSTRUCTION QUANTUM TECHNOLOGIES RULING NUMBERS

RULING NUMBER 1.1111

RULING NUMBER 1.3333

RULING NUMBER 5

RULING NUMBER 9

RULING NUMBER 10

RULING NUMBER 11.11

RULING NUMBER 12

RULING NUMBER 16

RULING NUMBER 22.5

4,500/

RULING NUMBER 27

RULING NUMBERS 37

RULING NUMBER 45

RULING SUN'S LIGHT NUMBER 144

 RULING NUMBER 150 SUNS AND 150 MOONS ON OUTER MOLTEN SEA RIM

150/ 12 / 1.111 =11.25   150 / 360 / 1.111 = .375
150 x 1.333 = 199.999   150 / 16 = 150 / 12

RULING NUMBER 300 TOTAL NUMBER OF SUNS AND MOONS ON MOLTEN SEA RIM

300 x 1.333 x 1.111 = 444.444 IN MUSIC A
300 / 12 / 1.111 = 22.5
300 / 16 = 18.753   300 / 12 = 25
LILLIES 4,800 lilies x 1,800 petals = 8,640,000

RULING NUMBER 360

RULING NUMBER 432

RULING NUMBER 450

RULING NUMBER 720

720 / 16 = 45   720 / 1.333 = 540   720 / 1.111 = 432 [432+234 = 666]

RULING NUMBER 2,520

MOLTEN SEA ARK ATOMIC RECONSTRUCTION QUANTUM TECHNOLOGIES RULING NUMBERS

2,520 = 3 x 4 x 5 x 6 x 7
2,520 = 3 x 7 x 10 x 12 (all perfect numbers)
2,520 x 2 =5,040
2,520

RULING NUMBER 4,500 VOLUME OF RIM OF THE MOLTEN SEA

4,500 / 0.125 = 36,000

RULING NUMBER 14,400

14,400 = 1+2+3+4+5

VOLUME OF THE MOLTEN SEA IS 3,000 BATHS x 72 = 216,000 LOGS

216,000 Logs = 300 Homer = 54,000 Cad

VOLUME OF THE MOULTEN SEA IS 3,000 BATHS X 72 = HANDS

VOLUME OF THE MOULTEN SEA IS 2,000 BATHS x 72 = 144,000
1,44,000 / 11.11 EQUALS 129,600 RESONANT FREQUENCY ENERGY UNITS [RFEU]

VOLUME OF THE MOULTEN SEA IS 2,000 BATHS x 72 = 144,000 logs
144,000 logs = 36,000 Cad = 200 homer = 18,000 Him = 36,000 Kab =144,000 Logs
1,44,000 / 11.11 EQUALS 129,600 RESONANT FREQUENCY ENERGY UNITS [RFEU]

VOLUME OF THE MOULTEN SEA IS 2,000 BATHS = 150 RITUAL BATHS

VOLUME OF THE MOLTEN SEA INNER SPHERE = 47,712 / 2 = 23,856 / 0.666 = 35,784

VOLUME OF THE MOULTEN SEA IS 2,000 BATHS x 72 = 129,600 (RFEU)

VOLUME OF THE MOULTEN SEA IS 129,600 x 1.333 = 172,800

172,800 RFE [1728 + 8271 = 9,999]

VOLUME OF THE MOLTEN SEA OUTER SPHERE = 56,684 / 2 = 28,342 / 0.666 = 42,513

 MINUS VOLUME OF THE INNER SPHERE = 8,972 / 2 = 4,486 / 0.6666 = 6,729

AREA OF TOP OF THE MOLTEN SEA =

CIRCUMFERANCE OF THE MOLTEN SEA =

RULING NUMBERS 129,600, 9,999 AND 36
MADE FROM REFLECTIVE NUMBER 3456 AND

259,200 [EARTHS EQUINOX] = 3 X 4 X 5 X 6 X 6 X 5 X 4 X 3 X 2 / 2 = 129,600 / 0.125 = 1,036,800

129,600 / 36 = 3,600 [36 = 3+4+5+6+6+5+4+3]   9,999 = 3,456 + 6,543

129,600 / 1.111 = 116,640   129,600 X 1.111 = 144,000
129,600 / 1.333 = 97,200.0000000024   129,600 x 1.333 = 172,800

129,600 / 360 = 129,600/180= 129,600/ 90 = 129,600/45 = 129,600/22.5 =

129,600 / 9,999=12.9612961   9,999/129,600 = 0.077152777

9,999 = 3,456 + 6,543   9,999/36 = 277.75 /0.125 = 2,222

9,999 / 3,456 = 2.8932291666 = 1/X = 0.3456345634563456

9,999 / 6,543 = 1.52819807427785   1/x = 0.654365436543

6,543 / 9,999 = 0.654365436543

9,999 / 720 = 13.8875   9,999 / 360 = 27.775   9,999 / 180 = 55.55   9,999 / 90 = 111.111
9,999 / 45 =222.2   9,999 / 11.25 = 888.8

144,000 / 1.111 = 129,600

360 X 360 = 129,600   180 x 720 = 129,600

129,600 x 7.5 = 972,000 / 360 = 2,700 / 9 = 300 / 1.333 = 225 / 1.111 =

202.5000000000070875 = 1/x = 0.00493827160494

RULING NUMBER 1,776 = 999 + 777 AND 48
MADE FROM REFLECTIVE NUMBER 789

789 + 987 = 1,776   999+ 777 = 1,776

7 + 8 + 9 +9 + 8 +7 = 48

7 x 8 x 9 x 9 x 8 x 7 = 254.016

RULING NUMBER 144,000, 7,777 AND 28
MADE FROM REFLECTIVE NUMBER 2345

At Base 9 the harmonic speed of light is 144,000 minutes of Arc
Where normal 86,400 (864,000 Sun 888) seconds in a day based on harmonic 8 are
expanded to into 9,720 (9,720 Earth 999) based on harmonic 9 splitting the earth into

27 grid hours the number of grid seconds then equals 97,200. Therefore the grid speed of light is 144,000 minutes of Arc per grid second.

2 to the 7th + 3 SQUARED + 5 CUBED = 144,000

144,000 = 2 X 3 X 4 X 5 X 5 X 4 X 3 X 2 X 10   144,000 / 0.125 = 1,152,000

144,000 / 28 = 5,142.85714285714   [28 = 2+3+4+5+5+4+3+2] 7,777 = 2,345 + 5,432

144,000 / 7,777= 18.5161373280185   7,777 / 144,000 = 0.054006944
144,000 x 144,000 =20,736,000,000 / 7,777 = 2,66,323.7752346663238

144,000 x 144,000 x 144,000 = 2,985,984,000,000,000 / 7,777 = 383,950,623,633.792

144,000 SQUARE ROOT = 379.473319220205 / 7,777 = 0.0487943061875

144,000 CUBE ROOT = 52.4148278841779 / 7,777 = 52.4148278841779

7,777 / 1.5241579027586648377 = 5,102.48970000042

7,777 = 2,345 + 5,432   7,777/28 = 277.75 /0.125 = 2,222

7,777 / 2,345 = 3.31641791044776 =   1/x = 0.30153015

7,777 / 5,432 = 1.43170103092784

5,432 / 7,777 = 0.698469846984

2,345 / 7,777 = 0.30153015301

144,000 /1.111 = 129,600

360 x 400 = 144,000   180 x 800 = 144,000   200 x 720 = 144,000

RULING NUMBER 1,728,000 [Suns Resonant Frequency Energy units]
9999 AND 36 and 48 MADE FROM REFLECTIVE NUMBER 1728

1,728 + 8,271 = 9,999   1+7+2+8+8+2+7+1= 36

1,728 / 36 = 48

1 x 7 x 2 x 8 x 8 x 2 x 7 x 1 = 12,544   / 36 = 348.444

SANSKRIT LONG 100 YR RULING NUMBER   3,110,400,000,000

311,040,000,000,000 / 25,920 = 12,000,000,000,000

3,110,400, 000,000 / 25,920 = 120,000,000

3,110,400,000,000 / 864,000 = 3,600,000 / 266.666 = 13,500 / 22.5 =

3,110,400,000,000 / 7.5 =

THE RIGHT ANGLE TRIANGLE DISPLAYS THE RELATIONSHIPS BETWEEN
AETHER, MATTER, TIME AND SPACE FOR THE IMPLODED SPHERE TORUS
WITH RATIOS 3/4, 4/3, 3/5, 5/3, 4/5, 5/4 AND A x A + B x B = C x C
THE SCAFFOLDING FOR THE TORUS IS ITS SKIN Pi WHICH IF THE RADIUS AND
DEGREES ARE 22.5 THEN THE AREA OF THE TORUS IS EXACTLY

Pi x 100 = 314.1592653589793

HARMONIC SPEED OF LIGHT 144 (288) 72 x 3 = 216, 72 x 4 = 360 and 72 x 5 = 288

Area = TIME = 38,880/144 = 270
/24/60/60 x 7.5 x 144 = 486/22.5 = 216

MATTER 216 / 3 = 72   Light 72 x 5 = 288 [144 x 2]

Matter 360 = 72 x 4

LIGHT 144,000 A [720 x 200   900 x 160]

AT   = 13,824,000,000 / 4.444 = 3,110,400,000
/24/60/60 = 160,000/ 720 = 222.222

Light 144,000 / 3 = 48,000   Aether 48,000 x 5 = 240,000
[144,000 / 266.666 = 540]   [240,000 / 266.666 = 900]
[240,000 / 6.666 = 36,000]
[240,000 / 5.555 = 43,200]
[240,000 / 3.333 = 72,000]

Matter 48,000 x 4 = 192,000
[192,000/266.666 = 720   192,000/5.555 = 34,560   192,000/3.333=57,600   192,000/144,000=1.333 (16/12)]

LIGHT 144,000 C Hyp [720 x 200   900 x 160]

A 4,976,640,000/4.444= 1,119,744,000/360= 3,110,400
4,976,640,000/144= 34,560,000/25,920= 1,333.333
/24/60/60 = 57,600/720/22.5 = 3.555
Aether 28,800 x 3 = 86,400   Light 144,000 / 5 = 28,800
[144,000 / 266.666 = 540]   [144,000 / 8.888 = 162,200]
[144,000 / 6.666 = 21,600]
[144,000 / 5.555 = 25,920]
[144,000 / 4.444 = 32,400]
Matter 28,800 x 4 = 115,200   [144,000 / 3.333 = 43,200]

[115,200 / 266.666 = 432 115,200 / 8.888 = 12,960 115,200 / 6.666 = 17,280 115,200 / 5.555 = 20,736 115,200 / 4.444 = 25,920
[115,200 / 3.333 = 34,560   115,200 / 2.222 = 518,400   115,200 / 1.111 = 103,680   115,200/144 = 800   115,200 / 720 = 160]

LIGHT 144,000 B [720 x 200   900 x 160]

A = 7,776,000,000/4.444=1,749,600,000/360=4,860,000
7,776,000,000/144=54,000,000/900=60,000/8=7,500
/24/60/60=90,000/720=125/22.5= 5.555
Light 36,000 x 3 = 108,000   Aether 36,000 x 5 = 180,000
[108,000/5.555 = 19,440] [108,000/266.666 = 405]   [180,000/ 266.666 = 675]
[180,000/ 5.555 = 32,400]
[180,000 / 4.444 = 40,500]
[180,000 / 3.333 = 54,000]
144,000 / 4 = 36,000

144,000 / 8.888 = 16,200   144,000 / 7.777 = 18,514.2857   144,000 / 6.666 = 216,000   144,000 / 5.555 = 25,920
144,000 / 4.444 = 32,400   144,000 / 3.333 = 43,200   144,000 / 2.222 = 64,800   144,000 / 1.111 = 129,600
144,000 / 1.333 = 108,000   144,000 /   900 = 160   144,000 / 7.5 = 19,200   144,000 / 266.666 = 540

LIGHT CUBED 144,000 x 144,000 x 144,000 = 2,985,980,000,000,000

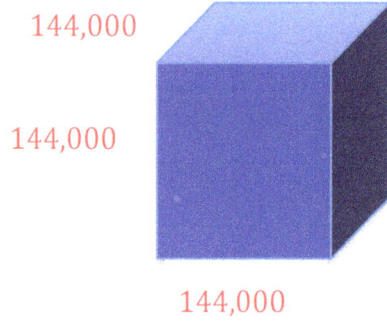

144,000

144,000

144,000

THE SUN 864,000 A [720 x 1,200]

[900 x 960]
Area = TIME = 497,664,000,000
/ 24 / 60 / 60 / 7.5 / 1,440 = 266.666

Aether 864,000 / 3 = 288,000

Light 288,000 x 5 = 1,440,000

Matter 288,000 x 4 = 1,152,000

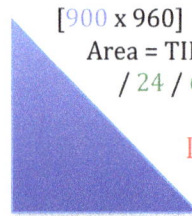

THE SUN 864,000 B [720 x 1,200   900 x 960]

Area = TIME = 165,888,000,000
/ 24 / 60 / 60 / 7.5 / 1,440 = 177.777

Aether 216,000 x 3 = 648,000

Light 216,000 x 5 = 1,080,000

864,000 / 4 = 216,000

THE SUN 864,000 C [720 x 1,200   900 x 960]

Area = TIME = 179,159,040,000
/ 24 / 60 / 60 / 7.5 / 1,440 = 192
/24/60/60x7.5x144=/16/22.5= 43,200

Matter 172,800 x 3 = 518,400

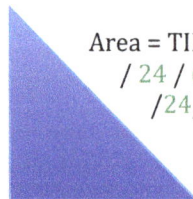

Light 864,000/ 5 = 172,800

691,200 = 172,800 x 4

SUN CUBED 864,000 x 864,000 x 864,000 = 644,972,540,000,000,000

(720 x 895,795,200,000,000)
(900 x 716,636,160,000,000)

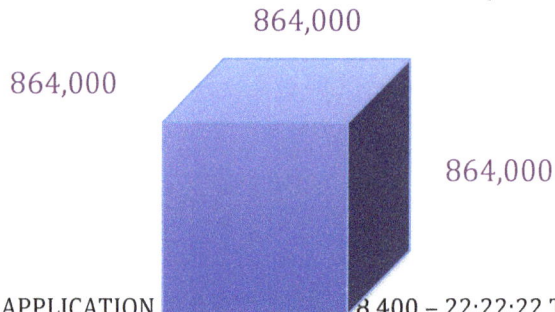

864,000

864,000

864,000

555

THE MOON'S 2,160 D CUBE = 10,077,696,000

(720 x 13,996,800) (900 x 11,197,440)

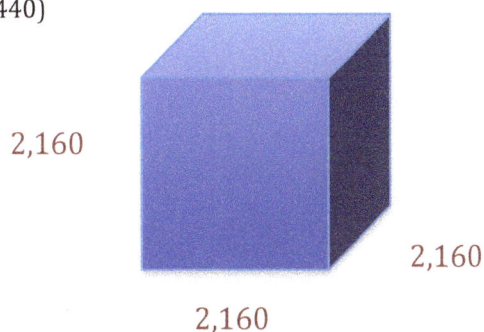

2,160

2,160

2,160

THE MOON = 2,160 D [720 x 3 900 x 2.4]

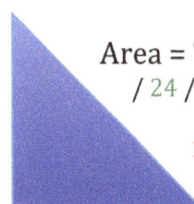

Area = TIME = 3,110,400
/ 24 / 60 / 60 = 36

Aether 2,160 / 3 = 720

matter 720 x 5 = 3,600

[720 x 720 = 518,400]

Light 2,880 = 720 x 4

EARTH = 7,960 D [720 x 11] [900 x 8.8]

Area = TIME = 49,766,400,000

Aether 7,920 / 3 = 2,640

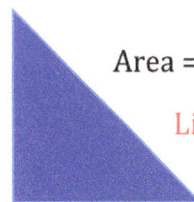

Light 2,640 x 5 = 13,200

Matter 2,640 x 4 = 10,560 [OCTAVE 660]

THE EARTH'S 7,920 SQUARED = 31,680 [720 x 44] [900 x 35.2]

THE EARTH'S 7,920 CUBED = 496,793,088,000 [720 x 689,990,400] [900 x 551,992,320]

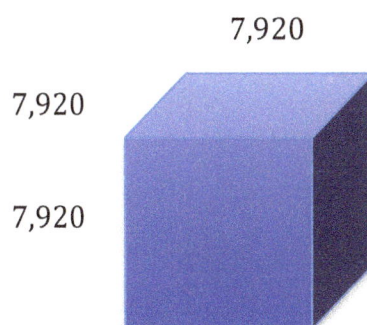

7,920

7,920

7,920

7,920

EARTH'S EQUINOX A 25,920 YRS [720 x 36]
[900 x 28.8]

Area = TIME = 447,897,600
/24/60/60 = 5,184/.75/16/ = 432

Time 25,920 YRS / 3 = 8,640   Light 8,640 x 5 = 43,200

Matter 8,640 x 4 = 34,560 = 3 x 4 x 5 x 6 = 360 x 360 = 129,600
[34,560 / 25,920 = 1.333 (16/12)   25,920 / 34,560 =   0.75 (12 / 16)]

EARTH'S EQUINOX B 25,920 YRS [720 x 36]
[900 x 28.8]

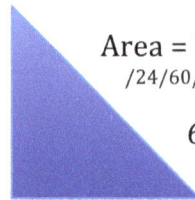

Area = TIME = 251,942,400
/24/60/60 = 2,916/.75 /16 = 243

6,480 x 3 = 19,440   6,480 x 5 = 32,400

25,920 / 4 = 6,480
[34,560 / 25,920 = 1.333 (16/12)   25,920 / 34,560 = 0.75 (12 / 16)]

EARTH'S EQUINOX C HYP 25,920 YEARS   [720 x 36] [900 x 28.8]
Area = TIME = 161,243,136
/24/60/60 =1,866.24 / .75/720/16 = 0.216

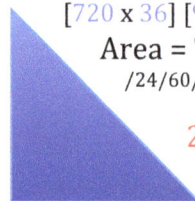

5,184 x 3 = 15,552   25,920 / 5 = 5,184

5,184 x 4 = 20,736 / 24 = 864 / 60 = 14.4 / 60 = 0.24
[20,736 / 25,920 = 0.8   25,920 / 20,736 = 1.25]

EARTH'S EQUINOX 25,920 SQUARED = 103,608   (720 x 144)   (900 x 115.12)

THE EARTH'S EQUINOX 25,920 CUBED = 17,414,258,688,000   (720 x 24,186,470,400)
(900 x 19,349,176,320)

25,920

25,920

25,920

THE SUN 432,000 R [720 x 600   900 x 480]

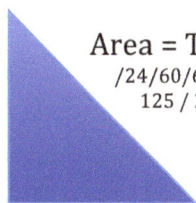

Area = TIME = 124,416,000,000
/24/60/60/ = 1,440,000 / 720 = 2,000 / 16 = 125
125 / 22.5 = 5.555 / .125 = 44.444 / 11.111 = 4

432,000 / 3 = 144,000

144,000 x 5 = 720,000

144,000 x 4 = 576,000

THE MOON 1,080 R [720 x 1.5   900 x 1.2]

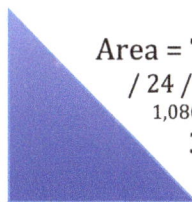

Area = TIME = 777,600
/ 24 / 60 / 60 = 9   777,600 / 720 = 1,080
1,080 /16 = 67.5/22.5 = 3 / 0.125 = 24 / 9 = 2.666
360 x 5 = 1,800

1,080 / 3 = 360
[720 x 1.5   900 x 1.2]

360 x 4 = 1,440

THE EARTH 3,960 R [720 x 5.5   900 x 4.4]

AT = 10,454,400 / 720 = 14,520
/24/60/60 = 121

3,960 / 3 = 1,320

1,320 x 5 = 6,600

1,320 x 4 = 5,280

1 DAY 43,200 Seconds [720 x 60   900 x 48]

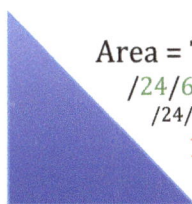

Area = TIME = 1,244,160,000
/24/60/60 = 14,400/.75/720/16 = 1.666
/24/60/60=14,400/16=900/22.5 = 40/9 = 4.444
14,400 x 5 = 72,000

43,200 Sec / 3 = 14,400

14,400 x 4 = 57,600
[57,600 / = 0.75   115,200 / 86,400 = 1.333 (16/12)]

6 DAYS 518,400s [720 x 720   900 x 576]

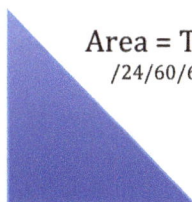

Area = Time = 178,744,320,000
/24/60/60 = 2,068,800

518,400 / 3 = 172,800

172,800 x 5 = 864,000

172,800 x 4 = 689,600

PART 17 OF 20 – MSAART PATENT APPLICATION NOTES DRAFT 518,400 – 22:22:22 THUR 22ND SEPT 2022
© STRIKE FOUNDATION GUARANTEE LIMITED | MALCOLM BENDALL 2022 | GRAPHICS - STEVE EARL

7 DAYS 604,800 Sec [720 x 840] [900 x 672]

Area - Time = 243,855,360,000
/24/60/60/7.777/360/16/22.5 = 0.0875
/24/60/60/720/16/ =

604,800 / 3 = 201,600

201,600 x 5 = 1,008,000

201,600 x 4 = 806,400

30 DAYS 2,592,000 Sec [720 x 3,600] [900 x 2,880]

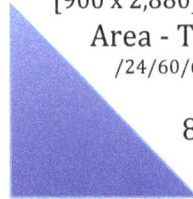

Area - Time = 4,478,976,000,000
/24/60/60=51,840,000/.75/720/16 /22.5 = 266.666

2,592,000 / 3 = 864,000

864,000 x 5 = 4,320,000

864,000 x 4 = 3,456,000

360 DAYS 31,104,000s [720 x 432,000] [900 x 34,560]

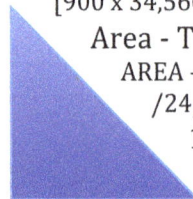

Area - Time = 86,400
AREA – TIME = 644,972,544,000,000
/24/60/60/7.5/720/16 = 864,000

360 Days/ 3 = 120
31,104,000 Seconds / 3 = 10,368,000

120 x 5 = 600
10,368,000 x 5 = 51,840,000

120 x 4 = 480
10,368,000 x 4 = 41,472,000

360 DAY'S ONE YEAR SQUARE = 1,440 (720 x 2)   (900 x 1.6)

360 DAY ONE YEAR CUBE = 46,656,000 (720 x 64,800)   (900 x 51,840)

46,656,000

46,656,000

46,656,000

50 yrs 1,555,200,000 Sec [720 x 2,160,000] [900 x 1,728,000]
50 x 360 = 18,000 x 24 = 432,000 x 60 x 60 = 1,555,200,000

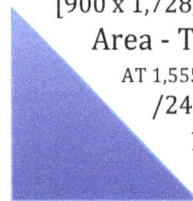

Area - Time 50 yrs = 1,666.666
AT 1,555,200,000 = 6,449,725,440,000,000,000
/24/60/60/7.5/720/16 = 864,000,000

50 yrs / 3 = 16.666
1,555,200,000 Seconds / 3 = 518,400,000

16.666 x 5 = 83.333
51,840,000 x 5 = 2,592,000,000

PART 17 OF 20 – MSAART PATENT APPLICATION NOTES DRAFT 518,400 – 22:22:22 THUR 22ND SEPT 2022
© STRIKE FOUNDATION GUARANTEE LIMITED | MALCOLM BENDALL 2022 | GRAPHICS - STEVE EARL

16.666 x 4 = 66.666
518,400,000 x 4 = 8,294,400,000

50 YEARS SQUARE 1,555,200,000 Sec x 4 = 6,220,800,000 [720 x 8,640,000] [900 x 6,912,000]
[266.666 x 23,328,000]

6,220,800,000

6,220,800,000

50 YEARS SQUARE 1,555,200,000 x 1,555,200,000 x 1,555,200,000 =
376,147,987,661,000,000,000,000,000 / 11.1111 = 338533188898100,000,000,000,000
376,147,987,661,000,000,000,000,000 / 266.666 = 14,105,549,537,320,000,000,000,000
376,147,987,661,000,000,000,000,000 / 720,000 =
376,147,987,661,000,000,000,000,000 / 900,000 =

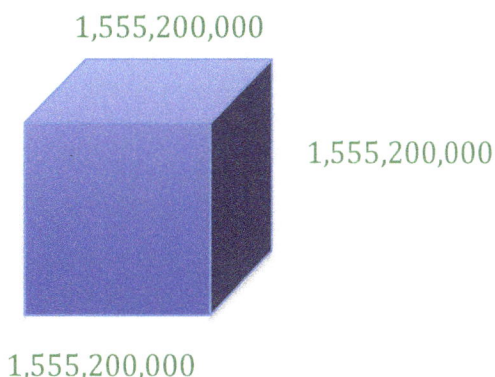

1,555,200,000

1,555,200,000

1,555,200,000

PHI 1.680339887499

1.0082039324994/ 3 = 0.336067977499

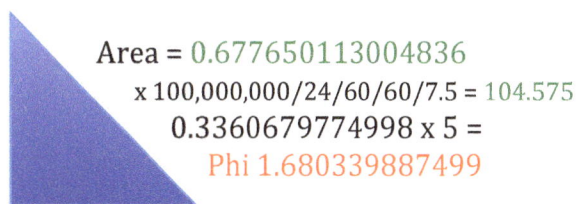

Area = 0.677650113004836
x 100,000,000/24/60/60/7.5 = 104.575
0.3360679774998 x 5 =
Phi 1.680339887499

0.3360679774998 x 4 = 1.3442719099992

PIE 3.141592653589793

Area - Time = 2.36870505626128
x 100,000,000/24/60/60/7.5 = 365.54
0.62831853071794 x 5 =
PIE 3.1415926535897

1.88495559215382/3 = 0.62831853071

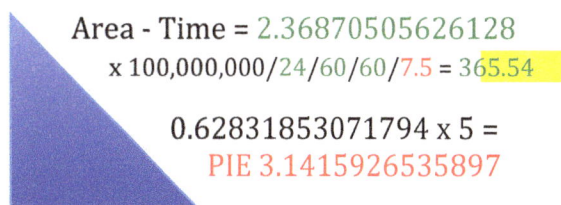

0.628318530717959 x 4 = 2.51327412287176

SPEED OF LIGHT 186,282 MILES Per/Sec

Area - Time = 23,133,989,016
/ 24 / 60 / 60 / .75 / 720 / 16 = 30.99010

186,282 / 3 = 62,094   62,094 x 5 = 310,470

62,094 x 4 = 248,376

HYDROGEN IN WATER 11.111

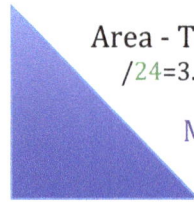

Area - Time = 82.3
/24=3.42916/60/60/7.5=.00012700617

11.111 / 3 = 3.70370370   Matter 3.70370 x 5 = 18.518

14.814 = 3.70370 x 4

OXYGEN IN WATER 88.888

Area - Time = 5,267.48971193373
/24 / 60 / 60 / 7.5 /720 /16 = 0.00000529

88.888 / 3 = 29.6296   Matter 29.6296 x 5 = 148.148

29.6296 x 4 = 118.518518

SANSKRIT TIME A 311,040 [720 x 432]
[900 x 345.6]

Area - Time = 21,499,084,800
/24 / 60 / 60 / 7.5 /720 /16 = 2.88

311,040 / 3 = 103,680   Matter 103,680 x 5 = 518,400

103,680 x 4 = 414,720

SANSKRIT TIME B 311,040 [720 x 432]

[900 x 345.6]

Area - Time = 36,279,705,600
/24 / 60 / 60 / 7.5 / 720/16 = 4.86

77,760 x 3 = 233,280   77,760 x 5 = 388,800

311,040 / 4 = 77,760

SANSKRIT TIME C Hyp 311,040 [720 x 432]

[900 x 345.6]

Area - Time = 23,219,011,584
/24/60/60 = /7.5/720/16 = 3.1104

62,208 x 3 = 186,624

311,040 / 5 = 62,208

62,208 x 4 = 248,832

SANSKRIT LONG YEAR 311,040 x 4 = 1,244,160 (720 x 1,728[SUN SQUARE]) (900 x 1,382.4)

311,040

311,040

SANSKRIT LONG YEAR CUBED   311,040 x 311,040 x 311,040 = 30,091,839,012,860,000
(720 x 41,794,220,851,200) (900 x 33,435,376,680,960)
30,091,839,012,860,000/24/60/60/7.5/720/16/22.5=179,159.04/144=1,244.16/144 = 8.64/144 = 0.06

311,040

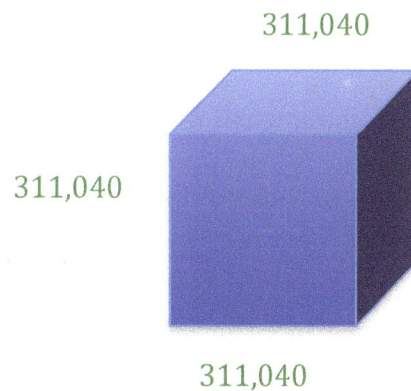

311,040

311,040

311,040

RESONANT MATTER – ALTERNATING CURRENT

PYTHAGOREAN, KEPLER 2,160 SQUARE RESONANCE MOON PARABLE

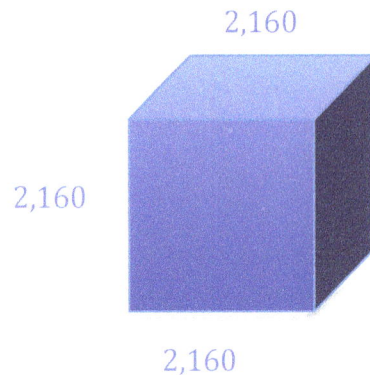

2,160

2,160

2,160

2,160

Pythagoras (whose names Greek Gematria adds up to 864 the diameter of the Sun) proved that all shapes resonate at the sum of their internal angles, a cubes internal angles add up to 2,160 degrees the diameter in miles of the Moon. When the length of the cube squaring the Moon equals the sum of the internal angles you get a natural harmonic resonance feedback system. To calculate the base frequency one must divide 2,160 by 11.111 that equals 194.4 as laid out below.

2,160 Moons Diameter / 11.111 = 194.4 (POS)

194.4 / 0.0432 = 450

432,000 / 194.4 = 2,222.222 – 2D

Aether (DC) to Matter (AC) conversion by the Sun Torus is 864,000 / 194.4 = 4,444.444

Time, based on one 24 hour day, is 24 hours x 60 minutes x 60 seconds = 86,400 Seconds / 194.4 = 444.444 which equals A in music which is 40 x 11.111 = 444.444 so under the Pythagorean system A in music equates to a day in seconds and to Red in color which is 444 Billion vibrations per second.

DIMENSIONAL NUMBERS

144,000 / 1.111 = 129,600.000000001296 / 1.111 = 116,640.0000000023328

DIMENSIONAL NUMBERS SUN, MOON AND EARTH

311,040 x 5.555 (AETHER) = 1,728,000 SUN SQUARE REPRESENTING FINITE MATTER / 2 = 864,000

16 D – AETHER SUN - 864,000 / 16.161616 = 53,460

16 D – AETHER SUN - 864,000 / 16 = 54,000

12D – AETHER SUN - 864,000 / 12.121212 = 71,280

12 D – AETHER SUN - 864,000 / 12 = 72,000

11D – AETHER SUN - 864,000 / 11.111 = 77,760

10D – AETHER SUN - 864,000 / 10 = 86,400

9D – AETHER SUN - 864,000 / 9.999 = 86,400.000000000864

8D – AETHER SUN - 864,000 / 8.888 = 97,200

7D – AETHER SUN - 864,000 / 7.777 = 111,085.7142857143

6D – AETHER SUN - 864,000 / 6.666 = 129,600 (360 x 360)

5D – AETHER SUN - 864,000 / 5.555 = 155,520.0000000015552

4D – TIME - 864,000 / 4.444 = 194,400.000000001944

3D – MATTER - 864,000 / 3.333 = 259,200.000000002592

2D – FLATLAND - 864,000 / 2.222 = 388,800.000000003888

1D – SPIRAL - 864,000 / 1.111 = 777,600.00000000777

11.111 AND 5.555 AETHER OCTAVE SEQUENCE

186,413,511.111
93,206,755.555
46,603,337.777
23,301,688.888
11,650,844.444
5,825,422.222
2,912,711.111
1,456,355.555
728,177.777
364,088.888
182,044.444
91,022.222
45,511.111
22,755.555
11.377.777
5,688.888
2,844.444
1,422.222
711.111
355.555
177.777
88.888
44.444
22.222

11.111

3,456,000/11.111= 311,040 1,728,000/11.111= 155,520 864,000/11.111= 77,760 432,000/11.111= 38,880
311,040 x 11.111 = 3,456,000 311,040/11.111 = 27,993.6 518,400 /11.111 = 46,656 518,400 x 11.111= 576
144,000 x 11.111 = 1,600,000 144,000 / 11.111 = 129,600   7.5 x 11.111 =83.333   11.111 / 7.5 = 1.48148

5.555

3,456,000/5.555 = 622,080   1,728,000/5.555 = 311,040   864,000/5.555 = 155,520   432,000/5.555 = 77,760
311,040/5.555= 55,987.2 311,040 x 5.555= 1,728,000 518,400/5.555 = 93,312 518,400 x 5.555 = 2,880,000
144,000 / 5.555 = 25,920 / 3.333 = 7,776 x 1.333 = 10,368/ 24 = 432/60 = 7.2/60 = 0.12/60 = 0.002
144,000 x 5.555 = 800,000 / 24 = 33,333.333 / 60 = 555.555 / 60 = 9.2592 / 60 = ***0.154320987654321***
1x5x4x3x2x9x8x7x6x5x4x3x2x1 = 43,545,600 x 43,545,600 = 1,896,219,279,360,000
7.5 / 5.555 = 1.35   5.555 x 7.5 = 41.666

2.777
1.3888
0.69444
0.347222
0.1736111
0.08680555
0.0434027888
0.0217013888
0.01085069444
0.005425347222
0.0027126736111

12, 24, 48 AETHER TO MATTER OCTAVE SEQUENCE

402,653,184
201,326,592
100,663,296
50,331,648
25,165,824
12,582,912
6,291,456
3,145,728
1,572,864
786,432
393,216
196,608
98,304
49,152
23,576
12,228
6,144
3,072
1,536
768
384
192
96
48
24

3,456,000/24= 144,000 1,728,000/24= 72,000 864,000/24 = 36,000 432,000/24= 18,000
31,104,000,000 / 24 = 12,96,000,000   518,400 / 24 = 21,600   518,400 x 24 = 12,441,600
144,000/24= 6,000(MOLTEN SEA SPHERE VOLUME IN BATHS) 12/7.5 = 1.6 7.5/24 = 0.3125 24/7.5= 3.2
24 / 11.111 = 2.16   24 x 11.111 = 266.666   12 x 7.5 = 90   24 x 7.5 = 180   48 x 7.5 = 360

12
6
3
1.5
0.75 (LIGHT TRAVELS 7.5 x AROUND THE EARTH IN 1 SECOND)
0.375
0.1875
0.09375
0.046875
0.0234375
0.01171875
0.005859375
0.0029296875
0.00146484375

SUN OCTAVE 864,000 SEQUENCE (POS)

452,984,832,000
226,492,416,000
113,246,208,000
56,623,104,000
28,311,552,000
14,155,776,000
7,077,888,000
3,538,944,000
1,769,472,000
884,736,000
442,368,000
221,184,000
110,592,000
55,296,000
27,684,000
13,824,000
6,912,000
3,456,000
1,728,000

864,000 SUN AETHER OCTAVES

864,000 x 7.5 = 6,480,000 x 7.5 = 48,600,000

432,000
216,000
108,000
54,000
27,000
13,500
6,750
3,375
1,687.5
843.75
421.875
210.9375
105.46875
52.734375
26.3671875
13.18359375
6.591796875
3.2958984375
1.64794921875
0.823974609375
0.4119873046875
0.20599365234375
0.102996826171875

ALL SIX 864,000 COMBINATIONS

864,000	846,000	684,000	648,000	486,000	468,000
56,623,104*	55,443,456*	44,826,624*	42,467,328*	31,850,496*	30,670,848*
28,311,552*	27,721,728*	22,413,312*	21,233,664*	15,925,248*	15,335,424*
14,155,776*	13,860,864*	11,206,656*	10,616,832*	7,962,624*	7,667,712*
7,077,888*	6,930,432*	5,603,328*	5,308,416*	3,981,312*	3,833,856*
3,538,944*	3,465,216*	2,801,664*	2,654,208*	1,990,656*	1,916,928*
1,769,472*	1,732,608*	1,400,832*	1,327,104*	995,328*	958,464*
884,736*	866,304*	700,416*	663,552*	497,664*	479,232*
442,368*	433,152*	350,208*	331,776*	248,832*	239,616*
221,184*	216,576*	175,104*	165,888*	124,416	119,808*
110,592*	108,288*	87,552*	82,944*	62,208*	59,904*
55,296*	54,144*	43,776*	41,472*	31,104*	29,952*
27,684*	27,072*	21,888*	20,736*	15,552*	14,976*
13,824*	13,536*	10,944*	10,368*	7,776*	7,488*
6,912,000	6,768,000	5,472,000	5184,000	3,888,000	3,744,000
3,456,000	3,384,000	2,736,000	2,592,000	1,944,000	1,872,000
1,728,000	1,692,000	1,368,000	1,296,000	972,000	936,000

864,000	846,000	684,000	648,000	486,000	468,000
432,000	423,000	342,000	324,000	243,000	234,000
216,00	211,500	171,000	162,000	121,500	117,000
108,000	105,000	85,500	81,000	60,750	58,500
54,000	52,875	42,750	40,500	30,375	29,250
27,000	26,437.5	21,375	20,250	15,187.5	14,625
13,500	13,218.75	10,687.5	10,125	7,593.75	7,312.5
6,750	6,609.375	5,343.75	5,062.5	3,796.875	3,656.25
3,375	3,304.6875	2,671.875	2,531.25	1,898.43	1,828.1
1,687.5	1,652.34375	1,335.9375	1,265.62	949.2187	914.06
843.75	826.171875	667.96875	632.812	474.6093	457.03
421.875	413.0859375	333.984375	316.406	237.3046	228.51
210.9375	206.54296875	166.99218	158.2031	118.6523	114.25
105.46875	103.271484375	83.496093	79.10156	59.32617	57.128
52.734375	51.6357421875	41.7480468	39.55078	29,66308	28.564
26.3671875	25.8187109375	20.8740234	19.77539	14.83154	14.282

864,000	846,000	684,000	648,000	486,000	468,000

5D THE SUN / AETHER

5D – AETHER SUN - 864,000 / 5.555 = 155,520.0000000015552

1 / x = 0.00000643004115552

864,000 x 864,000 x 864,000 = 644,972,540,000,000,000 / 5.555 = 116,095,057,920,001,160.95 / 311,040,000 / 311,040,000 = 1.2
1.2 x 7.5 x 1.333 = 12 the Aether number so proving my Area Time Volume Molten Sea observations.

155,520.0000000015552 / 266.666 = 583.20145800365084113

583.20145800365084113 x 1.333 = 777.601 x 1.111 = 864.0002160005378 x 11.111 = 9,600.0240000596641492 x 8 = 76,800.1920005 x 7.5 = 576,001.440004 /144 = 4,000.01000002 /2 = 2,000 / 16 = 125.0003125007687694 / 22.5 = 5.55556944447897 x 400 = 2222.227777779159 x 0.125 = 17,777.82222332711387

5D - GAP = 4D 194,400 – 5D 155,520 = 38,880

38,880 / .2666 = 145,800/ 22.5 = 6,480 x 1.111 = 7,200 x 11.111 = 9,600 / 16 = 600 x 400 = 240,000

38,880 / 22.5 = 1,728 / 16 = 108 x 400 = 43,200 / 1.111 = 38,880.0000000003888 x 11.111 = 432,000 / 1.333 = 324,000.000000008

38,880 / 16 = 2,430 / 22.5 = 22.5 x 400 = 9,000 / 7.5 = 1,200 x 1.333 = 1,600 / 1.111 = 1,440

864,000 / 266.666 = 3,240.000000000081   SUN RADIUS SHUFFLE
155,520 / 266.666 = 583.201458003645
864,000 x 266.666 = 230,400
155,520 x 266.666 = 41,471,999.89632

155,520 / 864,000 [SUN] = 0.18
155,520 / 2,160 [MOON] = 72 / 0.18 = 400
155,520 / 7,920 [EARTH] = 19.636363 / 0.18 = 109.0909055555

864,000 / 1.111 = 777,600 POS
155,520 / 1.111 = 139,968
864,000 / 1.333 = 648,000.000000016 SUN D SHUFFLE
155,520 / 1.333 = 116,640

AETHERIAL MUSIC 11.111 BASE CODE SEQUENCE PSLAM 144 DAVID'S HARP'S 10 STRINGS
3 x 7 =   21 x 11.111 = 233.333
3 x 8 =   24 x 11.111 = 266.666 C

3 x 9 = 27 x 11.111 = 300
3 x 10 = 30 x 11.111 = 333.333
3 x 11 = 33 x 11.111 = 366.666
3 x 12 = 36 x 11.111 = 400
3 x 13 = 39 x 11.111 = 433.333
3 x 14 = 42 x 11.111 = 466.666
3 x 15 = 45 x 11.111 = 500
3 x 16 = 48 x 11.111 = 533.333 (2 x 266.666) C Sharp Atherial Music

864,000 / 24 = 36,000 / 11.111 = 3,240 SUN	155,520 / 24 = 6,480 SUN
864,000 / 27 = 32,000 / 11.111 = 2,880 LIGHT	155,520 / 27 = 5,760 LIGHT
864,000 / 30 = 28,800 / 11.111 = 2,590 EQUX	155,520 / 30 = 5,184 TIME
864,000 / 32 = 27,000 / 11.111 = 2,430 SUN	155,520 / 32 = 4,860 SUN SHUFFLE
864,000 / 36 = 24,000 / 11.111 = 2,160 MOON	155,520 / 36 = 4,320 SUN R
864,000 / 40 = 21,600 / 11.111 = 1,944 POS	155,520 / 40 = 3,888 POS
864,000 / 45 = 19,200 / 11.111 = 1,728 SUN x 2	155,520 / 45 = 3,456 SUN x 4
864,000 / 48 = 18,000 / 11.111 = 1,620 MOON SHUFFLE	155,520 / 48 = 3,240 SUN R SHUFFLE

864,000 / 11.111 = 77,760.0000000007776 TIME
155,520 / 11.111 = 13,996.8

5D 16 SECTORS OF THE MOLTEN SEA 22.5 DEGREES EACH

155,520 / 1 = 155,520...........[TIME 155,520 / 24 = 6,480 / 60 = 108 / 60 = 1.8 / 60 = 0.03]
155,520 x 1 = 155,520.......[LIGHT 155,520 / 864 = 180, 155,520 / 144 = 1,080   155,520 / 1.333 = 116,640]

155,520 / 2 = 77,760..........[77,760 / 864 = 90   77,760 / 144 = 540   77,760 / 1.333 = 58,320]
155,520 x 2 = 311,040........[SANSKRIT 100 YEARS 311,040,000,000]

155,520 / 3 = 51,840..........[6 DAYS OF CREATION IN SECONDS = 518,400 x 60 ARC SECONDS = 31,104,000]
155,520 x 3 = 466,560........[466,560 / 864 = 540   466,560 / 144 = 3,240   466,560 / 1.333 = 349,920]

155,520 / 4 = 38,880 ....[GAP BETWEEN DIMENSIONS ¼ QUADRITURE CONFIRMS MY MODEL]
155,520 x 4 = 622,080........[622,080 / 864 = 720   622,080 / 144 = 4,320   622,080 / 1.333 = 466,560]

155,520 / 5 = 31,104 ....[SANSKRIT LONG YEAR 311,040,000,000]
155,520 x 5 = 777,600........[777,600 / 864 = 900   777,600 / 144 = 5,400   777,600 / 1.333 = 583,200]

155,520 / 6 = 25,920..........[EARTHS PROCESSION OF THE EQUINOX 25,920 YEARS]
155,520 x 6 = 933,120........[933,120 / 864 = 1,080   933,120 / 144 = 6,480   933,120 / 1.333 = 933,120]

155,520 / 7 = 22,217.14285714286
155,520 x 7 = 1,088,640.....[1,088,640 / 864 = 1,260 1,088,640 / 144 = 7,560   1,088,640 / 1.333 = 816,480]

155,520 / 8 = 19,440.............[PLANET OCTAVE 864 SEQUENCE (POS)]
155,520 x 8 = 1,244,160.....[1,244,160 / 864 = 1,440   1,244,160 / 144 = 8,640   1,244,160 / 1.333 = 933,120]

155,520 / 9 = 17,280...........[GOLDEN NUMBER 864,000 x 2]
155,520 x 9 = 1,399,680......[1,399,680 / 864 = 1,620   1,399,680 / 144 = 9,720   1,399,680 / 1.333 = 1,049,760]

155,520 / 10 = **15,552**..............[PLANET OCTAVE 864 SEQUENCE (POS)]
155,520 x 10 = **1,555,200**......[1,555,200 / 864 = 1,800   1,555,200 / 144 = 10,800   1,555,200 /1.333 = 1,166,400]

155,520 / 11 = **14,138.1818181**
155,520 x 11 = **1,710,720**...[1,710,720 / 864 = 1,980   1,710,720 / 144 = 11,880   1,710,720 / 1.333 = 1,283,040]

155,520 / 12 = **12,960**..........[RESONANT FREQUENCY ENERGY UNIT RFEU 129,600]
155,520 x 12 = **1,866,240** ....[1,866,240 / 864 = 2,160   1,866,240 / 144 = 12,960   1,866,240 / 1.333 = 1,283,040]

155,520 / 13 = **11,963.0769230769**..............[
155,520 x 13 = **2,021,760**.....[2,021,760 / 864 = 2,340   2,021,760 / 144 = 14,040   2,021,760 / 1.333 = 1,516,320]

155,520 / 14 = **11,108.571428571428571**...[
155,520 x 14 = **2,177,280**......[2,177,280 / 864 = 2,520 2,177,280 / 144 = 15,120   2,177,280 / 1.333 = 1,632,960]

155,520 / 15 = **10,368**.............[TIME NUMBER 5,184 X 2]
155,520 x 15 = **2,332,800**....[2,332,800 / 864 = 2,700   2,332,800 / 144 = 16,200   2,332,800 / 1.333 = 1,749,600]

155,520 / 16 = **9,720**.............[EARTH'S DIAMETER 7,920]
155,520 x 16 = **2,488,320**......[2,488,320 / 864 = 2,880 2,488,320 / 144 = 17,280   2,488,320 / 1.333 = 1,866,240]

5D - MUSIC OCTAVE SCALE BASED ON C = 266.666 VIBRATIONS PER SEC

266.666 x .666 (Molten Sea 2/3 full) = 177.777 / 2 = 88.888 1/3 air cavity of Molten Sea
155,520 / 24 = 6,480 x 11.111 = 72,000 x 1.333 = 96,000 / 266.666 = 360
360 x 8 = 2,880 / 16 = 180 / 22.5 = 8 x 400 = 3,200

155,520 / 27 = 5,760 x 11.111 = 64,000 x 1.333 = 85,333 / 266.666 = 320
320 x 8 = 2,560 / 16 = 160 / 22.5 = 7.111 x 400 = 2,844.444

155,520 / 30 = 5,184 x 11.111 = 57,600 x 1.333 = 76,800 / 266.666 = 288
288 x 8 = 2,304 / 16 = 144 / 22.5 = 6.4 x 400 = 2,560

155,520 / 32 = 4,860 x 11.111 = 54,000 x 1.333 = 72,000 / 266.666 = 270
270 x 8 = 2,160 / 16 = 135 / 22.5 = 6 x 400 = 2,400

155,520 / 36 = 4,320 x 11.111 = 48,000 x 1.333 = 64,000 / 266.666 = 240
240 x 8 = 1,920 / 16 = 120 / 22.5 = 5.3 x 400 = 2,133.333

155,520 / 48 = 3,240 x 11.111 = 36,000 x 1.333 = 48,000 / 266.666 = 180
180 x 8 = 1,440 / 16 = 90 / 22.5 = 4 x 400 = 1,600

5D – AETHER

COMPASS DIVIDED INTO 16 SEGMENTS 22.5 DEGREES PER SEGMENT

155,520 / 5.625 = 27,648 x 1.333 = 36,864 (864 SUN)
       27,648 x 11.111 = 307,200 x 1.333 = 409,600 / 266.666 = 1,536
155,520 / 11.25 = 13,824 x 1.333 = 18,432, (432 SUNS RADIUS)
       13,824 x 11.111 = 153,600 x 1.333 = 204,7800 / 266.666 = 768

CARDINAL COMPASS ANGLES

155,520 / 22.5 = 6,912 x 1.333 = 9,216 ....(RFEU)
6,912 x 11.111 = 76,800 x 1.333 = 102,400 / 266.666 = 384
155,520 / 45 = 3,456 x 1.333 = 4,608 ....(468 SUNS DIAMETER MIRROR)
155,520 / 67.5 = 2,304 x 1.333 = 3,072
155,520 / 90 = 1,728 x 1.333 = 2,304 ....(432 SUNS RADIUS MIRROR)
1,728 x 11.111 = 19,200 x 1.333 = 25,600 / 266.666 = 96

155,520 / 112.5 = 1,382.4 x 1.333 = 1,843.1999
155,520 / 135 = 1,152 x 1.333 = 1,536
155,520 / 157.5 = 987.42857142857142857 x 1.333 = 1,316.5714285713956571
155,520 / 180 = 864 x 1.333 = 1,152 [144 OCTAVE]
864 x 11.111 = 9,600 x 1.333 = 12,800 / 266.666 = 48

155,520 / 202.5 = 768 x 1.333 = 1,023
155,520 / 225 = 691.2 x 1.333 = 921.59999999997696
155,520 / 247.5 = 628.3636 x 1.333 = 837.818181
155,520 / 270 = 576 x 11.111 = 6,400 x 1.333 = 8,533.333/266.666 = 32

155,520 / 292.5 = 531.692 x 11.111 = 1,728,000 x 1.333 = 2,304,000 / 266.666 = 8,640
155,520 / 315 = 493.714 x 11.111 = 5,485.714 x 1.333 = 7314/266.666 = 27.428571428571154286
155,520 / 337.5 = 460.8 x 11.111 = 5,119.9999999999488 x 1.333 = 6,826.6666666664277333
6,826.6666666664277333 / 266.666 = 25.6
155,520 / 360 = 432 x 11.111 = 4,800 x 1.333 = 6,400/266.666 = 24

155,520 DIVIDED BY OCTAVES OF 720

155,520 / 540 = 288 x 11.111 = 3,200 x 1.333 = 4,266.666 / 266.666 = 16   [MATTER]
155,520 / 360 = 432 x 11.111 = 4,800 x 1.333 = 6,400 / 266.666 = 24
155,520 / 720 = 216 x 11.111 = 2,400 x 1.333 = 3200 / 266.666 = 12   [AETHER]
155,520 / 1,440 = 108 x 11.111 = 1,199.99999999988 x 1.333 = 1,600/266.666 = 6   [LIGHT]
155,520 / 2,880 = 54 x 11.111 = 600 x 1.333 = 800 / 266.666 = 3
155,520 / 5,760 = 27 x 11.111 = 300 x 1.333 = 400 / 266.666 = 1.5
155,520 / 11,520 = 13.5 x 11.111 = 150 x 1.333 = 200 / 266.666 = 0.75
155,520 / 23,040 = 6.75 x 11.111 = 75 x 1.333 = 100 / 266.666 = 0.375
155,520 / 46,080 = 3.375 x 11.111 = 37.5 x 1.333 = 50 / 266.666 = 0.1875

155,520 / 1,296 = 120 x 11.111 = 1,333.333 x 1.333 = 1,777.7777777777155556 / 266.666 = 6.666
155,520 / 144 = 1,080 x 11.111 = 12,000 x 1.333 = 15,999.99999999944 / 266.666 = 60

5D GAP = 4D 194,400 – 5D 155,520 = 38,880

38,880 / 1.111 = 34,992.00000000034992
38,880 / 1.333 = 29,160.000000000729
38,880 / 11.111 = 3,499.200000000003

16 SECTORS OF THE MOLTEN SEA 22.5 DEGREES EACH

38,880 / 2 = 19,440...........[POS]
38,880 / 3 = 12,960...........[RFEU NUMBER 129,600]
38,880 / 4 = 9,720.............[EARTH'S DIAMETER NUMBER 7,920]
38,880 / 5 = 7,776...... ......[POS]
38,880 / 6 = 6,480.............[SUN'S DIAMETER NUMBER 864,000]
38,880 / 7 = 5,554
38,880 / 8 = 4,860.............[SUN'S DIAMETER NUMBER 864,000]
38,880 / 9 = 4,320.............[SUN'S RADIUS NUMBER 423,000]
38,880 / 10 = 3,888.............[POS]
38,880 / 11 = 3,534.545
38.880 / 12 = 3,240.............[SUN'S RADUIS NUMBER 432,000]
38,880 / 13 = 2,990.76923076923
38,880 / 14 = 2,777.14285714286
38,880 / 15 = 2,592.............[EARTH'S PROCESSION OF THE EQUINOX NUMBER 25,920]
38,880 / 16 = 2,430.............[SUN'S RADIUS NUMBER 432,000]

DEGREES OF THE MOLTEN SEA

38,880 / 5.625 = 6,912..............[RFEU NUMBER 129,600]
38,880 / 11.25 = 3,456..............[SUN'S DIAMETER SQUARE NUMBER 3,456,000]

38,880 / 22.5   = 1,728..............[SUN DIAMETER x 2]
38,880 / 45   = 864.................[THE SUN'S DIAMETER NUMBER 864,000]
38,880 / 67.5   = 576.................[LIGHT 144 x 4]
38,880 / 90   = 432.................[GOLDEN NUMBER SUN'S RADIUS 432,000]

38,880 / 112.5 = 345.6..............[SUN DIAMETER x 4]
38,880 / 135   = 288................[LIGHT SPEED DIAMETER NUMBER 144,000 x 2 = 288,000]
38,880 / 157.5 = 246.8
38,880 / 180   = 216................[MOON'S DIAMETER 2,160]

38,880 / 202.5   = 192
38,880 / 225   = 172.8
38,880 / 247.5   = 157.090909
38,880 / 270   = 144................[LIGHT SPEED RADIUS NUMBER]

38,880 / 292.5   = 132.9230
38,880 / 315   = 123.42857142857142857
38,880 / 337.5   = 115.2
38,880 / 360   = 108................[MOON'S RADIUS 1,080 AND EARTH'S + MOON'S DIA = 10,080]

38,880 / 720   = 54
38,880 / 900   = 432................[GOLDEN NUMBER SUN'S RADIUS 432,000]

38,880 / 1,296   = 30
38,880 / 144   = 270

4D TIME OCTAVE 311,040 - 518,400 SEQUENCE (TOS)

5,218,385,264,640
2,609,192,632,320
1,304,596,316,160
652,298,158,080
326,149,076,040
163,074,539,520
81,537,269,760
40,768,634,880
20,384,317,440
10,192,158,720
5,096,079,360
2,548,039,680
1,274,019,840
637,009,920
318,504,960
159,252,480
79,626,240
39,813,120
19,906,560
9,953,280
4,976,640
2,488,320
1,244,160
622,080

311,040 TIME OCTAVES (311,040 / 864 = 360 Degrees)

311,040 x 7.5 (The one second light takes to travel 7.5 times around the Earths equator) = 2,332,800

155,520
77,760
38,880
19,440
9,720
4,860

2,430

1,215
607.5
303.75
151.875
75.9375
37.96975
18.984375
9.4921875
4.74609375
2.373046875
1.1865234375
0.59326171875
0.296630859375
0.1483154296875
0.07415771484375
0.037078857421875

4D RESONANCE 129,600 RFEU OCTAVES

543,581,798,400
271,790,899,200
135,895,449,600
67,947,724,800
33,973,862,400
16,986,931,200
8,493,465,600
4,246,732,800
2,123,366,400
1,061,683,200
530,841,600
265,420,800
132,710,400
66,355,200
33,177,600
16,588,800
8,294,400
4,147,200
2,073,600
1,036,800
518,400
259,200

129,600 RFEU (129,600 / 864)
(129,600 x 7.5 = 972,000 x 7.5 = 7,290,000, 129,600 / 7.5 = 17,280 / 7.5 = 2,304)

64,800
32,400
16,200
8,100
4,050
2,025
1,012.5
506.25
253.125
126.5625
63.28125
31.640625
15.8203125
7.91015625
3.955078125
1.9775390625
0.98876953125
0.494384765625
0.2471923828125

4D – TIME 864,000 / 4.444 = 194,400.000000002
1/x = 0.00000514403292

864,000 x 864,000 x 864,000 = 644,972,540,000,000,000 / 4.444
= 145,118,822,400,001,451.19 / 311,040,000 / 311,040,000 = 1.5
Therefore 1.5 is 2 x 0.75 x (1.333 x lights standard 7.5) = 7.5 and 150 Moon's and 150
Sun's to keep time 311,040,000 being all Sanskrit time.

194,400.000000002 / 266.666 = 729.000000000018225 (EARTH'S D)

4D GAP = 3D 259,200 (Earth's Eqnx) – 4D 194,400 = 64,800 (Sun)

64,800 / .266666 = 243.000000006 / 22.5 = 10.8 x 1.111 = 12 x 11.111 = 133.333 / 16 =
8.333 x 400 = 3,333,333
64,800 / 22.5 = 2,880 / 16 = 180 x 400 = 72,000 / 720 = 100 x 7.5 = 750 x 1.333 = 1,000
/9 = 111.111 / 266.666 = 0.416
864,000/3 = 21,600/22.5 = 960 / 6.25 = 153.6 / 16 = 9.6 (6.25 x 16 = 100)

194,400.000000002 / 1.111 = 174,960.0000000197496
194,400.000000002 / 1.333 = 145,800.000000018645
194,400.000000002 / 11.111 = 17,496.00000000197496

4D 16 SECTORS OF THE MOLTEN SEA 22.5 DEGREES EACH

194,400.000000002 / 2 = 97,200 [EARTH'S DIAMETER 7,920]
194,400.000000002 x 2 = 388,800.[388,800 / 24 = 16,200 / 60 = 270 / 60 = 4.5 / 60 = 0.075]

194,400.000000002 / 3 = 64,800 [SUN'S DIAMETER 864,000]
194,400.000000002 x 3 = 583,200.[583,200 / 24 = 24,300 / 60 = 405.0 / 60 = 6.75 / 60 = 0.1125]

194,400.000000002 / 4 = 48,600 [SUN'S DIAMETER 864,000]
194,400.000000002 x 4 = 777,600.[777,600 / 24 = 32,400 / 60 = 540 / 60 = 9 / 60 = 0.15]

194,400.000000002 / 5 = 38,880...................................[PLANET OCTAVE 311,040 SEQUENCE (POS)]
194,400.000000002 x 5 = 972,000.[972,000 / 24 = 40,500 / 60 = 675 / 60 = 11.25 / 60 = 0.1875]

194,400.000000002 / 6 = 32,400.0000000033......[SUN'S RADIUS 432,000]
194,400.000000002 x 6 = 1,166,400.[1,166,400 / 24 = 48,600 / 60 = 810 / 60 = 13.5 / 60 = 0.22]

194,400.000000002 / 7 = 27,771.428571431428571
194,400.000000002 x 7 = 1,360,800.[1,360,800 /24 = 56,700 /60 = 945 /60 = 15.75 /60 = 0.2625]

194,400.000000002 / 8 = 24,300.0000000025......[SUN'S RADIUS 432,000]
194,400.000000002 x 8 = 1,555,200...[1,555,200/ 24 = 64,800 / 60 = 1,080 / 60 = 18 / 60 = 0.3]

194,400.000000002 / 9 = 21,600.0000000022......[MOON'S DIAMETER 2,160]
194,400.000000002 x 9 =1,749,600[1,749,600/24 = 72,900/60 = 1,215/60 = 20.25/60 = 0.3375]

194,400.000000002 / 10 = 19,440....................[PLANET OCTAVE 311,040 SEQUENCE (POS)]
194,400.000000002 x 10 = 1,944,000.[1,944,000/24 = 81,000 /60 = 1,350 /60 = 22.5 /60 = 0.375]

194,400.000000002 / 11 = 17,672.7272727
194,400.000000002 x 11 = 2,138,400.[2,138,400/24= 89.100/60= 1,485/60= 24.75/60= 0.4125]

194,400.000000002 / 12 = 16,200.000000017............[MOON'S DIAMETER 2,160]
194,400.000000002 x 12 = 2,332,800...[2,332,800 /24 = 97,200 /60 = 1,620 /60 = 27 /60 = 0.45]

194,400.000000002 / 13 = 14,953.846153847692308
194,400.000000002 x 13 =2,527,200[2,527,200/24 = 105,300/60 =1,755/60 =29.25/60 = .4875]

194,400.000000002 / 14 = 13,885.714285715714286
194,400.000000002 x 14 = 2,721,600...[2,721,600/24 = 113,400/60 = 1,890/60 = 31.5/60 = 31.5]

194,400.000000002 / 15 = 12,960.0000000013............[RFEU 129,600]
194,400.000000002 x 15 = 2,916,000[2,916,000/24= 121,5000/60= 2,025/60= 33.75/60= .5625]

194,400.000000002 / 16 = 12,150.0000000012
194,400.000000002 x 16 = 3,110,400..[3,110,400 / 24 = 129,600 / 60 = 2,160 / 60 = 36 /60 = 0.6]

CARDINAL DEGREES OF THE MOLTEN SEA

194,400.000000002 / 5.625   = 34,560.0000000036..............[SUN'S SQUARE 3,456,000]
194,400.000000002 / 11.25   = 17,280.0000000018..............[SUN'S D x 2]

194,400.000000002 / 22.5   = 8,640.000000000888   [SUN'S DIAMETER 864,000]
194,400.000000002 / 45   = 4,320.00000000044...............[SUN'S RADIUS 432,000]
194,400.000000002 / 67.5   = 2,880.0000000002962963....[LIGHT SPEED 144,000]
194,400.000000002 / 90   = 2,160. 00000000022...............[MOON'S DIAMETER]

194,400.000000002 / 112.5 = 1,728.......................................[SUN'S D x 2]
194,400.000000002 / 135   = 1,440.00000000015.................[LIGHT SPEED 144,000]
194,400.000000002 / 157.5 = 1,234.28571428571
194,400.000000002 / 180   = 1,080.00000000011...............[MOON'S RADIUS]

194,400.000000002 / 202.5 = 960.0000000000099
194,400.000000002 / 225   = 864.0000000000888.............[SUN'S DIAMETER 864,000]
194,400.000000002 / 247.5 = 785.454545
194,400.000000002 / 270   = 720.0000000000740740

194,400.000000002 / 292.5   = 664.615384615453
194,400.000000002 / 315   = 617.14285714292063492
194,400.000000002 / 337.5   = 576.00000000005925926
194,400.000000002 / 360   = 540.0000000000555

194,400.000000002 / 540   = 360.000000000037
194,400.000000002 / 720   = 270.00000000002777
194,400.000000002 / 900   = 216.......................................[MOON'S RADIUS 2,160]

194,400.000000002 / 1,296 [RFEU] = 150.0000000000154321
194,400.000000002 / 144   = 1,350.00000000013888
194,400.000000002 / 144,000 = 1.35000000000013888
194,400.000000002 / 266.666 = 729.000000000093225
194,400.000000002 / 22.5   = 8,640.000000000888......[SUN'S DIAMETER 864,000]
864,000 / 1.111   = 77,600
864,000 / 1.333   = 648,000.0000000162............................[SUN'S DIAMETER 864,000]
864,000 / 11.111 = 77,760.0000000007776

16 SECTORS OF THE MOLTEN SEA 22.5 DEGREES EACH

864,000 / 2 = 432,000...[SUN'S RADIUS 432,000]
864,000 / 3 = 288,000...[LIGHT SPEED 144,000]
864,000 / 4 = 216,000...[MOON'S RADIUS 2,160]
864,000 / 5 = 172,800
864,000 / 6 = 144,000..[LIGHT SPEED 144,000]
864,000 / 7 = 123,428.57142857142857
864,000 / 8 = 108,000..[EARTH'S + MOON'S DIAMETER = 10,080]

864,000 / 9   = 96,000
864,000 / 10 = 86,400..[SUN'S DIAMETER 864,000]
864,000 / 11 = 78,545.454545.
864,000 / 12 = 72,000
864,000 / 13 = 66,461.538461538461538
864,000 / 14 = 61,714.285714285714286
864,000 / 15 = 57,600
864,000 / 16 = 54,000

DEGREES OF THE MOLTEN SEA

864,000 / 5.625 = 153,600
864,000 / 11.25 = 76,800

864,000 / 22.5 = 38,400
864,000 /   45 = 19,200
864,000 / 67.5 = 12,800
864,000 /   90 = 9,600

864,000 / 112.5 = 7,680
864,000 /   135 = 6,400
864,000 / 157.5 = 5,485.714285714286
864,000 /   180 = 4,800

864,000 / 202.5 = 4,266.666
864,000 /   225 = 3,840
864,000 / 247.5 = 3,490
864,000 /   270 = 3,200

864,000 / 292.5 = 2,953.84615385
864,000 / 315   = 2,742.85714285714

864,000 / 337.5 = 2,560
864,000 / 360   = 2,400

864,000 / 540   = 1,600
864,000 / 720   = 1,200
864,000 / 900   = 960

864,000 / 1,296 = 666.666
864,000 / 144   = 6,000

16 SECTORS OF THE MOLTEN SEA 22.5 DEGREES EACH

64,800 / 2 = 32,400.....................................[SUN'S RADIUS MIRROR 432,000]

64,800 / 3 = 216,000..................................[MOON'S DIAMETER 2,160]

64,800 / 4 = 16,200...................................[MOON'S DIAMETER 2,160]

64,800 / 5 = 12,960...............[RESONANT FREQUENCY RESONANT UNIT (RFEU) 129,600]

64,800 / 6 = 10,800...................................[MOON'S RADIUS   1,080]

64,800 / 7 = 9,257.1428571428571429

64,800 / 8 = 8,100.....................................[MOON'S RADIUS   1,080]

64,800 / 9 = 7,200

64,800 / 10 = 6,480...................................[SUN'S RADIUS 432,000 SHUFFLE]

64,800 / 11 = 5,890.909090

64,800 / 12 = 5,400..................................[PLANET OCTAVE 864 SEQUENCE (POS)]

64,800 / 16 = 4,050..............[RESONANT FREQUENCY RESONANT UNIT (RFEU) 129,600]

DIVIDED BY MUSIC OCTAVE ROOTS

64,800 / 24 = 2,703.333
64,800 / 27 = 2,402.962962962
64,800 / 30 = 2,162.666...[MOON DIAMETER 2,160]
64,800 / 32 = 2,027.5
64,800 / 36 = 1,802.222
64,800 / 40 = 1,622
64,800 / 44 = 1,474.545454
64,800 / 48 = 1,351.666
64,800 / 266.666 = 243.3000000000060825..............[SUN'S RADIUS 432,000]
64,800 / 11.111 = 5,832

4D GAP 3D - COMPASS ANGLES FOR 16 SECTORS

64,800 / 5.625 = 11,520
64,800 / 11.25 = 5,760
64,800 / 22.5 = 2,880...[LIGHT SPEED 144,000]
64,800 / 45 = 1,440...[LIGHT SPEED 144,000]
64,800 / 67.5 = 960
64,800 / 90 = 720......................................[TORUS GOLDEN NUMBER]
      TOTAL = 6,000

64,800 / 112.5 = 576......................................[LIGHT SPEED 144,000]
64,800 / 135 = 480
64,800 / 157.5 = 411
64,800 / 180 = 360
      TOTAL = 1,827

64,800 / 202.5 = 320
64,800 / 225 = 288......................................[LIGHT SPEED 144,000]
64,800 / 247.5 = 261.81
64,800 / 270 = 240
      TOTAL = 1,109.81

64,800 / 292.5 = 221.538
64,800 / 315 = 205.71
64,800 / 337.5 = 192
64,800 / 360 = 180
      TOTAL = 870.21
  GRAND TOTAL = 9,807.02

64,800 / 540 = 120
64,800 / 720 = 90
64,800 / 900 = 72
      TOTAL = 282
64,800 / 1,296 = 50
64,800 / 144 = 450
64,800 / 266.666 = 243.0000000000006075............[SUN'S RADIUS 432,000 SHUFFLE]
64,800 / 11.111 = 5,832

3D - 135 DEGREES SOLOMANS MOLTEN SEA

2,264,924,160
1,132,462,080
566,231,404
283,115,520
141,557,760
70,778,880
38,389,440
17,604,720
8,847,360
4,432,680
2,211,840
1,105,920
552,960
276,480
138,240
69,120
34,560
17,280
8,640
4,320
2,160
1,080
540
270

135 DEGREES MOLTEN SEA SOUTH EAST CORNER OF SOLOMON'S TEMPLE
(864,000 / 900 = 960,   311,040 / 900 = 345.6)

67.5
33.75
16.875
8.4375
4.21875
2.109375
1.0546875
0.52734375
0.263671875
0.1318359375
0.06591796875
0.032958984375
0.0164794921875
0.00823974609375
0.004119873046875
0.0020599365234375
0.00102996826171875

3D - VAJRA TORUS OCTAVE 16 SEQUENCE (VOS)

268,435,456
134,217,728
67,108,864
33,554,432
16,777,216
8,388,608
4,194,304
2,097,152
1,048,576
524,288
262,144
131,072
65,536
32,768
16,384
8,192
4,096
2,048
1,024
512
256
128
64
32

16 VAJRA SECTIONS OCTAVES   (864 / 16 = 54)
16 x 7.5 = 120 (MOLTEN SEA BATH VOLUME) x 7.5 = 900

8
4
2
1
0.5
0.25
0.125
0.0625
0.03125
0.015625
0.0078125
0.00390625
0.001953125
0.0009765625
0.00048828125
0.000244140625
0.0001220703125
0.00006103515625
0.000030517578125

3D - MATTER-MUSIC <mark>266.666</mark> OCTAVE SEQUENCE

1,118,481,066.666
559,240,533.333
279,620,266.666
139,810,133.333
69,905,066.666
34,952,533.333
17,476,266.666
8,738,133.333
4,369,066.666
2,184,533.333
1,092,266.666
546,133.333
273,066.666
136,533.333
68,266.666
34,133.333
17,066.666
8,533.333
4,266.666
2,133.333
1,066.666
533.333

<mark>266.666</mark> MATTER ELEMENTAL OCTAVES

266.666 x 7.5 = 2,000 x 7.5 = 15,000

<mark>133.333</mark>
66.666
33.333
16.666
8.333
4.1666
2.08333
1.041666
0.5208333
0.26041666
0.130208333
0.0651041666
0.03255208333
0.016276041666
0.0081380208333
0.00406901041666
0.002034505208333
0.0010172526041666

3D - ANGLES 720, 360, 22.5 OCTAVE SEQUENCE (AOS)

12,079,595,520
6,039,797,760
3,019,898,880
1,509,949,440
754,974,720
377,487,360
188,743,680
94,371,840
47,185,920
23,592,960
11,796,480
5,898,240
2,949,120
1,474,560
737,280
368,640
184,320
92,160
46,080
23,040
11,520
5,760
2,880
1,440
720

2,880 1,440, 720, 360,180, 90, 45, 22.5 DEGREES ANGLE OCTAVES

360 x 7.5 = 2,700 x 7.5 = 20,250   360 / 7.5 = 48 / 7.5 = 6.4

180
90
45
22.5
11.25
5.625
2.8125
1.40625
0.703125
0.3515625
0.17578125
0.087890625
0.0439453125
0.02197265625
0.010986328125
0.0054931640625
0.00274658203125
0.001373291015625

3D - 7,920 D 3,960 R OCTAVE SEQUENCE

1,063,004,405,760
531,502,202,880
265,751,101,440
132,875,550720
66,437,775,360
33,218,887,680
16,609443,840
8,304,721,920
4,152,360,960
2,076,180,480
1,038,090,240
519,045,120
259,522,560
129,761,280
64,880,640
32,440,320
16,220,160
8,110,080
4,055,040
2,027,520
1,013,760
506,880
253,440
126,720
63,360
31,680
15,840
7,920
3,960
1,980
990
495
247.5
123.75
61.875
30.9375
15.46875
7.734375
3.8671875
1.93359375
0.966796875
0.4833984375
0.24169921875
0.120849609375
0.0604248046875

3D - 388,880 OCTAVE SEQUENCE

6,524,323,758,080
3,262,161,879,040
1,631,080939,520
815,540,469,760
407,770,234,880
203.885 117,440
101,942,558,720
50,971,279,360
25,485,639,680
12,742,819,840
6,371,409,920
3,185,704,960
1,592,852,480
796,426,240
398,213,120
199,106,560
99,553,280
49,776,640
24,888,320
12,444,160
6,222,080
3,111,040
1,555,520
777,760
388,880
194,440
97,220
48,610
24,305
12,152.5
6,076.25
3,038.125
1,519.0625
759.53125
379.765625
189.8828125
94.94140625
47.470703125
23.7353515625
11.86767578125
5.933837890625

3D - 25,920 OCTAVE SEQUENCE

434,865,438,720
217,432,719,360
108,716,359,680
54,358,179,840
27,179,089,920
13,589,544,960
6,794,772,480
3,397,386,240
1,698,693,120
849,346,560
424,673,280
212,336,640
106,168,320
53,084,160
26,542,080
13,271,040
6,635,520
3,317,760
1,658,880
829,440
414,720
207,360
103,680
51,840 GOLDEN MIRROR NUMBER FOR TIME
25,920
12,960 RFEU
6,480 SUN D
3,240 SUN R
1,620 MOON D
810
405
202.5
101.25
50.625
25.3125
12.65625
6.328125
3.1640625
1.58203125
0.791015625
0.3955078125
0.19775390625
0.098876953125
0.0494384765625

3D – MATTER 864,000 / 3.333 = 259,200.000000003
1/x = 0.00000385802469

864,000 x 864,000 x 864,000 = 644,972,540,000,000,000 / 3.333 =
193,491,763,200,001,934.92 / 2 / 31,104,000 / 31,104,000 = 0
193,491,763,200,001,934.92 / 31,104,000 / 31,104,000 = 2
31,104,000 x 31,104,000 x 2 = 193,491,763,200,001,934.92

259,200 / 266.666 = 972.00000000003402 (7,920 EARTH'S DIAMETER)
972.00000000003402 / 11.111 = 87.48 x 1.333 = 116.64 x 1.111 = 129.6
/16 = 8.1 / 22.5 = 0.36 x 400 = 144 x 8 = 1,080 x .666 = 720 x .75 = 540 x 0.8
= 432 x .8333 = 360 x .5 = 180 x .5 = 90 x .5 = 45 x .5 = 22.5 x 400 = 9,000

3D - GAP = 2D 388,880 – 3D 259,200 = 129,680 / 3 = 43,226.666

43,226.666 / 3 = 14,408.888
43,226.666 / 22.5 = 1,921.18511851851
129,680 / 266.666 = 486.000000000012
129,680 / 6.666 = 19,452
129,680 / 5.555 = 23,342.4
129,680 / 4.444 = 29,178
129,680 / 3.333 = 38,906
129,680 / 2.222 = 58,356
129,680 / 1.111 = 116,712
129.680 / 1.333 = 172,906
129,680 / 144 = 900.555
3D - 259,200.000000003 / 1.111 = 233,280. 0000000023328
3D - 259,200.000000003 / 1.333 = 194,400.00000000486
3D - 259,200.000000003 / 11.111 = 23,328.00000000023328

3D 16 SECTORS OF THE MOLTEN SEA 22.5 DEGREES EACH

259,200.000000003 / 2 = 129,600.........[RESONANT FREQUENCY RESONANT UNIT (RFEU) 129,600]
259,200.000000003 x 2 = 518,400.......[518,400 / 24 = 21,600 / 60 = 360 /60 = 6 /60 = 0.1]

259,200.000000003 / 3 = 86,400   [SUN'S DIAMETER 864,000]
259,200.000000003 x 3 = 777,600.......[777,600 / 24 = 32,400 / 60 = 540 /60 = 9 /60 = 0.15]

259,200.000000003 / 4 = 64,800   [SUN'S DIAMETER 864,000 SHUFFLE]

259,200.000000003 x 4 = 1,036,800.......[1,036,800 / 24 = 43,200 / 60 = 720 /60 = 12 /60 = 0.2]

259,200.000000003 / 5 = 51,840..............[TIME OCTAVE 311,040 AND 518,400 SEQUENCE (TOS)]

259,200.000000003 x 5 = 1,296,000.......[1,296,000/ 24 = 54,000 / 60 = 900 /60 = 15 /60 = 0.25]

259,200.000000003 / 6 = 43,200..............[SUN'S RADIUS 432,000]

259,200.000000003 x 6 = 1,555,200...[1,555,200 / 24 = 64,800 / 60 = 1,080 /60 = 18 /60 = 0.3]

259,200.000000003 / 7 = 37,028.5714285714
259,200.000000003 x 7 = 1,814,400.[1,814,400 / 24 = 75,600 / 60 = 1,260 /60 = 21 /60 = 0.35]

259,200.000000003 / 8 = 32,400.............[SUN'S RADIUS 432,000 SHUFFLE]
259,200.000000003 x 8 = 2,073,600...[2,073,600 / 24 = 86,400 / 60 = 1,440 /60 = 24 /60 = 0.4]

259,200.000000003 / 9 = 28,800.............[LIGHT PLASMOID OCTAVE 144 SEQUENCE (POS)]
259,200.000000003 x 9 = 2,332,800...[2,332,800 / 24 = 97,200 / 60 = 1,620 /60 = 27 /60 = 0.45]

259,200.000000003 / 10 = 25,920...........[EARTH'S PROCESSION OF THE EQUINOX NUMBER 25,920]
259,200.000000003 x 10 = 2,592,000...[2,592,000 / 24 = 108,000 / 60 = 1,800 /60 = 30 /60 = 0.5]

259,200.000000003 / 11 = 23,563.6363
259,200.000000003 x 11 = 2,851,200.[2,851,200 / 24 = 118,800 / 60 = 1,980 /60 = 33 /60 = 0.55]

259,200.000000003 / 12 = 21,600............[MOON'S DIAMETER 2,160]
259,200.000000003 x 12 = 3,110,400...[3,110,400 / 24 = 129,600 / 60 = 2,160 /60 = 36 /60 = 0.6]

259,200.000000003 / 13 = 9,938.14615384615
259,200.000000003 x 13 = 3,369,600.[3,369,600 / 24 = 140,400 / 60 = 2,340 /60 = 39 /60 = 0.65]

259,200.000000003 / 14 = 18,514.2857142857
259,200.000000003 x 14 = 3,6828,800.[3,628,800 / 24 = 151,200 / 60 = 2,520 /60 = 42 /60 = 0.7]

259,200.000000003 x 15 = 17,280 [SUN'S DIAMETER 864,000 x 2 = 1,728,000 SUN SQUARE]
259,200.000000003 x 15 = 3,888,000...[3,888,000 /24 = 162,000 /60 = 2,700 /60 = 45 /60 = 0.75]

259,200.000000003 / 16 = 16,200............[MOON'S DIAMETER 2,160 SHIFTED]
259,200.000000003 x 16 = 4,147,200.[4,147,200 / 24 = 172,800 / 60 = 2,880 /60 = 48 /60 = 0.8]

DEGREES OF THE MOLTEN SEA

259,200.000000003 / 11.25 = 23,040...................[SUN'S RADIUS MIRROR 432,000]

259,200.000000003 / 22.5   = 11,520
259,200.000000003 / 45   = 5,760....................[LIGHT SPEED 144,000]
259,200.000000003 / 67.5   = 3,840
259,200.000000003 / 90   = 2,880....................[LIGHT SPEED 144,000]

259,200.000000003 / 112.5 = 2,304.................... [SUN'S RADIUS 432,000]
259,200.000000003 / 135   = 1,920
259,200.000000003 / 157.5 = 1,645
259,200.000000003 / 180   = 1,440....................[LIGHT SPEED 144,000]

259,200.000000003 / 202.5 = 1,280
259,200.000000003 / 225   = 1,152
259,200.000000003 / 247.5 = 1,047
259,200.000000003 / 270   = 960

259,200.000000003 / 292.5 = 886.153846153846
259,200.000000003 / 315   = 822.85714285714285714

259,200.000000003 / 337.5 = 768
259,200.000000003 / 360   = 720

259,200.000000003 / 540   = 480
259,200.000000003 / 720   = 360
259,200.000000003 / 900   = 288..[LIGHT SPEED 144,000 x 2]

259,200.000000003 / 129,600 = 2
259,200.000000003 / 144 = 1,800
259,200.000000003 / 144,000 = 1.8
259,200.000000003 / 266.666 = 972.00243 000607501519............[EARTH'S DIA 7,920]

16 SECTORS OF THE MOLTEN SEA 22.5 DEGREES EACH

129,680 / 2 = 64,840
129,680 / 3 = 43,226.666
129,680 / 4 = 32,420
129,680 / 5 = 25,936
129,680 / 6 = 21,613.333
129,680 / 7 = 18,525.714285714385714
129,680 / 8 = 16,210
129,680 / 9 = 14,408.888
129,680 / 10 = 12,968
129,680 / 11 = 11,789.090909
129,680 / 12 = 10,806.666
129,680 / 16 = 8,105

3D GAP 4D - COMPASS ANGLES FOR 16 SECTORS

129,680 / 5.625 = 23,045.222
129,680 / 11.25 = 11,527.111

129,680 / 22.5   = 5,763.555
129,680 / 45   = 2,881.777
129,680 / 67.5   = 1,921.18518515
129,680 / 90   = 1,440.888

129,680 / 112.5 = 1,152.7111
129,680 / 135   = 960.592592
129,680 / 157.5 = 823.365
129,680 / 180   = 720.444

129,680 / 202.5 = 640.39506172839506173
129,680 / 225   = 576.3555
129,680 / 247.5 = 523.9595959
129,680 / 270   = 480.296296296

129,680 / 292.5 = 443.350427350427350427

590

129,680 / 315   = 411.68253968254
129,680 / 337.5 = 384.23703703703
129,680 / 360   = 360.222

129,680 / 540   = 240.148148148
129,680 / 720   = 180.111
129,680 / 1,296 = 100.0617283950617283
129,680 / 144   = 900.555
129,680 / 1,440   = 90.0555
129,680 / 266.666 = 486.3000000000121575

PART 17 OF 20 – MSAART PATENT APPLICATION NOTES DRAFT 518,400 – 22:22:22 THUR 22ND SEPT 2022
© STRIKE FOUNDATION GUARANTEE LIMITED | MALCOLM BENDALL 2022 | GRAPHICS - STEVE EARL

2D - 388,800 OCTAVE SERIES

3,261,490,790,400
1,630,745,395,200
815,372,697,600
407,686,348,800
203,843,174,000
101,921,587,200
50,960,793,600
25,480,396,800
12,740,198,400
6,370,099,200
3,185,049,600
1,592,524,800
796,262,400
398,131,200
199,065,600
99,532,800
49,766,400
24,883,200
12,441,600
6,220,800

3,110,400 4D - TIME

1,555,200 = 5D [60 x 25,920 = 1,555,200 5D]

777,600 = 1D [30 x 25,920 = 777,600 1D]

388,800 = 2D [15 x 25,920 = 388,800 2D]

194,400 = 4D – TIME [7.5 x 25,920 = 194,400 4D]

97,200 3D [3.75 x 25,920 = 97,200 3D]

48,600
24,300
12,150
6,075
3,037.5
1,518.75
759.375
379.6875
189.84375
94.921875
47.4609375
23.73046875
11.865234375
5.9326171875
296630859375
1.483154296875
0.7415771484375
0.370.78857421875

2D – FLATLAND   864,000 / 2.222 = 388,800.000000003888
$$1 / x = 0.00000257201646$$

864,000 x 864,000 x 864,000 = 644,972,540,000,000,000 / 2.222 =
290,237,644,800,000,000 / 311,040,000 / 311,040,000 = 30

864,000 x 4 = 3,456,000 / 2.222 = 1,555,200
388,800 / 266.666 = 1,458.00000000005103

2D - GAP = 3D 777,600 – 2D 388,800 = 388,800
GAP = OCTAVE VALUE SAME AS PLANET SCALE

Octaves of Octave gap up

388,800 x 2 = 777,600 x 2 =1,555,200 x 2 = 3,110,400 x 2 = 6,220,800 x 2 = 12,441,600 x
2 = 24,833,200 x 2 = 49,766,400 x 2 = 99,532,800 x 2 = 199,065,600 x 2 = 398,131,200 x
2 = 796,262,400

Octaves of octave gap down

388,800 / 2 = 194,400 / 2 = 97,200 / 2 = 48,600 / 2 = 24,300 / 2 = 12,150 / 2 = 6,075 / 2
= 3,037 / 2 = 1518.75 / 2 = 759.375 / 2 = 378.6875 / 2 = 189.84375 / 2 = 94.921875 / 2
= 47.4609375 / 2 = 23.730 / 2 = 11.86 / 2 = 5.93261719 / 2 = 2.96630859 x 8 =
11.8652344

388,800 / 3   = 129,600
388,800 / 266.666 = 1,458

2D - 388,800.000000004 / 1.111 = 34,992.00000000069984
34,992.00000000069984 / 1.333 = 466,560
388,800.000000003888 x 11.111 = 4,320,000 / 266.666 = 16,200

2D 16 SECTORS OF THE MOLTEN SEA 22.5 DEGREES EACH

388,800.0000000388 / 2 = 194,400.000000001944...........[SANSKRIT 100 YEARS 311,040 OCTAVES]
388,800.0000000388 x 2 = 777,600...[777,600 / 24 = 32,400 / 60 = 540 /60 = 9 / 60 = 0.15]

388,800.0000000388 / 3 = 129,600.000000001296.........[RFEU 129,600]
388,800.0000000388 x 3 = 1,166,400[1,166,400 /24 = 48,600 /60 = 810 /60 = 13.5 /60 = .225]

388,800.0000000388 / 4 = 97,200.0000000000972.........[SUN'S RADIUS 432,000]
388,800.0000000388 x 4 = 1,555,200...[1,555,200/ 24 = 64,800 / 60 = 1,080 /60 = 18 / 60 = 0.3]

388,800.00000000388 / 5 = 77,76.0000000007776............[SANSKRIT 100 YEARS 311,040,000,000]
388,800.0000000388 x 5 = 1,944,000[1,944,000/24= 81,000 /60= 1,350 /60= 22.5 /60= 0.375]

388,800.0000000388 / 6 = 648,00.000000000648............[SUN'S RADIUS 432,000]
388,800.0000000388 x 6 = 2,332,800...[2,332,800/24= 97,200 /60 = 1,620 /60 = 27 / 60 = 0.45]

388,800.000000038 / 7 = 55,542.8571428577
388,800.000000038 x 7 = 2,721,600[2,721,600/24= 133,400 /60= 1,890 /60= 31.5 /60 = 0.525]

388,800.0000000388 / 8 = 48,600.000000000486.................[SUN'S RADIUS 432,000]
388,800.0000000388 x 8 = 3,110,400...[3,110,400/24= 129,600 /60= 2,160 /60= 36 /60= 0.6]

388,800.0000000388 / 9 = 43,200.000000000432.................[SUN'S RADIUS 432,000]
388,800.0000000388 x 9 = 3,499,200[3,499,200/24= 145,800/60= 2,430/60 = 40.5/60= .675]

388,800.0000000388 / 10 = 38,880.0000000003888....[SANSKRIT 100 YEARS 311,040 OCTAVES]
388,800.0000000388 x 10 = 3,888,000...[3,888,000/24= 162,000/60 = 2,700 /60 = 45 /60= 0.75]

388,800.0000000388 / 11 = 35,345.454545454898909
388,800.0000000388 x 11 = 4,276,800.[4,276,800/24= 178,200/60= 2,970/60= 49.5/60= 0.825]

388,800.0000000388 / 12 = 32,400.000000000324...[SUN'S RADIUS 432,000]
388,800.0000000388 x 12 = 4,665,600...[4,665,600/24= 194,400/60= 3,240/60= 54/60= 0.9]

388,800.0000000388 / 13 = 29,907.692307692606769
388,800.0000000388 x 13 = 5,054,400.[5,054,400/24= 210,600/60= 3,510/60= 58.5 /60= 0.975]

388,800.0000000388 / 14 = 27,771.428571428849143
388,800.0000000388 x 14 = 5,443,200.[5,443,200/ 24 = 26,800 / 60 = 3,780 /60 = 63 / 60 = 1.05]

388,800.0000000388 / 15 = 25,920.0000000002592...[EARTH'S EQUX 25,920 Yrs]
388,800.0000000388 x 15 = 5,832,000.[5,832,000/24= 243,000/60= 4,050/60= 67.5/60= 1.125]

388,800.0000000388 / 16 = 24,300.000000000243...[SUN'S RADIUS 432,000]
388,800.0000000388 x 16 = 6,220,800...[6,220,800/24= 259,200/60= 4,320/60= 72/60= 1.2]

CARDINAL COMPASS ANGLES FOR THE MOLTEN SEA

388,800.000000003888 / 5.625 = 69120.0000000006912........[RFEU 129,600]
388,800.000000003888 / 11.25 = 34,560.0000000003456.......[SUN'S SQUARE 864,000]

388,800.000000003888 / 22.5 = 17,280.0000000001728
388,800.000000003888 / 45 = 8,640.0000000000864.........[SUN'S DIA 864,000]
388,800.000000003888 / 67.5 = 5,760.0000000000576
388,800.000000003888 / 90 = 4,320.0000000000432.........[SUN'S RADIUS 432,000]

388,800.000000003888 / 112.5 = 3,456.....................[SUN'S SQUARE DIA 864,000 x 4]
388,800.000000003888 / 135 = 2,880.0000000000288..........[LIGHT SPEED 144,000]
388,800.000000003888 / 157.5 = 2,468.571428571428
388,800.000000003888 / 180 = 2,160.0000000000216......[MOON'S DIA 2,160]

388,800.000000003888 / 202.5 = 1,920.0000000000192
388,800.000000003888 / 225 = 1,728.00000000001728.......[SUN'S DIA 864,000 x 2]
388,800.000000003888 / 247.5 = 1,570.9090909091066182
388,800.000000003888 / 270 = 1,440.0000000000144..........[LIGHT SPEED 144,000]

388,800.000000003888 / 292.5 = 1,329.2307962307825231
388,800.000000003888 / 315 = 1,234.2857142857266286

388,800.000000003888 / 337.5 = 1,152.00000000001152
388,800.000000003888 / 360   = 1,080.0000000000108......[EARTH + MOON 10,080]

388,800.000000003888 / 540   = 720.0000000000072
388,800.000000003888 / 720   = 540.0000000000054
388,800.000000003888 / 900   = 432.0000000000432......... [SUN'S RADIUS 432,000]

SO 900 DEGREES IN THE TORUS PLANE AND 900, 432 LOG VOLUME UNITS IN THE SACRED TORUS BLOWN UP 100 PI SPHERE

388,800.000000003888 /   1.333 = 29,166.00000000292
388,800.000000003888 /   11.111 = 3,499.920000000035
388,800.000000003888 / 1,29,600 = 300.00000000000300
388,800.000000003888 /   144,000 = 2.700000000000027
388,800.000000003888 /   266.666 = 145.8300000000146
388,800.000000003888 /   7.5 = 5,185.0666

ALL SIX 864,000 COMBINATIONS

864,000	846,000	684,000	648,000	486,000	468,000
56,623,104*	55,443,456*	44,826,624*	42,467,328*	31,850,496*	30,670,848*
28,311,552*	27,721,728*	22,413,312*	21,233,664*	15,925,248*	15,335,424*
14,155,776*	13,860,864*	11,206,656*	10,616,832*	7,962,624*	7,667,712*
7,077,888*	6,930,432*	5,603,328*	5,308,416*	3,981,312*	3,833,856*
3,538,944*	3,465,216*	2,801,664*	2,654,208*	1,990,656*	1,916,928*
1,769,472*	1,732,608*	1,400,832*	1,327,104*	995,328*	958,464*
884,736*	866,304*	700,416*	663,552*	497,664*	479,232*
442,368*	433,152*	350,208*	331,776*	248,832*	239,616*
221,184*	216,576*	175,104*	165,888*	124,416	119,808*
110,592*	108,288*	87,552*	82,944*	62,208*	59,904*
55,296*	54,144*	43,776*	41,472*	31,104*	29,952*
27,684*	27,072*	21,888*	20,736*	15,552*	14,976*
13,824*	13,536*	10,944*	10,368*	7,776*	7,488*
6,912,000	6,768,000	5,472,000	5184,000	3,888,000	3,744,000
3,456,000	3,384,000	2,736,000	2,592,000	1,944,000	1,872,000
1,728,000	1,692,000	1,368,000	1,296,000	972,000	936,000

864,000	846,000	684,000	648,000	486,000	468,000
432,000	423,000	342,000	324,000	243,000	234,000
216,00	211,500	171,000	162,000	121,500	117,000
108,000	105,000	85,500	81,000	60,750	58,500
54,000	52,875	42,750	40,500	30,375	29,250
27,000	26,437.5	21,375	20,250	15,187.5	14,625
13,500	13,218.75	10,687.5	10,125	7,593.75	7,312.5
6,750	6,609.375	5,343.75	5,062.5	3,796.875	3,656.25
3,375	3,304.6875	2,671.875	2,531.25	1,898.43	1,828.1
1,687.5	1,652.34375	1,335.9375	1,265.62	949.2187	914.06
843.75	826.171875	667.96875	632.812	474.6093	457.03
421.875	413.0859375	333.984375	316.406	237.3046	228.51
210.9375	206.54296875	166.99218	158.2031	118.6523	114.25
105.46875	103.271484375	83.496093	79.10156	59.32617	57.128
52.734375	51.6357421875	41.7480468	39.55078	29,66308	28.564
26.3671875	25.8187109375	20.8740234	19.77539	14.83154	14.282

864,000	846,000	684,000	648,000	486,000	468,000

1D - 864,000 / 1.111 = 777,600.000000007776

1/x = 0.00000128600823

864,000 x 864,000 x 864,000 = 644,972,540,000,000,000 / 1.111 =
580475289600005804.75 / 311,040,000 / 311,040,000 = 0
Therefore 1D has no Time as 311,040,000 is all Area Volume Time (AVT)

777,600 / 266.666 = 2,916.00000000010206........................[RFEU 129,600]

1D - GAP = 1D 777,600 – 0D 7,783.783783783783 = 769,816.216216216216224076

1D 16 SECTORS OF THE MOLTEN SEA 22.5 DEGREES EACH

769,816.2162162162163 / 2 = 384,908.108108112038
769,816.2162162162163 / 3 = 256,605.405405405333
769,816.2162162162163 / 4 = 192,454.0540540540525
769,816.2162162162163 / 5 = 153,963.243243242
769,816.2162162162163 / 6 = 128,302.7027027016666
769,816.2162162162163 / 7 = 109,973.74517374428571
769,816.2162162162163 / 8 = 96,227.02702702625
769,816.2162162161263 / 9 = 85,535.1351353444
769,816.2162162162163 / 10 = 76.981.621621621
769,816.2162162162163 / 11 = 69,983.2923832918182
769,816.2162162162163 / 12 = 64,151.3513508333
769,816.2162162162163 / 16 = 48,113.513513125

769,816.2162162162163 / 360 = 2,138.37837837836111
769,816.2162162162163 / 720 = 1,069.1891891891805556
769,816.2162162162163 / 144 = 5,345.945945945902777

864,000 / 1.333 = 648,000.000000016
864,000 x 1.333 = 1,151,999.9999999712

0D –TRINITY FACTOR 864,000 / 111 = 7,783.7837833783783783
                    1/X = 0.00012847222222
77,600.0000000007776 / 1.1111 = 69,840.00000000006984
77,600.0000000007776 / 1.3333 = 58,200.000000001455
77,600.0000000007776 / 11.111 = 6,984.00000000006984

1D 16 SECTORS OF THE MOLTEN SEA 22.5 DEGREES EACH

77,600.000007776 / 2 = 38,800...[SANSKRIT 100 YEARS 311,040,000,000]

77,600.000007776 x 2 = 155,200...[155,200/24= 6,466/60= 107.777/60= 1.79629/60= 0.2993]

77,600.000007776 / 3 = 25,866.666
77,600.000007776 x 3 = 232,800...[232,800/24 = 9,700/60 = 161.666/60 = 2.694/60 = 0.0449]

77,600.000007776 / 4 = 19,400...[SANSKRIT 100 YEARS 311,040,000,000]

77,600.000007776 x 4 = 310,400.[310,400/24 = 12,933.33 /60= 215.555 /60= 185.5 /60= 3.0925]

77,600.000007776 / 5 = 15,520...[SANSKRIT 100 YEARS 311,040,000,000]

77,600.000007776 x 5 = 388,000.[388,000/24= 16,166.6/60= 269.4 /60= 4.49/60= 0.074845679]

77,600.000007776 / 6 = 12,933.333
77,600.000007776 x 6 = 465,600...[465,600/ 24 = 19,400 /60 = 323.3 / 60 = 5.38 /60 = 0.898148]

77,600.000007776 / 7 = 11,085.714285714285714
77,600.000007776 x 7 = 543,200...[543,200/24= 22,633/60= 377.2/60= 6.287/60= 0.1047839]

77,600.000007776 / 8 = 9,700
77,600.000007776 x 8 = 620,800.[620,800/24= 25,866/60= 431.1/60= 7.185/60= 0.1197530864]

77,600.000007776 / 9 = 8,622.222
77,600.000007776 x 9 = 698,400...[698,400/24= 29,100/60= 485 /60= 8.83 /60= 0.1347222]

77,600.000007776 / 10 = 77,760
77,600.000007776 x 10 = 776,000...[776,000/ 24 = 32,333 /60 = 538 / 60 = 8.981 /60 = 0.14969]

77,600.000007776 / 11 = 7,054.5454
77,600.000007776 x 11 = 853,600.[853,600/24= 35,566.6 /60= 592.7 /60= 9.879 /60= 0.164660]

77,600.000007776 / 12 = 6,466.666
77,600.000007776 x 12 = 931,200...[931,200 /24= 38,800 /60= 646.6 /60= 10.7 /60= 0.179629]

77,600.000007776 / 13 = 5,969.230769230077
77,600.07776 x 13 = 1,008,800 [1,008,800/24= 42,033.3 /60= 700.5 /60= 11.67
11.67/60 = 0.194598765432099]

77,600.000007776 / 14 = 5,542.8571428571428571
77,600.000007776 x 14 = 1,086,400 [1,086,400/ 24 = 42,266.666 /60 = 754.444 / 60 = 12.574
12.5740740740/60 = 0.209567901234568]

77,600.000007776 / 15 = 5,173.333
77,600.000007776 x 15 = 1,164,000 [1,164,000/ 24 = 48,500/60 = 808.333 / 60 = 13.472
13.47222 /60 = 02245370370370]

77,600.000007776 / 16 = 4,850
77,600.000007776 x 16 = 1,241,600 [1,241,600/24= 51.733.33 /60= 862.222 /60=
14.370370370 /60 = 0.239506172839506]

CARDINAL COMPASS ANGLES FOR THE MOLTEN SEA

77,600.0000000007776 / 5.625 = 13,795.555
77,600.0000000007776 / 11.25 = 6,897.777

77,600.0000000007776 / 22.5 = 3,448.888
77,600.0000000007776 / 45 = 1,724.444
77,600.0000000007776 / 67.5 = 1,149.629629
77,600.0000000007776 / 90 = 862.222

77,600.0000000007776 / 112.5 = 689.777
77,600.0000000007776 / 135 = 574.814814814
77,600.0000000007776 / 157.5 = 492.698
77,600.0000000007776 / 180 = 431.111

77,600.0000000007776 / 202.5 = 383.20987654321
77,600.0000000007776 / 225 = 344.888
77,600.0000000007776 / 247.5 = 313.535353535
77,600.0000000007776 / 270 = 107.77

77,600.0000000007776 / 292.5 = 265.299145299145
77,600.0000000007776 / 315 = 246.349206349206
77,600.0000000007776 / 337.5 = 229.925925925
77,600.0000000007776 / 360 = 215.555

77,600.0000000007776 / 540 = 143.703703703
77,600.0000000007776 / 720 = 107.777
77,600.0000000007776 / 1,296 [360 x 360] = 59.87654320987654321
77,600.0000000007776 / 144 = 538.888

THE MOON
2,160 Miles in Diameter

2,160 / 8.888 = 243.00000000000243 x 11.111 = 2,700 / 266.666 = 10.125 x 1.333 = 13.5

2,160 x 8.888 = 19,200 x 11.111 = 213,333.333 / 266.666 = 800

2,160 / 7.777 = 277.71428 x 11.111 = 3,085.71428571429 / 266.666 = 11.5714 x 1.333 = 15.428571

2,160 x 7.777 = 16,800 x 11.111 = 186,666.666 / 266.666 = 700

2,160 / 6.666 = 324.00000000000324 x 11.111 = 3,600 / 266.666 = 13.5 x 1.333 = 18

2,160 x 6.666 = 14,400 x 11.111 = 160,000 / 266.666 = 600

2,160 / 5.555 = 388.80000000003888 x 11.111 = 4,320/ 266.666 = 16.2 x 1.333 = 21.6
$\qquad$ 1/x = 0.00257201646091

2,160 x 5.555 = 12,000 x 11.111 = 133,333.3333333306666 / 266.666 = 500
$\qquad$ 1 / x = 0.00008333

2,160 / 4.444 = 486.0000000000486 (SUN 864) x 11.111 = 5,400 / 266.666 = 20.25 x 1.333 = 27
$\qquad$ 1/x = 0.00205761316872

2,160 x 4.444 = 96,000 [9,599.999] x 11.111 = 106,666.66666666453333 / 266.666 = 400

2,160 / 3.333 = 648.000000000006480 (Sun 864) x 11.111 = 72,000/ 266.666 = 27 x 1.333 = 36
$\qquad$ 1/x = 0.00154320987654

2,160 x 3.333 = 72,000 [7,199.999] x 11.111 = 80,000 / 266.666 = 300
$\qquad$ 1 / x = 0.00333

2,160 / 2.222 = 972.00000000000972 (EARTH 7,920) / 266.666 = 3.645 x 1.333 = 4.86
$\qquad$ 1/x = 0.00102880658436

2,160 x 2.222 = 4,800 [4,799.999999999952] x 11.111 = 53,333.33333333226666 / 266.666 = 200
$\qquad$ 1/x = 0.005

2,160 / 1.111 = 1,944.00000000001944 x 11.111 = 21,600 / 266.666 = 81.000000000002025
$\qquad$ 1/x = 0.00051440329218

2,160 x 1.111 = 2,400 [399.999999999976] x 11.111 = 26,666.666666666133333 / 266.666 = 100

2,160 / 111 = 19.459459459 x 11.111 = 216.216216216 / 266.666 = 0.810810810810822973 1/x = 1.23

2,160 x 111 = 239,760 x 11.111 = 2,664 [1,663,999.99999997336] / 266.666 = 9,990
$\qquad\qquad\qquad$ 1/x = 0.0001001001001

2,160 x 111 = 239,760 / 16 = 14,980 / 22.5 = 666 x 400 = 266,400 / 360 = 740
740 x 1.111 = 822.222/266.666 = 3.083 x 11.111 = 34.2555 x 1.333 = 45.674074074072475481 x 8 =
365.392 x 7.5 = 2,740.4444444443485289

2,160 / 1.333 = 1,957.5 / 22.5 = 87 x 400 = 34,800 / 11.111 = 3,132
2,160 x 1.333 = 2,880 / 22.5 = 128 x 400 = 51,200 / 11.111 = 4,608

THE EARTH - GIA
7,920 Miles in Diameter

7,920 / 22.5 = 352

7,920 x 22.5 = 178,200

7,920 / 16 = 495 / 22.5 = 22

7,920 x 16 = 126,720 / 22.5 = 5,632

7,920 / 12 = 660

7,920 x 12 = 95,040

7,920 / 8.888 = 891 x 11.111 = 9,900 / 266.666 = 37.125 x 1.333 = 49.5

7,920 x 8.888 = 70,400 x 11.111 = 782,222.2 / 266.666 = 2,933.333 x 1.333 = 3,911.111

7,920 / 7.777 = 1,018.28 x 11.111 = 11,314.285714 / 266.666 = 42.42 x 1.333 = 56.56

7,920 x 7.777 = 61,600 x 11.111 = 684,444.4/ 266.666 = 2,566.666 x 1.333 = 3,422.222

7,920 / 6.666 = 1,188 x 11.111 = 13,200 / 266.666 = 49.5 x 1.333 = 66

7,920 x 6.666 = 52,800 x 11.111 = 586,66.666 / 266.666 = 2,200 x 1.333 = 2,933.333

7,920 / 5.555 = 1,425.6000000001 x 11.111 = 1,4256 = 15,840 / 266.666 = 59.4000000000015 x 1.333 x 1.111 = 88
    1/x = 0.00070145903479

7,920 x 5.555 = 44,000 [1/x = 0.000022727272] x 11.111 = 488,888.888888879111111 x 1.333 = 651,851.851

7,920/4.444 = 1,782.00000000001782 x 11.111 = 19,800/266.666 = 74.25 x 1.333 = 99
    1/x = 0.00056116722783

7,920 x 4.444 = 35,199.999999999648 [35,200 1 / x = 0.000028409090] x 11.111 = 391,111.111111102888 / 266.666 = 1,466.666666666674 x 1.333 = 1,955.5555555555164444

7,920 / 3.333 = 2,376.00000000002376 x 11.111 = 26,400 / 266.666 = 99.000000000002475 x 1.333 = 132
    1/x = 0.00042087542088

7,920 x 3.333 = 26,399.999999999736 [26,400] x 11.111 = 29,333.3333332746667 x 1.333 = 391,111.111111093511111

7,920 / 2.222 = 3,564.0000000000564 x 11.111 = 39,600/ 266.666 = 148.5x 1.333 = 198
        1/x =
7,920 x 2.222 = 17599.999999999824 x 11.111 = 195,555.55555555164444 / 266.666 = 733.333 x 1.333 =

7,920 / 1.111 = 7,128.712871287128 x 11.111 = 79,200 / 266.666 = 297.000000000007425 x 1.333 = 396
      1/x =

7,920 x 1.111 = 8,799.12 [8,800] x 1.333 = 11,733.333 x 11.111 = 130,370.3703703645037 / 266.666 = 488.890111114144467425 /16 = 30.555 /22.5 = 1.3580 x 400 = 543.21123457127186028 x 360 = 195,556.0444456578697

7,920 / 111 = 71.351351351 x 11.111 = 792.792792792 x 1.333 = 1,057.0570570570200601
      1/x = 0.01401515

7,920 x 111 = 879,120
      1 / x = 0.00000113750114

7,920 / 1.333 = 5,940.00000000001485

7,920 x 1.333 = 10,560 / 7.5 = 1,408

7,920 / 194.4 = 0.020618556701031 [1/x = 48.5]

7,920 x 194.4 = 1,539,648

THE STARS

11 / 9.999 = 9.9

11 x 9.999 = 109.000

11 / 8.888 = 1.2375

11 x 8.888 = 97.777

11 / 7.777 = 1.414285714285714

11 x 7.777 = 85.555

11 / 6.666 = 1.65

11 x 6.666 = 73.333

11 / 5.555 = 1.9800000000000198 x 11.111 = 22 = 29.333

11 x 5.555 = 61.111

11 / 4.444 = 2.475

11 x 4.444 = 48.888

11 / 3.333 = 3.3

11 x 3.333 = 36.666

11 / 2.222 = 4.95

11 x 2.222 = 24.444

11 / 1.111 = 9.9

11 x 1.111 = 12.222

11 x 111 = 12

1D - ONE DIMENSION

1.111

864,000 x 864,000 x 864,000 = 644,972,540,000,000,000 / 1.111 =
580,475,289,600,005,804.75 / 311,040,000 / 311,040,000 = 6.00000000000006

1D = 1.111 / 1.333 x 10 = 8.333   1/x = 11.999999999999458333 [12]

2D =123456790123454321 / 1.333 = 0.92592592592593005556 1/x = 10.8 [POS]

3D = 1.371742112 / 1.333 x 10 = 1.02880658436213477   1/x = 9.72 {Earths Diameter}

4D = 1.5241579027586648377 / 1.333 x 10 =   11.431184270690272062
1/x = 0.87480000000001 {TIME}

5D = 1.6935087808429439957 / 1.333 x 10 = 12.70131585632243975
        1/x = 0.78732000000002 {AETHER}

5D = 1.6935087808429439957 x 1.333 x 10 = 22.5580117077905355439
        1/x = 0.44286750000003 {AETHER}

5D = 1.6935087808429439957 / 1.111 x 10 = 15.241579027586648377
        1/x = 0.65610000000003 {AETHER}

5D = 1.6935087808429439957 x 1.111 x 10 = 18.8167642315881
        1/x = 5.3144100000003188646 {AETHER}

1.111 x 1.333 = 1.481481481

1.111 x 7.5   = 8.33333

1.111 x 9   = 9.9999

1.111 / 2.222 = 0.5

1.111 / 3.333 = 0.33333

1.111 / 4.444 = 0.25

1.111 / 5.555 = 0.2

COMPASS ANGLES

1.111 x 5.625 = 6.25   [6.25] /

1.111 x 11.25 = 12.5   [12.5] /

1.111 x 22.5 = 25   [25] / 360 =0.069444 x 9 = 0.625 x 400 = 250

1.111 x 45   = 50   [50] /

1.111 x 67.5 = 75   [75] /

1.111 x 90   = 99.999   [100]

1.111 x 180 = 199.999   [200]

1.111 x 270 = 299.999   [300]

1.111 x 360 = 399.999   [400]

1.111 x 400 = 444.444   [444]

1.111 x 450 = 499.999   [500]

1.111 x 540 = 599.999   [600]

1.111 x 630 = 699.999   [700]

1.111 x 720 = 799.999   [800]

1.111 x 810 = 899.999   [900]

1.111 x 900 = 999.999   [1,000]

1.111 x 1,440 = 1,599.999   [1,600]

1.111 x 1,800 =1,999.999   [2,000]

1.111 x 2,700 = 2,999.999   [3,000]

1.111 x 4,500 = 4,999.999   [5,000]

1.111 x 8,100 = 8,999.999   [9,000]

1.111 x 9,000 = 9,999.999   [10,000]

1.111 x 129,600 = 143,999.999   [144,000]

1.111 x 144,000 = 159,999.9   [160,000]

2D - FLAT LAND – THE SECOND DIMENSION

864,000 x 864,000 x 864,000 = 644,972,540,000,000,000 / 2.222 =
290,237,644,800,002,902.38 / 311,040,000 / 311,040,000 = 3.00000000000003

1.111 x 1.111 = 1.23456790123454321
1.111 + 1.111 = 2.222   1.021 + 1.201 = 2.222

1 + 2 +3 + 4 + 5 + 6 + 7 + 9 + 0 + 1 + 2 + 3 + 4 + 5 + 4 + 3 + 2 + 1 = 62

1 x 2 x 3 x 4 x 5 x 6 x 7 x 9 x 0 x 1 x 2 x 3 x 4 x 5 x 4 x 3 x 2 x 1 = 130,636,800
                                                1 / x = 7.654810895552

*130,636,800 / 24 = 5,442,200 / 60 = 90,720 / 60 = 1,512 / 7.5 = 201.6 / 16 = 16
16 / 22.5 = 0.7111

Octaves down 0.7111 // 0.3555 // 0.1777 // 0.0888 // 0.0444 // 0.0222 // 0.0111 //
0.00555 // 0.002777 // 0.0013888 // 0.00069444 // 0.000347222 // 0.0001736111 //
0.00008680555 // 0.000043402777 // 0.0000217013888 // 0.00001085069444 //

Octaves up 0.7111 // 0.14222 // 0.28444 // 0.56888 // 1.13777 // 2.27555// 4.55111
// 9.10222 // 18.20444 // 36.40888 // 72.81777 // 145.6333 // 291.27111 //
582.54222 // 1,165.08444 // 2330.16888 //4,660.33777// 932.67555 //
18,641.35111 // 37,282.70222 // 74,565.40444 / 149,130.80888 // 298,261.61777

0.7111 / 0.125 = 5.6888 x 400 = 2,275 Octaves down

*130,636,800 /720 = 181,440 /16 = 11,340 /22.5 = 504 / 11.111 = 45.36 /1.333 = 34.02

130,636,800 / 62 = 2,107,045.1612903225806   1 / x = 0.00000047459828

          123456790123454321
     + 123454321097654321
TOTAL   246811111221108642

1D = 1.111 / 1.333 x 10 = 8.333 1/x = 11.999999999999458333 [12]

2D = 123456790123454321 / 1.333 = 0.92592592592593005556
                              1/x = 10.8 {AETHER}

3D = 1.371742112 / 1.333 x 10 = 1.02880658436213477   1/x = 9.72 {Earths Diameter}

4D = 1.5241579027586648377 / 1.333 x 10 = 11.431184270690272062
                              1/x = 0.87480000000001 {TIME}

5D = 1.6935087808429439957 / 1.333 x 10 = 12.70131585632243975
                              1/x = 0.78732000000002 {AETHER}

5D = 1.6935087808429439957 x 1.333 x 10 = 22.5580117077905355439
     1/x = 0.44286750000003 {AETHER}

5D = 1.6935087808429439957/ 1.111 x 10 = 15.241579027586648377
     1/x = 0.65610000000003 {AETHER}
5D = 1.6935087808429439957 x 1.111 x 10 = 18.8167642315881

$$1/x = 5.3144100000003188646 \ \{\text{AETHER}\}$$

123,456,790,123,454,321
+ 123,454,321,097,654,321
246,901,111,221,108, 642

(12,345,679,012,345,432,099)
+ (99,032,454,321,097,654,321)
Sum = (111,378,133,333,443,086,420)

(111,378,133,333,443,086,420 / 10,581,580,800 = ?

(10,581,580,800 / 111,378,133,333,443,086,420 = ?

(111,378,133,333,443,086,420 / 79 = ?

(1+2 + 3 + 4 + 5 + 6 + 7 + 9 + 0 + 1 + 2 + 3 + 4 + 5 + 4 + 3 + 2 + 0 + 9 + 9 = 79 ***)

(1 x 2 x 3 x 4 x 5 x 6 x 7 x 9 x 0 x 1 x 2 x 3 x 4 x5 x 4 x 3 x 2 x 0 x 9 x 9 = 10,581,580,800)

(10,581,580,800 / 79 = 133,944,060.759494)

CORRECT NUMBER IS PROBABLY BEST

123,456,790,123,454,321
+ 123,454,321,097,654,321
Sum = 246,901,111,221,108,642

1 + 2 + 3 + 4 +5 + 6 +7 + 9 + 0 + 1 + 2 + 3 + 4 + 5 + 4 + 3 + 2 +1 = 62

1 x 2 x 3 x 4 x 5 x 6 x 7 x 9 x 1 x 2 x 3 x 4 x 5 x 4 x 3 x 2 x 1 = 130,636,800

130,636,800 / 62 = 2,107,045.16129032

246,901,111,221,108,642 / 130,636,800 = 2,027,767.9124190733392

2.222 x 1.333 = 2.962962

1.333 / 2.222 = .6

2.222 x 7.5 = 16.666

2.222 x 9 = 20

2D - COMPASS ANGLES FOR 2.222

2.222 x 11.25   = 25

2.222 x 22.5   = 50 / 360 = 1.3888 x 9 = 1.25 x 400 = 500

2.222 x 45   = 100

2.222 x 90   = 200

2.222 x 135   = 300

2.222 x 180   = 400

2.222 x 225   = 500

2.222 x 270   = 600

2.222 x 315   = 700

2.222 x 360   = 800

2.222 x 405   = 900

2.222 x 450   = 1,000

2.222 x 4,500   = 10,000

2.222 x 45   = 100

2.222 x 450   = 1,000

2.222 x 4,500   = 10,000

2.222 x 400   = 888.888

2.222 x 720   = 1,600

2.222 x 1,440   = 3,200

2.222 x 4,500   = 10,000

2.222 x 72,000   = 160,000

2.222 x 129,600   = 288,000

2.222 x 144,000   = 320,000

2.222 x 259,200   = 576,000

2.222 x 288,000   = 640,000

2.222 x 576,000   = 1,280,000

2.222 x 1,152,000 = 2,600,000

2D - TWO DIMENSIONAL NUMBERS

1.111 x 1.111 = 1.<mark>23456790123454321</mark> x 1.333   = 1.6460905349793497942

1.111 x 1.111 = 1.23456790123454321 x 7.5   = 9.259259

1.111 x 1.111 = 1.23456790123454321 x 9   = 11.111

1.111 x 1.111 = 1.23456790123454321 x 11.25   = 13.888

1.111 x 1.111 = 1.23456790123454321 x 22.5   = 27.777 / 360 = 277.777X 9 X 400 =

1.111 x 1.111 = 1.23456790123454321 x 45   = 55.55

1.111 x 1.111 = 1.23456790123454321 x 90   = 111.111

1.111 x 1.111 = 1.23456790123454321 x 180   = 222.222

1.111 x 1.111 = 1.23456790123454321 x 360   = 444.444

1.111 x 1.111 = 1.23456790123454321 x 400   = 493.82716049381728395

1.111 x 1.111 = 1.23456790123454321 x 720   = 888.888

1.111 x 1.111 = 1.23456790123454321 x 1,440 = 1,777.777

1.111 x 1.111 = 1.23456790123454321 x 129,600 = 160,000

1.111 x 1.111 = 1.23456790123454321 x 144,000 = 177,777.777

1.111 x 1.111 = 1.23456790123454321 x 432,000 = 533.333.333

1.111 x 1.111 = 1.2345679012345432099 x 864,000 = 1,066,666.666

1.111 x 1.111 = 1.2345679012345432099 x 1,728,000 = 2,133,333.333

1.2345679012345432099 x 311,040,000,000 = 384,000,000,000.

311,040,000,000 / 1.2345690123454321 = 251,942,400,000

311,040,000,000 / 130,636,800 = 2380.9523809523809523

311,040,000,000 x 7.5 x 1.333 / 720 / 16 / 22.5 / 9 / 400 = 3,333.333

3,333.333 x 11.111 = 37,037.0370 x 400 x 400 = 5,925,925,925.9257185185

3D - MATTER – THE THIRD DIMENSION

1.111 x 3 = 3.333

864,000 x 864,000 x 864,000 = 644,972,540,000,000,000 / 3.333 =
193,491,763,200,001,934.92 / 311,040,000 / 311,040,000 = 2.00000000000002

1.111 x 1.111 x 1.111 = 1.3717421124828120713

1 + 3 +7 + 1 + 7 + 4 + 2 + 1 + 1 + 2 + 4 + 8 + 2 + 8 + 1 +2 + 0 + 7 + 1 + 3 = 65

1 x 3 x 7 x 1 x 7 x 4 x 2 x 1 x 1 x 2 x 4 x 8 x 2 x 8 x 1 x 2 x 0 x 7 x 1 x 3 = 50,577,408
1/x = 0.000000001977167

50,577,408 / 65 = 778,113.969230769   1/x = 0.00000128515878

          13717421124828120713
        + 31702182842112471731
TOTAL   44419603966940592444

1D = 1.111 / 1.333 x 10 = 8.333 1 / x = 11.999999999999458333 [12]

2D =123456790123454321 / 1.333 = 0.92592592592593005556
                    1 / x = 10.8   {AETHER}

3D = 1.371742112 / 1.333 x 10 = 1.02880658436213477 1 / x = 9.72 {Earths Diameter}

4D = 1.5241579027586648377 / 1.333 x 10 = 11.431184270690272062
1 / x = 0.87480000000001 {TIME}

5D = 1.6935087808429439957 / 1.333 x 10 = 12.70131585632243975
      1 / x = 0.78732000000002 {AETHER}

5D = 1.6935087808429439957 x 1.333 x 10 = 22.5580117077905355439
      1/x = 0.44286750000003 {AETHER}

5D = 1.6935087808429439957/ 1.111 x 10 = 15.241579027586648377
      1/x = 0.65610000000003 {AETHER}

5D = 1.6935087808429439957 x 1.111 x 10 = 18.8167642315881
      1/x = 5.3144100000003188646 {AETHER}

1 + 3 + 7 + 1 + 7 + 4 + 2 + 1 + 1 + 2 + 4 + 8 + 2 + 8 + 1 + 2 + 0 + 7 + 1 + 3 = 65

1 x 3 x 7 x 1 x 7 x 4 x 2 x 1 x 1 x 2 x 4 x 8 x 2 x 8 x 1 x 2 x 0 x 7 x 1 x 3 = 50,577,408

        13,717,421,124,828,120,713
      + 31,702,182,842,112,471,731
Sum = 44,419,603,966,940,402,444

3.333 x 1.333 = 4.444
3.333 x 7.5 = 25
3.333 x 9 = 30

3D - COMPASS ANGLES FOR 3.333

3.333 x 5.625 = 18.75

3.333 x 11.25 = 37.5

3.333 x 22.5 = 75 / 360 = 277.777 x 9 x 400 =

3.333 x 45   = 150

3.333 x 90   = 300

3.333 x 180   = 600

3.333 x 270   = 900

3.333 x 360   = 1,200

3.333 x 400   = 1,333.333

3.333 x 720   = 2,400

3.333 x 1,440   = 4,800

3.333 x 129,600 = 432,000

3.333 x 144,000 = 480,000

3.333 x 1,728,000 = 5,760.000

311,040,000,000 / 1,728,000 = 180,000

311,040,000,000 / 129,600 = 2,400,000

311,040,000,000 / 144,000 = 2,160,000

1,728,000 / 3.333 = 518,400

1,728,000 / 360 = 4,800

1,728,000 / 129,600 = 13.333

1,728,000 / 144,000 = 12

3.333 x 45 = 149.999   = [150]

3.333 x 450 = 1,499.99   = [1,500]

3.333 x 4,500 = 14,999.999   = [15,000]

1.371742112482812 0713 / 1.111 = 1.2345679012345323456   1 / x = 0.81

1.371742112482812 / 11.111 = 0.123456790123454321 1 / x = 8.10000000000016

1.3717421124828 / 1.333 = 1.0288065843621347737   1 / x = 0.972 {Earths Diameter}

1.3717421124828120713 x 129,600 = 1777,777.777777777244444

1.3717421124828120713 x 144,000 = 197,530.8641972493827 [Sun 864 Earth 972]

1.3717421124828120713 x 45   = 61.72839506172654321

1.3717421124828120713 x 90   = 123.45679012345308642

1.3717421124828120713 x 180 = 246.91358024690617284

1.3717421124828120713 x 360 = 493.82716049381234568

1.3717421124828120713 x 720 = 987.65432098762469136

1.3717421124828120713 / 129,600 = 0.00001058844298802686   1/x = 94,478.4

1.3717421124828120713 / 144,000 = 0.00000952598689   1/x =104,976

1.3717421124828120713   1/x = 0.72900000000002 (EARTH 7,920)

1.3717421124828120713 / 3.1415 = 0.43663907566979164145

1.3717421124828120713 x 3.1415 = 4.3094549431957469742

1.3717421124828120713 / 0.125 = 10.9739368998625

16 SECTORS OF THE MOLTEN SEA 22.5 DEGREES EACH

1.3717421124828120713 / 5.625 = 0.2438652644413888127 1/ x = 4.100625000012

1.3717421124828120 / 11.25 = 0.12193226322206944063 1/x = 8.20125000000025

1.3717421124828120713 / 22.5 = .0609663161103472032   1/x = 16.4025000000005

1.3717421124828120713 / 45 = 0.0304831580551736016 1/x = 32.805000000001

1.3717421124828120713 / 67.5 = 0.0203221053701157344 1/x =

1.3717421124828120713 / 90 = 0.01524157900275868008   1/x = 65.610000000002

1.3717421124828120713 / 112.5 = 0.01219326322207

1.3717421124828120713 / 135 = 0.01016105268506

1.3717421124828120713 / 157.5 = 0.00870947373005

1.3717421124828120713 / 180 = 0.0076207895137934004 1/x = 131.220000000004

1.3717421124828120713/ 202.5 = 0.0067740351233719115 1/x = 147.62250000004

1.3717421124828120713 / 225 = 0.0060966316110347203 1/x = 164.025000000005

1.3717421124828120713 / 247.5 = 0.00554239237367   1/x = 180.427500000005

1.3717421124828120713 / 270 = 0.00508052634252893336 1/x = 196.830000000006

1.3717421124828120713 /292.5 = 0.00468897166238728618 1/x=213.232500000006

1.3717421124828120713 / 315 = 0.00435473686502   1 /x = 229.635000000007

1.3717421124828120713 / 337.5 = 0.00406442107402 1 / x = 246.037500000007

1.3717421124828120713 / 360 = 0.0038103947568967002   1/x = 262.440000000008

1.3717421124828120713 / 720 = 0.019051973784483501 1/x = 524.880000000016

1.37174211248281207 / 144,000 = .000009525986922418 1/x = 104,976.000000003

1.111 x 1.111 x 1.111 = 1.3717421124828120713 x 129,600 = 177,777.777

1.111 x 1.111 x 1.111 = 1.3717421124828120713 x 144,000 = 197,530.864197525

4D TIME – FOURTH DIMENSION – 4D

1.111 x 4 = 4.444

864,000 x 864,000 x 864,000 = 644,972,540,000,000,000 / 4.444 =
145,118,822,400,001,451.19 / 311,040,000 / 311,040,000 = 1.500000000000015

1.111 x 1.111 x 1.111 x 1.111 = 1.5241579027586648377

1 + 5 + 2 + 4 + 1 +5 + 7 + 9 + 0 + 2+ 7+ 5 + 8 + 6 + 6 + 4 + 8 + 3 + 7 + 7 = 97

1 x 5 x 2 x 4 x 1 x 5 x 7 x 9 x 0 x 2 x 7 x 5 x 8 x 6 x 6 x 4 x 8 x 3 x 7 x 7 = 1,194,891,264,000

1,194,891,264,000 / 266.666 = 4,480,842,240.00011 / 864 = 5,186,160
5,186,160 / 216 = 24,010 / 792 = 30.315656

1,194,891,264,000/97 = 12,318466,639.175257732 / 266.666 = 46,194,249.8969084

   15241579027586648377
 + 77384668572097514251
 = 92625247599584162638

1.5241579027586648377

4D = 1.5241579027586648377 / 1.333 x 10 = 11.431184270690272062
1 / x = 0.87480000000001 {}

1D = 1.111 / 1.333 x 10 = 8.333 1 / x = 11.999999999999458333 [12]

2D = 123456790123454321 / 1.333 = 0.92592592592593005556
                           1 / x = 10.8   {POS}

3D = 1.371742112 / 1.333 x 10 = 1.02880658436213477 1 / x = 9.72 {Earths Diameter}

4D = 1.5241579027586648377 / 1.333 x 10 = 11.431184270690272062
                                    1 / x = 0.87480000000001 {TIME}

5D = 1.6935087808429439957 / 1.333 x 10 = 12.70131585632243975
                                    1 / x = 0.78732000000002 {AETHER}

5D = 1.6935087808429439957 x 1.333 x 10 = 22.5580117077905355439
                                    1 / x = 0.44286750000003 {AETHER}

5D = 1.6935087808429439957 / 1.111 x 10 = 15.241579027586648377
                                    1 / x = 0.65610000000003 {AETHER}

5D = 1.6935087808429439957 x 1.111 x 10 = 18.8167642315881
                                    1 / x = 5.3144100000003188646 {AETHER}

4D - TIME = 1.5241579027586648377/ 3.1415 = 0.0485154528553235609 1/x = 2.061

1.5241579027586648377 x 1.333 = 2.03221 (3221+1223 = 4444) 05370115023116
1.5241579027586648377 x 111 = 169.181527206212
1.5241579027586648377 x 11.111 = 16.935087808429439957
1.5241579027586648377 x 7.5 = 11.431184270689986283

1.5241579027586648377 x 9 = 13.717421124828

16 SECTORS OF THE MOLTEN SEA 22.5 DEGREES EACH

1.5241579027586648377 x 11.25 = 17.146776406034979424   1 / x = 0.05832

1.5241579027586648377 x 22.5 = 34.293552812069958848   1 / x = 0.02916

34.293552812069958848 x 360 = 12,345.679012345185185 x 360 =
4,444,444.444444266666 / 24 = 185,185.185185 / 60 = 3,086.4197530862962963 /
60 = 51.440329218104938272   1 / x = 0.002592

1.5241579027586648377 x 45 = 68.587105624139917695   1 / x = 0.01458 (REF)

1.5241579027586648377 x 90 = 137.17421124827983539   1 / x = 0.00729

1.5241579027586648377 x 180 = 137.17421124827983539   1 / x = 729

1.5241579027586648377 x 360 = 548.69684499311934156   1 / x = 0.0018225

1.5241579027586648377 x 400 = 609.66316110346593507   1 / x = 0.00164025

1.5241579027586648377 x 720 = 1,097.3936899862386831   1 / x = 0.00091125

 1.5241579027586648377 x 1,440 = 2,194.7873799724773663   1 / x = 0.000455625

1.5241579027586648377 x 129,600 = 197,530.86419752296296 1 /x = 0.0000050625

1.5241579027586648377 x 144,000 = 219,478.737997248   1 / x = 0.00000455625

1.5241579027586648377 x 311,040 = 474,074.074074055111111 1/x = 0.0-02109375

1.111 x 1.111 x 1.111 x 1.111 = 1.5241579027586648377 x 1.111 = 1.69350878084294
                                        1 / x = 0.590490000000003

1.111 x 1.111 x 1.111 x 1.111 = 1.5241579027586648377 x 1.333 =
20322105370115023116   1 / x = 0.492075

4D – TIME = 1.5241579027586648377 / 1.333 = 1.14311842706903

1.5241579027586648377 / 1.111 = 1.37174211248281   1 / x = 0.729

1.5241579027586648377 / 7.5 = 0.20332210537011553117 1 / x = 4.92075
1.5241579027586648377 / 9 = 0.1693508780842960931   1 / x = 5.90490000000024

16 SECTORS OF THE MOLTEN SEA 22.5 DEGREES EACH

1.5241579027586648377 / 11.25 = 0.1354807024674368745 1/ x = 7.381125000003

1.5241579027586648377 / 22.5 = 0.06774035123372 1 / x   = 14.7622500000006

1.5241579027586648377 / 45 = 0.0338701756168592186 1 / x = 29.5245000000012

1.5241579027586648377 / 90 = 0.0169350878084296093 1 / x = 59.0490000000024

1.5241579027586648377 / 180 = 0.0084675439042148047 1 /x = 118.098000000005
1.5241579027586648377 / 360 = 0.00423377719521074023 1 /x = 236.196000000009

1.5241579027586648377 / 400 = 0.0038103947568966621 1 / x = 262.44000000001
1.5241579027586648377 / 720 = 0.00211688597605   1 / x = 472.392000000019

1.5241579027586648377/1,440 = 0.0010584429880268506 1/x =944.784000000038

5D - AETHER – FITH DIMENSION

1.111 x 5 = 5.555

864,000 x 864,000 x 864,000 = 644,972,540,000,000,000 / 5.555 =
116,095,057,920,001,160.95 / 311,040,000 / 311,040,000 = 1.200000000000012

1.111 x 1.111 x 1.111 x 1.111 x 1.111 = 1.6935087808429439957
1 / x = 0.590490000000003

1 + 6 + 9 + 3 + 5 + 0 + 8 + 7 + 8 + 0 + 8 + 4 + 2 + 9 + 4 + 3 + 9 + 9 + 5 + 7 = 107

1x 6 x 9 x 3 x 5 x 0 x 8 x 7 x 8 x 0 x 8 x 4 x 2 x 9 x 4 x 3 x 9 x 9 x 5 x 7 = 7,110,822,297,600

*****7,110,822,297,600 / 266.666 = 26,666,666.6 [266.666 x 100,000]

7,110,822,297,600 / 5.555 = 1279,948,013,568.0127995 / 11.111 = 1.151953212 / 24
= 479,980,505,088 / 60 = 7,999,675.0848 / 60 = 133,327.91808 / 7.5 =
17,777.055744000799968

 17,777.055744000799968 / 720 = 24.690355200001111066
24.690355200001111066 / 16 = 1.543147200000694416 / 22.5 =
0.0685843200000030863 / .125 = 0.54867456 x 400 = 219.4698240000987614

7,110,822,297,600 / 31,104,000,000 = 17,777.055744000799968
*****17,777.055744000799968 / 1.2 / 24 / 60 / 60 / 7.5 = 9,144,576
9,144,576 / 720 = 12,700 / 16 = 793.8 / 22.5 = 35.28 / 0.125 = 282.24
282.24 x 400 = 266.666 = 423.36

1D = 1.111 / 1.333 x 10 = 8.333   1 / x = 1.1999999999999458333 [12]

2D =123456790123454321 / 1.333 = 0.92592592592593005556
    1/x = 10.8   {POS}

3D = 1.371742112 / 1.333 x 10 = 1.02880658436213477   1/x = 9.72 {Earths Diameter}

4D = 1.5241579027586648377 / 1.333 x 10 = 11.431184270690272062
1/x = 0.87480000000001 {}

5D = 1.6935087808429439957 / 1.333 x 10 = 12.70131585632243975
    1/x = 0.78732000000002 {AETHER}

5D = 1.6935087808429439957 x 1.333 x 10 = 22.5580117077905355439
    1/x = 0.44286750000003 {AETHER}

5D = 1.6935087808429439957 / 1.111 x 10 = 15.241579027586648377
    1/x = 0.65610000000003 {AETHER}

5D = 1.6935087808429439957 x 1.111 x 10 = 18.8167642315881
    1/x = 5.3144100000003188646 {AETHER}

144,000 = 2 x 3 x 4 x 5 x 5 x 4 x 3 x 2 x 10   144,000 / 0.125 = 1,152,000

PART 17 OF 20 – MSAART PATENT APPLICATION NOTES DRAFT 518,400 – 22:22:22 THUR 22ND SEPT 2022
© STRIKE FOUNDATION GUARANTEE LIMITED | MALCOLM BENDALL 2022 | GRAPHICS - STEVE EARL

144,000 / 28 = 5142.85714285714   [28 = 2+3+4+5+5+4+3+2]   7,777 = 2,345 + 5,432

144,000 / 7,777 = 18.5161373280185   7,777 / 144,000 = 0.054006944

144,000 / 5,555 = 25.922592259   5,555 / 144,000 = 0.0385763888

144,000 x 144,000 = 20,736,000,000 / 7,777 = 2,66,323.7752346663238

144,000 x 144,000 x 144,000 = 2,985,984,000,000 / 7,777 = 3,842,965,250,965.250965

144,000 SQUARE ROOT = 379.473319220205 / 7,777 = 48.78942675688357

144,000 CUBE ROOT = 52.4148278841779 / 7,777 = 6.739049299394309

7,777 / 1.5241579027586648377 =

7,777 = 2,345 + 5,432   7,777 / 28 = 277.75 / 0.125 = 2,222

7,777 / 2,345 = 3.31641791044776 = 1 / x = 0.30153015

7,777 / 5,432 = 1.43170103092784

5,432 / 7,777 = 0.698469846984

2,345 / 7,777 = 0.30153015301

144,000 / 1.111 = 129,600 / 266.666 = 486.00000000001215

360 x 400 = 144,000   180 x 800 = 144,000   200 x 720 = 144,000   900 x 144 = 129,600

5.555 x 3.14159265358979 (Pi) = 17.4515471906913

5.555 x 1.333 = 7.40666

5.555 x 7.5 = 41.6625

5.555 x 9 = 49.995

5D –AETHER - ANGLES OF THE COMPASS FOR 5.555

5.555 x 11.25 = 62.4999 [62.5]   1 / x = 0.016 x 432,000 = 6.91200000000007
                                                1/x = 0.00014467592593

5.555 x 22.5 = 124.9875 [125]   1 / x = 0.008 x 432,000 = 3,456

[3 + 4 + 5 + 6 + 6 + 5 + 4 + 3 = 36]

[3 x 4 x 5 x 6 x 6 x 5 x 4 x 3 = 129,600]..[RESONANT FREQUENCY RESONANT UNIT (RFEU) 129,600]
[129,600 / 36 = 3,600]
[1 + 2 + 3 + 4 + 5 + 5 + 4 + 3 + 2 + 1 = 30]
[1 x 2 x 3 x 4 x 5 x 5 x 4 x 3 x 2 x 1 = 14,400].....[LIGHT PLASMOID OCTAVE 144 SEQUENCE (POS)]
[14,400 / 30 = 480]

5.555 / 266.666 = 0.0208333333333336458   1 / x   = [48] 47.999 x 432 =

5.555 / 22.5 = 0.24691358024691   1 / x = 4.050

5.555 / 9 = 0.617283950617278   1 / x = 1.62............[MOON'S DIAMETER 2,160 SHIFTED]

5.555 / 12 = 0.46296296296   1 / x = 2.16............[MOON'S DIAMETER 2,160]

5.555 / 15 = 0.370370370   1 / x = 2.70000000000003*x 432,000 = 11,664

5.555 / 16 = 0.925925925   1 / x = 1.08....[PLANET OCTAVE 864 SEQUENCE (POS)]

5.555 / 18 = 0.308641975308643888   1 / x = 3.24...........[SUN'S RADIUS 432,000 SHUFFLE]

5.555 / 24 = 0.23148148148   1 / x = 4.23............[SUN'S RADIUS 432,000 SHUFFLE]

5.555 / 27 = 0.205771316872428   1 / x = 4.86   [SUN'S DIAMETER 864,000 SHUFFLE]

5.555 / 30 = 0.185185185   1 / x = 5.4

5.555 / 36 = 0.154320987654319444   1 / x = 6.48   [SUN'S DIAMETER 864,000 SHUFFLE]

5.555 / 37 = 0.150150150   1 / x = 6.66...[MUSIC 266.666 OCTAVE 11.111 SEQUENCE (MOS)]

5.555 / 38 = 0.146198830409366   1 / x = 6.84   [SUN'S DIAMETER 864,000 SHUFFLE]

5.555 / 39 = 0.142450142450142450   1 / x = 7.02

5.555 / 40 = 0.13888   1 / x = 7.2

5.555 / 72 = 0.077160493827159722   1 / x = 12.960

5D - 16 SECTORS OF THE MOLTEN SEA 22.5 DEGREES EACH

5.555 x 45   = 250   1 / x = 0.004

5.555 x 90   = 500   1 / x = 0.002

5.555 x 180   = 1,000   1 / x = 0.001

5.555 x 270   = 1,500   1 / x = 0.000666

5.555 x 360   = 2,000   1 / x = 0.0005

5.555 x 400   = 2,222.2   1 / x = 0.00045

5.555 x 720   = 4,000   1 / x = 0.00025

5.555 x 1,440   = 8,000   1 / x = 0.000125

5.555 x 1,776   = 9,866.666   1 / x = 0.00010135135

5.555 x 2,520   = 14,000   1 / x = 0.0000071428571429

5.555 x 25,920   = 144,000   1 / x = 0.0000069444

5.555 x 129,600 = 72,000   1 / x = 0.00000138

5.555 x 144,000 = 800,000   1 / x = 0.00000125

5.555 x 1,728,000 = 9,600,000.999   1 / x = 0.00000010416667

1.6935087808429439957 x 129,600 = 219,478.73799724554184

1.6935087808429439957 x 144,000 = 243,865.26444138393538 1 / x = 0.004100625

1.111 x 1.111 x 1.111 x 1.111 x 1.111 = 1.6935087808429439957

1.6935087808429439957 x 1.333 = 2.2580117077905355439

1.6935087808429439957 x 7.5 = 12.701315856322079967   1 / x =0.078732

1.6935087808429439957 x 9 = 15.241579027586495961   1 / x = 0.06561

16 SECTORS OF THE MOLTEN SEA 22.5 DEGREES EACH

1.6935087808429439957 x 11.25 = 19.051973784483119951   1 / x = 0.052488

1.6935087808429439957 x 22.5 = 38.103947568966239902   1 / x = 0.026244

1.6935087808429439957 x 45 = 76.2078951379325   1 / x = 0.013122

1.6935087808429439957 x 90 = 152.41579027586495961   1 / x = 0.0006561

1.6935087808429439957 x 180 = 304.83158055173   1 / x = 0.0032805

1.6935087808429439957 x 360 = 609.66316110346   1 / x = 0.00164025

1.6935087808429439957 x 400 = 677.40351233717759827   1 / x = 0.001476225

1.6935087808429439957 x 720 = 1219.3263222069196769   1 / x = 0.000820125

1.6935087808429439957 x 1,440 = 2,438.6526444138393538   1 / x = 0.0004100625

1.6935087808429439957 x 129,600 = 219,418.7379972455418   1 / x =0.0000455625

1.6935087808429439957 x 144,000 = 243,865.2644413839353   1 / x = 0.0004100625

1.111 x 1.111 x 1.111 x 1.111 = 1.6935087808429439957 / 129,600 =

1.111 x 1.111 x 1.111 x 1.111 = 1.6935087808429439957 / 144,000 =

1.111 x 1.111 x 1.111 x 1.111 x 1.111 x 1.111 =1.6935087808429439957 x 129,600 =

1.111 x 1.111 = 1 + 2 + 3 + 4 + 5 + 6 + 7 + 9 + 1 + 2 + 3 + 4 + 5 + 4 + 3 + 2 + 9 + 9 = 79

1 x 2 x 3 x 4 x 5 x 6 x 7 x 9 x 0 x 1 x 2 x 3 x 4 x 5 x 4 x 3 x 2 x 0 x 9 x 9 = 10,581,580,800

10,581,580,800 / 79 = 133,944,060.759494

144,000 = 2 x 3 x 4 x 5 x 5 x 4 x 3 x 2 x 10   144,000 / 0.125 = 1,152,000

144,000 / 28 = 5,142.85714285714   [28 = 2 + 3 + 4 + 5 + 5 + 4 + 3 + 2]

7,777 = 2,345 + 5,432

144,000/7,777= 18.5161373280185   7,777 / 144,000 = 0.054006944

144,000 x 144,000 = 20,736,000,000 / 7,777 = 2,66,323.7752346663238

144,000 x 144,000 x 144,000 = 2,985,984,000,000 / 7,777 =

144,000 SQUARE ROOT = 379.473319220205 / 7,777 =

144,000 CUBE ROOT = 52.4148278841779 / 7,777 =

7,777 / 1.5241579027586648377 =

7,777 = 2,345 + 5,432   7,777 / 28 = 277.75 / 0.125 = 2,222

7,777/ 2,345 = 3.31641791044776 = 1/x = 0.30153015

7,777 / 5,432 = 1.43170103092784

5,432 / 7,777 = 0.698469846984

2,345 / 7,777 = 0.30153015301

144,000 / 1.1111 = 129,600

360 x 400 = 144,000   180 x 800 = 144,000   200 x 720 = 144,000

6D - ENERGY – SIXTH DIMENSION 6.666

1.111 x 6 = 6.666

864,000 x 864,000 x 864,000 = 644,972,540,000,000,000 / 6.666 =
96,745,881,600,000,967.459 / 311,040,000 / 311,040,000 = 1

1.111 x 1.111 x 1.111 x 1.111 x 1.111 x 1.111 = 1.88167642315588078451
$$1 / x = 0.53144100000003$$

1 + 8 +8 +1+ 6 + 7 +6 + 4 + 2 + 3 + 1 + 5 + 8 + 8 + 8 + 0 + 7 + 8 + 4 + 5 +1 = 101

1 x 8 x 8 x 1 x 6 x 7 x 6 x 4 x 2 x 3 x 1 x 5 x 8 x 8 x 7 x 8 x 4 x 5 x 1 = 138,726,604,800

1.88167642315588078451

138,726,604,800 / 266.666 = 520,224,786

1D = 1.111 / 1.333 x 10 = 8.333 1 / x = 1.1999999999999458333 [12]

2D = 123456790123454321 / 1.333 = 0.92592592592593005556
$$1 / x = 10.8 \quad \{POS\}$$

3D = 1.371742112 / 1.333 x 10 = 1.02880658436213477
$$1 / x = 9.72 \ \{Earths\ Diameter\}$$

4D = 1.5241579027586648377 / 1.333 x 10 =   11.431184270690272062
$$1 / x = 0.87480000000001 \ \{TIME\}$$

5D = 1.6935087808429439957 / 1.333 x 10 = 12.70131585632243975
$$1 / x = 0.78732000000002 \ \{AETHER\}$$

5D = 1.6935087808429439957 x 1.333 x 10 = 22.5580117077905355439
$$1 / x = 0.44286750000003 \ \{AETHER\}$$

5D = 1.6935087808429439957 / 1.111 x 10 = 15.241579027586648377
$$1/x = 0.65610000000003 \ \{AETHER\}$$

5D = 1.6935087808429439957 x 1.111 x 10 = 18.8167642315881
$$1 / x = 5.3144100000003188646 \ \{AETHER\}$$

7D - SPIRIT – SEVENTH DIMENSION 7.777

1.111 x 7 = 7.777

864,000 x 864,000 x 864,000 = 644,972,540,000,000,000 / 7.777 =
82,925,041,371,429,400.679 / 311,040,000 / 311,040,000 = 0.8571428571428657143

1.111 x 1.111 x 1.111 x 1.111 x 1.111 x 1.111 x 1.111 = 2.0907515812875433648
1 / x = 0.47829690000003

2.0907515812875433648

2 + 0 + 9 + 0 + 7 + 5 + 1 + 5 + 8 + 1 + 2 + 8 + 7+ 5 + 4 + 3 + 3 + 6 + 4 + 8 =

2 x 0 x 9 x 0 x 7 x 5 x 1 x 5 x 8 x 1 x 2 x 8 x 7 x 5 x 4 x 3 x 3 x 6 x 4 x 8 =

2.0907515812875433648 / 266.666 =

1D = 1.111 / 1.333 x 10 = 8.333   1 / x = 1.1999999999999458333 [12]

2D = 123456790123454321 / 1.333 = 0.92592592592593005556
    1 / x = 10.8   {AETHER}

3D = 1.371742112 / 1.333 x 10 = 1.02880658436213477 1 / x = 9.72 {Earths Diameter}

4D = 1.5241579027586648377 / 1.333 x 10 = 11.431184270690272062
1 / x = 0.87480000000001 {}

5D = 1.6935087808429439957 / 1.333 x 10 = 12.70131585632243975
    1 / x = 0.78732000000002 {AETHER}

5D = 1.6935087808429439957 x 1.333 x 10 = 22.5580117077905355439
    1 / x = 0.44286750000003 {AETHER}

5D = 1.6935087808429439957 / 1.111 x 10 = 15.241579027586648377
    1 / x = 0.65610000000003 {AETHER}

5D = 1.6935087808429439957 x 1.111 x 10 = 18.8167642315881
    1/x = 5.3144100000003188646 {AETHER}

8D – NEW BEING – EIGHTH DIMENSION 8.888

1.111 x 8 = 8.888

864,000 x 864,000 x 864,000 = 644,972,540,000,000,000 / 8.888 =
72,559,411,200,000,725.594 / 311,040,000 (ATV) / 311,040,000 (ATV) = 0.75

1.111 x 1.111 x 1.111 x 1.111 x 1.111 x 1.111 x 1.111 x 1.111 = 2.3230573125416916192
1 / x = 0.43046721000003

2.3230573125416916192

2 + 3 + 2 + 3 + 0 + 5 + 7 + 3 + 1 + 2 + 5 + 4 + 1 + 6 + 9 + 1 + 6 + 1 + 9 + 2 =

2 x 3 x 2 x 3 x 0 x 5 x 7 x 3 x 1 x 2 x 5 x 4 x 1 x 6 x 9 x 1 x 6 x 1 x 9 x 2 =

2.3230573125416916192

2.3230573125416916192 / 266.666 = 520,224,786

1D = 1.111 / 1.333 x 10 = 8.333 1/x = 1.1999999999999458333 [12]

2D = 123456790123454321 / 1.333 = 0.92592592592593005556
1/x = 10.8 {AETHER}

3D = 1.371742112 / 1.333 x 10 = 1.02880658436213477   1/x = 9.72 {Earths Diameter}

4D = 1.5241579027586648377 / 1.333 x 10 = 11.431184270690272062
1/x = 0.87480000000001 {}

5D = 1.6935087808429439957 / 1.333 x 10 = 12.70131585632243975
1/x = 0.78732000000002 {AETHER}

5D = 1.6935087808429439957 x 1.333 x 10 = 22.5580117077905355439
1/x = 0.44286750000003 {AETHER}

5D = 1.6935087808429439957 / 1.111 x 10 = 15.241579027586648377
1/x = 0.65610000000003 {AETHER}

5D = 1.6935087808429439957 x 1.111 x 10 = 18.8167642315881
1/x = 5.3144100000003188646 {AETHER}

9D - ENERGY – NINTH DIMENSION 9.999

1.111 x 9.999 = 11.1111111111110888

864,000 x 864,000 x 864,000 = 644,972,540,000,000,000 / 9.999 =
64,497,254,400,000,644.973 / 311,040,000 / 311,040,000 = 0.666

1.111 x 1.111 x 1.111 x 1.111 x 1.111 x 1.111 x 1.111 x 1.111 x 1.111 = 2.5811747917129648763
1 / x = 0.38742048900003

2.5811747917129648763

1 + 8 +8 +1+ 6 + 7 +6 + 4 + 2 + 3 + 1 + 5 + 8 + 8 + 8 + 0 + 7 + 8 + 4 + 5 +1 =

1 x 8 x 8 x 1 x 6 x 7 x 6 x 4 x 2 x 3 x 1 x 5 x 8 x 8 x 7 x 8 x 4 x 5 x 1 =

2.5811747917129648763

138,726,604,800 / 266.666 = 520,224,786

1D = 1.111 / 1.333 x 10 = 8.333 1/x = 1.1999999999999458333 [12]

2D =123456790123454321 / 1.333 = 0.92592592592593005556
1 / x = 10.8   {AETHER}

3D = 1.371742112 / 1.333 x 10 = 1.02880658436213477 1 / x = 9.72 {Earths Diameter}

4D = 1.5241579027586648377 / 1.333 x 10 =   11.431184270690272062
1/ x = 0.87480000000001 {}

5D = 1.6935087808429439957 / 1.333 x 10 = 12.70131585632243975
1 / x = 0.78732000000002 {AETHER}

5D = 1.6935087808429439957 x 1.333 x 10 = 22.5580117077905355439
1 / x = 0.44286750000003 {AETHER}

5D = 1.6935087808429439957 / 1.111 x 10 = 15.241579027586648377
1 / x = 0.65610000000003 {AETHER}

5D = 1.6935087808429439957 x 1.111 x 10 = 18.8167642315881
1 / x = 5.3144100000003188646 {AETHER}

10 D – UNIVERSAL LAW – TENTH DIMENSION
10.101010

1.111 x 10.101010 = 11.223344556677665544332211

6,350,400 x 6,350,400 = 40,327,580,160,000 / 24 / 60 / 60 / 7.5 = 62,233,920 / 720 = 86,436 / 16 = 5,402.25 / 22.5 = 240.1

40,327,580,160,000 / 311,040,000 = 129,654

40,327,580,160,000 / 51,840,000,000,000 = 0.777924

40,327,580,160,000 / 259,248,729,600 = 155.555 // 77.777 //38.888 //19.444 //9.7222 // 4.86111 // 2.430555 // 1.215277

864,000 x 864,000 x 864,000 = 644,972,540,000,000,000 / 10.101010 = 63,852,281,856,000,638.523 / 311,040,000 / 311,040,000 = 66.0000000000066

1.111 x 1.111 x 1.111 x 1.111 x 1.111 x 1.111 x 1.111 x 1.111 x 1.111 x 1.111 = 2.8679719907921545161

1 / x = 0.34867844010003

2.8679719907921545161 / 0.660000000000066 = 4.34541210726071

2 + 8 +6 +7+ 9 + 7 +1 + 9 + 9 + 0 + 7 + 9 + 2 + 1 + 5 + 4 + 5 + 1 + 6 + 1 = 99

2 x 8 x 6 x 7 x 9 x 7 x 1 x 9 x 9 x 7 x 9 x 2 x 1 x 5 x 4 x 5 x 1 x 6 = 259,248,729,600

259,248,729,600 / 266.666 = 972,182,736.000024

1D = 1.111 / 1.333 x 10 = 8.333   1 / x = 1.1999999999999458333 [12]

2D =123456790123454321 / 1.333 = 0.92592592592593005556
    1 / x = 10.8   {AETHER}

3D = 1.371742112 / 1.333 x 10 = 1.02880658436213477 1 / x = 9.72 {Earths Diameter}

4D = 1.5241579027586648377 / 1.333 x 10 =   11.431184270690272062
    1 / x = 0.87480000000001 {}

5D = 1.6935087808429439957 / 1.333 x 10 = 12.70131585632243975
    1 / x = 0.78732000000002 {AETHER}

5D = 1.6935087808429439957 x 1.333 x 10 = 22.5580117077905355439
    1 / x = 0.44286750000003 {AETHER}

5D = 1.6935087808429439957 / 1.111 x 10 = 15.241579027586648377
    1 / x = 0.65610000000003 {AETHER}

5D = 1.6935087808429439957 x 1.111 x 10 = 18.8167642315881
    1 / x = 5.314410000003188646 {AETHER}

11 D - ENTROPY – ELEVENTH DIMENSION
11.111

1.111 x 11.111 = <mark>12.3456790123454321</mark>

864,000 x 864,000 x 864,000 = 644,972,540,000,000,000 / 11.111 =
58,047,528,960,000,580.475 / 311,040,000 / 311,040,000 = 0.600000000000006

1.111 x 1.111 x 1.111 x 1.111 x 1.111 x 1.111 x 1.111 x 1.111 x 1.111 x 1.111 x 1.111 = 3.18663554532458422627
  1 / x = 0.31381059609003

3.18663554532458422627 / 0.600000000000006 = 5.31105924220745

1 + 8 +8 +1+ 6 + 7 +6 + 4 + 2 + 3 + 1 + 5 + 8 + 8 + 8 + 0 + 7 + 8 + 4 + 5 +1 =

1 x 8 x 8 x 1 x 6 x 7 x 6 x 4 x 2 x 3 x 1 x 5 x 8 x 8 x 7 x 8 x 4 x 5 x 1 =

3.18663554532458422627

3.18663554532458422627 / 266.666 = 520,224,786

1D = 1.111 / 1.333 x 10 = 8.333 1/x = 1.1999999999999458333 [12]

2D =123456790123454321 / 1.333 = 0.92592592592593005556
    1/x = 10.8   {AETHER}

3D = 1.371742112 / 1.333 x 10 = 1.02880658436213477   1/x = 9.72 {Earths Diameter}

4D = 1.5241579027586648377 / 1.333 x 10 =   11.431184270690272062
1/x = 0.87480000000001 {}

5D = 1.6935087808429439957 / 1.333 x 10 = 12.70131585632243975
    1/x = 0.78732000000002 {AETHER}

5D = 1.6935087808429439957 x 1.333 x 10 = 22.5580117077905355439
    1/x = 0.44286750000003 {AETHER}

5D = 1.6935087808429439957 / 1.111 x 10 = 15.241579027586648377
    1/x = 0.65610000000003 {AETHER}

5D = 1.6935087808429439957 x 1.111 x 10 = 18.8167642315881
    1/x = 5.3144100000003188646 {AETHER}

12 D - GOVERNMENT – TWELTH DIMENSION
12.12121212121212121212121212121212

1.111 x 12.121212 = 13.46801346800013198653

864,000 x 864,000 x 864,000 = 644,972,540,000,000,000 / 12.121212 =
53,210,234,880,000,532.102 / 311,040,000 / 311,040,000 = 0.5500000000000055

1.111 x 1.111 x 1.111 x 1.111 x 1.111 x 1.111 x 1.111 x 1.111 x 1.111 x 1.111 x 1.111 x 1.111 = 3.54070616147172489848
1 / x = 0.282242953648103

3.54070616147172489848 / 0.5500000000000055 = 6.43764756631212
This number octaves up to the Planetary Octave Sequence (POS)

3 + 5 +4 + 0 + 7 + 0 + 6 + 1 + 6 + 1 + 4 + 7 + 1 + 7 + 2 + 4 + 8 + 8 + 4 + 8 = 86

3 x 5 x 4 x 0 x 7 x 0 x 6 x 1 x 6 x 1 x 4 x 7 x 1 x 7 x 2 x 4 x 8 x 8 x 4 x 8 = 48,554,311,680

48,554,311,680 / 86 = 564,585,019.53488372093

48,554,311,680 / 144 = 337,182,720 / 7.5 = 44,957,696 / 1.333 = 33,718,272 / 16 =
2,107,392 /

48,554,311,680 / 9.999 = 4,855,431,168.00004855431168

48,554,311,680 / 8.888 = 5,462,360,064.0000546236

48,554,311,680 / 7.777 = 6,242,697,216.000062426

48,554,311,680 / 5.555 = 8,739,776,102.40008739776

48,554,311,680 / 6.666 = 8,739,776,102.4

48,554,311,680 / 3.333 = 14,566,293,504.00014566293504

48,554,311,680 / 2.222 = 21,849,440,256.00021849440256

48,554,311,680 / 1.111 = 43,698,880512.00043698880512

48,554,311,680 / 266.666 = 182,078,668.80000455197

48,554,311,680 / 311,040,000 = 156.103111

3.54070616147172489848 / 266.666 = 520,224,786

1D = 1.111 / 1.333 x 10 = 8.333   1 / x = 1.1999999999999458333 [12]

2D = 123456790123454321 / 1.333 = 0.92592592592593005556
    1 / x = 10.8   {AETHER}

3D = 1.371742112 / 1.333 x 10 = 1.02880658436213477 1 / x = 9.72 {Earths Diameter}

4D = 1.5241579027586648377 / 1.333 x 10 = 11.431184270690272062
    1 / x = 0.87480000000001 {}

5D = 1.6935087808429439957 / 1.333 x 10 = 12.70131585632243975
    1 / x = 0.78732000000002 {AETHER}

5D = 1.6935087808429439957 x 1.333 x 10 = 22.5580117077905355439
    1 / x = 0.44286750000003 {AETHER}

5D = 1.6935087808429439957 / 1.111 x 10 = 15.241579027586648377
    1 / x = 0.65610000000003 {AETHER}

5D = 1.6935087808429439957 x 1.111 x 10 = 18.8167642315881
    1 / x = 5.3144100000003188646 {AETHER}

16 D – TORUS SECTORS – SIXTEENTH DIMENSION
16.161616161616161616161616161616

1.111 x 16.161616 = 17.95735129684264871

864,000 x 864,000 x 864,000 = 644,972,540,000,000,000 / 16.161616 =
39,907,676,160,000,399.077 / 311,040,000 / 311,040,000 = 0.4125

0.4125 / 0.125 = 3.300000000000033

1.111 x 1.111 x 1.111 x 1.111 x 1.111 x 1.111 x 1.111 x 1.111 x 1.111 x 1.111 x 1.111 x 1.111 x 1.111 x 1.111 x 1.111 x 1.111 =

= 5.396595277353426698 / 0.4125 = 13.0826552178265
1 / x = 0.18530201888521

= 5.396595277353426698 x 0.4125 = 2.22609555190829

13.0826552178265 x 17.95735129684264871 = 234.92983564197325961

13.0826552178265 / 17.95735129684264871 = 0.00425659004642

SECTION C - IMPLOSIVE VORTEX TORUS MATHS

SECTION D – DIMENSIONAL NUMBERS

AETHER (DC) TO MATTER (DC) EQUATIONS

144,000 x 7.5 = 1,080,000 / 360 = 3,000 / 9 = 333.333 / 1.333 = 250 / 1.111 = 225.00000000000625 = 1 / x = 0.0044444

144,000 x 7.5 = 1,080,000 / 360 = 3,000 / 16 = 187.5 / 22.5 = 8.333 x 1.333 = 11.111 / 9 = 1.2345679012345679 x 400 = 493.827160493815 x 360 = 177,777.777

288,000 x 7.5 = 2,160,000 x 1.333 = 2,879,999.999 x 10 = 28,799,999.999.99999928 / 720 = 39,999.999 / 32 = 1,249.999 / 22.5 = 55.555
55.555 / 9 = 6.1728395061726851852 x 11.111 =68.58 x 400 = 27,432

266,666.666 x 7.5 = 1,999,999.999 / 360 = 5,555.555 / 9 = 617.283950617269 / 1.333 = 462.962962 / 1.111 = 416.666   1/x = 0.0024

266,666.666 x 7.5 = 1,999,999.9999 [2,000,000] x 1.333 = 2,666,666.666

129,600 x 7.5 = 972,000 / 360 = 2,700 / 22.5 = 120 / 16 = 7.5

129,600 +144,000 = 273,600 (EARTH)

7,777 + 9,999 = 1,776 / 2 = 888 / 2 = 444 / 2 = 222 / 2 = 111 / 2 = 55.5 /2 = 27.75 / 2 = 13.875 / 2 = 6.9375 / 0.125 = 55.5

REFLECTIVE MIRROR NUMBERS

They can only exist if mirror image exists in other polarity.

11.11 = 5.555 + 5.555

1,110 = 456 + 654

1,111 = 555.5 + 555.5 = 1,111

2,222 = 1,021 +1,201 = 1,111 + 1,111 = 2,222

3,333 = 2,031 + 1,302 = 1,122 + 2,211 = 3,030 + 0303

4,444 = 2,042 + 2,402 = 3041 + 1403 = 2,222 + 2,222 = 4,040 + 0404

5,555 = 1,324 + 4,231 = 1,234 + 4,321 = 2,413 + 3,142
         = 3,322 + 2,233 = 5,050 + 0505 =

6,666 = 3,333 + 3,333 = 2,244 + 4,422= 2,424 + 4,242

7,777 = 2,345 + 5,432 = 2,435 + 5,342   = 3,344 + 4,433 =

8,888 = 4,444 + 4,444 = 4,534 + 4,354 = 6,622 + 2,266 = 4,084 + 4,804

9,999 = 3,456 + 6,543 = 3,546 + 6,453

11,110 = 5,555 + 5,555 = 3,737 + 7,373

266.666 RESONANT FREQUENCIES

MARS 4,222 / 266.666 C in Music Resonant Base Frequency of Matter = 15.8325
MARS 4,222 x *11.111* x *1.333* = 3,940,000 / *266.666* = 1,440
1,440 x (7.5 EARTH) = *10,800* (Shift Moon 2,160 + Earth 7,920 Diameter + 720)
10,800 / *5.555* = *1,994*
1,944 / 2 = // *972* // *486* // *243* //

4,222 / 5.555	= 759.9600000000001
4,222 / 3.333	= 1,266.6
4,222 / 144	= 29.31944
4,222 / 266.666	= 15.8325
4,222 / 1.333	= 3,166.5
4,222 / 7.5	= 562.9333
4,222 / 11.111	= 379.98
4,222 / 2,160	= 1.9546296296
4,222 / 7,920	= 0.53308080

687 Days Orbit x 24 x 60 x 60 = 59,356,800
59,356,800 / 144 = 412,200
59,356,800 / 720 = 824,400

59,356,800 / 266.666 = 222,588
59,356,800 / 11.111 = 5,342,112.005342112
59,356,800 / 1.333 = 44,517,600.111294

MARS RIGHT ANGLE TRIANGLE NUMBERS
4,222 / 3 = 1,407.333

1,407.333 x 4 = 5,629.333
1,407.333 x 5 = 7,036.666
AREA 3 X 4 = 23,767,045.333 / 2 = 11,883,522.666

JUPITER RESONANT FREQUENCIES

JUPITER 88,840 x 11.111 x 1.333 = 1316237 / 266.666 = 4935.888 x 7.5 = 37,019.16
4,925.888 / 5.555 = 888.4598408884598

88,840 / 5.555 = 333.1500000000001
88,840
Day = 10 hours x 60 x 60 = 36,000
Year = 4,333 x 24 = 103,992 / 10 = 10,933.2
Year = 4,333 earth days x 24 x 60 x 60 = 374,371,200

JUPITER RIGHT ANGLE TRIANGLE NUMBERS

MARS 4,222 x 11.111 x 1.333 = 3, 940, 000 / 266.666 = 1440 x 7.5 =
10,800 (MOON plus EARTH + 720 diameters).

4,222 / 6.666 = 633.3000006333
4,222 / 5.555 = 759.9600000000001
4,222 / 3.333 = 1,266.6
4,222 / 144 = 29.31944444444444
4,222 / 266.666 = 15.8325
4,222 / 1.333 = 3,166.50000791625
4,222 / 7.5 = 562.9333333333333
4,222 / 11.111 = 379.98000037998
4,222 / 216 = 19.5462962962963
4,222 / 792 = 5.330808080808081
4,222 / 864 = 4.886574074074074

88,840 / 4,222 = 21.04216011369019
687 Days Orbit x 24 Hours x 60 Minutes x 60 seconds = 59,356,800 / 7.5 = 7,914,240
7,914,240 / 144 = 54,960 / 16 = 3,435 / 22.5 = 152.666

59,356,800 / 144 = 412,200
59,356,800 / 720 = 824,400
59,356,800 / 266.666 = 222,588
59,356,800 / 11.111 = 5,342,112.005342112
59,356,800 / 1.333 = 44,517,600.111294

JUPITER 88,840 x 11.111 x 1.333 = 1,316,237 / 266.666 = 4,935.888 x 7.5 = 37,019.16
4,925.888 / 5.555 = 888.4598408884598
88,840 / 3.333 = 26,652
88,840 / 4.444 = 19,989
88,840 / 5.555 = 333.1500000000001
88,840 / 6.666 = 13,326.000013326

Day = 10 Hours x 60 x 60 = 36,000
Year = 4,333 x 24 = 103,992 / 10 = 10,933.20
Year = 4,333 Earth Days x 24 x 60 x 60 = 374,371,200 x 60 Arc Sec = 22,462,272,000

EARTH DIAMETER IN MILES 7,920 REVERSE MIRROR NUMBERS

[88 x 11.111 = 977.777]
[88 / 11.111 = 7.92]
[33 / 11.111 = 2.97]
[33 x 11.111 = 366.666]
[791 / (977.777 - 366.666) = 1.296]

  792 = [99 x 8]   and   [9 x 88]
- 297 = [99 x 3 #] and   [9 x 33]
= 495 = [99 x 5 #] and   [9 x 55]
- 594 = [99 x 6.01010]   [9 x 66]
= 99 root number EARTH DIAMETER

SUN DIAMETER IN MILES 864,000 REVERSE MIRROR NUMBERS

36 x 24 x 60 x 60 = 3,110,400 / *864* = 3,600 / 144 = 25 / 22.5 = 1.111
360 x 24 x 60 x 60 = 31,104,000 / *864* = 36,000 / 144 = 250 / 22.5 = 11.111

3,110,400 / 7.5 = 414,720
3,110,400 x 7.5 = 23,328,000 / 360 = *64,800*
23,328,000 / 144,000 = 162
23,328,000 / 266.666 = 87,480.00000000002

26.8 x 24 x 60 x 60 = 2,315,520

[96 x 11.111 = 1,066.666 (266.6 x 3)]
[96 / 11.111 = *8.64*]
[52 / 11.111 = *4.68*]
[52 x 11.111 = 577.777 (144.444 x 4)]
[864 / (1,066.666 - 577.777) = 488.8]

864 = [99 x 8.727272]	[9 x 96]
- *468* = [99 x 4.727272]	[9 x 52]
= 369 = [99 x 4 #] and	[9 x 44]
- 963 = [99 x 7] and	[9 x 77]
= 297 = [99 x 3 #] and	[9 x 33]
- *792 = EARTH* [99 x 8]	[9 x 88]
= 495 = [99 x 5 #]	[9 x 55]
- 594 = [99 x 6.101010]	[9 x 66]
99 = [99 x 1]	[9 x 11]

= 99 root number SUN DIAMETER

SUN RADIUS IN MILES 432,000 REVERSE MIRROR NUMBERS

[48 x 11.111 = 533.333] Music C #
[48 / 11.111 = *4.32*]
[26 / 11.111 = *2.34*]
[26 x 11.111 = 288.888 light #]

432 = [99 x 4.363636]	[9 x 48]
- *234* = [99 x 2.363636]	[9 x 26]
= 198 = [99 x 2] and	[9 x 22]
- 891 = [99 x 9] and	[9 x 99]
- 693 = [99 x 7] and	[9 x 77]
= 297 = [99 x 3 #] and	[9 x 33]
- *792 = EARTH* [99 x 8]	[9 x 88]
= 495 = [99 x 5 #] and	[9 x 55]
- 594 = [99 x 6.101010]	[9 x 66]

= 99 root number SUNS RADIUS

© STRIKE FOUNDATION GUARANTEE LIMITED | MALCOLM BENDALL 2022 | GRAPHICS - STEVE EARL

MOON DIAMETER IN MILES 2,160 REVERSE MIRROR NUMBERS

2,160 x 4 = **8,640**

[24 x 11.111 = **266.666**] Music C
[24 / 11.111 = **2.16**]
[68 / 11.111 = **6.12**]
[68 x 11.111 = 755.555]

  216 = [99 x 2.1818]   [9 x 24]
- **612** = [99 x 6.2828]   [9 x 68]
= 396 = [99 x 4 #] and   [9 x 44]
- 693 = [99 x 7] and   [9 x 77]
= 297 = [99 x 3 #] and   [9 x 33]
- **792 =** [99 x 8]   [9 x 88] EARTH
= 495 = [99 x 5 #] and   [9 x 55]
- 594 = [99 x 6.101010] [9 x 66]

= 99 ROOT NUMBER FOR THE MOON

JUPITER DIAMETER IN MILES 88,840 REVERSE MIRROR NUMBERS

  8884
- **4888**
= 3996 = 999 x 4   [9 x 444]
- 6993 = 999 x 7   [9 x 777]
= 2997 = 999 x 3   [9 x 333]
- **7992 = 999 x 8**   [9 x 888]
= 4995 = 999 x 5   [9 x 555]
- 5994 = 999 x 6   [9 x 666]
= 999 = 999 x 1   [9 x 111]

999 = ROOT NUMBER FOR MARS

999 code = 473,856 / 8884 = 53.33813597478613
473,856 / 266.666 =1,776.960000000001
1,776.96// 888.48// 444.24// 222.12// 111.06// 55.53// 27.765

MARS DIAMETER IN MILES 4,222 REVERSE MIRROR NUMBERS

Radius = 2,111 Circle = 16,888
4,222 / **266.666** = 15.8325
4,222 / **6.666** = 633.3000000000001
4,222 / **5.555** = 759.6

  4222
- **2224**
= 1998 / 999 = 2   [9 x 222]
  8991 / 999 = 9   [9 x 999]
= 6993 / 999 = 7   [9 x 777]
- 3996 / 999 = 4   [9 x 444]

= 2997 / 999 = 3   [9 x 333]
- 7992 / 999 = 8   [9 x 888]
= 4995 / 999 = 5   [9 x 555]
- 5994 / 999 = 6   [9 x 444]
   999 / 999 = 1   [9 x 111]

= 999 ROOT NUMBER FOR JUPITER

165,834,792 / 4,222 = 39,278.7285646613
297,438,561 / 2.666 = 1,115,394.60375
1,115,394.60375

29,743,856 / 2.666 = 111,539.46
29,743,856 / 4,222 = 7,044.968261487447
29,743,856 / 266.666 =111,539.46
29,743,856 / 5.555 =5,353,894.080000001
29,743,865 / 4.444 = 6,692,367.600000001
29,743,856 / 3.333 =8,923,156.800000001
29,743,856 / 2.222 = 13,384,735.2
29,743,866 / 1.111 = 26,769,470.4
29,743,866 / 11.111 = 2,676,947.04
29,743,866 / 111 = 267,962.666

   4,222
+ 2,224
= 6,446
+ 6,446
= 12,892
+ 29,821
= 42,713
+ 31,724
= 74,437
+ 73,447
= 147,884
+ 488,741
= 636,625
+ 526,636
= 1,163,261
+ 1,623,611
= 2,786,872
+ 2,786,872
= 5,573,744
   4,473,755
   10,047,499
   99,474,001
   109,521,500
   005,125,901
   114,647,401

SUMMARY TABLE OF REVERSE MIRROR NUMBERS FOR THE SUN, MOON, EARTH, JUPITER AND MARS.

SUN REVERSE MIRROR NUMBERS OF THE 864 DIAMETER AND RADIUS OF 432

864 = [99 x 8.727272]	[9 x 96]	**432** = [99 x 4.3636]	[9 x 48]	{48 x 11.111 = 533.333 Music C }
- 468 = [99 x 4.727272]	[9 x 52]	- 234 = [99 x 2.3636]	[9 x 26]	{ 26 x 11.111 = 288.888 }
= *396* = [99 x 4] and	[9 x 44]	= 198 = [99 x 2]	[9 x 22]	{ 22 x 11.111 = 244.444 }
- 693 = [99 x 7] and	[9 x 77]	- 891 = [99 x 9]	[9 x 99]	{ 99 x 11.111 = 1,100 }
= 297 = [99 x 3] and	[9 x 33]	= 693 = [99 x 7]	[9 x 77]	{ 77 x 11.111 = 855.555 }
		- 396 = [99 x 4]	[9 x 44]	{ 44 x 11.111 = 488.888 }
- *792* = [99 x 8] EARTH	[9 x 88]	= 297 = [99 x 3]	[9 x 33]	{ 33 x 11.111 = 366.666 }
= 495 = [99 x 5]	[9 x 55]	- 792 = [99 x 8]	[9 x 88]	{ 88 x 11.111 = 977.777 }
- 594 = [99 x 6]	[9 x 66]	= 495 = [99 x 5]	[9 x 55]	{ 55 x 11.111 = 611.111 }
99 = [99 x 1]	[9 x 11]	- 594 = [99 x 6]	[9 x 66]	{ 66 x 11.111 = 733.333 }
		= 99 = [99 x 1]		

= 99 ROOT NUMBER FOR THE SUN'S DIAMETER AND 9 TIMES 11, 33, 44, 52, 55, 66, 77, 88 AND 96

MOON REVERSE MIRROR NUMBERS HAVE THE 2,160 DIAMETER AND RADIUS 1,080

216 = [99 x 2.1818]	[9 x 24]	{ 24 x 11.111 = 266.666 C}	1080 = [9 x 120]	{ 120 x 11.11 = 1,333.333 }
- 612 = [99 x 6.1818]	[9 x 68]	{ 68 x 11.111 = 755.555 }	801 = [9 x 89]	{ 89 x 11.111 = 988.888 }
= 396 = [99 x 4] and	[9 x 44]	{ 44 x 11.111 = 488.888 }	279 = [9 x 31]	{ 31 x 11.111 = 344.444 }
- 693 = [99 x 7] and	[9 x 77]	{ 77 x 11.111 = 855.555 }	972 = [9 x 108]	{ 108 x 11.11 = 1,199.999 }
= 297 = [99 x 3] and	[9 x 33]	{ 33 x 11.111 = 366.666 }	693 = [9 x 77]	{ 77 x 11.111 = 855.555 }
- 792 = [99 x 8] EARTH	[9 x 88]	{ 88 x 11.111 = 977.777 }	396 = [9 x 44]	{ 44 x 11.111 = 488.888 }
= 495 = [99 x 5] and	[9 x 55]	{ 55 x 11.111 = 611.111 }	297 = [9 x 33]	{ 33 x 11.111 = 366.666 }
- 594 = [99 x 6]	[9 x 66]	{ 66 x 11.111 = 733.333 }	792 = [9 x 88]	{ 88 x 11.111 = 977.777 }
99 = [99 x 1]	[9 x 11]	{ 11 x 11.111 = 122.222 }	495 = [9 x 55]	{ 55 x 11.111 = 611.111 }
			594 = [9 x 66]	{ 66 x 11.111 = 733.333 }
			99 = [9 x 11]	

= 99 ROOT NUMBER FOR THE MOON'S DIAMETER AND 9 TIMES 11, 24, 33, 44, 55, 66, 68, 77, AND 88

EARTH REVERSE MIRROR NUMBERS HAVE THE 7,920 DIAMETER AND RADIUS 3,960

792 = [99 x 8] and	[9 x 88]	396 = [9 x 41]	{ 41 x 11.111 = 455.555 }
- 297 = [99 x 3] and	[9 x 33]	693 = [9 x 77]	{ 77 x 11.111 = 855.555 }
= 495 = [99 x 5] and	[9 x 55]	297 = [9 x 33]	{ 33 x 11.111 = 366.666 }
- 594 = [99 x 6] and	[9 x 66]	792 = [9 x 88]	{ 88 x 11.111 = 977.777 }
99 = [99 x 1] and	[9 x 11]	495 = [9 x 55]	{ 55 x 11.111 = 611.111 }
		594 = [9 x 66]	{ 66 x 11.111 = 733.333 }
		= 99 = [9 x 11]	

= 99 ROOT NUMBER FOR THE EARTH'S DIAMETER AND 9 TIMES 11, 33, 55, 66, AND 88

JUPITER REVERSE MIRROR NUMBERS OF THE 88,840 DIAMETER AND RADIUS 44,420

8884 = [999 x 8.892]	[9 x 987.111]	4442 = [9 x 493.555]
- 4888 = [999 x 4.892]	[9 x 543.111]	2444 = [9 x 271.555]
= 3996 = [999 x 4]	[9 x 444]	1998 = [9 x 222]
- 6993 = [999 x 7]	[9 x 777]	8991 = [9 x 999]
= 2997 = [999 x 3]	[9 x 333]	6993 = [9 x 777]
- 7992 = [999 x 8]	[9 x 888]	3996 = [9 x 444]
= 4995 = [999 x 5]	[9 x 555]	2997 = [9 x 333]
- 5994 = [999 x 6]	[9 x 666]	7992 = [9 x 888]
= 999 = [999 x 1]	[9 x 111]	4995 = [9 x 555]
		5994 = [9 x 666]
		999 = [9 x 111]

= 999 ROOT NUMBER FOR JUPITERS DIAMETER AND 9 TIMES 111, 333, 444, 555, 666, 777 AND 888

MARS REVERSE MIRROR NUMBERS OF THE 4,222 DIAMETER AND 2,111 RADIUS

4222 = [999 x 4.226]	[9 x 469.111]	2111 = [9 x 234.555]
- 2224 = [999 x 2.226]	[9 x 247.111]	1112 = [9 x 123.555]
= 1998 / 999 = 2	[9 x 222]	999 = [9 x 111]
8991 / 999 = 9	[9 x 999]	
= 6993 / 999 = 7	[9 x 777]	
- 3996 / 999 = 4	[9 x 444]	
= 2997 / 999 = 3	[9 x 333]	
- 7992 / 999 = 8	[9 x 888]	
= 4995 / 999 = 5	[9 x 555]	
- 5994 / 999 = 6	[9 x 666]	
999 / 999 = 1	[9 x 111]	

= 999 ROOT NUMBER FOR MARS DIAMETER AND 9 TIMES 111, 222, 333, 444, 555, 666, 777, 888 AND 999 CODE

1, 2, 3, 4, 5, 6, 7, 8, 9 NUMBER SEQUENCE 9 BASE DECODED

518,400 / **576 (4 x 144)** = **900** Degrees in **1** of **8** Sacred Torus Cardinal planes total = **7,200**
24 x **24** = **576** /2 = **288** /2 = **144** /2 = **72** /2 = **36** /2 = **18** /2 = **9** /2 = **4.5** /2 = **2.25** / **2** = **1.125**

SAT   = SANSKRIT 4,800 x 360 = *144,000* / 720 = 200
TRETAA   = SANSKRIT 3,600 x 360 = *129,600* / 720 = 180
DWAAPAR = SANSKRIT 2,400 x 360 = *864,000* / 720 = 1,200
KALI   = SANSKRIT 1,200 x 360 = *432,000* / 720 = 600

12,345,678 + 87,654,321 = 99,999,999
1,234,567 + 7,654,321 = 8,888,888
123,456 + 654,321 = 777,777
12345 + 54321 = 66,666
1234 + 4321 = 5,555   (Tungsten's Boiling point 11.111 x 500)
123 + 321 = 444
12 + 21 = 33

518,400 = **5 x 1 x 8 x 4** = [160] x **4 x 8 x 1 x 5** [160] = 25,600 = 160 x 160
25,600 + 160 +160 = *25,920* = the Earth's Equinox so clearly the square [time] plus the product = The Equinox *25,920*
144,000 = 1 x 4 x 4 [16] x 1 x 4 x 4 = **256** / 16 = 16 So the same for the speed of light of
144,000 if you add 256 [square product] + 16 +16 [square time] you get 288 twice 144 yet we only added 4 x 4 to 4 x 4 to equal 256 [4 x 4 x 4 x 4]

8 x 9 x 10 [720] x 10 x 9 x 8 = 518,400 / 9 = 57,600 / 144 = 600
again **518,400** + *720* + *720* = 519,840 [/ *518,400* = 1.002777] 519,840 / *24* / *60* / *60* / *7.5* = 0.80222
518,400 / *144* = 36,000   518,400 / *266.666* = 1944   *518,400* / *720* = 720
8 + 9 + 10 = [27] + 10 + 9 + 8 = 54 x *11.111* = 600   54 / *11.111* = 4.86 [Suns Diameter]
again 518,400 / 54 = 9,600 / *266.666* = 36
518,400 / 54 = 9,600

891 + 198 = 1,980 [990 + 990] / 9 = 220
891 − 198 = 693 / 9 = 77
891 x 198 = 8,722 / 9 = 969.111

518,400 / 54 = 9,600 / 22.5 = 426.666 / 16 = 26.666 x 10 = 266.666

518,400 = *24* Hours x *60* Minutes x *60* Seconds x *6* Days of Creation and *518,400* / 60 = **8,64**3.3

518,400 / 720 = 720 / 16 = 45 / 22.5 = 2 / 0.125 = 16 x 400 = 6,400 / 144 = 44.444 TIME
518,400 // 259,200 // 129,600 // **64,800** // **32,400** // **16,200** // 8,100 // 4,050 // 2 025 // 1,012.5

1 x 2 x 3 x 4 x 5 x 6 x 8 x 9 x 10 = [518,400] 10 x 9 x 8 x 6 x 5 x 4 x 3 x 2 x 1 = *268,738,560,000 TIME*
268,738,560,000 / 24 = 11,197,440,000 / 60 = 186,624,000 / 60 = 3,110,400 (SANSKRIT ALL TIME)
3,110,400 (SANSKRIT ALL TIME) / 60 = 51,840
3,110,400 (SANSKRIT ALL TIME) / 7.5 = 414,720 / 16 = 25,920 (EARTHS EQU) / 22.5 = 1,152 / 0.125 = 9,216
9,216 x 400 = 3,686,400 / 144 = 25,600   3,686,400 / 3,456 = 1,066.666   3,686,400 / 864 = 4,266.666
3,686,400 / 432 = 8,533.333   3,686,400 / 900 = 4,096

12,345,689 + 98,654,321 = 111,000,010
1 + 2 + 3 + 4 + 5 + 6 + 8 + 9 [38] + 6 + 5 + 4 + 3 + 2 + 1 [1,444] = 1,444

2,687,385,600 x 1,444 = 3,880,584,806,400
2,687,385,600 / 1,444 = 1,861,070.360110803
1 x 2 x 3 x 4 x 5 x 6 = [720] x 6 x 5 x 4 x 3 x 2 x 1 = 518,400 / 144 = 3,600 / 9 = 400
1 + 2 + 3 + 4 + 5 + 6 [21] + 6 + 5 + 4 + 3 + 2 + 1 [21] = 42 x 11.111 = 466.666
8 x 9 = [721,] x 9 x 8 = 5,184 / 144 = 36
8 + 9 + 9 + 8 = 34 / 11.111 = 34 x
89 x 98 = 8,722
89 + 98 = 187

(518,400 / 54 = 9,600 / 22.5 = 426.666 / 16 = 26.666 x 10 = 266.666 (RFEU)
{518,400 = 24 Hours x 60 Minutes x 60 Seconds x 6 Days of Creation and 518,400 / 60 = 8,643.3}
{518,400 // 259,200 // 129,600 // 64,800 // 32,400 // 16,200 // 8,100 // 4,050 // 2 025 // 1,012.5}

518,400 / 34 = 15,247.058823529411765

7 x 8 x 9 = [504 (72 x 7)] x 9 x 8 x 7 = 254,016 / 144 = 1,764   254,016 / *266.666* = 952.56
7 + 8 + 9 + 9 + 8 + 7 = 48 x *11.111* = 533.333 {*266.666* x 2} 48 / *11.111* = 4.32 [suns radius]
705,600 / 48 =
789 + 987 = 1776
789 x 987 = 778,743

254,016

6 x 7 x 8 x 9 = [3,024 (72 x 42)] x 9 x 8 x 7 x 6 = 9,144,576
9,144,576 / 144 = 63,504 9,144,576 / 266.666 = 34,292.16
6 + 7 + 8 + 9 = [30] + 9 + 8 + 7 + 6 = 60 x *11.111* = 666.666   60 / *11.111* = 5.4
6789 + 9876 = 16,665
6789 x 9876 = 67,048,164

9,144,576

5 x 6 x 7 x 8 = [2,680 (72 x 37.222)] x 8 x 7 x 6 x 5 = 2,822,400 / 144 = 19,600 2,822,400 /
266.666 = 10,584
5 + 6 + 7 + 8 = [26] + 8 + 7 + 6 + 5 = 52 x *11.111* = 577.777   52 / *11.111* = 4.68 [Suns diameter]

5678 + 8765 = 14,443
5678 x 8765 = 49,767,670

2,822,400 / 52 = 54,276.9230

49,767,670 / 144 = 345,608.819444
3,456.0881944 / 864 = 4.000102076903292

4 x 5 x 6 x 7 = [840] x 7 x 6 x 5 x 4 = 705,600 / *144* = 4,900   705,600 / 266.666 = 2,646
4 + 5 + 6 + 7 + 7 + 6 + 5 + 4 = 44 x *11.111* = 488.888   44 / *11.111* = 3.96
705,600 / 44 = 16,036.3636

4567 + 7654 = 12,221
4567 x 7654 = 34,955,818

705,600 x *1.111* = 783,999.999216
705,600 / *144* = 4,900
705,600 / *22.5* = 31,360 / 16 = 1,960
705,600 / *5.555* = 127,008.000127008

3 x 4 x 5 x 6 x 7 = [2,520 (72 x 35)] x 7 x 6 x 5 x 4 x 3 = 6,350,400
3 + 4 + 5 + 6 + 7 + 7 + 6 + 5 + 4 + 3 = 50
6,350,400 / 50 = 127,008

3 x 4 x 5 x 6 = [360 (72 x 5)] x 6 x 5 x 4 x 3 = 129,600 + 360 + 360 = 13,320

129,600 / *360* = *1,440* / *360* = 4 x *129,600* = *518,400*
129,600 / *144* = 900   *129,600* / *266.666* = *486*

3 + 4 + 5 + 6 + 6 + 5 + 4 + 3 = 36 x *11.111* = 400 *36* / *11.111* = *3.24* 36 / *266.666* = *0.135*
129,600 / 36 = 3,600

3456 + 6543 = 9999
3456 − 6543 = 3,087 / 9 = 343 x *11.111* = 3,811.111 x 1.333 = 5,081.481463696
3456 x 6543 = 22,612,608 / 9 = 2,512,512

129,600 x *1.111* = 144,000
129,600 / *144* = 900
129,600 / *5.555* = 23,328.000023328
129,600 / *6.666* = 19,440.00001944
129,600 / *22.5* = 5,760 / 16 = 360

2 x 3 x 4 x 5 = [120] x 5 x 4 x 3 x 2 = 14,400 / 120 = 120

14,400 / *144* = 100   *14,400* / *266.666* = 54 [1-9 sum]
2 + 3 + 4 + 5 + 5 + 4 + 3 + 2 = 28 x *11.11* = 311.11 28 / *11.11* = 2.52 28 / *266.666* = 0.105

2345 + 5432 = 7,777
2345 − 5432 = 3,087
2345 x 5432 = 12,738,040

14,400 x 1.111 = 16,000
14,400 / 22.5 = 640
144,000 x 90 = 12,960,000
14,400 x 9 = 129,600

5 x 6 x 7 = [210] x 7 x 6 x 5 = 44,100 / *144* = 306.25   44,100 / *266.666* = 165.375
5 + 6 + 7 + 7 + 6 + 5 = 36 x *11.11* = 400   *36* / *11.111* = *3.24*   36 / *266.666* = *0.135*
44,100 / 36 = 1,225

567 + 765 = 1332 (2 x 666)
567 − 765 = 198
567 x 765 = 433,755

44,100 x 1.111 = 49,000
44,100 / 1.111 = 39,690
44,100 x 11.111 = 489,999.999
44,100 / 11.111 = 3,969
44,100 x 1.333 = 58,800
44,100 / 1.333 = 33,075
44,100 / 22.5 = 1,960 / 16 = 122.5
44,100 x 22.5 = 992,250
44,100 x 9 = 4,900
44,100 / 9 =

1 x 2 x 3 x 4 = [24] x 4 x 3 x 2 x 1 = 576
1 + 2 + 3 + 4 + 4 + 3 + 2 + 1 = 20
576 / 20 = 28.8

576 = 144 x 4
1234 + 4321 = 5,555
1234 x 4321 = 5,332,114 / 9 = 592,457.111

1 x 2 x 3 [6] x 3 x 2 x 1 = 36
1 + 2 + 3 + 3 + 2 + 1 = 12
36 / 12 = 3

36 = 144 / 4
123 + 321 = 444 / 9 = 49.333   444 / 37 = 12   444 / 37 = 12
123 x 321 = 39,483 / 9 = 4,387

1 x 2 x 2 x 1 = 4
1 + 2 + 2 + 1 = 6
4 = 144 / 36
12 + 21 = 33 / 9 = 3.666
12 x 21 = 252 / 9 = 28

4 x 5 x 6 x 7 = [840] x 7 x 6 x 5 x 4 = 705,600 / 144 = 4,900
4 + 5 + 6 + 7 + 7 + 6 + 5 + 4 = 44
705,600 / 44 = 16,036.363636

4567 + 7654 = 12,221 / 9 = 1357.888
4567 x 7654 = 34,955,818 / 9 = 3883979.777
34,955,818 / 12,221 = 2,860.30750034776204893

705,600 / 22.5 = 31,360
705,600 / 5.555 = 127,008.000127008

3 x 9 x 6 = [162] x 6 x 9 x 3 = 26,244 product
3 + 6 + 9 + 9 + 6 + 3 = 36 sum
26,244 / 36 = 729

396 + 693 = 1,089 / 9 = 121
396 − 693 = 297 / 9 = 33
396 x 693 = 274,428

3 x 6 x 9 = [162] x 9 x 6 x 3 = 26,244
3 + 6 + 9 + 9 + 6 + 3 = 36

369 + 963 = 1,332 / 9 = 148
369 − 963 = 594 / 9 = 66
369 x 963 = 355,347 / 9 = 39,483
39,483 / 148 = 266.7770270270270
6 x 3 x 9 = [162] x 9 x 3 x 6 = 26,244
6 + 3 + 9 + 9 + 3 + 6 = 36
639 + 936 = 1,575 / 9 = 175
639 − 936 = 297 / 9 = 33
639 x 936 = 598,104 / 9 = 66,456 / 175 = 379.748571

9 x 6 x 3 = [162] x 3 x 6 x 9 = 26,244
9 + 6 + 3 + 3 + 6 + 9 = 36

963 + 369 = 1,332 / 9 = 148
963 − 369 = 594 / 9 = 66
963 x 369 = 355,347 / 9 =

3 x 9 x 6 = [162] x 6 x 9 x 3 = 26,244
3 + 9 + 6 + 6 + 9 + 3 = 36

396 + 693 = 1,089 / 9 = 121
396 − 693 = 297 / 9 = 33
396 x 693 = 274,428 / 9 = 30,492
3 x 6 x 9 = [162] x 9 x 6 x 3 = 26,244
3 + 6 + 9 + 9 + 6 + 3 = 36

369 + 963 = 1,332/ 9 = 148
369 x 963 = 355,347

9 x 3 x 6 = [162] x 6 x 3 x 9 = 26,244
3 + 6 + 9 + 9 + 6 + 3 = 36

5 x 6 x 7 x 9 x 9 x 7 x 6 x 5 = 3,572,100
5 + 6 + 7 + 9 + 9 + 7 + 6 + 5 = 54

5679 + 9765 = 15,444 / 9 = 1,716
5679 x 9765 = 55,455,435 / 9 = 6,161,715   55,455,435 / 22.5 = 2,464,686

2 x 3 x 4 x 5 x 6 x 7 x 9 x 9 x 7 x 6 x 5 x 4 x 3 x 2 = 2,057,529,600
2 + 3 + 4 + 5 + 6 + 7 + 9 + 9 + 7 + 6 + 5 + 4 + 3 + 2 = 72

2,057,529,600 / 144 = 14,288,400
2345679 + 9765432 = 12,111,111 / 9 = 1,345,679

2345679 x 9765432 = 22,906,568,768,328 / 9 =
22,906,568,768,328 / 12,111,111 = 1,891,368. 08079193

518,400 / 6 = *86,400*
518,400 x 6 = *3,110,400*

2 x 3 x 4 x 5 x 6 x 6 x 5 x 4 x 3 x 2 = *518,400*

8 x 10 x 10 x 9 x 8 = *518,400*
2 x 3 = 6 x 4 = 24 x 5 = 120

150 / 9 = 16.6666 x 2 = 33.333 x 2 = 66.666 x 2 = 133.333 x 2 = 266.666
300 / 9 = 33.333 x 2 = 66.666 x 2 =133.333 x 2 = 266.666

225 / 9 = 25
22.5 / 9 = 2.5
2,400 / *9* = *266.666*

2,800 / 9 = 3.111 x 100 = 311.111

54 / 9 = 6	45 / 9 = 5
540 / 9 = 60	450 / 9 = 50
5,400 / 9 = 600	4,500 / 9 = 500
5,994 / 9 = 666 Power	4,995 / 9 = 555 AETHER Grace

518,400 / 5,994 = 86.4864864   Power of the Sun   518,400 / 666 = 778.37837

518,400 / 4,995 = 103.7837837   518,400 / 555 = 934.054054054
(Gematria 1-9 = 45 to 10-90 = 450 to 100 – 900 = 4,500 = 4,995)
(Gematria 1-10-100 = 111 thru to 9-90-900 = 999)

518,400 / 999 = 518.918918
518,400 / 888 = 583.7837837
518,400 / 777 = 667.181467
518,400 / 666 = 778.378378
518,400 / 555 = 934.0540540
518,400 / 444 = 1,167.567567
518,400 / 333 = 1,556.756756
518,400 / 222 = 2,335.135135
518,400 / 111 = 4,670.270270270

2 x 3 x 4 x 5 x 6 x 6 x 5 x 4 x 3 x 2 = *518,400*

8 x 9 x 10 x 10 x 9 x 8 = *518,400*

8 x 9 x 8 x 9 = *5,184*

6 DAYS x *24 HOURS* x *60 MINUTES* x *60 SECONDS* = *518,400* / *86,400* = 6

518,400 x 6 = *3,110,400*   *3,110,400* x *144* = *447,897,600* / *864* = *518,400* / *720* = *720*
3,110,400 / *144* = *21,600*

MOONS 2,160 DIAMETER AND RADIUS 1,080

MOON'S DIAMETER = 2,160 Miles

2,160 = 2 x 1 x 6 (12) x 2 x 1 x 6 = *144*

2,160 / 24 Hrs = 90 [0.0111] / 60 Min = 1.5 [0.666] / 60 Sec = 0.025 [40] / 60 Arc sec = 0.00041667 [2,400]

2,160 / 360 = 6 / 16 = 0.375 [2.666] / 22.5 = 0.01666 /.125 = 0.1333 [7.5] x 400 = 3,000 (Volume in Baths of a full Solomons Molten Sea)

144 / 24 Hrs = 6 [1/x = 0.1666] / 60 Min = 0.1 [1/x =10] / 60 Sec = 0.001666 [1/x = 600] / 60 Arc Sec = 0.00002777 [1/x = 36,000]

144 / 16 = 9 / 22.5 = 0.4 [2.5] / 0.125 = 3.2 x 400 = 1,280

144 / 7.5 = 19.2

144 / 12 = 12

2,160 = 2 x 1 x 6 (12) x 2 x 1 x 6 = *144*

MOON'S RADIUS = 1,080 Miles

1,080 = 1 x 8 (8) x 1 x 8 = 64   1+ 8 (= 9) +1+ 8 =18

1,080 / 24 Hrs = 45 [0.0222] / 60 Min = 0.75 [1.333] / 60 Sec = 0.0125 [80] / 60 Arc sec = 0.000208333 [4,800]

1,080 / 16 = 67.5 / 22.5 = 3 [0.333] /.125 = 24 x 400 = 9,600 / 144 = 66.666 x 7.5 = 500 [0.002]

64 / 24 hrs = 2.666 [1/x = 0.375] / 60 Min = 0.0444 [1/x = 22.5] / 60 Sec = 0.000740 [1/x = 1,350] / 60 Arc Sec = 0.000012345679012 [1/x = 81,000]

64 / 16 = 4 / 22.5 = 0.1777 [5.625] /.125 = 1.4222 x 400 = 568.888 [0.0017578125]

64 / 7.5 = 8.5333   *144* / 64 = 2.25 (22.5)

PART 17 OF 20 – MSAART PATENT APPLICATION NOTES DRAFT 518,400 – 22:22:22 THUR 22ND SEPT 2022
© STRIKE FOUNDATION GUARANTEE LIMITED | MALCOLM BENDALL 2022 | GRAPHICS - STEVE EARL

MOON'S RELATIONSHIP TO TIME 518,400 AND 311,040

The Moon is moving at one Moon's Diameter per hour, therefore 51,840 Miles (2,160 x 24) is travelled in 24 hours. The Earth is about 15 billion inches from the Moon. The distance between the Earth and the Moon is 108 times the Moon's diameter of 2,160 miles being Equal to 233,280 Miles. The distance from the Sun to the Earth is 108 times the Suns 864,000 Mile Diameter which is a distance of 93,312,000 Miles (972 Earth's D).

51,840 Miles x 63,360 inches per Mile = 3,284,582,400 Inches

3,284,582,400 / 24 Hrs = 136,857,600 / 60 Min = 2,280,960 / 60 Sec = 38,016 / 60 Arc sec = 633.6

3,284,582,400 / 16 = 205,286,400 / 22.5 = 9,123,840 / 0.125 = 72,990,720
72,990,720 x 400 = 29,196,288,000 / 900 = 32,440,320

63,360 / 24 Hrs = 2,640 / 60 Min = 44 / 60 Sec = 0.7333 [1.363636] / 60 Arc sec = 0.01222 [81.818181]

63,360 / 16 = 3,960 / 22.5 = 176 /.125 = 1,408 x 400 = 563,200 x 900 = 506,880,000

51,840 Miles x 5,280 Feet (Sun's shadow's speed =   ) = 273,715,200 Feet

273,715,200 / 24 Hrs = 11,404,800 / 60 Min = 190,080 / 60 Sec = 3,168 / 60 Arc Sec = 52.8

273,715,200 / 360 = 760,320 / 16 = 47,520 / 22.5 = 2112 /.125 = 16,896 x 400 = 6,758,400 / 900 = 7,509.333

5,280 / 24 Hrs = 220 / 60 Min = 3.666 / 60 Sec = 0.06111 [16.363636] / 60 Arc Sec = 0.00101851851   [981.817181]

51,840 Miles x 1,760 Yards = 91,238,400 Yards

91,238,400 / 24 Hrs = 3,801,600 / 60 Min = 63,360 / 60 Sec = 1,056 / 60 Arc Sec = 17.6

91,238,400 / 360 = 253,400 / 16 = 15,840 / 22.5 = 704 /.125 = 5,632 x 400 = 2,252,800 x 900 = 2,027,520,000

91,238,400 / 16 = 5,702,400 / 22.5 = 253,440 /.125 = 2,027,520 x 400 = 811,008,000

1,760 / 24 Hrs = 73.333 / 60 Min = 1.222 / 60 Sec = 0.02037037 [49.090909] / 60 Arc Sec = 0.00033950617284 [2,945.454545]

In six days this equals 6 days x 51,840 Miles = 311,040 Miles (Sanskrit all time) that

Is 311,040 x 63,360 inches per mile = 19,707,494,400 inches

19,707,494,400 / 24 Hrs = 821,145,600 / 60 Min = 13,685,760 / 60 Sec = 228,096 / 60 Arc Sec = 3,801.6

3,801.6 Inches x 7.5 = 28,512 x 7.5 = 213,840

19,707,494,400 / 360 = 54,743,040 / 16 = 3,421,440 / 22.5 = 152,064 /.125 = 1,216,512 x 400 = 486,604,800

19,707,494,400 / 16 = 1,231,718,400 / 22.5 = 54,743,040 /.125 = 437,944,320 x 400 =175,177,728,000

311,040 Miles x 5,280 Feet (Sun's shadow's speed =   ) = 1,642,291,200 Feet

1,642,291,200 / 24 Hrs = 68,428,800 / 60 Min =1,140,480 / 60 Sec =19,008 / 60 Arc Sec = 316.8 / 60 Arc Arc Sec = 5.28

1,642,291,200 / 360 = /16 = / 22.5 = /.125 = x 400 =

316.8 x 7.5 = 2,376 x 7.5 = 17,820 x 7.5 = 133,650   5.28 x 7.5 = 39.6 x 7.5 = 297

5,280 Feet / 24 Hrs = 220 / 60 Min =3.666 / 60 Sec = 0.06111 [16.363636] / 60 Arc sec = 0.0010185185185 [981.818181]

311,040 x 1,760 Yards = 547,430,400 Yards

547,430,400 / 24 Hrs = 22,809,600 / 60 Min = 380,160 / 60 Sec = 6,336 / 60 Arc sec = 105.6

547,430,400 / 360 = 1,520,640 / 16 = 95,040 / 22.5 = 4,224 /.125 = 33,792 x 400 = 13,516,800

1,760 / 24 Hrs = 73.333 [0.01363636] / 60 Min = 1.222 [0.818181] / 60 Sec = 0.02037037 / 60 Arc sec = 0.00033951 [2,945.454545]

1,760 / 360 = 4.888 [0.20454545] / 16 = 0.30555 [3.272727] / 22.5 = / 0.125 = x 400 =

The Suns 27.332 Average Solar day x 51,840 = 1,416,890.88

The Suns Polar 24.3 x 34.2 Equitorial Solar day = 831.06

The seconds in six days are 6 days x 24 Hrs x 60 Min x 60 Sec = *518,400 Sec*

518,400 / 24 Hrs = 21,600 / 60 Min = 360 / 60 Sec = 6 / 60 Arc sec = 0.1[10]

518,400 / 360 = 1,440 / 16 = 90 / 22.5 = 4 /.125 = 32 x 400 = 12,800

311,040 / 24 Hrs = 12,960 / 60 Min = 216 / 60 Sec = 3.6 / 60 Arc sec = 0.06 [16.666]

311,040 / 360 = 864 / 16 = 54 / 22.5 = 2.4 /.125 = 19.2 x 400 = 7,680

EARTHS DIAMETER 7,920 AND RADIUS 3,960

EARTH'S DIAMETER = 7,920 Miles

7,920 = 7 x 9 x 2 [126] x 7 x 9 x 2 = 15,876

7,920 / 24 Hrs = 330 / 60 Min = 5.5 / 60 Sec = 0.091666 [10.909090] / 60 Arc Sec = 0.000152777 [654.545454]

7,920 / 1.111 = 7,128 - // 3,456 / 1,782 / 864 / 432 / 216 / 108 / 54 / 27 / 13.5

7,920 / 266.666 = 29.7000001 x 24 = 712.800002 x 11.111 = 7,920.00001

7,920 / 24 = 330   7,920 / 9 = 880   7,920 / 40 = 198   7,920 / 30 = 264   7,920 / 20 = 396

7,920 / 12 = 660   7,920 / 36 = 220   7,920 / 48 = 165   7,920 / 144 = 55

*7,920 / 16 = 495 + // 990 / 1,980 / 3,960 / 7,920 / 15,840 / 31,680*** / 63,360*

3168

7,920 / 50 = 158.4   7,920 x 70 = 554,400   7,920 / 72 = 110

15,876 / 24 hrs = 661.5 [1/x = 0.00151171579743] / 60 min = 11.025 [1/x = 0.090702947845805] / 60 sec = 0.18375 [1/x = 5.442176870748299] / 60 arc sec = 0.0030625 [1/x = 326.530612244898]

15,876 / 7.5 = 2,116.8

15,876 x 7.5 = 119,070

15,876 / 266.666 = 59.535000000000001

EARTH'S RADIUS = 3,960 Miles

3,960 / 9 = 440   3,960 / 266.666 = 14.85   3,960 / 144 = 27.5   3,960 / 11.111 = 356.4

3,960 / 1.111 = 3,564   3,960 / 24 = 165   3,960 / 7.5 = 528   3,960 x 7.5 = 29,700

3,960 / 16 = / 22.5 = /.125 = x 400 =

3,960 / 24 hrs = 165 / 60 min = 2.75 [0.0458333] / 60 sec = 0.0458333 [21.818181] / 60 arc sec = 0.000763888

3,960 = 3 x 9 x 6 [162] x 6 x 9 x 3 = 26,224

26,224 / 24 hrs = 1,093.5 / 60 min = 18.225 / 60 sec = 0.30376 / 60 arc sec = 0.0050625

26,224 / 7.5 = 3,499.2

26,224 / 266.666 = 98.3400002

NOTE OCTAVE DOWN REVERSES THE NUMBERS 15,876 / 26,224 = 0.605399633923

EARTHS CIRCUMFERENCE 24,920 AND 7.5 TIME

EARTHS CIRCUMFERENCE = 24,920 Miles

24,920 MILES = 2 x 4 = 8 x 9 = 72 x 2 = **144**   24,920 / **7.5** = **3,322.666**

144 x 2 = **288** x 4 = 1,152 x 9 = 10,368   10,368 x 2 = 20,736   **144** x **144** = **20,736**

24,920 / 24 Hrs = 1,038.333 / 60 Min = 17.30555 / 60 Sec = 0.28842592592592 / 60 Arc Sec = 0.004807098765432

24,920 / 360 = 69.222 / 16 = 4.3263888 / 22.5 = 0.192283950617284 [5.200642] / .125 = 1.538271604938272 x 400 = 615.3086419753086

24,920 / 266.666 = 93.4500002   24,920 / 144 = 173.0555   24,920 / 7.5 =186,900

20,736 / 24 Hrs = **864** / 60 Min = **14.4** / 60 Sec = **0.24** / 60 Arc sec = 0.004 [1/x = 250]

20,736 / **266.666** = 77.76

20,736 / **11.111** = 1,866.24   20,736 / 144 = 144   144 x 144 = 20,736   20,736 / 9 = 2,304

PROCESSION OF THE EQUINOX = 25,920 YEARS

25,920 / 24 Hrs = 1,080 (Moon's D) [0.00092592592592] / 60 Min = 18 [0.0555] / 60 Sec = 0.3 [3.333] / 60 Arc sec = 0.005 [200]

25,920 / *266.666* = **97.20000000000002 [Earths D 792]**

25,920 = 2 x 5 = 10 x 9 = 90 x 2 = 180

180 x 2 = 360   360 x 5 =1,800   1,800 x 16,200   16,200 x 2 = 32,400

180 x 180 = 32,400

32,400 / 24 Hrs = 1,350 / 60 Min = 22.5 [0.0444] / 60 Sec = 0.375 [**2.666**] / 60 Arc Sec = 0.00625 [160]

32,400 / 7.5 = 4,320 / 7.5 = 576 (576 / 2 = 288 / 2 = 144)

32,400 / 266.666 = 121.5 // **243** Sun R / **486** SUN D / **972** / 1,944 / 3,888 / 7,776 / 15,552 [18 x 864] / 31,104 [36 x 864] / 62,208

SUN'S DIAMETER 864,000 AND RADIUS 432,000

SUN'S DIAMETER = 864,000 Miles

864,000 Miles = 8 x 6 x 4 [*192*] x 8 x 6 x 4 = 36,864

864,000 / 24 Hrs = 36,000 [0.00002777] / 60 Min = 600 [0.001666] / 60 Sec = 10 [0.1] / 60 Arc sec = 0.1666 [6]

864,000 / 16 = 54,000 [+ // 108,000 / 216,000 / 432,000 / 864,000 / 1,728000 / 3,456,000 / 6,912,000 / 13,824,000] / 22.5 = 2,400 / 0.125 = 192 x 400 = 7,680,000

36,864 / 24 Hrs = 1,536 [0.000651041666] / 60 Min = 25.6 [0.0390625] / 60 Sec = 0.42666 [2.34375] / 60 Arc sec = 0.007111 [140.652]

36,864 / *266.666* = 138.24 / 16 = 8.64 / 22.5 = 0.384 [2.6041666] / 0.125 = 3.072

SUN'S RADIUS = 432,000 Miles

432,000 = 4 x 3 x 2 = [*24*] x 2 x 3 x 4 = *576* // 288 / *144* / 72 / 36 / 18 / 9

432,000 / 24 Hrs = 1,800 / 60 Min = 300 / 60 Sec = 5 / 60 Arc sec = 0.08333 [12]

576 / 24 Hrs = 24 / 60 Min = 0.4 [2.5] / 60 Sec = 0.00666 [150] / 60 Arc sec = 0.000111 [9,000 (Degrees in Plasmoid sector)]

576 / *266.666* = *2.16* / 16 = *0.135*

36,864 / *576* = *64*

36,864 / *266.666* = 138.24 // down // *69.12* // *34.56* // *17.28* // *8.64* // *4.32* // *2.16* // *1.08* // .54 // .27 // .13

LIST OF ALL GEMATRIA REVERSE NUMBERS MULTIPLIED

1 (=1) x 1 = 1 / 144 = 0.0069444

4 [2 x 2]

1 x 2 (= 2)
x 2 x 1 = 4 / 144 = 0.02777

9 x [3 x 3]

1 x 2 x 3 (= 6)
x 3 x 2 x 1 = 36 / 144 = 0.25

16 x [4 x 4]

1 x 2 x 3 x 4 (= 24)
x 4 x 3 x 2 x 1 = 576 / 144 = 4

25 x [5 x 5]

1 x 2 x 3 x 4 x 5 (= 120)
x 5 x 4 x 3 x 2 x 1 = 14,400 / 144 = 100

36 x [6 x 6]

1 x 2 x 3 x 4 x 5 x 6 (= 720)
x 6 x 5 x 4 x 3 x 2 x 1 = 518,400 / 144 = 3,600

49 x [7 x 7]

1 x 2 x 3 x 4 x 5 x 6 x 7 (=5,040)
x 7 x 6 x 5 x 4 x 3 x 2 x 1 = 25,401,600 / 144 = 176,400

64 x [8 x 8]

1 x 2 x 3 x 4 x 5 x 6 x 7 x 8 (= 40,320)
x 8 x 7 x 6 x 5 x 4 x 3 x 2 x 1 = 1,625,702,400 / 144 = 11,289,600

81 x [9 x 9]

1 x 2 x 3 x 4 x 5 x 6 x 7 x 8 x 9 (= 362,880)
x 9 x 8 x 7 x 6 x 5 x 4 x 3 x 2 x 1 = 131,681,894,400 / 144 = 914,457,600

100 [10 x 10]

1 x 2 x 3 x 4 x 5 x 6 x 7 x 8 x 9 x 10 (= 3,628,800)
x 10 x 9 x 8 x 7 x 6 x 5 x 4 x 3 x 2 x 1 = 13,168,189,440,000 / 144 = 91,445,065,555.555

121 [11 x 11]

1 x 2 x 3 x 4 x 5 x 6 x 7 x 8 x 9 x 10 x 11 (= 39,916,800)
x 11 x 10 x 9 x 8 x 7 x 6 x 5 x 4 x 3 x 2 x 1 = 1,593,350,922,240,000 / 144 = 11,064,936,960,000

144 [12 x 12]

1 x 2 x 3 x 4 x 5 x 6 x 7 x 8 x 9 x 10 x 11 x 12 (= 479,001,600)
x 12 x 11 x 10 x 9 x 8 x 7 x 6 x 5 x 4 x 3 x 2 x 1 = 22,944,253,280,256 / 144 = 159,335,092,224

256 [16 x 16]

1 x 2 x 3 x 4 x 5 x 6 x 7 x 8 x 9 x 10 x 11 x 12 x 13 x 14 x 15 x 16 (= 20,922,789,888,000)
x 16 x 15 x 14 x 13 x 12 x 11 x 10 x 9 x 8 x 7 x 6 x 5 x 4 x 3 x 2 x 1 = 437,763,136,697,390,000,000,000,000

20,922,789,888,000 / 31,104 = 672,672,000 /

LIST OF ALL GEMATRIA REVERSE NUMBERS ADDED

1
+ 1 = 2 / 144 = 0.13888

1 + 2 = 3 [3 x 1]
+ 2 + 1 = 6 / 144 = 0.041666

1 + 2 + 3 = 6 [3 x 2]
+ 3 + 2 + 1 = 12 / 144 = 0.08333

1 + 2 + 3 + 4 = 10 [2 x 5]
+ 4 + 3 + 2 + 1 = 20 / 144 = 0.13888

1 + 2 + 3 + 4 + 5 = 15 [3 x 5}
+ 5 + 4 + 3 + 2 + 1 = 30 / 144 = 0.208333

1 + 2 + 3 + 4 + 5 + 6 = 21 [3 x 7]
+ 6 + 5 + 4 + 3 + 2 + 1 = 42 / 144 = 0.2777

1 + 2 + 3 + 4 + 5 + 6 + 7 = 28 [4 x 7]
+ 7 + 6 + 5 + 4 + 3 + 2 + 1 = 56 / 144 = 0.3888

1 + 2 + 3 + 4 + 5 + 6 + 7 + 8 = 36 [4 x 9]
+ 8 + 7 + 6 + 5 + 4 + 3 + 2 + 1 = 72 / 144 = 0.5

1 + 2 + 3 + 4 + 5 + 6 + 7 + 8 + 9 = 45 [5 x 9]
+ 9 + 8 + 7 + 6 + 5 + 4 + 3 + 2 + 1 = 90 / 144 = 0.625

1 + 2 + 3 + 4 + 5 + 6 + 7 + 8 + 9 + 10 = 55 [5 x 11]
+ 10 + 9 + 8 + 7 + 6 + 5 + 4 + 3 + 2 + 1 = 110 / 144 = 0.763888

1 + 2 + 3 + 4 + 5 + 6 + 7 + 8 + 9 + 10 + 11 = 66 [6 x 11]
+ 11 + 10 + 9 + 8 + 7 + 6 + 5 + 4 + 3 + 2 + 1 = 132 / 144 = 0.91666

1 + 2 + 3 + 4 + 5 + 6 + 7 + 8 + 9 + 10 + 11 + 12 = 78 [6 x 13]
+ 12 + 11 + 10 + 9 + 8 + 7 + 6 + 5 + 4 + 3 + 2 + 1 = 156 / 144 = 1.08333

2 + 6 + 12 + 20 + 30 + 42 + 56 + 90 + 110 + 132 + 156 = 656 [16 x 41]
656 / 144 = 4.555   656 / 266.666 = 2.46   656 / 11.111 = 59.04

1 + 2 + 3 + 4 + 5 + 6 + 7 + 8 + 9 + 10 + 11 +12 +13 +14 +15 + 16 = 136

+16 +15 + 14 + 13 + 12 + 11 + 10 + 9 + 8 + 7 + 6 + 5 + 4 + 3 + 2 + 1 = 272 / 144 = 1.888

653

9ᵀᴴ PLACE EVEN NUMBERS 2 to 18 MIRRORED

2 x 4 x 6 x 8 x 10 x 12 x 14 x 16 x 18 x 18 x 16 x 14 x 12 x 10 x 8 x 6 x 4 x 2 = 34,519,618,525,594,000
34,519,618,525,594,000 / 518,400 = 66,588,770,304 / 9 = 7,398,752,256 / 8 = 924,844,032

34,519,618,525,594,000 / 24 = / 60 = / 60 = / 7.5 = 5,327,101,624.3201

34,519,618,525,594,000 / 360 = 95,887,829,237,761 / 24 = 399,532,621,824 / 60 = 66588770304001 / 60 = 110,981,283.84 / 7.5 = 14,797,504.51200017 / 11.111 = 1,331,775.406080016 / 1.333 = 998,831.554556001122 / 999 = 999.8313859459582

66,588,770,304 / 16 = 4,161,798,144 / 22.5 = 185,013,250.84444

2 + 4 + 6 + 8 + 10 + 12 + 14 + 16 + 18 = [90] + 18 + 16 + 14 + 12 + 10 + 8 + 6 + 4 + 2 = (2 x 90) 180 [6 x 90 = 540]

180 / 162 = 1.111   180 / 16 = 11.25   162 / 12 = 13.5   11.25 / 13.5 = .8333   13.5 / 11.25 = 1.2

9ᵗʰ PLACE ODD NUMBERS 1 TO 17 MIRRORED

(180 / 162 = 1.111 AETHER / MATTER RATIO)

1 x 3 x 5 x 7 x 9 x 11 x 13 x 15 x 17 = [34,459,425] x 17 x 15 x 13 x 11 x 9 x 7 x 5 x 3 x 1 = 1,187,451,971,330,625

1,187,451,971,330,625 / 518,400 = 299,060.9522087191

1,187,451,971,330,625 / 360 / 24 / 60 / 60 / 7.5 = 5,090,243.361328125 / 11.111 = 579,156.578 / 1.333 = 434,367.4335 / 16 = 27,147.96459375001 / 22.5 = 1206.5762041666 / 9 = 134.0641226851852 / 1.111 = 120.65762041666

1 + 3 + 5 + 7 + 9 + 11 + 13 + 15 + 17 = [8] + 17 + 15 + 13 + 11 + 9 + 7 + 5 + 3 + 1 = (2 x 81) = 162 [81 x 6 = 486]

9ᵀᴴ PLACE FIBANACCI NUMBERS 1 TO 55 MIRRORED

1 x 2 x 3 x 5 x 8 x 13 x 21 x 34 x 55 = [122,522,400] x 55 x 34 x 21 x 13 x 8 x 5 x 3 x 2 x 1 = 15,011,738,501,760,000

15,011,738,501,760,000 / 360 / 24 / 60 / 60 / 7.5 = 6,435,073.0888

15,011,738,501,760,000 / 518,400 = 2,895,782,890

1 + 2 + 3 + 5 + 8 + 13 + 21 + 33 + 55 = [141] + 55 + 33 + 21 + 13 + 8 + 5 + 3 + 2 + 1 = (2 x 141) = 282 [141 x 6 = 846]

9ᵀᴴ PLACE OCTAVE 2 TO 512 SEQUENCE

2 x 4 x 8 x 16 x 32 x 64 x 128 x 256 x 512 = [35,184,372,088,832] x 512 x 256 x 128 x 64 x 32 x 8 x 4 x 2 = 7,888,240,972,855,000,000,000,000

7,888,240,972,855,000,000,000,000 / 9 = 8,764,712,192,206,100,000,000,000
8,764,712,192,206,100,000,000,000 / 360 = 24,346,422,756,128,000,000,000
/ 24 = 1014434281505300000000 / 60 = 16,907,238,025,089,000,000 / 60 = 28,178,700,418,150,000
/ 7.5 = 37,571,640,055,753,000   47 octaves down = 266.9625591220851

7,888,240,972,855,000,000,000,000 / 311,040,000,000,000 =
7,888,240,972,855,000,000,000,000 / 518,400 = 152,165,142,225,800,000,000

7,888,240,972,855,000,000,000,000 / 266.666 =
7,888,240,972,855,000,000,000,000 / 144,000 =
7,888,240,972,855,000,000,000,000 / 9 = 8,764,712,192,206,100,000,000,000 x 11.111 = 973,137,003,524,691,400,000,000

2 + 4 + 8 + 16 + 32 + 64 + 128 + 256 + 512 = [1,022] 512 + 256 + 128 + 64 + 32 + 16 + 8 + 4 + 2 = 2,044 [1,022 x 6 = 6,132]

REFLECTIVE COMPASS NUMBERS BASED ON 16 SEGMENTS OF 22.5 DEGREES PRODUCING 4 QUADRITURES AND 8 AXIS LINES

OPPOSITE ANGLES ADDED   720 / x   360 / x   x / 11.111   x / 16   x / 1.333 ADDED TO REFLECTION

1). 22.5 + 202.5	= 225	3.2	1.6	20.2520	14.0625	168	225 + 522 = 747
2). 45 + 225	= 270	2.66*	1.33	24.3024	16.875	202.5	270 + 072 = 342
3). 67.5 + 247.5	= 315	2.2857	1.14	28.3528	19.6875	236.25	315 + 513 = 828
4). 90 + 270	= 360	2	1	32.4032	22.5	270	360 + 036 = 396
5). 112.5 + 292.5	= 405	1.77*	.888	36.4536	25.3125	303.75	405 + 504 = 909
6). 135 + 315	= 450	1.6	.8	40.5040	28.125	337.5	450 + 054 = 504
7). 157.5 + 337.5	= 495	1.45*	.727	44.5544	30.9375	371.25	495 + 594 = 1,089
8). 180 + 360	= 540	1.33*	.666	48.6048	33.75	405	540 + 045 = 585
AXIS TOTAL	= 3,060	16.3158	8.151	275.4272	191.25	2,294.25	TOTAL = 5,400

3,060 + 0603 = 3,663   136 x 22.5 = 3,060   3060 / 37 = 99   360 x 8.5 = 3,060 / 720 = 4.25

1 TO 10 = 45   10 TO 90 = 450   100 TO 900 = 4,500 = SUM = 4,995

SECTION E - ELEMENTS

Every Element that occurs on an individual segment of the Torus Vajra model shares the same octave as the colour and sounds of that segment. Simply stated each sound has a corresponding octave shape (figures of chaldi), color and octave element. The molten sea geometry is designed to enhance the photon generation through Aether (DC) to Matter (AC) transmutation through sacred resonant geometry. The Molten Sea dimensions must be reduced first to hands to understand the basic geometric relationships and then divide the hands by .125 (one eighth) to calculate all the atomic elemental resonances.

RESONANT FREQUENCY, CHARGE DENSITY AND GRAVITY CALCULATIONS

The Molten Sea contains 2,000 baths of water (90 cubic meters), x 72 (7+2=9) equals, 144,000 (1 + 4 + 4 = 9) logs divided by 1.111 equals 129,600 (1 + 2 + 9 + 6 = 1 + 8 = 9) resonant Logs, another Bendall Translate. The Bendall Translate is a way of describing proportionately the use and application of the HS along with quadrature, spiral Fibonacci scaled orbital energy both Male (Anti – clockwise spin) and Female (clockwise spin),

As the speed of light is about 300,000,000 meters per second, a light foot, that is it takes 1ns to travel one foot. So therefore as there are 3 light hands in a light foot therefore there are 10 light hands per meter = 3 billion light hands per second (.333 ns / hand) therefore as 3 light hands = 1 ns/foot, a light foot, therefore the circumference of the molten sea rim being (4.5 times 30 cubits = 135 hands times 1.111) equals 150 resonant hands is equal to 50 light feet.

Therefore there are 2/3 of a light foot between the 300 knops, 300 knops in two rows of 150 per row, the top row spheres representing the Sun (8.3 light minutes from the Earth) representing photons travelling at light speed and the lower spheres the Moon representing (1.3 light seconds from Earth) tide pull, gravity and light being their respective two most powerful observed effects.

To ensure the least amount of interference between the frequencies, they are stepped up in octaves, a progressive doubling of their frequencies, are wrapped in harmonic packages in quadrature we describe as knops, some more commonly known as nodes or Elements. They have a circular 3D spiral vortex trajectory that ensured, so as at the least effort in the most efficient manor, the desired length and speed of travel over and around the Torus Vajra, back to the singularity point G of propagation was achieved. That is the complex vector based geometry of the spiral paths achieves both Entropy, expansive, counterclockwise, explosive force and negative entropy, contracting, clockwise, implosive force, octave, frequency and octave/degree/charge density harmonic, nodes and direct latitudinal discharge and charge to and from G and angular separation in all three dimensions.   This geometry and resultant charge density creates access to the Forth Dimension, Time (AC) and Fifth Dimension, HS Aether (DC) energy that appear to be vector less (from our resonant point of reference) dimensions. They are in truth, bound by the same vectors, demonstrably bound by the same sacred geometry as the first three dimensions and are hence accessible and able to be influenced. In doing so the Key Of David TORUS VAJRA was cast, to open and close the door to a Forth Dimension, Time (the fabric of which is Entropy and at 90 degrees (quadrature) to its Centropy) and the fifth Dimension, Aether energy (thought to contain at least 95 percent of the universe's energy). The Tesla based Torus Vajra is described below by way of mapping both the longitudinal, 9 based, counter rotating vortex spirals and intersecting latitudinal segments based on 16 and its octaves 32, and 64 as well as the cardinal points of 360 and 720 degrees and the Supreme Cardinal 144,000/ 1.111 = 126,900 = 360 times 360 in this new context part of the Bendall Translate along with the square root of 3 and 7.5 and 1.333, the numbers 22.5 and 45 and 9.

The 8 planes create 16 equatorial points of referance, because of the circular 360-degree form of the Torus Vajra geometry, separate the segments by 22.5 degrees. The introduction of degrees brings the vortex ruling number 9 it is also a key to the underlying fabric of the torus as 9 represents the ultimate paradox of the singularity point G the alpha and omega, zero and infinity
nothing and everything. Nine is the end of one series of ten, and the beginning of another, as 9 times 10 is 90 degrees that octaves down, in internal circular octaves to 45 (4+5=9) to 22.5 (2+2+5=9) and 11.25 (1+1+1+2+5=9), then up in 90 degree octaves to 180, 270, 360 and 720 as the molten sea one sphere inverted is effectively two spheres even 720 (80 times 9 or 90 times 8 or even 90 divided by 0.125 equals 720) rules.

THEN COLOUR OCTAVES TIME AND AETHER GEOMETRY
DIMENSIONS TRANSLATION
SANSKRIT TIME

311,040,000,000,000/360/360/16/22.5/11.111 = 600,000 times 400 = 240,000,000
311,040,000,000,000 in the Sanskrit texts is 100 years, therefore :-

311,040,000,000,000 / 100 yrs = 3,110,400,000,000
3,110,400,000,000 / 360 Days in a Year = 8,640,000,000 Days
8,640,000,000 / 24 Hours = 360,000,000 Hours
360,000,000 / 60 Minutes = 6,000,000 minutes
6,000,000 / 60 Seconds = 100,000 Seconds
100,000 Seconds / 60 Arc Sec = 1,6666.666 / 7.5 = 222.222

Time To Degrees to Music to Greek/Hebrew/Sanskrit language

100,000 / 16 = 6,250 / 22.5 = 277.777 / 11.111 = 25.0025
15 x 15 = 225   24 x 24 = 576   36 x 36 = 1,296   37 x 37 =1,369   37x 73 = 2,701

TIME = Seconds

Sari = 70 eclipses = 1,250 years

1 year = 518,400 min / yr times 60 sec = 31,104,000 sec
1 year = 1,440 min / day times 360 days = 518,400 Min yr
1 year = 360 days
1 year = 12 months
1 year = 52 weeks
1 month = 4 weeks
Week = 7 days
Day = 24 hours
24 hours = 1,440 Minutes
60 min hour times 60 Sec / Min = 3,600
60 sec min times 518,400 Min / yr = 31,104,000

360 DEGREES = ANGLES = OCTAVES =COLOR = ELEMENTS = TIME = ENERGY

0.02197265625/0.125 = 0.17578125
0.0439453125/0.125 = 0.3515625
0.087890625/0.125 = 0.703125
0.17578125/0.125=1.40625
0.3515625/0.125 = 2.8125
0.703125/0.125 = 5.625
1.40625/0.125 =11.25
2.8125/0.125 = 22.5
5.625/0.125 = 45
11.25/0.125 = 90
22.5/0.125 = 180
45/0.125 = 360

3,110,400,000,000 /360 days = 8,640,000,000 3,110,400,000,000 /720= 4,320,000,000

8,640,000,000 / 24 hours = 360,000,000   8,640,000,000 / 720 = 12,000,000

8,640,000,000 / 144,000 Min/day = 60,000   8,640,000,000 / 360 = 18,001.8001

360,000,000 / 60 Minutes = 6,000,000   360,000,000 / 16 = 22,500,000

6,000,000 / 60 Seconds = 100,000   6,000,000 / 22.5 = 266,666.66

[266,666.666 = 266.666 ONE OCTAVE span times 1,000]

266,666.6666 / 1.3333=200,000.000000005 200,000.000000005/11.11=18,001.80018

18,001.80018 / 7.5 = 2,400.24002400

TIME creates Music by imposing intervals between NODES {Notes} creating MUSIC Time creates elements, energy and colour through the same matrix explaining that there are 3 primary sounds and colours and every dis-chord in sound is a dis-chord in colour and if you spin all the colours of the rainbow together you get white a singularity the same effect employed on the imploding vortex motor to create a singularity point 360 and zero every thing and nothing every colour but no colour every frequency AC but only DC.

Each note reports to an angle defined by :-

111 / Angle Times 720 = Musical Note or Frequency Node

OCTAVES

1	111 / 360 times 720 = 222 A Flat	[37 x 6]	(1 sect)
2	111 / 180 times 720 = 444 A	[37 x 12]	(2 sect)
3	111 / 90 times 720 = 888 A Sharp	[37 x 24]	(4 sect)
4	111 / 45 times 720 = 1,776	[37 x 48]	(8 sect)
5	111 / 22.5 times 720 = 3,552	[37 x 96]	(16 sect)
6	111 / 11.25 times 720 = 7,104	[37 x 192]	(32 sect)
7	111 / 5.625 times 720 = 14,208	[37 x 384]	(64 sect)
8	111 / 2.8125 times 720 = 28,416	[37 x 768]	(128 sect)
9	111 / 1.40625 times 720 = 56,832	[37 x 1,536]	(256 sect)
10	111 / 0.703125 times 720 = 113,664	[37 x 3,072]	(512 sect)
11	111 / 0.3515625 times 720 = 227,328	[37 x 6,144]	(1,024 sect)
12	111 / 0.17578125 times 720 = 454,656,	[37 x 12,288]	(2,048 sect)
13	111 / 0.087890625 times 720 = 909,312	[37 x 24,576]	(4,096 sect)
14	111 / 0.0439453125 times 720 = 1,818,624	[37 x 49,152]	(8,192 sect)
15	111 / 0.02197265625 times 720 = 3,637,248	[37 x 98,304]	(16,384 sect)
16	111 / 0.010986328125 times 720 = 7,274,496	[37 x 196,608]	(32,768 sect)
17	111 / 0.0054931640625 times 720 = 14,548,992	[37 x 393,216]	(65,536 sect)
18	111 / 0.00274658203125 times 720 = 29,097,984	[37 x 786,432]	(131,072 sect)

311,040,000,000,000/100 = 3,110,400,000,000 311,040,000,000,000 /24= 12,960,000,000,000

3,110,400,000,000 / 5,760 = 540,000,000   3,110,400,000,000 / 16 = 194,400,000,000

3,110,400,000,000 / 2,880 = 1,080,000,000   3,110,400,000,000 / 18= 172,800,000,000

3,110,400,000,000 / 1,440 = 2,160,000,000   3,110,400,000,000 / 20= 155,520,000,000

3,110,400,000,000 / 720 = 4,320,000,000   3,110,400,000,000 / 22 = 141,381,818,181.81

3,110,400,000,000 / 360 = 8,640,000,000 3,110,400,000,000 /24 = 129,600,000,000

3,110,400,000,000/180 = 17,280,000,000   3,110,400,000,000/27 = 115,200,000,000

3,110,400,000,000/90 = 34,560,000,000   3,110,400,000,000/30 = 103,680,000,000

3,110,400,000,000/45 = 69,120,000,000   3,110,400,000,000/32 = 97,200,000,000

3,110,400,000,000/22.5 = 138,240,000,000   3,110,400,000,000/36 = 86,400,000,000

3,110,400,000,000/11.25 = 276,480,000,000   3,110,400,000,000/40 = 77,760,000,000

3,110,400,000,000/5.625 = 552,960,000,00   3,110,400,000,000/44 =70,690,909,090

3,110,400,000,000/ 2.8125 = 1,105,920,000   3,110,400,000,000/48 = 64,800,000,000

3,110,400,000,000 / 266.6666 = 11,664,000,000.0002916
[129,600,000,000 / 8,640,000,000 = 15,129,600,000,000 / 266.666666 = 486,000,000.00001215]

8,640,000,000/24 Hours = 360,000,000   8,640,000,000 / 27 = 320,000,000

8,640,000,000/144,000 Min / Day = 60,000   8,640,000,000 / 30 = 288,000,000

360,000,000/60 Minutes = 6,000,000   360,000,000 / 32 = 270,000,000

6,000,000/60 Seconds = 100,000   6,000,000 / 36 = 240,000,000

266,666.6666 / 1.3333 = 200,000.000000005

200,000.000000005 / 11.11 = 18,001.80018

18,001.80018 / 7.5 = 2,400.24002400

311,040,000,000,000/100=3,110,400,000,000/16=194,400,000,000/22.5=8,640,000,0
00/.125=69,120,000,000/10=6,912,000,000/1.3333=5,184,000000/7.5=691,200,000/
1.111 = 622,080,000.000006

100,000/7.5 times light around the earth in one second = 13,333.333 to Degrees
13,333.333/16 = 833.333/22.5 = 37.037037/11.111 = 3.333 times 400 = 1,333.333

311,040,000,000,000 in the Sanskrit texts is 100 years, therefore :-
311,040,000,000,000 / 100 Years = 3,110,400,000,000
3,110,400,000,000 / 360 Days in a Year = 8,640,000,000
8,640,000,000 / 30 Days in a Month = 288,000,000
288,000,000 / 24 Hours in a Day = 12,000,000
12,000,000 / 60 Minutes in an Hour = 200,000
200,000 / 60 Seconds in a Minute = 3,333.333

TO DEGREES

3,333.3333 / 16 = 208.333 / 22.5 = 9.25925925

9.25925925 times 400 = 3,703.703703 / 11.111 = 333.333

9.25925925 / 11.111 = 0.8333 times 400 = 333.333

333.333 / 1.111 = 300 / 1.333 = 225

225 / 9 = 25 / 7.5 = 3.333

311,040,000,000,000 in the Sanskrit texts is 100 years therefore :-

311,040,000,000,000 / 100 Yrs = 3,110,400,000,000
3,110,400,000,000 / 360 Days in a Year = 8,640,000,000
8,640,000,000 / 12 Months in a Year = 720,000,000
720,000,000 / 30 Days in a Month = 24,000,000
24,000,000 / 24 Hours in a Day = 1,000,000
1,000,000 / 60 Minutes in an Hour = 16,666.666
16,666.666 / 60 Seconds in a Minute = 277.777

DEGREES TO RESONANT FREQUENCY ENERGY UNITS (RFEU)

277.777 / 16 = 17.36111 / 22.5 = 0.77160 / 11.111 = 0.069444 x 400 = 27.777

Sanskrit language to Hebrew to Greek

ALEPH	1	YOD	10	KOPH	100	= 111 / 37 = 3 x 11.111	= 33.33
BETH	2	LAPH	20	RESH	200	= 222 / 37 = 6 x 11.111	= 66.66 = Abb
GIMEL	3	LAMED	30	SHIN [PHI]	300	= 333 / 37 = 9 x 11.1111	= 99.99
DALETH	4	MEM	40	TAU	400	= 444 / 37 = 12 [3] x 11.111 = 133.33 = Cb
HE	5	NUN	50	KOPPH	500	= 555 / 37 = 15 [6] x 11.111 = 166.66
VAU	6	SAMECH	60	MEM	600	= 666 / 37 = 18 [9] x 11.111 = 199.99
ZAYIN	7	AYIN [PI]	70	NUN	700	= 777 / 37 = 21 [3] x 11.111 = 233.33
CHETH	8	PE	80	PE	800	= 888 / 37 = 24 [6] x 11.111 = 266.66 = C
TETH	9	TSADDI	90	TSADDI	900	= 999 / 37 = 27 [9] x 11.111 = 299.99

= 4,995 / 37 = 135 [DEGREES MOLTEN SEA]

TOTAL = 45   TOTAL = 450   TOTAL = 4,500   = GRAND TOTAL = 4,995

(9 x 5)   (9 x 50)   (9 x 500)   = 9 x 555 = 4,995

So the first and last

111 + 999 = 1,110 / 37 = 30

222 + 888 = 1,110 / 37 = 30

333 + 777 = 1,110 / 37 = 30

444 + 666 = 1,110 / 37 = 30

555 + 555 = 1,110 / 37 = 30

PHI AND PI AS THEY RELATE TO GREEK AND HEBREW LANGUAGE

PHI

The 21 st [7+7+7] Greek letter is PHI (SHIN in Hebrew) this letter has a Gematria value as per the table above of 300 in the 333 line and the 4,500 line.
300 x 11.111 = 3,333.333

The Sun relates to Phi as the radius of the Sun 432,000 divided by 4 = 108,000, 108,000 squared equals 11,664 which if taken to its 32nd reducing octave = 1.618

PI

6/5 = 180 Degrees Sun Day / 150 Suns x Phi squared = About Pi (3.14)

The 16th [8+8] letter Greek letter is Pi (Ayin) this letter has a Gematria number of 70 in the 777 line and 4,500 line. It is significant Pi is the 16th letter as it confirms the 22.5 degree division of 360 degrees producing 16 segments. 70 / Phi =

The Sun relates to Pi as the radius of the Sun 432,000, [4+3+2+2+3+4 = 18 and 4 x 3 x 2 x 2 x 3 x 4 = 576 / 18 = 32 and also 432 plus 234 equals 666]

The Suns radius 432,000 divided by 4 = 108,000, 108,000 which if taken to its 36th reducing octave = 3.14, from 432,000 would be the 38th reduction.

THE MOLTEN SEA NUMBERS SAME BASE AS THE SANSKRIT

311,040,000,000,000 / 518,400 = 600,000,000 (6,000 Baths in 100 Pie sphere)

and

311,040,000,000,000 / 720 = 432,000,000,000 [432,000 Logs in 100 pie sphere]

and

311,040,000,000,000 / 144,000 = 2,160,000,000 [216,000 Logs full Molten Sea]

314.1592653589793 / 311.040 = 1.010028502311533

The Molten Sea full sphere has an area of 100 pie = 314.1592 and contains 6,000 Baths of water times 72 = 432,000 divided by 360 equals 1,200 divided by 16 equals 75 divided by 1.111 equals 67.5 divided by 22.5 equals 3 times 100 = 300 Ritual Baths (300 = 150 Suns and 150 Moons on out side of the Molten Sea)

Full the molten sea HALF SPHERE contains 3,000 Baths of water times 72 equals 216,000 logs (2+1+6=9) divided by 360 equals 600 divided by 16 equals 37.5 divided by 1.111 equals 33.75 divided by 22.5 equals 1.5 times 100 = 150 Ritual Baths.

216,000 logs divided by 16 segments equals 13,500 per segment divided by 360 equals 37.5 divided by 1.111 = 33.75. 13,500 per segment divided by 22.5 degrees = 1.5 216,000 logs divided by 1.111 equals 194,400 log volumes of Resonant potential photon energy units (RPFEU). 194,400 divided by 16 = 12,150 divided by 360 = 33.75

311,040,000,000,000 / 144,000 = 2,160,000,000

311,040,000,000,000 / 129,600 = 2,400,000,000

311,040,000,000,000 / 266.666 = 1,166,400,000,000

311,040,000,000,000 / 1,728,000 = 180,000,000

311,040,000,000,000 / 8.888 = 34,992,000,000,000.34992

311,040,000,000,000 / 7.777 = 39,990,857,142,857.542766

311,040,000,000,000 / 6.666 = 46,656,000,000,000.46656

311,040,000,000,000 / 5.555 = 55,987,200,000,000.559872

311,040,000,000,000 / 4.444 = 6.9984,000,000,000.69984

311,040,000,000,000 / 3.333 = 93,312,000,000,000.93312

311,040,000,000,000 / 2.222 = 139,968,000,000,001.399698

311,040,000,000,000 / 1.111 = 279,936,000,000,002.79936

1,728,000 / 8.888 = 194,400

1,728,000 / 7.777 = 222,171.428

1,728,000 / 6.666 = 259,200

1,728,000 / 5.555 = 311,040

1,728,000 / 4.444 = 388,800

1,728,000 / 3.333 = 518,400

1,728,000 / 2.222 = 777,600

1,728,000 / 1.111 = 1,555,200

1,728,000 / 1.333 = 1,296,000 / 100 = 129,600 [RFEU]

1,728,000 / 129,600 = 13.3333

The crossing over from the 5TH DIMENSION AETHER and 4TH DIMENSION TIME through the Rim of the Molten Sea (MS) into Creation the 3rd and 2nd and 1st dimensions is described in part mathematically by

144,000/9 = 16,000/7.5 = 2,133.333/1.111 = 1,920/1.333 = 1,440/360 = 360/16 = 22.5

144,000 / 360 / 16 / 22.5 = 1.111 and 144,000 / 1.111 = 129,600

Note:- 7.5 times 1.333 = 10 times 9 = 90 and 10/9/7.5/1.333 = 0.1111

Where 144,000 = Sacred Golden Number
Where 129,600 = 2,000 Baths times 72 and also 150 Ritual Baths Volume in Logs of the Molten Sea.
Where 9 = The Ruling number and Base Root for Circles, Spheres, Spirals, Pi and Torus.
Where 7.5 = It takes light one second to circle the earth 7.5 times.
Where 1.111 = The base unit of resonant frequency the Rod (Universal Constant).
Where 1.333 = Width in hands of the Molten Sea Rim = 16/12 ratio of Below (Earth AC) to Above (Heaven DC).

518,400 / 86,400 Seconds in a Day = 6 times 7.5 = 45 1 / x = 0.0222

144,000 / 86,400 Seconds in a Day = 1.666 times 7.5 = 12.5   1/x = 0.08

129,600 / 86,400 Seconds in a Day = 1.5 times 7.5 = 11.25   1/x = 0.0888

1,440 / 86,400 Seconds in a Day = 0.01666 times 7.5 = 0.125 {Base for Octaves all}

144,000 / 1.111 = Volume, Molten Sea   = 129,600 / 16 / 22.5 = 360 times 360
= 129,600 Logs which is 150 Ritual Baths.

16 segments times 100 = 1,600 / 1.111 = 1,440

1,440 / 16 = 90 = One quadrature

90 / 22.5 = 4   90 times 1,440 = 129,600

1,440 = Minutes in a Day

1,440 Times 360 = 518,400 / 4 = 129,600

1,440 Minutes times 7 = 10,080 Minutes in one Week

1,440 Minutes times 52 = 524,160 Minutes in 52 Weeks

86,400 Seconds in a day / 1,440 = 60 Minutes

864,000 [Sun's diameter] / 25,920 = 33.333 [3 x 11.111 and 3 Octaves up is 266.666]

432,000 / 25,920 = 16.666

25,920 [Equinox] / 2,160 [Moon] = 12

25,920 [Equinox] / 7,920 [Earth] = 3.272727

25,920 / 144 = 180

25,920 / 1,440 = 1,440 that is 1,440 x 1,440 = 25,920 the Equinox

86,400 / 25,920 = 3

129,600 / 25,920 = 5

3,456,000 / 25,920 = 133.333 [Next Octave up 266.666]

311,040,000 / 25,920 = 12,000

Aether, Light, Time, Matter, spirals and the Sacred Torus have the same digital root 9, 144,000 / 1.111 = 129,600 / 24 = 5,400 / 60 = 90 / 60 = 1.5 / 7.5 = 0.2 x 720 = 144 / 16 = 9 / 22.5 = 0.4 / 0.125 = 3.2 x 400 = 1,280 and therefore although unseen is bound by the same sacred geometry to have a similar vector field symmetry and form as the circle based Torus vortex, which makes it susceptible to be forced to the same torus geometry causing positive and negative time variance by the effective contraction and expansion, decreasing and increasing charge density, with the Torus field.

The Saros, 70 full consecutive Eclipses of the Moon, is 1,260 years times 2 = 2,520 times 2 = 5040 times 2 = 10,080 (minutes in a week) / 144 = 70 Saros.

360 / 7 = 51.4285714285 weeks in a 360 day year 144,000 / 51.428571428571 = 2,800

The 50 year Jubalee Sirus orbit 144,000 / 9 = 16,000 / 7.5 = 2,133.333 / 1.111 = 1,920 / 1.333 = 1,440 times 22.5 = 32,400 / 648 = 50

The 50 year Jubalee Sirus orbit 144,000 / 2,880 (2 times 1,440, 4 times 720) = 50

The 360 day year 144,000 / 9 = 16,000 / 7.5 = 2,133.333 / 1.111 = 1,920 / 1.333 = = 1,440 / 360 = 360

The 360 day year = 144,000 times 7.5 (time, one second, it takes for light to circle the earth 7.5 times and the degrees between 12 outside and 16 inside segments) = 972 / 360 = 2,700 / 9 = 300 times 1.333 = 400 / 1.111 = 360

The 12 months in a year 144,000 / 1.111 = 129,600 / 360 = 360 / 30 = 12
The 40 Days = 144,000 / 60 = 2,400 / 60 = 40 days and 40 years in the wilderness

The 30 day month= 144,000 / 7.5 (Speed of light to travel around the Earth 7.5 times) =
= 19,200 / 720 = 26.666 / 1.111 = 24 times 1.333 = 30 days
Week 144,000 / 9 = 16,000 /7.5 = 2,133.333 /1.111 = 1,920 /1.333 = 1,440 x 7 = 10,080

Day = 144,000 / 7.5 (SPEED OF LIGHT AROUND THE EARTH 7.5 IN ONE SECOND)
= 19,200 / 720 = 26.666 / 1.111 = 24 Hours

Day = 144,000 / 100 = 1,440 Minutes in a Day

Day = 144,000 / 9 = 16,000 / 7.5 = 2,133.333 / 1.111 = 1,920 / 1.333 = 1,440 / 60 = 24

Day = 144,000 / 7.5 (SPEED OF LIGHT AROUND THE EARTH 7.5 IN ONE SECOND)
= 19,200 / 720 = 26.666 / 1.111 = 24 Hours

The Hour= 144,000 / 100 = 1,440 MINUTES IN A DAY / 24 = 60 minutes

The Hour = 144,000 / 2,400 = 60 Minutes

The Second = 144,000 / 1.111 = 129,600 / 24 = 5,400 / 90 = 60 Seconds

PI - TRANSENDENTAL NUMBER BROUGHT BACK TO EARTH BY 9

Pi describes the relationships of the circles, spheres, vortex spirals and is the area of the Sacred Torus, Pi x 100. For thousands of years Pi has been the subject of study of every apt inquiring mind and is the very basis of modern society maths. Pi is simply a product of 9 times the octave roots of the elements as the number NINE rules the circles, spheres and the vortex spirals by cause that the resonant frequency of the cardinal cross section has internal angles totaling 900.

Pi = 9 times 10 to the 12th/(the Product from the Octave roots of the Elements)

PHI - TRANSENDENTAL NUMBER BROUGHT BACK TO EARTH BY 7

ALL ELEMENTAL FREQUENCIES ARE NOW ABLE TO BE EXACTLY CALCULATED FOR ALL EIGHT TOROIDAL SPHERES (ELEMENTS), ENABLING THE PRODUCTION OF GEOPOLYMERS FROM ROCK, THE SELECTIVE EXTRACTION THROUGH THE GALVANIC SERIES OF ANY ELEMENT AND THE FIRST SAFE NON TOXIC, NON PLANET DESTROYING, TRANSMUTATION OF ANY ELEMENT OR EATHER INTO POWER WITH AN ABOVE UNITY EFFECT.

Taking one Molten Sea segment a 1/16th (6.25 Pi) of 360 Degrees having an inside angle of 22.5 degrees and having a Radius of 22.5 Hands one simply draws at centres defined by 144,000 / 1.111 = 129,600 times .889 (Ratio of Oxygen to Hydrogen in water, H2O) = 115,214 and going down to the 14th Octave 7 then multiplying 7 (56/8) by 1,440 = 10,080 minutes in a week. [56 is the sacred number of the 3rd Dimension, as applied to the Sphere, the product of the root numbers of 1.111 cubed.]

SECTION G APPLICATIONS OF ALIEN MATHS AND CHEMISTRY

PROPRIETARY INFORMATION

THE BENDALL ENGINE
CHANGING THE PARADIGM, WHITE PAPER ON TOROIDAL VORTEX IMPLOSION BASED JET ENGINE TECHNOLOGY

APRIL 2017 - DRAFT 14

ABSTRACT

The breakthrough general unification model has been constructed and the first application of this wave guide design has been the application for replacement of existing jet engines as well as the implementation of new applications applying new fuels hitherto not conceived. This white paper outlines the overwhelming evidence in favour of replacing all axil based turbine technology from propellers in water to jet engine fans and helicopter blades in air. The key inputs, which led to the conceptualization of the implosive turbine design on a macro level, also led to the microscopic and atomic level uses and applications of the basic form and function to provide novel mechanisms to concentrate or disperse extreme charge densities and heat. This technology is original and inverts current thinking and has generated inexplicable yet repeatable and well-documented results. This novel technology promises to provide performance parameters that exceed by orders of magnitude current self limited self-constraining axil based constructs.

1). ABOUT MALCOLM ROY BENDALL

He has been involved in hi-tech and mechanical / technical innovation in the mining, mineral processing, smelting and oil and gas industries as a practical problem solving inventor. He has been a mine manager, drilling supervisor, plant manager, geologist, geophysicist and inventor and implementer of problem solving solutions to actual real world problems. He has built and flown kites, gliders and powered model planes and light planes and helicopters.

"I don't mind schooling as long as it does not interfere with my education." Mark Twain

2). THE CONCEPTUALIZATION OF THE BENDALL IMPLOSIVE TURBINE ENGINE
(BENDS ALL THE LAWS AND RULES BUT BRAKES NONE)

The 10 keys were at first glance unrelated but a common geometric based energy multiplier system well known and well documented, but to date untapped by modern mechanisms of our current civilisation. The use of these implosive forces, inherent within the operation of nature, uses principals well understood yet never implemented in modern times.

CONCEPTUALIZATION - CONTRIBUTIONS FROM NATURE

2A). BUTTERFLY EFFECT – TOROIDAL AIRFLOW

Having studied climatology in my third year at university I realised the inherent multiplier in force by a Toroidal structuring of airflow, the concept of the "butterfly effect". That effect is the concept that a butterfly flapping its wings in Africa creates a self-perpetuating expanding self organizing fractal system, incorporating both centipedal imploding and centrifugal exploding feed back system that by the time it reaches the USA has the force and destructive power of 10,000 atomic bombs (NASA calculation). My mind while trying to reconcile a butterfly creating the destructive power of 10,000 atomic bombs pondered on the truth that one could through extreme, acumen, cunning and rare craft, create a machine with geometry that could harness this truth in a controllable, sustainable and repeatable way. The implosive vortex turbine was the manifestation of this craft and cunning, which for good measure, also harvests the extreme electrical thunderstorm charges inescapably produced by the "butterfly effect".

2B). DOLPHIN EFFECT – FLUID TORIDAL VORTEX BUBBLES

Having studied Dolphins blowing Toroidal air bubbles that whilst in the spinning Toroidal shape become free of the normally understood laws of bouncy and inversely defying gravity. I became sure that the dolphins rings and the "butterfly effect" where the same cause with the same multiplier effect. I was right.

2 C). MANTIS SHRIMP – WEAPONISED TOROIDAL CAVITATION IMPLOSIVE BLAST

Having studied the Mantis shrimp, that utilizes the force concentrating multiplying effects of the imploding cavitation bubble made by first creating a vacuum with its claw then through intelligent design focusing that inherent force, I concluded high pressures and temperatures were involved. This conclusion was reached by observing burn marks around the holes drilled through cockle shells by the shrimps, documented temperatures of 10,000 Degrees C can be generated on the outside of an imploding vortex bubble, I determined to investigate the theory. With scuba gear at night I observed shrimps killing shellfish with the shock waves generated and recovered the shells of those same kills confirming my previous observations of washed up dead shellfish.

2 D). ARCHEAE EXTREMOPHILES

Archaea where first identified by Robert Oppenheimer who led expeditions to the most inhospitable environments on the planet that curiously excluded all other conventional life except shrimp. These Archaea are unique and not understood, the best way to allude to their extraordinary abilities is their close association with shrimp. Shrimp are now feeding on fluids from the Abyssal trench black smokers at miles depth in the ocean, also shrimp are feeding in hyper saline Flamingo lakes and shrimp are also feeding at the outflows from deep natural springs. Two obvious questions arise what are they feeding on as the fluids come out of rock and how can they survive in environments too extreme for life, the answer to these two questions is the same, Archaea. The shrimp at the black smoker vents should be well done at measured temperatures of over 400 degrees C so they must have a protective mechanism related also to the fact of what are they harvesting and feeding on.

The hyper saline Flamingo ephemeral lakes produce salt layers when they dry up and layers of Flamingo feces when they are filled with water. When the lakes are full of water a half bacteria half algae life form can only function with the help and synergistic association with Achaea that has the unique ability to extract protons and electrons from matter creating matter, Archea is the pre- matter to matter base terraforming entity that lives of electrons and protons by its geometry alone this trick allows it to protect the shrimp from the heat of the vents as it act as a Toroidal electron vacuum removing heat from the shrimp the same effect as shrimp in a hot shallow lake. Flamingos feeding off the shrimp produce faeces that have up the food chain concentrated the Archea to the extent that they are very alkaline PH 11 Natron ($Na_2CO_3.10H_2O$) salts. Natron salts preserve dead flesh and where used in China, Cambodia, Egypt, Tibet, and South Americans for mummification due to their ability to exclude bacteria and other destroying agents away. Archaea derived Flamingo lake deposits are near all ancient civilizations and are also the Catalysts (Salt) and reagents for producing Geopolymers (rock plastics) plastics that are seen in polygonal block constructions of Pre-History constructions found on all continents.

The Archaea have an implosive form similar to modern day diatoms and indeed the substrate to preserve and distribute them is diatomaceous earth. The effective simple function of my implosive vortex copies the Archaea form and function. Archaea terraform planets by converting the free electrons of higher charge density Toriodal form photons and Toroidal form energy into lower charge density frequency bound (figures of Chaldae) matter the inescapable conclusion being that the crust of the earth is simply Archaea shit!!! Reference material is on the Schuman cavity data base for anyone with a connection to down load it?

2 E). WHIRLWIND FIRE TORNADO

As a forest fire fighter in Australia recalled I witnessed fire tornadoes which sucked in fuel and air reaching up to 300 feet into the sky and literally exploding trees and houses on contact. The recorded temperatures reached by fire tornadoes are 2,000 degrees F (1,090 degrees C) by these structured airflows, however few have lived to tell the real temperatures generated. I have seen melted iron bolts on power poles along with turning the ceramic insulators to liquid along with liquefying the concrete poles and bushfire proof concrete rooms housing electrical transformers liquefying the metal doors and metal hinges along with the transformers themselves copper, aluminum and steel all melted into a pool. They generate winds of up to 100Mph and as they are a self-organizing system can last up to several hours as long as there is a source of fuel, I have observed that sand, dust or a water spout can extinguish them, otherwise there is no defense against them, many bush fire crews have been killed by them because of the speed they travel at and because the fire trucks, that normally provide shelter for the normal event of a passing fire front literally explode as the petrol tanks vaporize.

CONCEPTUALIZATION - CONTRIBUTIONS FROM MAN

2 F). IMPLODING HYDRODYNAMIC CAVITATION BUBBLES

Speed boat propellers are a good example of devises that generate cavitation bubbles that then implode from the North and South poles into their equatorial plane center. Running a speed boat at full throttle on a moonless night with dead calm waters one can observe a purple blue glow in the wash thrown up by the propeller. This ultraviolet light created by the imploding cavitation bubble has a measured frequency this frequency is the color of the 1% Argon gas dissolved in the water being taken up to higher energy atomic orbits then releasing the Violet to lavender blue light whilst it returns to its normal state. The light remains for several seconds before disappearing. By firing a gun underwater and filming it with a high speed camera one can observe the imploding and rebound reforming imploding sequence explaining the bounce down and therefore slow fading away effect.

2 G). STAR IN THE JAR

The creation of imploding cavitation bubbles by means of focusing within a spherical or cylindrical container sound waves that create and reform cavitation bubbles causing high temperatures and pressures that emit ultraviolet light.

2 H). NANOKEY BALL MILL

In 1998 I financed the construction of a NANOKEY BALL MILL designed to replace the current ball mills which are cylinders filled with steel spheres a design not changed for 5,000 years. Sad to say 60% of materials for our modern society from inks, to ceramics, to mineral processing and paint manufacture are processed still by this primitive and ineffectual base technology. The Ball Mill is always the bottle neck in mineral processing as well as all processing taking up to 90% of the overall process time and usually 60% of the Power consumption. The application of cavitation bubbles to break down

2 I). KEN SHOULDERS – EXOTIC VACUUM OCCURENCES (EVO's)

Ken Shoulders worked on studying Exotic Vacuum Occurrences (EVO's) with a high definition electron microscope and mass spectrometer to record the shape and elemental composition of impact craters caused by firing Plasmoids into high purity copper sheets. The impact creators and splashes, frozen in time as if by a freeze frame, recorded molten metal caught, as it were, half way through a splash out of the copper sheet target, being 99.999 % pure copper. Ken examined the composition of the splash material and found the elemental composition had changed. The target sheets were tested with highest purity metals and both the impact creators, frozen splash and a change in composition were recorded.

2 J). WAVEGUIDE FORCE MULTIPLYER

2 K). MAN'S OBSERVATIONS OF THE STRUCTURE AND FORM OF ELECTRICAL, GRAVITATIONAL, MAGNETIC, GALACTIC AND CRYSTALINE FORMS

Man has observed the natural movement of the Heavens and the rising and falling of the Sun and Moon since time began. Recent observations

By means of building and testing my theory I can confirm that indeed we have a lot to thank the butterfly, Mantus shrimp and dolphin for.

3). NO STRAIGHT LINES IN THE UNIVERSE LIGHT, ENERGY AND MATTER CURVE.

It is accepted that there are no straight lines in nature but what has not been realised is that one can then observe that there are then only two options with natural forces and that is clockwise implosive spin demonstrably generating a negative charge and anti-clockwise explosive spin generating a positive charge. This simple observation then can be applied fractally and in doing so one can summon toroidal vortex forces on both an atomic and universal scale and though simple geometry channel those force for effect.

4). OBSERVE AND COPY NATURAL IMPLOSIVE FORCES THEN CAPTURE THEM

Implosive self structuring forces increase the charge density and force of the input energy medium whilst reducing friction, concentrating heat and therefore the greater the force input the less the resistance offered against it. A system through self-structuring increases in power by multiplying input energy is an implosive system.

5). OBSERVE AND AVOID MANS EXPLOSIVE TURBINE AXIL BEARING BLADE CREATIONS

Explosive disorganising forces used by man, in his ignorance of a better way, decreases the charge density of the input energy medium whilst dissipating heat and increasing friction. A system decreasing in power by destroying, dissipating and obliterating input energy is an explosive system. An explosive system such as a wheel or classical jet engine spins around an central axil which is self limiting as the inertial forces work as a centrifuge creating forces literally wanting to pull the device apart and stored kinetic energy that makes the stopping of the device difficult especially as large engine weights caused by depths of section that are required to obtain the mechanical strength to achieve high performance characteristics add weight.

6). IMPLOSIVE TOROIDAL TURBINE TECHNOLOGY V's EXPLOSIVE TURBINE TECHNOLOGY

6 A). IMPLOSIVE CONCENTRATION OF FORCE V's EXPLOSIVE DISIPATION OF FORCE

The natural concentrating of force, reduction of friction, the generation of a centralised focused heat away from component parts, natural cooling of the turbine blades, higher burn temperatures, orders of magnitude higher compression ratios and orders of magnitude less weight of an implosion based jet engine is obviously a more efficient and effective way to implement energy applications than by the dissipation and dispersal of energy and creation and dispersal of both friction and heat onto the component parts caused by machines designed to operate under explosive conditions.

6 B). IMPLOSIVE ENGINE HAS NO MOVING PARTS, NO INERTIA, NO BEARINGS, NO AXIL, NO FRICTION AND MOVES INTO A VACUUM AND IS PUSHED FORWARD BY A PRESSURE WAVE IT SURFS ON V's CURRENT EXPLOSIVE JET ENGINES MOVING PARTS, INERTIA, BEARINGS, AXIL, STRESS FRACTURES, HEAT IN THE WRONG PLACES, VIBRATION, FRICTION CAUSING BALANCE PROBLEMS UNDER HIGH PERFORMANCE CONDITIONS AND TO ADD INSULT TO INJURY PUSHES INTO HIGH PRESSURE AND HAS BACKPULL

The basic principals of the implosive turbine are the exact opposite to those forces generated and propagated within the conventional explosive axil based turbine. The implosive turbine is stationary and uses the air itself to create a vacuum for it to move into and the air self compresses and self heats simply by the natural effects of a certain state of the art geometric structuring following strict constraining parameters.

6 C). THE BENDALL IMPLOSIVE TURBINE OPERATES IN AIR, WATER AND SPACE SEAMLESSLY V'S AIR BOUND JET ENGINE

The

6 D). THE BENDALL IMPLOSION OPERATES IN THREE MODES AT ONCE OR SEPERATLY V'S ONE MODE IN AIR FOR JET ENGINE

The

6 E). IMPLOSIVE TURBINE MOVES INTO A VACCUUM AND RIDES A PRESSSURE WAVE V's

An implosive engine generates a vacuum in front of both the wing and the engine and a pressure wave behind them so as it both pulls and pushes the wing and jet engine forward. In contrast the current jet engine effectively constricts air flow and creates drag once certain constraining speeds are reached where the blades become obstructions and being a primitive use of an axil bearing and blades suffers from the inherent design flaw of both weight and centrifugal forces creating inertial, heat and stress on the turbine blades at high work loads.

7). HOW THE IMPLOSIVE TURBINE CONCEPT WORKS

The

8). BENEFITS OF THE BENDALL IMPLOSIVE TURBINE TECHNOLOGY REPLACING EXISTING EXPLOSIVE JET TURBINE TECHNOLOGY

8 A) COMPRESSION RATIO   CONVENTIONAL JET ENGINE SET AT 40:1, BENDALL ENGINE DYNAMICALLY VARIABLE UP TO AND OVER 100,000:1

The

8 B) W E I G H T

The

8 C) F U E L C O N S U M P T I O N

The

8 D) R E L I A B I L I T Y

The

8 E) N U M B E R O F P A R T S

The

8 F) N U M B E R O F M O V I N G P A R T S

The

9). IMPLIMENTATION OF THE TECHNOLOGY FOR EXISTING AIRCRAFT

The

10). IMPLIMENTATION OF THE TECHNOLOGY FOR EXISTING LANDCRAFT

The

11). IMPLIMENTATION OF THE TECHNOLOGY FOR EXISTING SEACRAFT,

The

12). IMPLIMENTATION OF THE TECHNOLOGY FOR EXISTING SPACECRAFT

The
13). CURRENT STATUS OF PROTOTYPE

The

14). CURRENT STATUS OF PATENTS

The

15). REQUESTED INFORMATION ON TEST BEDS TO ESTABLISH OPERATIONAL FEASIBILITY

The

16). INDUSTRY PLAYERS AND RELEVANT CONTACTS

The

17). SUGGESTED IMPROVEMENTS TO THE PRESENTATION AND PACKAGING OF THE CONCEPTS IN NORMAL INDUSTRY TERMS AND FORMATS

The

18). CONCLUSION

The

PATENT APPLICATION

A Patent application for THE REVOLUTIONARY BENDALL ITP ENGINE

The new technology will replace all axial-based 'explosive' turbine mechanisms, from propellers in water to jet engine fans and helicopter blades in air, with an innovative 'Implosive Tornado Plasmoid' (ITP) Engine.

This device is a complete break from all previous engine developments.

The Bendall ITP Engine is based on a function that mimics or replicates the natural occurring forces in a tornado. Hurricane, typhoon, whirlwind or cyclone. In relation to the fluid state of matter, it is similar to the dynamics of a whirlpool or waterspout.

The ITP employs Plasma Magnetic Energy or Plasmoid, which takes the shape of a doughnut when emitted.

Basic Physics of the Bendall ITP Engine.

There are two options for electron spin in nature.

One: A clockwise up-spin that generates negative ions of high charge density of low gravity and energy. The up-spin is implosive.

Two: An anti-clockwise explosive down-spin generating a positive low charge density of high gravity and energy. The down-spin is explosive.

As in the eye of a tornado, the stillness, or lack of spin energy at rest in a Plasmoid's centre represents a point which has all potential energy and/or no applied kinetic energy at all; that is, all force and no force at the one junction.

These Tornado Plasmoid forces can be employed from a sub-atomic up to a universal or macro scale, such as a Galaxy.

The Plasmoid's doughnut or lifebuoy shape is technically a 'Torus' or surface generated by imploding a sphere. On implosion, the sphere's north and south poles meet at its equatorial plane's centre and create both its form and pivot point. The resultant doughnut or Torus uses natural implosive force causeways and patterns of energy.

Impact of the Bendall ITP Engine

The industrial and communication revolutions of the 19th and 20th Century followed by the current internet-based information explosion have obvious material advantages for the world. But the unfettered, unguided misuse of technology has led to hazards caused by the aforementioned anti-clockwise, spin-down, positive ion charge, which has resulted in widespread pollution in the atmosphere, rivers, lakes, and oceans which cover 70 per cent of the earth. Positive ion pollution means these water areas are becoming more acidic every day. An instance is the East China Sea that is now acidic. (A classic but typical example of pollution occurs in the airline industry. Flying at 35,000 feet, planes burn sulphur-infused fuel, which produces carbon dioxide, other noxious gases and positive ions.)

The atmosphere's net negative charge is maintained by the alkaline ocean's net negative charge. But the waterways and seas are being heavily polluted by explosive, positive ion technologies. If this 'old' or current technology continues it is inevitable that it will destroy human life and the planet itself. Then British scientist Stephen Hawking's prediction, that the earth is doomed and humankind must escape it inside 100 years, will come true.

The mission of this patent is to supersede and reverse all the destructive results of technology developed over the last two hundred years. The Bendall ITP Engine will produce clockwise, spin-up negative ions, as a source for energy production. The new spin-up, implosive technology led to the Bendall ITP Engine.

The first of many applications is a Torus (Toroidal) Wave Guiding Thrust device. The ubiquitous Torus has the capacity to catapult all societies, not only into a new industrial revolution, but also into a renaissance in the foundations of quantum and geometric thought. As stated, this unique model can be applied from quantum particles up to the size of galaxies. [1]

Summary: ITP Engine Versus Explosive Turbine Technology

The imploding natural spin-up Torus invokes:

i: concentration of forces;

ii: reduction of friction;

iii: generation of self-destructing centralized and focused heat that moves away from components;

iv: natural cooling of the flow and wave guide blades;

v: higher burn temperatures;

vi: higher compression ratios:

vii: less weight;

viii: a more efficient and effective way to implement energy applications that move into vacuums, and not into self-inflicted pressure waves.

By contrast, the conventional Brayton Cycle engine causes the dissipation and dispersal of energy. It also causes the dissipation and dispersal of both friction and heat onto the component parts, which are designed to operative under explosive conditions. This means that the Bendall ITP Engine is not subject to the same temperature limitations as the Brayton Cycle engine because the critical components are not exposed to high heat.

The Bendall ITP Engine has only one component shape. It has no moving parts, no inertia, no bearings, no axle, and no friction. It moves into a vacuum it has created itself and it is pushed forward by a pressure wave that if 'surfs' on, and which originates at the electrically assisted zero point.

The creator, Malcolm Bendall, has made and tested more than a hundred different prototypes in order to implement and validate various aspects of the Toriodal theory; to test materials suitability; and viable industrial scale manufacturing.

Again by contrast, the ITP is superior to the Brayton because the latter has moving parts; inertia, bearings, axil and potential stress fractures. In addition the Brayton has destructive heat that causes complete or partial failures, which contribute to balance, vibration and friction problems. These downsides features of the Brayton show up under high load performance conditions. Furthermore, it creates a pressure zone in front of it. This impedes its progress because of drag on the engine.

The Bendall ITP, on the other hand, creates a vacuum in front of itself that propels its forward motion.

<u>Description of the Bendall ITP Engine</u>

In a jet plane, the ITP thruster can operate similar to a normal jet engine (except for being implosive, not explosive), and/or a pulse motor. It can also work as an electro-static drive, either internally, or as applied differentially to the wing. In for instance, a close combat situation needing extremely high performance, the ITP can operate in wire combinations of the three modes. [RP: This is unclear and needs to be re-written].

The encapsulated ITP is a self-structuring, self-organizing, homeostatic electron, air or fluid flow system, As stated, when considering the gaseous state of matter, it is similar to a tornado, hurricane, typhoon, whirlwind or cyclone. In relation to a fluid state of matter, it acts like a whirlpool or waterspout. In the vacuum of outer-space it works in a different way to the Bussard Ram-jet proton drive but with the same effect.

In air, the ITP engine initiates the inward, centripetal implosive flow into the device by generating an imploding vortex tornado.

This is spun in a clockwise spiral motion around its central pole.

The open centre causes the air inflow to spiral into ever faster, tighter orbits around the imagined central pole.

The increasing speed of the tornado's natural rotation and contraction generates a greater and greater vacuum. This creates a higher suction, and therefore increased speed of incoming gases. The result is a dynamic, self-perpetuating positive feedback system. The nose cone faces into the airflow. This directs the flow, first to the outside of the device's flow guide bowl, then the outside of the engine. The device's central vacuum causes the air to enter the engine's periphery. It is then compressed and directed into a linear stream. The air is sent tangentially to the Tornado's axis and in the same spin direction, while being compressed and accelerated.

The external peripheral flow is forced over the engine's blades before coming into contact with the fuel. The fuel is fed in by the nose cone assembly down the pole of the engine's central axis. The device's static blades are cooled and are only subject to the combustion's radiant heat. The temperature is minimalized by the implosive concentration of air layers spiraling into the vortex's centre, which creates a heat shield. The vortex generates a natural Kelvin-Joule cooling effect, which further protects the blades form overheating. This 'squeezing' effect builds higher compression of fuel burn. This leads to far greater fuel efficiencies and cleaner exhausts than current technologies.

The ITP's 'zero point'---or transition from an implosion to an explosion---is outside and behind the device's wing and engine. The zero point creates a pressure which manifests as a standing wave. This forces the wing and engine forward.

In the case of a jet plane, the implosive engine flow can be instantly shut, and compression ratios and flow controlled in flight. This applies in both dense atmosphere near ground and near zero atmosphere conditions at extreme heights. This means maximum performance at all levels in a similar way to an Aero Spike engine.

The Bendall ITP Engine's different blades are electrically isolated and have their own command and control system. They regulate an individual capacitor discharge. The blades are also designed to be an electrostatic generator and controlled discharge system.

The air flow over one material creates a negative charge (in a prototype polycarbonate), and over a metallic material causes a positive charge. This difference in charges creates an electrical charge differential. If applied at the zero point an effect is created that has similarities in principal to the Bussard Ramjet proton drive, which is a by-product of the ITP's operation. This greatly reduces the plane's overall weight by replacing, or minimalizing the electrical generator and battery systems.

The ITP's wave guide thruster components can be of the one material. This generates the same air-over-metal, positive charge, with the applied implosive vortex collision photon resonance effect. This generates the electrical current, which still invokes a vortex collision pressure wave effect, and most importantly, Electrically Assisted Combustion (EAC).

Operation on and Under Water of the Bendall IPT Engines

The capacity bank of planes or boats can be fully charged by a free flight operation, replenished by jet fuel, or with stored charge from a previous operation by a plasmoid battery or base station charge. Once fuelled or charged, the planes and boats can submerge into water. They can operate in either a slow stealth mode or in a fast Squid Jet propulsion mode by using a novel configuration of the ITP Engine, which enables an effective one way valve pulse jet effect.

The underwater system uses both the combined potential difference spin effect, and the ITP's cavitation effect. This generates electricity which acts as a power source for a plasma explosion that mimics the simple Ram Jet engine of the World War 11 German-designed V1 and V2 rockets.

Operation in Outer Space of the Bendall IPT Engines

The IPT Engine is capable of vacuuming in both bound, and free atomic and intermolecular ions, in the form of protons and electrons, by means of its concentrated charge density and implosive structure. This characteristic is similar to the operation of the Bussard Ramjet space engine, though fundamentally different. The electrons are encapsulated by spiral orbital movement in and out of the IPT's centre. This prevents them from interacting with the environment.

The net effect of the dynamic pathways within the ITP shape insulate the electrons from 'normal' influences of matter or magnetic and electrical forces. This is because the central point of both implosion [up-spin electrons attracted to higher charge densities] and explosion [spin-down electrons attracted to lower charge densities] acts as a focus of osmotic gravitational force much higher than its low-charge density environment.

The ITP's geometric, geographic and gravity centres represent all frequencies [360], and therefore no frequencies [0]. This means the Engine can travel through matter as it neutralizes the frequencies through which it passes.

[1] The elements of the Bendall ITP Turbine engine are subject to the same forces that determine herd coherent harmonic frequencies and knops and nodes. These are pre-determined by the application in Time/Area/Space of a 100 Pi derived surface area of a Toroidal template. This is connected by a Fibonacci curve. These elemental nodes are created by dividing the Torus---the doughnut---into 16 equal segments or doughnut slices. Once that is done, the Toroidal structures can be predicted geometrically.

WATER AS A PARABLE TO UNLOCK THE SECRETS OF MATTER

7,920 EARTHS OCTAVE SEQUENCE

132,875,550,720
66,437,775,360
33,218,887,680
16,609443,840
8,304,721,920
4,152,360,960
2,076,180,480
1,038,090,240 [9,720 SPLICED COMPOSITE SHARED]
519,045,120 [518,400 SPLICED COMPOSITE SHARED]
259,522,560 [25,920 Years EARTHS EQUINOX]
129,761,280 [129,600 RFEU]
64,880,640 [648-864 MIRROR NUMBER SUNS DIAMETER]
32,440,320 [324-432 MIRROR NUMBER SUNS RADIUS]
16,220,160 [162-216 MIRROR NUMBER MOON'S D 2,160]
8,110,080 [162-216 MIRROR NUMBER MOON'S R 1,080]
4,055,040 [045-540 MIRROR NUMBER MOONS R/2= 540]
2,027,520 [297 SHUFFLED COMPOSITE 7,920]
1,013,760 [297 SHUFFLED COMPOSITE 7,920]
506,880 [468 SHUFFLED COMPOSITE SUN'S DIA 864]
253,440 [297 SHUFFLED COMPOSITE 7,920]
126,720 [972 SHUFFLED COMPOSITE 7,920]
63,360 [036-360 MIRROR CIRCLE NUMBER]
31,680 [468 MIRROR SUN'S DIAMETER 864]
15,840 [684 COMPOSITE SUN'S DIAMETER 864]

7,920 EARTH'S DIAMETER IN MILES

3,456,000 /7,920 = 436.3636 1,728,000 /7,920 = 225 864,000 /7,920 = 109.0909 432,000 /7,920 = 54.5454
311,040 / 7,920 = 39.2727 311,040 x 7,920 = 2,463,436,800
518,400 / 7,920 = 65.4545 518,400 x 7,920 = 4,105,728,000
144,000 / 7,920 = 18.181818 x 3.333 = 60.6060 x 1.333 = 80.8080 / 24 = 3.367003367/ 60 = 0.0561167227
144,000 x 7,920 = 1,140,480,000 / 24 = 47,520,000 / 60 = 7,92,000 / 60 = 13,200 / 60 = 220 / 7.5 = 29.333
7.5 / 7,920 = 0.00094696969 7,920x 7.5= 1,056 7,920/266.66= 29.70/7.5= 3.96 7,920 x 266.66= 2,111,100

3,960
1,980
990
495
247.5
123.75
61.875
30.9375
15.46875
7.734375
3.8671875
1.93359375
0.966796875
0.4833984375
0.24169921875
0.120849609375
0.0604248046875
0.03021240234375

11.111 AND 5.555 AETHER OCTAVE SEQUENCE

186,413,511.111
93,206,755.555
46,603,337.777
23,301,688.888
11,650,844.444
5,825,422.222
2,912,711.111
1,456,355.555
728,177.777
364,088.888
182,044.444
91,022.222
45,511.111
22,755.555
11.377.777
5,688.888
2,844.444
1,422.222
711.111
355.555
177.777
88.888
44.444
22.222

11.111

3,456,000/11.111= 311,040 1,728,000/11.111= 155,520 864,000/11.111= 77,760 432,000/11.111= 38,880
311,040 x 11.111 = 3,456,000 311,040/11.111 = 27,993.6 518,400 /11.111= 46,656 518,400 x 11.111= 576
144,000 x 11.111 = 1,600,000 144,000 / 11.111 = 129,600   7.5 x 11.111 =83.333   11.111 / 7.5 = 1.48148

5.555

3,456,000/5.555 = 622,080   1,728,000/5.555 = 311,040   864,000/5.555 = 155,520   432,000/5.555 = 77,760
311,040/5.555= 55,987.2 311,040 x 5.555= 1,728,000 518,400/5.555 = 93,312 518,400 x 5.555 = 2,880,000
144,000 / 5.555 = 25,920 / 3.333 = 7,776 x 1.333 = 10,368/ 24 = 432/60 = 7.2/60 = 0.12/60 = 0.002
144,000 x 5.555 = 800,000 / 24 = 33,333.333 / 60 = 555.555 / 60 = 9.2592 / 60 = ***0.154320987654321***
1x5x4x3x2x9x8x7x6x5x4x3x2x1 = 43,545,600 x 43,545,600 = 1,896,219,279,360,000
7.5 / 5.555 = 1.35   5.555 x 7.5 = 41.666

2.777
1.3888
0.69444
0.347222
0.1736111
0.08680555
0.0434027888
0.0217013888
0.01085069444
0.005425347222
0.0027126736111

PART 17 OF 20 – MSAART PATENT APPLICATION NOTES DRAFT 518,400 – 22:22:22 THUR 22ND SEPT 2022
© STRIKE FOUNDATION GUARANTEE LIMITED | MALCOLM BENDALL 2022 | GRAPHICS - STEVE EARL

12, 24, 48 AETHER TO MATTER OCTAVE SEQUENCE

402,653,184
201,326,592
100,663,296
50,331,648
25,165,824
12,582,912
6,291,456
3,145,728
1,572,864
786,432
393,216
196,608
98,304
49,152
23,576
12,228
6,144
3,072
1,536
768
384
192
96
48
24

3,456,000/24= 144,000 1,728,000/24= 72,000 864,000/24 = 36,000 432,000/24= 18,000
31,104,000,000 / 24 = 12,96,000,000   518,400 / 24 = 21,600   518,400 x 24 = 12,441,600
144,000 /24= 6,000 (MOLTEN SEA SPHERE VOLUME IN BATHS) 12/7.5 = 1.6 7.5/24 = 0.3125 24/7.5= 3.2
24 / 11.111 = 2.16   24 x 11.111 = 266.666   12 x 7.5 = 90   24 x 7.5 = 180   48 x 7.5 = 360

12
6
3
1.5
0.75 (LIGHT TRAVELS 7.5 x AROUND THE EARTH IN 1 SECOND)
0.375
0.1875
0.09375
0.046875
0.0234375
0.01171875
0.005859375
0.0029296875
0.00146484375
0.000732421875

PLANET OCTAVE 864,000 SEQUENCE (POS)

452,984,832,000
226,492,416,000
113,246,208,000
56,623,104,000
28,311,552,000
14,155,776,000
7,077,888,000
3,538,944,000
1,769,472,000
884,736,000
442,368,000
221,184,000
110,592,000
55,296,000
27,684,000
13,824,000
6,912,000
3,456,000
1,728,000

864,000 SUN AETHER OCTAVES

8 / 6 / 4 = 0.333

3,456,000 / 864 = 4,000 1,728,000/ 864 = 2,000 864,000/ 864 = 1,000 432,000/ 864 = 500

311,040 /864= 360 311,040x 864= 268,738,560 518,400/864= 600 5184x 864= 4,478,976

144,000 / 864 = 166.666 x 1.333 = 222.222 / 16 = 13.888 /22.5 = 0.61728 /0.125 = 4.93827

144,000 x 864 = 124,416,000 /24 = 5,184,000 /60 = 86,400 /60 = 1,440 /60 = 24 /7.5 = 3.2

864,000 x 7.5 = 6,480,000 x 7.5 = 48,600,000

432,000
216,000
108,000
54,000
27,000
13,500
6,750
3,375
1,687.5
843.75
421.875
210.9375
105.46875
52.734375
26.3671875
13.18359375
6.591796875
3.2958984375
1.64794921875

LIGHT PLASMOID OCTAVE 144 SEQUENCE (POS)

603,979,776,000
301,989,888,000
150,994,944,000
75,497,472,000
37,748,736,000
18,874,368,000
9,437,184,000
4,718,592,000
2,359,296,000
1,179,648,000
589,824,000
294,912,000
147,456,000
73,728,000
36,864,000
18,432,000
9,216,000
4,608,000
2,304,000
1,152,000
576,000
288,000

144,000 LIGHT OCTAVES

3,456,000 / 144,000 = 24   1,728,000 / 144,000 = 12   864,000 / 144,000 = 6   432,000 / 144,000 = 3
311,040 / 144,000 = 2.16   311,040 x 144,000 = 44,789,760,000   518,400 / 144,000 = 3.6
518,400 x 144,000 = 74,649,600,000 / 3.333 = 22,394,880,000 x 1.333 = 29,859,840,000
29,859,840,000 / 24 = 1,244,160,000 /60 = 20,736,000 /60 = 345,600 /60 = 5,760 \\\1,440
144,000 / 144 = 1,000 / 3.333 = 300 x 1.333 = 400 / 24 = 16.666 /60 = 0.2777 /60 = 0.00462962
144,000 x 144 = 20,736,000 / 24 = 864,000 / 60 = 14,400 / 60 = 240 / 60 = 4 / 7.5 = 0.5333 \0.2666
144,000 / 7.5 = 19,200   7.5 / 144,000 = 0.00005208333 \9\ 0.02666   144,000 x 7.5 = 1,080,000

72,000
36,000
18,000
9,000
4,500
2,250
1,125
5,625
281.25
140.625
70.3125
35.15625
17.578125
8.7890625
4.39453125
2.197265625
1.0986328125

TIME OCTAVE 311,040 AND 518,400 SEQUENCE (TOS)

5,218,385,264,640
2,609,192,632,320
1,304,596,316,160
652,298,158,080
326,149,076,040
163,074,539,520
81,537,269,760
40,768,634,880
20,384,317,440
10,192,158,720
5,096,079,360
2,548,039,680
1,274,019,840
637,009,920
318,504,960
159,252,480
79,626,240
39,813,120
19,906,560
9,953,280
4,976,640
2,488,320
1,244,160
622,080

311,040 TIME OCTAVES (311,040 / 864 = 360 Degrees)

3,456,000/311,040=11.111 1,728,000/311,040=5.729166 864,000/311,040=2.777 432,000/311,040=1.388
311,040 / 311,040 = 1   311,040 x 311,040 = 96,745,881,600   518,400 / 311,040 = 1.666
518,400 x 311,040= 161,234,136,000/24= 6,718,464,000/60= 111,974,400/60= 1,866,240/60= 31,104
31,104 / 60 = 518.4/60 = 86.4 / 60 = 1.44 / 60 = 0.024 / 60 = 0.0004 / 60 = 0.00000666 / 7.5 = 0.00000888
144,000 / 311,040 = 0.462962962 / 3.333 = 0.13888 x 1.333 = 0.185185 / 24 = 0.0077160 /60 = 0.00012
144,000 x 311,040= 44,789,760,000/24=1,866,240/60= 31,104,000/60= 518,400 /60 = 8,640 x 7.5 = 64,800
44,789,760,000 / 7.5 = 5,971,968,000 / 24 = 248,832,000 / 60 = 4,147,200 / 60 = 69,120 / 60 = 1,152\144
7.5 / 311,040 = 0.000024112654321   311,040 x 7.5 = 2,332,800

311,040 x 7.5 (The one second light takes to travel 7.5 times around the Earths equator) = 2,332,800

155,520
77,760
38,880
19,440
9,720
4,860

2,430

1,215
607.5
303.75
151.875
75.9375
37.96975
18.984375
9.4921875

RESONANCE 129,600 RFEU OCTAVES

543,581,798,400
271,790,899,200
135,895,449,600
67,947,724,800
33,973,862,400
16,986,931,200
8,493,465,600
4,246,732,800
2,123,366,400
1,061,683,200
530,841,600
265,420,800
132,710,400
66,355,200
33,177,600
16,588,800
8,294,400
4,147,200
2,073,600
1,036,800
518,400
259,200

129,600 RFEU (129,600/864= 150 129,600/31,104= 4.1666 129,600/5,184= 25 129,600/7.5 = 1,728)
3,456,000/129,600= 26.666 1,728,000/129,600=13.333 864,000/129,600=6.666 432,000/129,600= 3.333
311,040 / 129,600 = 2.4   311,040 x 129,600 = 40,310,784,000   518,400 / 129,600 = 4
518,400 x 129,600= 67,184,640,000/24=2,799,360,000/60=46,656,000/60=777,600/60=12,960/7.5= 1,728
144,000 / 1,296 = 111.111 / 3.333 = 33.333 x 1.333 = 44.444 / 7.5 = 5.925922592
144,000 x 129,600 = 16,796,160,000 / 24 = 699,840,000 / 60 = 11,664,000 / 60 = 194,400 / 60 = 3,240
129,600 x 7.5 = 972,000 x 7.5 = 7,290,000, 129,600 / 7.5 = 17,280 / 7.5 = 2,304

64,800
32,400
16,200
8,100
4,050

2,025

1,012.5
506.25
253.125
126.5625
63.28125
31.640625
15.8203125
7.91015625
3.955078125
1.9775390625
0.98876953125

135 DEGREES SOLOMANS MOLTEN SEA

2,264,924,160
1,132,462,080
566,231,404
283,115,520
141,557,760
70,778,880
38,389,440
17,604,720
8,847,360
4,432,680
2,211,840
1,105,920
552,960
276,480
138,240
69,120
34,560
17,280
8,640
4,320
2,160
1,080
540
270

135 DEGREES OCTAVES S.E CORNER SOLOMON'S TEMPLE

864,000/135=6,400 864,000x135= 116,640,000 311,040/135=2,304 311,040x135=41,990,400/864= 48.6
864 x 9 = 7,776   864,000 / 900 = 960   31,104 x 9 = 279,936   311,040 / 900 = 345.6
3,456,000 / 135 = 25,600   1,728,000 / 135 = 12,800   864,000 / 135 = 6,400   432,000 / 135 = 3,200
311,040 / 135 = 2,304   311,040 x 135 = 41,990,400   518,400 / 135 = 3,840   518,400 x 135 = 69,984,000
144,000 / 135 = 1,066.666 x 3.333 = 3,555.555 x 1.333 = 4,740.740740740
4,740.740740740 / 24 = 197.530864 /60 = 3.292181 /60 = 0.054869684499314 /60 = 0.000914494
144,000 x 135 = 19,440,000 / 24 = 810,000 / 60 = 13,500 / 60 = 225 / 60 = 3.75 / 7.5 = 0.5 / 0.125 = 4
7.5 x 135 = 1,012.5   7.5 / 135 = 0.0555   135 x 7.5 = 1,012.5   135 / 7.5 = 18 / 7.5 = 2.4 / 7.5 = 0.32

67.5
33.75
16.875
8.4375
4.21875
2.109375
1.0546875
0.52734375
0.263671875
0.1318359375
0.06591796875
0.032958984375
0.0164794921875

VAJRA TORUS OCTAVE 16 SEQUENCE (VOS)

268,435,456
134,217,728
67,108,864
33,554,432
16,777,216
8,388,608
4,194,304
2,097,152
1,048,576
524,288
262,144
131,072
65,536
32,768
16,384
8,192
4,096
2,048
1,024
512
256
128
64
32

16 VAJRA SECTIONS OCTAVES 864 /16= 54 432 /16= 27 311,040 /16= 19,440 518,400 /16= 32,400
3,456,000 / 16 = 55,296,000 1,728,000 / 16 = 108,000 864,000 / 16 = 54,000 432,000 / 16 = 27,000
3,456,000 x16= 55,296,000 1,728,000 x16 = 27,684,000 864,000 x16 = 13,824,000 432,000 x16 = 6,912,000
311,040 / 16 = 19,440 311,040 x 16 = 4,976,640 518,400 / 16 = 32,400 518,400 x 16 = 8,294,400
144,000 / 16 = 9,000 x 3.333 = 30,000 x 1.333 = 40,000 / 24 = 1,666.666 /60 = 27.777 /60 = 0.4629
144,000 x 16 = 2,304,000 / 24 = 96,000 / 60 = 1,600 / 60 = 26.666 / 60 = 0.444 / 7.5 = 0.0592592
7.5 / 16 = 0.46875 7.5 x 16= 120 16 / 7.5 = 2.1333\\\0.2666
16 x 7.5 = 120 (AETHER TO LIGHT TO MATTER) x 7.5 = 900

8
4
2
1
0.5
0.25
0.125
0.0625
0.03125
0.015625
0.0078125
0.00390625
0.001953125
0.0009765625

MATTER MUSIC 266.666 OCTAVESEQUENCE (POS)

1,118,481,066.666
559,240,533.333
279,620,266.666
139,810,133.333
69,905,066.666
34,952,533.333
17,476,266.666
8,738,133.333
4,369,066.666
2,184,533.333
1,092,266.666
546,133.333
273,066.666
136,533.333
68,266.666
34,133.333
17,066.666
8,533.333
4,266.666
2,133.333
1,066.666
533.333

266.666 MATTER ELEMENTAL OCTAVES

3,456,000/266.666 = 12,960 1,728,000/266.666 = 6,480 864,000/266.666 = 3,240 432,000/266.666 = 1,620
311,040/266.6 = 1,166.4 311,040x266.6 = 82,944,000 518,400/266.666 = 1,944 518,400x266.6 = 138,240
144,000/266.666=540/3.333=162 x1.333=216/24= 9 /60= 0.15/60=0.0025/60=0.00041666\5\0.002666
144,000 x 266.666 = 38,400,000 / 24 = 1,600,000 / 60 = 26,666.666 / 60 = 444.444 / 60 = 7.4074074
7.4074074 / 7.5 = 0.98765432098765432 / 1.111 = 0.888 x 1.333 = 1.8518518518
7.5 / 266.666 = 0.028125   266.666 x 7.5 = 2,000
266.666 x 7.5 = 2,000 x 7.5 = 15,000 x 7.5 = 112,500

133.333
66.666
33.333
16.666
8.333
4.1666
2.08333
1.041666
0.5208333
0.26041666
0.130208333
0.0651041666
0.03255208333
0.016276041666
0.0081380208333
0.00406901041666

ANGLES 720, 360, 22.5 OCTAVE SEQUENCE (AOS)

12,079,595,520
6,039,797,760
3,019,898,880
1,509,949,440
754,974,720
377,487,360
188,743,680
94,371,840
47,185,920
23,592,960
11,796,480
5,898,240
2,949,120
1,474,560
737,280
368,640
184,320
92,160
46,080
23,040
11,520
5,760
2,880
1,440
720

2,880 1,440, 720, 360,180, 90, 45, 22.5 DEGREES ANGLE OCTAVES

3,456,000/ 360 = 9,600   1,728,000/ 360 = 4,800   864,000/ 360 = 2,400 432,000/ 360 = 1,200
311,040/ 360 = 864 311,040 x 360 = 111,974,400 518,400/ 360 = 1,440 518,400 x 360 = 186,624,000
144,000 / 360 = 864 / 3.333 = 259.2 x 1.333 = 345.6 / 24 = 14.4 /60 = .24 /60 = 0.004 /60 = 0.0000666
144,000 x 360 = 400 / 24 = 16.666 / 60 = 0.2777 / 60 = 0.004629629 / 60 = 0.000077160493827
7.5 / 360 = 0.0208333   360 x 7.5 = 2,700
360 x 7.5 = 2,700 x 7.5 = 20,250 / 7.5 = 151,875 360 / 7.5 = 48 / 7.5 = 6.4 / 7.5 = 0.85333\\\\\0.02666

180
90
45
22.5
11.25
5.625
2.8125
1.40625
0.703125
0.3515625
0.17578125
0.087890625
0.0439453125
0.02197265625

PLANET OCTAVE 864,000 SEQUENCE (POS)

452,984,832,000
226,492,416,000
113,246,208,000
56,623,104,000
28,311,552,000
14,155,776,000
7,077,888,000
3,538,944,000
1,769,472,000
884,736,000
442,368,000
221,184,000
110,592,000
55,296,000
27,684,000
13,824,000
6,912,000
3,456,000
1,728,000

864,000 SUN AETHER OCTAVES

8 / 6 / 4 = 0.333
3,456,000/ 864 = 4,000 1,728,000/ 864 = 2,000 864,000/ 864 = 1,000 432,000/ 864 = 500
311,040/864= 360 311,040x 864= 268,738,560 518,400/864= 600 5184x 864=4,478,976
144,000 / 864 = 166.666 x 1.333 = 222.222 / 16 = 13.888/22.5 = 0.61728/0.125 = 4.93827
144,000 x 864 = 124,416,000 /24 = 5,184,000/60 = 86,400 /60 = 1,440 /60 = 24 /7.5 = 3.2
864,000 x 7.5 = 6,480,000 x 7.5 = 48,600,000

432,000
216,000
108,000
54,000
27,000
13,500
6,750
3,375
1,687.5
843.75
421.875
210.9375
105.46875
52.734375
26.3671875
13.18359375
6.591796875
3.2958984375
1.6479492187

PLANET OCTAVE 846,000 SEQUENCE (POS)

1,774,190,592,000
887,095,296,000
443,547,648,000
221,773,824,000
110,886,912,000
55,443,456,000
27,721,728,000
13,860,864,000
6,930,432,000
3,465,216,000
1,732,608,000
866,304,000
433,152,000
216,576,000
108,288,000
54,144,000
27,072,000
13,536,000
6,768,000
3,384,000
1,692,000

846,000

8 / 4 / 6 = 0.333

3,456,000 / 846 = 4,085.1 1,728,000 / 846 = 2,042.5 864,000 / 846 = 1,021.2 432,000 / 846 = 510.6
311,040 /846 = 367.65 311,040 x 846 = 263,139,840 518,400 /846 = 612.76 518,400 x846 = 438,566,400
144,000 / 846 = 170.212 x 1.333 = 226.95 / 24 = 9.456 /60 = 0.1576 /60 = 0.02626 /60 = 0.00043779
144,000 x 846 = 121,824 / 24 = 5,076,000 / 60 = 84,600 / 60 = 1,410 / 60 = 23.5
7.5 / 846 = 0.00886524822695   846 x 7.5 = 6,345

423,000
211,500
105,750
52,875
26,437.5
13,218.75
6,609.375
3,304.6875
1,652.34375
826.171875
413.0859375
206.54296875

PLANET OCTAVE 684,000 SEQUENCE (POS)

1,434,451,968,000
717,225,984,000
358,612,992,000
179,306,496,000
89,653,248,000
44,826,624,000
22,413,312,000
11,206,656,000
5,603,328,000
2,801,664,000
1,400,832,000
700,416,000
350,208,000
175,104,000
87,552,000
43,776,000
21,888,000
10,944,000
5,472,000
2,736,000
1,368,000

684,000

6 / 8 / 4 = 0.1875

3,456,000/ 684 = 5,052.63 1,728,000/ 684 = 2,526.315 864,000/ 684 = 1,263.1578 432,000/ 684 = 631.57
311,040/ 684 = 454.73684 311,040 x 684 = 212,751,360 518,400/684 = 757.89 518,400 x 684= 354,585,600
144,000 / 684 = 210.526 x 1.333 = 280.70 / 24 = 11.6959 /60 = 0.1949/60 = 0.0032488 /60 = 0.0000541477
144,000 x 684 = 98,496,000 /24 = 4,104,000 /60 = 68,400 /60 = 1,140 /60 = 19 / 7.5 = 2.5333
7.5 / 684 = 0.01096491 684 x 7.5 = 5,130

342,000
171,000
85,500
42,750 (EARTH 792)
21,375
10,687.5
5,343.75
2,671.875
1,335.9375
667.96875
333.984375
166.9921875

PLANET OCTAVE 648,000 SEQUENCE (POS)

1,358,954,496,000
679,477,248,000
339,738,624,000
169,869,312,000
84,934,656,000
42,467,328,000
21,233,664,000
10,616,832,000
5,308,416,000
2,654,208,000
1,327,104,000
663,552,000
331,776,000
165,888,000
82,944,000
41,472,000
20,736,000
10,368,000
5,184,000
2,592,000 (EARTHS EQ)
1,296,000 (RFEU)

648,000

6 / 4 / 8 = 0.1875
3,456,000 /648 = 5,333.333 1,728,000/648 = 2,666.666 864,000/648 = 1,333.333 432,000/648 = 666.666
311,040 / 648 = 480   311,040 x 648 = 201,553,920   518,400 / 648 = 800   518,400 x 648 = 335,923,200
144,000/648=222.222 x 1.333=296.296/24=12.34567901234568/60=0.2057/60=0.00342/60= 0.0000571
144,000 x 648 = 93,312,000 / 24 = 3,888,000 / 60 = 64,800/ 60 = 1,080 / 60 = 18 / 7.5 = 2.4
7.5 / 648 = 0.0115740740740   648 x 7.5 = 4,860

324,000
162,000
81,000
40,500
20,250
10,125
5,062.5
2,531.25
1,265.625
632.8125
316.40625
158.203125

PLANET OCTAVE 486,000 SEQUENCE (POS)

1,019,215,872,000
509,607,936,000
254,803,968,000
127,401,984,000
63,700,992,000
31,850,496,000
15,925,248,000
7,962,624,000
3,981,312,000
1,990,656,000
995,328,000
497,664,000
248,832,000
124,416,000
62,208,000
31,104,000
15,552,000
7,776,000
3,888,000
1,944,000
972,000 (EARTH)

486,000

4 / 8 / 6 = 0.08333

3,456,000/486 = 7,111.111   1,728,000/486 = 3,555.555 864,000/486= 1,77.777 432,000/486= 888.888

311,040/486 = 640 311,040 x 486 = 151,165,440 518,400/486= 1,066.666 518,400 x 486 = 251,942,400

144,000 / 486 = 296.296 x 1.333 = 395/ 24 = 16.46 /60 = 0.265 /60 = 0.0044 /60 = 0.0000737

144,000 x 486 = 69,984 / 24 = 2,916,000 /60= 48,600/60= 810 / 60 = 13.5 / 7.5 = 1.8

7.5 / 486= 0.0154320987654321   486 x 7.5 = 3,645

243,000
121,500
60,750
30,375
15,187.5
7,593.75
3,796.875
1,898.4375
949.21875
474.609375
237.3046875
118.65234375

PLANET OCTAVE 468,000 SEQUENCE (POS)

1,962,934,272,000
981,467,136,000
490,733,568,000
245,366,784,000
122,683,392,000
61,341,696,000
30,670,848,000
15,335,424,000
7,667,712,000
3,833,856,000
1,916,928,000
958,464,000
479,232,000
239,616,000
119,808,000
59,904,000
29,952,000
14,976,000
7,488,000
3,744,000
1,872,000
936,000

468,000

4 / 6 / 8 = 0.08333
3,456,000/468= 7,384.6 1,728,000/468= 3,692.30 864,000/468= 1,846.153846 432,000/468= 923.07692
311,040/468= 664.61 311,040 x 468= 145,566,720 518,400/468= 1,107.692 518,400x468 = 242,611,200
144,000 / 468 = 307.692 x 1.333 = 410.25610/24= 17.094017/60 = 0.2849 /60 = 0.004748 /60 = 0.00079
144,000 x 468 = 67,392,000 / 24 = 2,808,000/60 = 46,800 /60 = 780 /60 = 13 / 7.5 = 1.7333
7.5 / 468 =0.0160256410   468 x 7.5 = 3,510

234,000
117,000
58,500
29,250
14,625
7,312.5
3656.25
1828.125
914.0625
457.03125
228.515625

ALL SIX 864,000 COMBINATIONS

864,000	846,000	684,000	648,000	486,000	468,000
56,623,104*	55,443,456*	44,826,624*	42,467,328*	31,850,496*	30,670,848*
28,311,552*	27,721,728*	22,413,312*	21,233,664*	15,925,248*	15,335,424*
14,155,776*	13,860,864*	11,206,656*	10,616,832*	7,962,624*	7,667,712*
7,077,888*	6,930,432*	5,603,328*	5,308,416*	3,981,312*	3,833,856*
3,538,944*	3,465,216*	2,801,664*	2,654,208*	1,990,656*	1,916,928*
1,769,472*	1,732,608*	1,400,832*	1,327,104*	995,328*	958,464*
884,736*	866,304*	700,416*	663,552*	497,664*	479,232*
442,368*	433,152*	350,208*	331,776*	248,832*	239,616*
221,184*	216,576*	175,104*	165,888*	124,416	119,808*
110,592*	108,288*	87,552*	82,944*	62,208*	59,904*
55,296*	54,144*	43,776*	41,472*	31,104*	29,952*
27,684*	27,072*	21,888*	20,736*	15,552*	14,976*
13,824*	13,536*	10,944*	10,368*	7,776*	7,488*
6,912,000	6,768,000	5,472,000	5184,000	3,888,000	3,744,000
3,456,000	3,384,000	2,736,000	2,592,000	1,944,000	1,872,000
1,728,000	1,692,000	1,368,000	1,296,000	972,000	936,000

864,000	846,000	684,000	648,000	486,000	468,000
432,000	423,000	342,000	324,000	243,000	234,000
216,00	211,500	171,000	162,000	121,500	117,000
108,000	105,000	85,500	81,000	60,750	58,500
54,000	52,875	42,750	40,500	30,375	29,250
27,000	26,437.5	21,375	20,250	15,187.5	14,625
13,500	13,218.75	10,687.5	10,125	7,593.75	7,312.5
6,750	6,609.375	5,343.75	5,062.5	3,796.875	3,656.25
3,375	3,304.6875	2,671.875	2,531.25	1,898.43	1,828.1
1,687.5	1,652.34375	1,335.9375	1,265.62	949.2187	914.06
843.75	826.171875	667.96875	632.812	474.6093	457.03
421.875	413.0859375	333.984375	316.406	237.3046	228.51
210.9375	206.54296875	166.99218	158.2031	118.6523	114.25
105.46875	103.271484375	83.496093	79.10156	59.32617	57.128
52.734375	51.6357421875	41.7480468	39.55078	29,66308	28.564
26.3671875	25.8187109375	20.8740234	19.77539	14.83154	14.282

864,000	846,000	684,000	648,000	486,000	468,000

© STRIKE FOUNDATION GUARANTEE LIMITED | MALCOLM BENDALL 2022 | GRAPHICS - STEVE EARL

3,456,000/ = 1,728,000/ = 155,520 864,000/ = 432,000/ =
311,040 x = 311,040/ = 518,400 / = 518,400 x =
144,000 x = 144,000 / = 7.5 x = / 7.5 =

3,456,000/ = 1,728,000/ = 864,000/ = 432,000/ =
311,040/= 311,040 x = 518,400/ = 518,400 x =
144,000 / = / 3.333 = x 1.333 = / 24 = /60 = /60 = /60 =
144,000 x = / 24 = / 60 = / 60 = / 60 =
7.5 / = x 7.5 =

3,456,000 / = 1,728,000 / = 864,000 / = 432,000 / =
31,104,000,000 / = 518,400 / =
144,000 / = 7.5 / =
x 7.5 = x 7.5 =

8,000,000,000,000 / 144,000 = 5,555,555.555/ 7.5 = 740,740.740 (1/x = 0.00000135)
740,740.740 / 720 = 1,028.80658436214 /16 = 64.3004115226337448559 /22.5 = 2.857796067672611
8,000,000,000,000 /129,600= 6,172,839,50617284/1.333/144,000=32.150205761316881/x=0.031104
8,000,000,000,000 / 266.666 = 3,000,000,000.
8,000,000,000,000 / 1.333 = 600,000,000,000
8,000,000,000,000 / 6.666 = 120,000,000,000
8,000,000,000,000 / 5.555 = 144,000,000,000
8,000,000,000,000 / 4.444 = 180,000,000,000
8,000,000,000,000 / 3.333 = 240,000,000,000
8,000,000,000,000 / 2.222 = 360,000,000,000
8,000,000,000,000 / 1.111 = 720,000,000,000 / 9 = 80,000,000,000

LEGEND OF TERMS

SLEF = STANDARD LIGHT EARTH FREQUENCY [LIGHT 7.5 TIMES AROUND EARTH IN 1 SEC]

B = BENDALL TRANSLATOR

D MIN = MINIMUM DISTANCE FROM G AFTER SPIRALLING IN FROM TORUS EQUATOR

D MAX = MAXIMUM DISTANCE FROM G AFTER SPIRALLING OUT FROM TORUS EQUATOR

G = SINGULARITY ZERO POINT OF A TORUS

KD = KEY OF DAVID - 8 FREQUENCY KNOP CODE

TG = TESLA - THE SPEED OF LIGHT AT G THAT POINT DEFINED BY

F = ELEMENTAL SURFACES - (FERMI) – KNOPS ON THE KNOPS

R = THE ROD = 1 / x = AC / DC = 1.111 = ENERGY AT REST / ENERGY IN MOTION –
WORK TO PUSH AND PULL ONE CYCLE

O = OBOLINSKY TIME VARIENCE (RATE OF INCREASE AND DECREASE OF LIGHT SPEED)

V = POTENTIAL ENERGY

R = THE ROD THE MEDIUM THROUGH WHICH TORTION IS APPLIED THE ROD IS THE POLE ONLY
CUTTING THRU G AND DEFINING NORTH AND SOUTH ON THE TORUS AND IS PROBAGATED FROM
THE POINT G.

TABLE OF NUMBER RELATIONSHIPS AND MEANINGS
COLOR CODED

....................[EARTH'S PROCESSION OF THE EQUINOX NUMBER 25,920]

     [EARTH'S DIAMETER 7,920 x 2 = 15,840 x 2 = 31,680]

     [EARTH'S DIAMETER 7,920]

     [EARTH'S RADIUS 3,960]

     [EARTH'S RADIUS 3,960 / 2 = 1,980 / 2 = 990]

....[SUN'S DAY 26.24 EARTH DAYS AT EQUATOR 31 EARTH DAYS AT ITS POLE]

     [SUN'S DIAMETER 864,000 x 2 = 1,728,000 x 2 = 3,456,000 SUN SQUARE]

     [SUN'S DIAMETER 864,000]

     [SUN'S DIAMETER 864,000 MIRROR]

     [SUN'S DIAMETER 864,000 SHUFFLE]

..................[SUN'S RADIUS 432,000]

..................[SUN'S RADIUS 432,000 SHUFFLE]

..................[SUN'S RADIUS MIRROR 432,000]

...[LIGHT SPEED DIAMETER NUMBER 144,000 x 2 = 288,000 LIGHT DIA EARTH 7.5]

...[SANSKRIT 100 YEARS 311,040 OCTAVES]

...[MOON'S DIAMETER 2,160 x 2 = 4,320 x 2 = 8,640]

...................[MOON'S DIAMETER 2,160]

...................[MOON'S DIAMETER 2,160 MIRROR]

...................[MOON'S DIAMETER 2,160 SHIFTED]

...................[MOON'S RADIUS 1,080]

...................[MOON'S RADIUS 1,080 AND EARTH'S + MOON'S DIA = 10,080]

...[RESONANT FREQUENCY RESONANT UNIT (RFEU) 129,600]

...[TIME OCTAVE 311,040 AND 518,400 SEQUENCE (TOS)]

...[VAJAR TORUS OCTAVE 16 SEQUENCE (VOS)]

...[PLANET OCTAVE 864 SEQUENCE (POS)]

...[MUSIC 266.666 OCTAVE 11.111 SEQUENCE (MOS)]

...[LIGHT PLASMOID OCTAVE 144 SEQUENCE (POS)]

THE END OF THE BEGINNING

PART 17 OF 20 – MSAART PATENT APPLICATION NOTES DRAFT 518,400 – 22:22:22 THUR 22ND SEPT 2022

SUN - 864,000 - 5D (5.555)

864,000 MILES IS THE SUN'S DIAMETER
86,400 SECONDS IN 1 EARTH DAY
8,640 MOON SQUARE [4 x 2,160]
864,000 D ÷ 4 = SUN SQ OF 3,456,000
3,456,000 / 16 = 216,000 [Moon]
3,456,000 / 259.2 H = 13,333 [16 / 12]
3,456,000 / 6.666 = 518,400 [TIME]
TOTAL 864 = 846 = 684 = 648 = 486 = 468
864,000 / 25,920 = 33.333 [GY]
864,000 / 518,400 = 1.666 [POS]
864,000 / 129,600 [RFEU] = 6.666

Aether [DC] to Matter [AC] converter

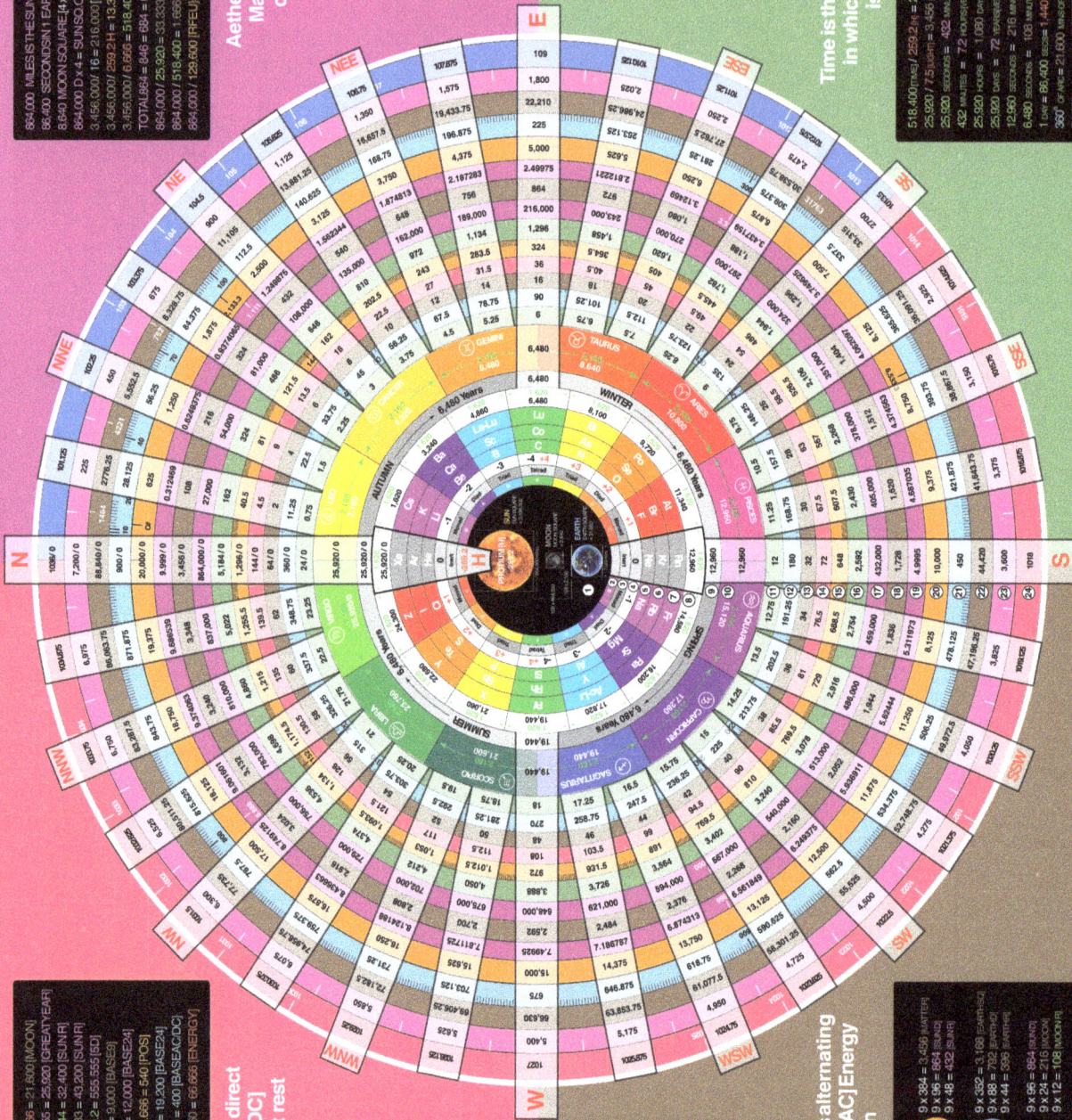

TIME - 518,400 - 4D (4.444)

Time is the mould in which Matter is formed

518,400 [TIME] / 259.2 H = 2,000 [BAR]
25,920 / 7.5 [WAR] = 3,456 [BANG]
25,920 DAYS = 432 [WHATEVER]
432 HOURS = 72 HOURS PER QUARTER
25,920 HOURS = 1,080 DAYS PER MOON
25,920 DAYS = 72 HOURS PER ...
12,960 SECONDS = 216 MINUTES PER ...
6,480 SECONDS = 108 MINUTES PER ...
1 DAY = 86,400 = 1,440 MINUTES PER DAY
360° OF ARC = 21,600 MIN OF ARC = 1,296,000 SEC OF ARC

AETHER / LIGHT / ENERGY - 144,000 - 6D (6.666)

144,000 / 6.666 = 21,600 [MOON]
144,000 / 5.555 = 25,920 [GREAT YEAR]
144,000 / 4.444 = 32,400 [SUN R]
144,000 / 3.333 = 43,200 [SUN R]
144,000 / 259.2 = 555.555 [SD]
144,000 / 16 = 12,000 [BASE 24]
144,000 / 12 = 12,000 [BASE 24]
144,000 / 266.666 = 540 [POS]
144,000 / 7.5 = 19,200 [BASE 24]
144,000 / 360 = 400 [BASE AC/DC]
144,000 / 2,160 = 66.666 [ENERGY]

Aether is direct current [DC] Energy at rest

MATTER - 3,456,000 - 3D (3.333)

Matter is alternating current [AC] Energy in motion

9 x 384 = 3,456 MATTER
9 x 96 = 864 [SUN]
9 x 48 = 432 [SUN R]
9 x 352 = 3,168 [EARTH SQ]
9 x 88 = 792 [EARTH R]
9 x 44 = 396 [EARTH R]
9 x 96 = 864 [SUN]
9 x 24 = 216 [MOON]
9 x 12 = 108 [MOON]

SUN
EARTH
MOON

OUR CURRENT POSITION ON THE GREAT YEAR
(25,920 Years) THE DAWNING OF THE AGE OF AQUARIUS

PLASMOID UNIFICATION MODEL

MOLTEN SEA ARK ATOMIC RECONSTRUCTION TECHNOLOGY (MSAART)

Plasmoids are doughnut or toroidal shaped clusters of net Protons or net Electrons that once captured and placed into a Toroidal orbit are capable of absorbing, storing and releasing enormous amounts of energy present within their self-generated and structured electro-magnetic containment field. Plasmoids, in effect, function as an atomic battery that can be self-charging due to its ability to convert matter to available clean energy. Plasmoids by their unique geometry cause a consequential electro-magnetic containment field to generate a Zero point naturally and casually, without much effort, have the ability to convert the nuclear Mass of Protium (Atoms) into energy.

The Plasmoid Unification Model (PUM) posits that Plasmoids are epoch-making and that knowledge of them has been hidden in plain sight for centuries. This PUM 'slide rule' reveals the algorithmic relationships between life's elements critical to mankind's existence and development. It starts with Protium [H] which has a melting point of -259.2°C and is the most abundant element in our Solar System. Protium determines the 25,920 Great Year frequency of our Solar System. The resonant frequencies of all other elements can then be calculated when 25,920 years is reduced from years to days, hours and seconds.

The PUM is evidence that the Universe is an intelligent design. That design is in perfect octave harmonic resonance with itself. Therefore, all of creation from Galaxies to Planets to Elements all resonate in unison with a collective chord 'As Above So Below'. This is interconnected with an Energy 'web'; the 24 components and laws of which are all based and governed on the same 16 sector Torus Plasmoid precepts shown. The concepts and ruling principles of the PUM can, and have, been applied to make Energy to Matter and Matter to Energy conversions. When applied to the modern hydrocarbon powered internal combustion engine, PUM technology removes exhaust toxic waste products and increases the engine power output by transforming waste energy back into fuel. Plasmoid employed in conjunction with the Plasmoid Toroidal Implosive Turbine provide a new novel Matter to Energy and Energy to Matter propulsion device for water, land, air and space travel.

LEGEND

Aether/Light   **Sun**   **Time**   **Matter**   **Frequency**   **Degrees**

1	Sun (864,000) [DC], Earth (7,920) [AC], Moon (2,160) [AC]
2	Music [AC] (Do = C = 24 x 11.111 = 266.666)
3	Elemental Crystal Forms +/- Monad, Diad, Triad, Tetrad
4	Elemental Valencies 0, -1, -2, -3, -4 and 0, +1, +2, +3, +4
5	Elements 1-16 (He - Cl)
6	Elements 17-32 (Ar - I)
7	Elements 33-48 (Xe - 'Z')
8	Elemental Frequencies (1,620 x 16 = 25,920 light frequency of -259.2 C)
9	Seasons of Great Year (25,920 / 0 years)
10	Zodiac Great Year (25,920 / 0 years)
11	Clock 24 Hour / 0 Hour [AC] / 0 Hour [DC] Clock
12	Compass 360° Degrees (AC) / 0° Degrees [DC]
13	Matter 64 / 0 64 Points being 32 Resonant Planes / 0
14	Light 144 / 0
15	Resonate Frequency Energy Unit (RFEU) - 1,296 (129,600)
16	All Time 5,184 (518,400 secs)
17	Aether - Sun (864,000 miles diameter, 432,000 miles radius)
18	Matter 3,456 (3,456,000 miles - Sun Square)
19	Dimensions 3D Matter = 3.33, 4D Time = 4.44, 5D Aether = 5.55, 6D Light = 6.666
20	Sound and Music (0 - 20,000 Hz)
21	Language 1-9, 10-90, 100-900 [111,222,333,444,555,666,777,888,999] [45,450,4,500]
22	Solar System Sun & Planetary Diameters and Radii in miles
23	Plasmoids 7,200 Degrees / 0, 32 Planes 64 Radial Points, One Zero Point
24	All Plasmoid Energy = All Alternating Current [AC] Frequencies

(= Non Ionizing   = Ionizing)

www.ingramcontent.com/pod-product-compliance
Lightning Source LLC
Chambersburg PA
CBHW050935210326

41597CB00036B/6192